Mathematics for Electricity and Electronics

Arthur Kramer

Delmar Publishers Inc.

I(T)P An International Thomson Publishing Company

Albany · Bonn · Boston · Cincinnati · Detroit · London · Madrid · Melbourne
Mexico City · New York · Pacific Grove · Paris · San Francisco · Singapore · Tokyo
Toronto · Washington

NOTICE TO THE READER

Delmar Staff:
Publisher: Michael A. McDermott
Administrative Editor: Wendy Welch
Developmental Editor: Mary E. Clyne
Project Editor: Barbara Riedell
Production Manager: Larry Main
Art & Design Coordinator: Lisa Bower

Copyright © 1995
by Delmar Publishers Inc.
The ITP trademark is used under license.

For more information,

Delmar Publishers Inc.
3 Columbia Circle, Box 15015
Albany, New York, 12212-5015

International Thomson Publishing
Berkshire House
168-173 High Holborn
London, WC1V7AA
England

Thomas Nelson Australia
102 Dodds Street
South Melbourne 3205
Victoria, Australia

Nelson Canada
1120 Birchmont Road
Scarsborough, Ontario
M1K 5G4, Canada

International Thomson Publishing GmbH
Konigswinterer Str. 418
53227 Bonn
Germany

International Thomson Publishing Asia
221 Henderson Bldg. #05-10
Singapore 0315

International Thomas Publishing Japan
Kyowa Building, 3F
2-2-1 Hirakawa-cho
Chiyoda-ku, Tokyo 102
Japan

Printed in the United States of America
published simultaneously in Canada by Nelson Canada,
a division of The Thomson Corporation

1 2 3 4 5 6 7 8 9 10   XXX   00 99 98 97 96 95 94

Library of Congress Cataloging-in-Publication Data
Kramer, Arthur D., 1940—
    Technical mathematics/Arthur Kramer.
        p.   cm.
    Includes index.
    ISBN 0-8273-5804-0
    1. Mathematics   I. Title.
QA39.2.P49   1995   92-12345
123-dc20   CIP

Contents

Preface

Introduction and Objectives

You may be wondering: Why a special math book for electricity and electronics? There are several good reasons.

- First and foremost, electricity and electronics use more mathematics and require more formulas than most other technical fields.
- Second, the need for intermediate and advanced mathematical ideas comes earlier in the study of electricity and electronics than in the other basic engineering technologies such as civil, chemical, or mechanical.
- Third, a general technical math course cannot adequately cover many of the important mathematical concepts as they apply to electricity and electronics. For some topics, such as complex numbers, the terminology and the symbolism are different from those used in general mathematics.
- A fourth very good reason is that seeing and understanding the applications that apply to electricity and electronics is the best way to get a good grasp of the mathematical ideas.

Mathematics for Electricity and Electronics is designed to serve these needs in the best possible way. There are two underlying themes throughout the book:

1. *Clarity and ease of understanding.* The language is down to earth and written with the student, and the students' needs, in mind. Concepts are presented in the most concrete way with many applications and illustrations.
2. *Developing and mastering the critical skill needed by all technical people: the ability to understand, analyze and solve problems.* Concepts and relationships are emphasized. Memorizing formulas is not stressed as much as the ability to use and adapt formulas to fit problems.

Applications to electronics are shown as often as possible and, at times, some practical applications are also shown to further motivate learning and reinforce ideas such as gas mileage, temperature conversion, test average, sales tax, and velocity.

All of the mathematical ideas needed for the first year of electronics, and most needed for the second year, are included in the text. Only accepted electrical and electronic notation is used. There is enough material in the text to prepare the student for the study of calculus, if required.

Special Features

A most outstanding feature of this text is the following:

- Every set of problems in the nonelectronic chapters, from the most elementary to the most advanced, includes a special *"Applications to Electronics"* section. This section contains realistic and useful applications to electricity and electronics. These problems not only demonstrate the need to learn the mathematics, but they help one to understand the ideas better when done in the context of an electronics application. The mathematics required to do these problems is thoroughly explained within each section. It is not necessary to understand all the electronic concepts to do these problems. However, the problems do expose one to these concepts and can help in understanding them when encountered later.

Another equally outstanding feature is the following:

Close the Circuit

- In almost all sections of the nonelectronics chapters, there are examples that "close the circuit" between theory and practice. They show how to apply the mathematical ideas to a problem in electricity or electronics and help one to do the Applications to Electronics at the end of the section.
    Close the circuit problems are marked with the icon shown in the margin.

Additional helpful features in this book are as follows:

- Every chapter contains an "Error Box" that points out a common misunderstanding and stresses the correct approach. There are practice problems in each error box to immediately reinforce the correct procedure.

- Every chapter concludes with Chapter Highlights and Review Questions. The Chapter Highlights summarize the important ideas and formulas but also refer the student back to specific examples and figures in the chapter that emphasize these key concepts. The Review Questions contain more than just further examples of the same types of problems found throughout the chapter. Where possible, they combine and link together material from several sections and help the student understand the interrelationships.

- The book contains two early chapters on arithmetic and elementary algebra to strengthen and reinforce the basic skills and/or to review fundamental concepts. There are special calculator sections at the end of these chapters to aid in the use of the calculator.

Unique to this book and an important feature is the following:

- Chapter 4 contains two special sections, one on formulas and one on verbal problem solving. These are designed to help the student develop the needed skill of analyzing and solving problems. The sections stress the logical steps of using a formula and setting up an equation, and they provide many different types of problems to sharpen one's reasoning. In addition, every problem set throughout the book contains verbal problems to help master this ability.

Other special features include the following:

- Every chapter begins with an introduction that gives an overview of the chapter and a list of chapter objectives.

- There are special sections on engineering notation and the metric system.
- There are six chapters devoted exclusively to the mathematics of electronic concepts.
- There are more than 350 examples, many of which show calculator steps, and more than 350 figures to help explain and illustrate important concepts.
- There are more than 2500 student questions and problems of many types and of varying difficulty to develop a wide range of skills.
- There is a special appendix on calculator functions, including an introduction to the Texas Instruments (TI) graphics calculator.

Organization

Although the chapters have been arranged in a very logical and progressive order, clearly it may not satisfy everybody's needs. To this end, care was taken to make as many topics and chapters as independent as possible and to allow for maximum flexibility in the order of presentation. One of the limitations is that much of the material in Chapters 1 through 4 should be understood before doing any subsequent chapters. The topics in Chapters 5 through 16 can then be covered in almost any sound pedagogic order to fit various needs and requirements.

Development

As an instructor for more than 30 years in both mathematics and electronics, and as an author of several mathematics texts, I am very aware and sensitive to the needs of electronics students and to their pitfalls in mastering mathematics. All the material in the text has been classroom tested and reviewed by six technical reviewers. In addition, the entire book, including every example and problem, has been checked for accuracy both by me and by other technical reviewers.

Supplements

- An *Instructor's Solutions Manual* contains solutions to every text problem with suggested course outlines and teaching tips.
- A *Student's Solutions Manual and Study Guide* contains solutions to every other odd problem and extra examples and problems sets to reinforce major concepts.
- A *computerized Test Bank* contains approximately 100 questions and answers for each chapter.

Acknowledgements

Many thanks to the editorial and production staffs at Delmar in helping to develop this book, with special recognition to Jim Corcoran for being instrumental in initiating the project and to Mary Clyne who worked closely with me over many of the rough spots.

Special thanks to my many students and colleagues whose help and needs were invaluable in the writing of this book.

Thanks to my wife Carol for her patience and understanding when authoring interfered with quality time together.

My appreciation to the companies who were so generous in providing photographs to help illustrate the text.

Finally, the publisher and I gratefully acknowledge the contributions of many reviewers who offered their valuable insights:

John Anagnost
Neil Catone
Susan Drake
Ken Exworthy
Arnie Garcia
Roger Harlow
Carl Jensen
David Longobardi
Byron Paul
Richard Stedman
Richard Steinmeier
Ames Stewart
Lewis Van Vliet
Stan Vittetoe

We also thank the following reviewers for their diligence in verifying the text's technical accuracy:

Wendell Johnson
David Schmidt
Henry Smith
George Wyatt

Mathematics for Electricity and Electronics

CHAPTER

1

Basic Arithmetic

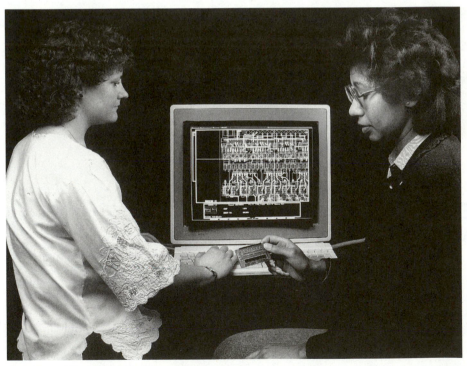

Courtesy of Department of Energy.

Researchers use an auto/router system to find the best approach for surface mounting 350 devices in a small 4"-by-2.3" circuit board.

Electricity and electronics are two of the most important fields in today's technical world. An understanding of mathematics is essential for mastering the ideas of electricity and electronics. This chapter is designed to reinforce your basic arithmetic skills and help you with calculations and the handling of numbers. The first three sections review the basic arithmetic operations and their application to fractions and decimals. Section 1.4 explains precision and accuracy of numbers, which are important in calculations and interpreting data. Section 1.5 discusses how to perform basic arithmetic operations on the calculator and, what is most important—how to check the results. The chapter concludes with a summary of the important ideas of the chapter and some comprehensive review questions to test your understanding and help strengthen these ideas.

Chapter Objectives

In this chapter, you will learn:

- The basic laws of arithmetic and the order of operations.
- How to add, subtract, multiply, and divide fractions.
- How to add, subtract, multiply, and divide decimals.
- How to convert between fractions, decimals, and percentages.
- What is meant by precision and accuracy of numbers and the rules for calculating with approximate numbers.
- How to use the calculator and its memory to perform basic arithmetic operations and check for correct results.

1.1
ARITHMETIC OPERATIONS

There are four basic operations in arithmetic: addition, subtraction, multiplication, and division. However, subtraction is the inverse of addition, and division is the inverse of multiplication. Therefore, the laws of arithmetic are defined only for addition and multiplication.

There are two types of laws that apply to addition and multiplication. The first type of law is the *commutative law.* This law says that it makes no difference in what *order* two numbers are added or multiplied. For example,

$$3 + 7 = 7 + 3 = 10 \quad \text{and} \quad 3 \times 5 = 5 \times 3 = 15$$

For any two numbers a and b the *commutative laws* are, therefore,

$$a + b = b + a$$

$$a \times b = b \times a \tag{1.1}$$

Observe, however, that subtraction and division are *not* commutative:

$$7 - 3 \text{ is not equal to } 3 - 7$$

$$2 \div 6 \text{ is not equal to } 6 \div 2$$

The second type of law is the *associative law.* This law states that if three numbers are to be added or multiplied together, it makes no difference if the operations start with the first and second numbers or with the second and third. For example, in addition, if you want to add $6 + 3 + 2$, you can do it two ways. Using parentheses to indicate which operation is to be done first, the two ways are as follows:

$$(6 + 3) + 2 = 9 + 2 = 11$$

$$6 + (3 + 2) = 6 + 5 = 11$$

Note that *operations in parentheses are always done first.* In multiplication, if you want to multiply $8 \times 4 \times 2$, you can also do it two ways:

$$(8 \times 4) \times 2 = 32 \times 2 = 64$$

$$8 \times (4 \times 2) = 8 \times 8 = 64$$

For any two numbers a and b the *associative laws* are, therefore,

$$(a + b) + c = a + (b + c)$$

$$(a \times b) \times c = a \times (b \times c) \tag{1.2}$$

Observe that subtraction and division are also *not* associative. The subtraction sign and the division sign apply only to the number that immediately follows the sign *moving from left to right.* For example,

$$6 - 3 - 2 \quad \text{means} \quad (6 - 3) - 2 = 3 - 2 = 1$$

$$8 \div 4 \div 2 \quad \text{means} \quad (8 \div 4) \div 2 = 2 \div 2 = 1$$

When the commutative and associative laws are applied together, it follows that three or more numbers can be added or multiplied in any order. For example, 2, 3, and 4 can be added (or multiplied) in any one of six different ways with the same result:

$$(2 + 3) + 4 = (3 + 2) + 4 = (4 + 2) + 3$$
$$= (2 + 4) + 3 = (3 + 4) + 2 = (4 + 3) + 2 = 9$$

Another important law of arithmetic which combines multiplication and addition is the *distributive law.* This law says that multiplication distributes over addition. For any three numbers a, b, and c,

$$a \times (b + c) = (a \times b) + (a \times c) \tag{1.3}$$

For example,

$$3 \times (2 + 4) = (3 \times 2) + (3 \times 4)$$

which correctly states that

$$3 \times 6 = 6 + 12 = 18$$

The distributive law is important in algebra.

The *order of operations* in arithmetic is, moving from left to right,

1. Perform operations in parentheses.
2. Do multiplication or division.
3. Do addition or subtraction.

Computers and scientific calculators are programmed to perform the operations in this order. It is called *algebraic logic* or the *algebraic operating system*. Study the following examples that illustrate the order of operations.

EXAMPLE 1.1

Calculate the following:

$$4 \times 19 - 36 + 6 \div 2$$

Solution   Apply the order of operations and perform multiplication and division first as shown by parentheses:

$$(4 \times 19) - 36 + (6 \div 2) = 76 - 36 + 3$$

Then subtract and add:

$$(76 - 36) + 3 = 40 + 3 = 43$$

Notice that you must move from left to right and subtract the 36 first before adding the 3. The subtraction sign applies only to the 36. If you add the 3 first, it will become subtracted from the 76 and the answer will not be correct:

$$76 - (36 + 3) = 76 - 39 = 37$$

You can check that your calculator uses algebraic logic as follows. Enter Example 1.1 exactly as it appears above:

$$4 \boxed{\times} 19 \boxed{-} 36 \boxed{+} 6 \boxed{\div} 2 \boxed{=} \rightarrow 43$$

You should get 43 when you press $\boxed{=}$. Section 1.5 contains more information on the basic calculator operations.

■

EXAMPLE 1.2

Calculate the following:

$$5 \times 17 - (12 + 4) \div 2 \times 3$$

Solution   Perform the operation in parentheses first:

$$5 \times 17 - (12 + 4) \div 2 \times 3 = 5 \times 17 - (16) \div 2 \times 3$$

Now do the multiplication and division:

$$5 \times 17 - 16 \div 2 \times 3 = 85 - 8 \times 3 = 85 - 24$$

EXAMPLE 1.2 (Cont.)

Note that you must move from left to right and divide the 2 into the 16 before you multiply by 3.
Then do the subtraction:

$$85 - 24 = 61$$

You can check Example 1.2 on the calculator using the parentheses keys as follows:

$$5 \;\boxed{\times}\; 17 \;\boxed{-}\; \boxed{(}\; 12 \;\boxed{+}\; 4 \;\boxed{)}\; \boxed{\div}\; 2 \;\boxed{\times}\; 3 \;\boxed{=}\; \rightarrow 61$$

The next example involves division with a fraction line. It is important to understand that a *fraction line is like parentheses*. It groups all the operations in the top, or numerator of the fraction, and all the operations in the bottom, or denominator of the fraction. Study the next example carefully.

EXAMPLE 1.3

Calculate the following:

$$\frac{5 \times 21 \times 12}{15 \times 7 \times 3}$$

Solution   The fraction line is like a division sign with parentheses. The example actually means

$$(5 \times 21 \times 12) \div (15 \times 7 \times 3)$$

The direct way to do this problem is to first multiply all the numbers in the top and all the numbers in the bottom:

$$\frac{5 \times 21 \times 12}{15 \times 7 \times 3} = \frac{1260}{315}$$

Then do the division:

$$\frac{1260}{315} = 4$$

Another way to do Example 1.3 involves the division of factors. A factor of a number is one that divides evenly into the number. It is also called a divisor. For example, all the ways to factor 12, not using 1 and the number itself, are

$$12 = 2 \times 6 = 3 \times 4 = 2 \times 2 \times 3$$

In the fraction in Example 1.3 all the numbers in the numerator *and* the denominator are multiplied together. These numbers are, therefore, factors. Common factors in the numerator and the denominator of a fraction can be divided out to simplify the fraction. For example,

$$\frac{12}{4} = \frac{3 \times \overset{1}{\cancel{4}}}{\underset{1}{\cancel{4}}} = 3$$

EXAMPLE 1.3 (Cont.)

Example 1.3 can then be done by first dividing all the common factors in the numerator and the denominator and then multiplying:

$$\frac{\overset{1}{\cancel{5}}\times\overset{3}{\cancel{21}}\times\overset{4}{\cancel{12}}}{\underset{3}{\cancel{15}}\times\underset{1}{\cancel{7}}\times\underset{1}{\cancel{3}}}=\frac{\overset{1}{\cancel{3}}\times 4}{\underset{1}{\cancel{3}}}=\frac{4}{1}=4$$

One must be very careful to divide only factors in a fraction. For example, in the following fraction,

$$\frac{3+4}{2-4}$$

you cannot divide out the 4s because they are not factors of the numerator and the denominator.

The calculator solution of Example 1.3 is shown in Section 1.5 as Example 1.25.

EXERCISE 1.1

In problems 1 through 18, test your understanding of arithmetic by doing the problem mentally or by hand. (Use the calculator only to check your results.)

1. $6 + 5 - 7 + 3 + 5 + 4$

2. $8 + 2 - 3 + 9 - 1$

3. $5 \times 2 \times 3 \times 4$

4. $12 \div 3 \times 2 \div 2$

5. $(800 + 20) \div 20$

6. $10 \div (5 + 40) \times 9$

7. $8 + 14 \div 2 \times 4$

8. $7 - 6 \div 3 + 8 \div 4$

9. $5 + (8 - 1) \times 6 \div 2$

10. $(5 - 1) \div 2 + 3 \times 4$

11. $\dfrac{9 \times 4}{3 \times 4} + \dfrac{18}{6}$

12. $\dfrac{8 \times 9}{4} - \dfrac{15}{3}$

13. $\dfrac{12 \times 15}{5 \times 3 \times 2}$

14. $\dfrac{8 \times 7 \times 6}{4 \times 28}$

15. $\dfrac{6 \times 27 \times 5}{9 \times 2 \times 15}$

16. $\dfrac{7 \times 10 \times 24}{6 \times 4 \times 14}$

17. $\dfrac{(3 + 5) \times 2}{13 - 11}$

18. $\dfrac{6 + 8 \times (4 - 1)}{4 - 1 \times 2}$

In problems 19 through 22, solve each applied problem by hand. (Use the calculator to check your results.)

19. One car travels 171 miles on 9 gallons of gasoline. A second car travels 210 miles on 10 gallons of gasoline. How many more miles per gallon does the car with the better gas mileage get?

20. An electronics technician earns $520 for a 40-hour week and a math teacher earns $420 for a 30-hour week. Who earns more per hour? How much more is earned per hour?

21. A Mariner space probe traveling at an average speed of 6000 mi/h takes 400 days to reach Mars. What is the total distance traveled by the space probe?

22. A bus route is 22 kilometers long. It takes the bus 50 minutes to travel the route in one direction and 70 minutes to travel the route in the other direction. What is the average speed of the bus in kilometers per hour for the total trip? (Note: Average rate = Total distance/Total time.)

Applications to Electronics   In problems 23 through 28, solve each applied problem by hand. (Check for accuracy with the calculator.)

23. Figure 1–1 shows three different resistances R_1, R_2, and R_3 connected in series to a battery. Each subscript identifies a different resistance. A series circuit has one current path, and the same current flows through each series resistor. If $R_1 = 12 \; \Omega$ (ohms), $R_2 = 2 \; \Omega$, $R_3 = 10 \; \Omega$, and the voltage of the battery source $V_B = 24$ V (volts), the current I in amps (A) is given by Ohm's law:

$$I = 24 \div (12 + 2 + 10)$$

Calculate the value of I. Note that the current shown in Figure 1–1 flows from negative to positive. This is electron current and is the current flow used throughout the text.

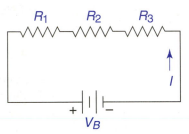

FIGURE 1–1   Series circuit for problems 23 and 24.

24. In problem 23, if $R_1 = R_2 = 20 \; \Omega$, $R_3 = 10 \; \Omega$, and $V_B = 100$ V, the current I in amps is given by Ohm's law:

$$I = 100 \div (2 \times 20 + 10)$$

Calculate the value of I.

25. Figure 1–2 shows two different resistors R_1 and R_2 connected in parallel to a battery. Each subscript identifies a different resistance. A parallel circuit has more than one current path, and a different current flows through each parallel resistor. If $R_1 = 20 \; \Omega$, $R_2 = 12 \; \Omega$ and the total current in the circuit $I_T = 2$ mA (milliamps), the voltage of the battery source V_B in millivolts (mV) is given by

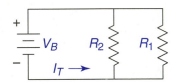

FIGURE 1–2   Parallel resistances for problems 25 and 26.

$$V_B = 2 \times \left(\frac{20 \times 12}{20 + 12} \right)$$

Calculate the value of V_B.

26. In problem 25, if $R_1 = 18$ kΩ (kilohms) and $R_2 = 9$ kΩ, the total power dissipated in milliwatts (mW) is given by

$$P_T = 2 \times 2 \times \left(\frac{18 \times 9}{18 + 19} \right)$$

Calculate the value of P_T.

27. In BASIC and most other computer languages, the following keyboard symbols are used for the arithmetic operations:

Addition	+
Subtraction	−
Multiplication	∗
Division	/

Applying the order of operations, calculate the following written in BASIC:

$$2 + 3 * 8 / 4 - 1$$

28. As in problem 27, calculate the following BASIC expression:

$$5 - 15 / 5 + 10 * (1/2)$$

Courtesy of Hewlett Packard Company.

Technician uses electronics to monitor the auto assembly process at Ferrari S.p.A.

 1.2
FRACTIONS

Calculations with fractions, decimals, and percentages are very important in technical work. Mistakes are often made because the concepts are not understood well enough. The calculator can prevent some of these mistakes, but it is not a substitute for clear understanding. Study this section and the next one thoroughly. The more problems you do correctly, the better you will grasp the concepts.

Reducing Fractions

Consider the following fraction:

$$\frac{6}{8}$$

The fraction line means division, but since 8 does not divide evenly into 6 the fraction cannot be simplified to a whole number. However, *it can be reduced to lowest terms* by dividing out common factors (divisors) in the numerator and the denominator:

$$\frac{6}{8} = \frac{\overset{1}{\cancel{2}} \times 3}{\underset{1}{\cancel{2}} \times 4} = \frac{3}{4}$$

To reduce a fraction into lowest terms, it is not necessary to show the factors before dividing. You can just divide the numerator and denominator of the preceding fraction by 2 to get the result.

The following examples with fractions are designed to be done without the calculator to reinforce your skills in arithmetic. You should use the calculator only to check results.

EXAMPLE 1.4

Simplify (reduce to lowest terms):

$$\frac{28}{42}$$

Solution   The common factors of 28 and 42 are 2 and 7 (or 14). You can divide the numerator and denominator by 2 and then 7 (or 14):

$$\frac{28}{42} = \frac{14}{21} = \frac{2}{3}$$

You can also first factor the numerator and the denominator into their smallest factors and then divide out the common factors:

$$\frac{28}{42} = \frac{\overset{1}{\cancel{2}} \times 2 \times \overset{1}{\cancel{7}}}{\underset{1}{\cancel{2}} \times 3 \times \underset{1}{\cancel{7}}} = \frac{2}{3}$$

The smallest factors of a number are numbers that have no other factors except one and the number itself. These smallest factors are called *prime numbers*. For example, 2, 3, 5, 7, 11, and so forth are prime numbers.

It is important to emphasize again that *you can only divide out factors* in the numerator and denominator of a fraction. If numbers are separated by a + or − sign, then they are not factors. Factors are always separated by multiplication signs. For example, in the fraction

$$\frac{5+1}{5 \times 7}$$

5 is *not* a factor in the numerator but *is* a factor in the denominator.

Multiplying and Dividing Fractions

To *multiply fractions* you multiply the numerators and the denominators:

$$\frac{A}{B} \times \frac{C}{D} = \frac{A \times C}{B \times D} \tag{1.4}$$

You then reduce the result to lowest terms. However, you can divide out common factors first that are in the numerator and denominator of *either* fraction and then multiply. This simplifies the multiplication. Study the next two examples, which show this procedure.

EXAMPLE 1.5

Multiply the following fractions:

$$\frac{3}{16} \times \frac{2}{9}$$

Solution Divide out common factors in any numerator and denominator and then multiply:

$$\overset{1}{\underset{8}{\cancel{\frac{3}{16}}}} \times \overset{1}{\underset{3}{\cancel{\frac{2}{9}}}} = \frac{1}{24}$$

EXAMPLE 1.6

Multiply the following:

$$4 \times \frac{3}{14} \times \frac{5}{9}$$

Solution Before multiplying, express the whole number 4 as a fraction with a denominator of 1. Then proceed in the same way you would for two fractions by dividing out all common factors that occur in *any* numerator and denominator:

$$\overset{2}{\cancel{\frac{4}{1}}} \times \overset{1}{\underset{7}{\cancel{\frac{3}{14}}}} \times \underset{3}{\cancel{\frac{5}{9}}} = \frac{10}{21}$$

Note that in this example a whole number is just a fraction with a denominator of 1. Therefore, when you multiply a whole number by a fraction, you can just multiply the numerator by the whole number. For example,

$$4 \times \frac{3}{14} = \frac{4 \times 3}{14} = \frac{12}{14} = \frac{6}{7}$$

To *divide fractions,* invert the divisor (the fraction after the division sign), and change the operation to multiplication:

$$\frac{A}{B} \div \frac{C}{D} = \frac{A}{B} \times \frac{D}{C} = \frac{A \times D}{B \times C} \tag{1.5}$$

EXAMPLE 1.7

Divide the following fractions:

$$\frac{5}{12} \div \frac{15}{16}$$

Solution Invert the fraction after the division sign, and change to multiplication. Then divide out common factors and multiply:

$$\frac{5}{12} \div \frac{15}{16} = \frac{5}{12} \times \frac{16}{15} = \frac{\overset{1}{\cancel{5}}}{\underset{3}{\cancel{12}}} \times \frac{\overset{4}{\cancel{16}}}{\underset{3}{\cancel{15}}} = \frac{4}{9}$$

Study the next example, which combines multiplication and division of fractions.

EXAMPLE 1.8

Calculate the following:

$$\frac{5}{4} \times \frac{8}{15} \div 2$$

Solution Invert the 2 to $\frac{1}{2}$ and change the division to multiplication. Then divide out common factors and multiply:

$$\frac{\overset{1}{\cancel{5}}}{\underset{1}{\cancel{4}}} \times \frac{\overset{2}{\cancel{8}}}{\underset{3}{\cancel{15}}} \times \frac{1}{2} = \frac{2}{6} = \frac{1}{3}$$

Adding Fractions

Adding fractions is not as straightforward as multipliying fractions. Calculators can add fractions as decimals, and some can add them as fractions. However, it is necessary for you to understand how to add fractions so that you can estimate results, recognize an incorrect answer, and troubleshoot for the error. More important, *to add fractions in algebra requires first understanding the process in arithmetic.*

Fractions can be added or subtracted *only if their denominators are the same.* When the denominators are the same, you add the fractions by adding the numerators over the common denominator:

$$\frac{1}{8} + \frac{5}{8} = \frac{1+5}{8} = \frac{6}{8} = \frac{3}{4}$$

Observe that you always reduce the result if possible.

The rule for adding fractions with the same denominator is then

$$\frac{A}{D} + \frac{B}{D} = \frac{A+B}{D} \tag{1.6}$$

When the denominators of fractions that are to be added are different, it is first necessary to change the fractions so the denominators are the same. A fraction can be changed to an equivalent fraction by dividing out common factors or by multiplying the numerator and denominator by the same factor. For example, the following fractions are all equivalent:

$$\frac{3}{10} = \frac{6}{20} = \frac{9}{30} = \frac{30}{100} \quad \text{and so forth}$$

Note that *a fraction is a ratio of two whole numbers.* The fraction $\frac{3}{10}$ is a ratio of 3 parts to 10 parts. The equivalent fractions $\frac{6}{20}$, $\frac{9}{30}$, etc., are all equal to the same ratio.

The process of adding two fractions whose denominators are different is done as follows. Suppose you wanted to add the fractions:

$$\frac{2}{3} + \frac{5}{6}$$

Since the denominators are not the same, you must first change one, or both, fractions to equivalent fractions with a common denominator. You look for the lowest common denominator (LCD), which is the smallest number that contains each denominator as a divisor. Since 3 divides into 6 exactly, the LCD equals 6.

Change the first fraction to an equivalent fraction by multiplying the numerator and denominator by 2:

$$\frac{2(2)}{3(2)} + \frac{5}{6} = \frac{4}{6} + \frac{5}{6}$$

Note that parentheses are used here instead of a × sign for multiplication. Parentheses or a dot (·) are used in algebra so as not to confuse the letter x with multiplication.

Now you can combine the numerators over the LCD and reduce the result:

$$\frac{4}{6} + \frac{5}{6} = \frac{4+5}{6} = \frac{9}{6} = \frac{3}{2}$$

Study the following examples carefully. They further explain the step-by-step process of adding fractions and finding the LCD when the denominators do not have common divisors.

EXAMPLE 1.9

Add the fractions

$$\frac{1}{10} + \frac{5}{6}$$

Solution   You must first find the LCD of 10 and 6. One method is to try multiples of the largest denominator until the other denominator divides into one of the multiples. That is, first try multiplying $10 \times 2 = 20$. Clearly, 6 does not divide into 20. Then multiply $10 \times 3 = 30$. Now, 6 divides into 30, so 30 is the LCD.

EXAMPLE 1.9 (Cont.)

Another method—which is important to know because it is used in algebra—is to completely factor each denominator into its smallest factors or primes. The prime factors of each denominator are

$$10 = (2)\ (5)$$

$$6 = (2)\ (3)$$

You then make up the LCD so that it contains the factors of every denominator by placing the same prime factors above each other as follows and then "bringing down" the necessary factors:

$$
\begin{array}{r}
10 = (2)\qquad (5) \\
6 = (2)\ (3)\quad \Big| \\
\downarrow\ \ \downarrow\ \ \downarrow \\
\hline
\text{LCD} = (2)\ (3)\ (5) = 30
\end{array}
$$

Note that if a factor appears twice in a denominator, it must appear twice in the LCD. To determine what to multiply each fraction by to change to the LCD, divide its denominator into the LCD. Since 10 divides into 30 three times, the numerator and denominator of the first fraction need to be multiplied by 3. Since 6 divides into 30 five times, the numerator and denominator of the second fraction need to be multiplied by 5:

$$\frac{1(3)}{10(3)} + \frac{5(5)}{6(5)} = \frac{3}{30} + \frac{25}{30} = \frac{3+25}{30} + \frac{28}{30} = \frac{14}{15}$$

Observe that the answer is reduced to lowest terms.

▪

EXAMPLE 1.10

Combine the fractions:

$$\frac{7}{15} + \frac{5}{12} - \frac{1}{3}$$

Solution   Use the method of prime factors. The prime factors of each denominator are

$$15 = (3)\ (5)$$
$$12 = (2)\ (2)\ (3)$$
$$3 = (3)$$

Place the prime factors above each other as follows:

$$
\begin{array}{r}
15 = \qquad\qquad (3)\ (5) \\
12 = (2)\ (2)\ (3)\quad \Big| \\
3 = \quad\big|\ \ \big|\ \ (3)\quad \Big| \\
\downarrow\ \ \downarrow\ \ \downarrow\ \ \downarrow \\
\hline
\end{array}
$$

Then the LCD $= (2)\ (2)\ (3)\ (5) = 60$

EXAMPLE 1.10 (Cont.)

Note that the factor of (2) must appear twice in the LCD because it appears twice in one of the denominators. Note also that there is only one (3) in the LCD because it appears no more than once in any denominator. The prime factors of each denominator should be able to divide into the factors of the LCD.

Now multiply the numerator and denominator of each fraction by the necessary factor to change each denominator into the LCD and combine:

$$\frac{7(4)}{15(4)} + \frac{5(5)}{12(5)} - \frac{1(20)}{3(20)} = \frac{28 + 25 - 20}{60} = \frac{33}{60} = \frac{11}{20}$$

▪

Close the Circuit

The following example serves to "close the circuit" between theory and practice. It shows how to apply the preceding mathematical ideas to a problem in electronics. Close the circuit examples are an important theme throughout the text. They are marked with the icon shown at left and help you to see the connection between mathematics and electronics. Study this application example.

EXAMPLE 1.11

Close the Circuit

Figure 1–3 shows a circuit containing three different resistances in parallel: R_1, R_2, and R_3. Each subscript identifies a different resistance. Each parallel resistance provides a separate path for the current to flow through. The equivalent resistance, R_{eq}, of the three resistances is given by the following formula:

$$\frac{1}{R_{eq}} = \frac{1}{R_1} + \frac{1}{R_2} + \frac{1}{R_3}$$

If $R_1 = 15\ \Omega$ (ohms), $R_2 = 10\ \Omega$, and $R_3 = 75\ \Omega$, find R_{eq}.

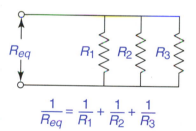

$$\frac{1}{R_{eq}} = \frac{1}{R_1} + \frac{1}{R_2} + \frac{1}{R_3}$$

FIGURE 1–3 Parallel circuit for Example 1.11.

Solution Substitute the given values for R_1, R_2, and R_3 in the formula and combine the fractions to calculate $\frac{1}{R_{eq}}$:

$$\frac{1}{R_{eq}} = \frac{1}{15} + \frac{1}{10} + \frac{1}{75}$$

The prime factors of each denominator are

$$
\begin{array}{rll}
15 = & (3)\ (5) \\
10 = (2) & (5) \\
75 = & (3)\ (5)\ (5) \\
\hline
& \downarrow\ \downarrow\ \downarrow\ \downarrow
\end{array}
$$

Then the LCD = (2) (3) (5) (5) = 150

The LCD must contain all the factors that are in each denominator.

Multiply each fraction by the necessary factor to change each denominator to the LCD:

$$\frac{1}{R_{eq}} = \frac{1(10)}{15(10)} + \frac{1(15)}{10(15)} + \frac{1(2)}{75(2)}$$

Add the fractions:

$$\frac{1}{R_{eq}} = \frac{10}{150} + \frac{15}{150} + \frac{2}{150} = \frac{10 + 15 + 2}{150} = \frac{27}{150}$$

EXAMPLE 1.11 (Cont.)

Then find R_{eq} by inverting the fraction and reducing the result:

$$R_{eq} = \frac{150}{27} = \frac{50}{9} \ \Omega \ \text{ or } 5.6 \ \Omega$$

Study the following example, which illustrates all the arithmetic operations with fractions.

EXAMPLE 1.12

Calculate by hand:

$$\frac{13}{8} - \frac{7}{5} \times \frac{15}{14} + 2 \div \frac{8}{15}$$

Solution First invert the last fraction and change the operation of division to multiplication:

$$\frac{13}{8} - \frac{7}{5} \times \frac{15}{14} + 2 \times \frac{15}{8}$$

Then, applying the order of operations, do the multiplication first:

$$\frac{13}{8} - \left(\frac{\overset{1}{\cancel{7}}}{\underset{1}{\cancel{5}}} \times \frac{\overset{3}{\cancel{15}}}{\underset{2}{\cancel{14}}} \right) + \left(\overset{}{2} \times \frac{15}{\underset{4}{\cancel{8}}} \right) = \frac{13}{8} - \frac{3}{2} + \frac{15}{4}$$

Now combine the fractions over the LCD, which is 8, and do the addition and subtraction:

$$\frac{13}{8} - \frac{3(4)}{2(4)} + \frac{15(2)}{4(2)} = \frac{13 - 12 + 30}{8} = \frac{31}{8}$$

See Example 1.26 in Section 1.5 for the calculator solution of Example 1.12.

EXERCISE 1.2

In problems 1 through 6 simplify each fraction (reduce to lowest terms).

1. $\dfrac{6}{10}$

2. $\dfrac{12}{36}$

3. $\dfrac{28}{35}$

4. $\dfrac{27}{54}$

5. $\dfrac{39}{52}$

6. $\dfrac{34}{51}$

In problems 7 through 40, *calculate each problem by hand,* and express the answer in terms of a fraction or a whole number.

7. $\dfrac{5}{9} \times \dfrac{6}{25}$

8. $\dfrac{2}{21} \times \dfrac{7}{16}$

9. $\dfrac{8}{9} \div \dfrac{2}{3}$

10. $\dfrac{3}{11} \div \dfrac{1}{22}$

11. $\dfrac{3}{5} \times \dfrac{15}{7} \times \dfrac{14}{9}$

12. $\dfrac{3}{25} \times \dfrac{4}{9} \times \dfrac{5}{12}$

13. $6 \times \dfrac{4}{5} \div \dfrac{8}{15}$

14. $\dfrac{5}{6} \div 2 \times \dfrac{3}{10}$

15. $\dfrac{3}{17} \div \left(\dfrac{1}{34} \times \dfrac{1}{2} \right)$

16. $\left(\dfrac{9}{8} \div \dfrac{3}{4} \right) \div \dfrac{3}{2}$

17. $\dfrac{6}{7} + \dfrac{1}{7}$

18. $\dfrac{3}{10} + \dfrac{5}{10}$

19. $\dfrac{3}{8} + \dfrac{1}{4}$

20. $\dfrac{4}{15} + \dfrac{5}{6}$

21. $\dfrac{9}{20} - \dfrac{1}{6}$

22. $\dfrac{5}{14} - \dfrac{4}{21}$

23. $\dfrac{3}{4} - \dfrac{1}{2} + \dfrac{7}{10}$

24. $\dfrac{1}{6} + \dfrac{11}{20} - \dfrac{2}{3}$

25. $2 + \dfrac{7}{8} + \dfrac{2}{3}$

26. $\dfrac{5}{2} + \dfrac{5}{3} + \dfrac{5}{6}$

27. $\dfrac{16}{9} \times \dfrac{1}{2} + \dfrac{1}{4}$

28. $\dfrac{1}{6} + \dfrac{3}{8} \div \dfrac{1}{4}$

29. $\dfrac{1}{2} + \dfrac{2}{3} \div \dfrac{2}{5}$

30. $\dfrac{3}{4} \times \dfrac{5}{3} - \dfrac{1}{8}$

31. $\dfrac{5}{6} + \dfrac{2}{5} - \dfrac{3}{7} \times \dfrac{7}{10}$

32. $\dfrac{5}{6} - \dfrac{2}{5} \div \dfrac{8}{15} \times \dfrac{2}{9}$

33. $3 \times \dfrac{1}{6} + \dfrac{7}{2} - \dfrac{4}{5} \div 8$

34. $\dfrac{3}{100} + \dfrac{7}{10} \times \dfrac{2}{35} - \dfrac{1}{50}$

35. $\left(\dfrac{1}{2} + \dfrac{1}{3} \right) \times \left(8 \div \dfrac{4}{3} \right)$

36. $\left(1 + \dfrac{3}{8} \right) \div \left(1 - \dfrac{3}{8} \right)$

37. Peter decides to give $1000 of his savings to charity. He gives two-fifths to the heart fund, one-third of what is left to the cancer fund, and divides the balance evenly between an AIDS charity and an environmental fund. How much is given to the environmental fund?

38. A $40,000 inheritance is distributed as follows: half to the spouse, three-fourths of what is left to the children, and the remainder to a private nurse. How much does the nurse inherit?

39. A bookcase will be 8 ft $3\frac{1}{2}$ in. high and will contain six equally spaced shelves and a top, each $\frac{1}{2}$-in. thick. How far apart should each shelf be?

40. A grocery clerk is given three orders for cheese: $\frac{2}{3}$ lb, $1\frac{1}{2}$ lb and $\frac{3}{4}$ lb. What is the total amount of cheese the clerk has to prepare?

Applications to Electronics In problems 41 through 48 calculate each by hand, and *express the answer in terms of a fraction or a whole number.*

41. The voltage of a circuit is given by Ohm's law:

$$V = I \times R$$

where I = current and R = resistance. Calculate V in volts (V) substituting $I = \frac{5}{6}$ mA (milliamps) and $R = \frac{3}{10}$ kΩ (kilohms) in Ohm's law.

42. The power of a circuit is given by the power formula:

$$P = I \times I \times R$$

Calculate P in watts (W) substituting $I = \frac{2}{3}$ mA (milliamps) and $R = \frac{3}{10}$ MΩ (megohms) in the formula.

43. In Example 1.11, find R_{eq} when $R_1 = 2\ \Omega$, $R_2 = 3\ \Omega$, and $R_3 = 6\Omega$.

44. In Example 1.11, if $R_{eq} = 5\ \Omega$, $R_1 = 10\ \Omega$, and $R_2 = 20\ \Omega$, then R_3 is given by

$$\frac{1}{R_3} = \frac{1}{R_{eq}} - \frac{1}{R_2} - \frac{1}{R_1}$$

Substitute the values for R_{eq}, R_2, and R_1, and calculate the value of $\frac{1}{R_3}$. Then invert the fraction to find the value of R_3.

45. Figure 1–4 shows a circuit containing three different resistances R_1, R_2, and R_3. Each subscript identifies a different resistance. R_1 is connected in series to two resistances R_2 and R_3, which are connected in parallel. The total current passing through R_1 is divided between R_2 and R_3. The total resistance of this series-parallel circuit is given by

$$R_T = R_1 + \frac{R_2 R_3}{R_2 + R_3}$$

If $R_1 = 5\ \Omega$, $R_2 = 3\ \Omega$, and $R_3 = 9\ \Omega$, substitute these values in the formula and calculate R_T. Note: $R_2 R_3$ means $R_2 \times R_3$.

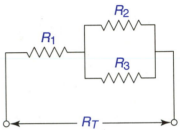

FIGURE 1–4   Series-parallel circuit for problems 45 and 46.

46. In problem 45, the total resistance of the series-parallel circuit is also given by

$$R_T = R_1 + 1 \div \left(\frac{1}{R_2} + \frac{1}{R_3} \right)$$

or

$$R_T = R_1 + \frac{1}{\dfrac{1}{R_2} + \dfrac{1}{R_3}}$$

If $R_1 = 3\ \Omega$, $R_2 = 10\ \Omega$, and $R_3 = 15\ \Omega$, substitute these values for R_1, R_2, and R_3, and calculate the value of R_T.

47. In most computer languages, as in arithmetic, operations in parentheses are performed first. Calculate the following BASIC expression (see Exercise 1.1, problem 27):

$$(1/10) * 3 / (3/2 - 1) * (20/3)$$

48. As in problem 47, calculate the following BASIC expression:

$$(4/5 - 1/10) / (7/10 + 1/5) * 45/100$$

☰ 1.3
DECIMALS AND PERCENTAGES

Decimals

Our number system is called the decimal system because it is based on the number ten. *Dec* means *ten* in Latin. For example, in expanded form, the number 5643 represents

$$5643 = 5000 + 600 + 40 + 3 = 5(1000) + 6(100) + 4(10) + 3(1)$$

From right to left, each digit represents a multiple of 1, 10, 100, 1000, and so forth. When a number is written with a decimal point, such as 56.43, the decimal digits 4 and 3 represent multiples of fractions whose numerators are 1 and *denominators* are 10, 100, 1000, etc., as follows:

$$56.43 = 5(10) + 6(1) + 4\left(\frac{1}{10}\right) + 3\left(\frac{1}{100}\right) = 50 + 6 + \frac{4}{10} + \frac{3}{100}$$

Decimals, then, are fractions with denominators of 10, 100, 1000, etc. The number of decimal places equals the number of zeros in the denominator as follows:

$$0.5 = \frac{5}{10}$$

$$0.21 = \frac{21}{100}$$

$$0.076 = \frac{76}{1000}$$

$$4.6 = 4 + \frac{6}{10} = \frac{46}{10}$$

To *add or subtract decimals,* line up the decimal points and the columns. Then add or subtract in the same way as whole numbers.

The following example shows how to add and subtract decimals.

EXAMPLE 1.13

Calculate the following:

$$7.74 + 5.05 - 10.4$$

Solution   Line up the decimal points and the columns. Add the first two numbers, bringing down the decimal point:

$$
\begin{array}{r}
7.74 \\
+\ 5.05 \\
\hline
12.79
\end{array}
$$

Then subtract as follows:

$$
\begin{array}{r}
12.79 \\
-\ 10.40 \\
\hline
2.39
\end{array}
$$

Note that a zero is added at the end of 10.4 to aid in the calculation, but it does not change the value of the decimal.

To *multiply decimals,* multiply the same way as with whole numbers. Then add the decimal places in all the numbers to determine the total number of decimal places in the answer. Study the following examples, which show how to multiply decimals.

EXAMPLE 1.14

Multiply:

$$0.1 \times 0.04$$

Solution   The number 0.1 has one decimal place, and 0.04 has two decimal places. Therefore, there are three decimal places in the answer:

$$0.1 \times 0.04 = 0.004$$

EXAMPLE 1.15

Multiply:

$$0.20 \times 0.31 \times 0.5$$

Solution   Multiply as if the numbers were 2, 31, and 5. *The zero at the end of 0.20 does not count as a decimal place.* It identifies the accuracy of the number. Then add the decimal places to determine the number of decimal places in the answer. The total decimal places is four, so there are *four* decimal places in the answer:

$$0.20 \times 0.31 \times 0.5 = 0.062 \times 0.5 = 0.0310$$

Note that the zero at the end can be dropped and the answer written as 0.031.

EXAMPLE 1.16

Calculate:

$$0.1 \times 0.03 + 0.013$$

Solution   First multiply, adding the decimal places:

$$0.1 \times 0.03 = 0.003$$

Then add, lining up the decimal point:

$$\begin{array}{r} 0.003 \\ +\ 0.013 \\ \hline 0.016 \end{array}$$

To *divide decimals,* move the decimal point in the divisor (the number you are dividing by) to the right as many places as there are in the number. Then do the same to the dividend (the number you are dividing into). For example, to divide 0.132 by 0.12, 0.12 is the divisor, so move the decimal point as follows:

$$0.132 \div 0.12 = 13.2 \div 12.$$

The number 0.132 is the dividend. Moving the decimal point to the right in the two numbers is the same as multiplying both numbers by 100. This does not change the quotient (the division result) of the two numbers. To better understand why, it is helpful to express the division as a fraction. Then multiplying both numbers by 100 is the same as multiplying numerator and denominator by 100, which does not change the value of the fraction:

$$\frac{0.132\,(100)}{0.12\,(100)} = \frac{13.2}{12.}$$

After you move the decimal point, then divide the numbers, keeping the decimal point in the same position as the numerator (or dividend):

$$\frac{13.2}{12} = 1.1$$

The 12 divides into the 13 once with one remaining. Then comes the decimal point and the 12 divides into 1.2, 0.1 time.

ERROR BOX

Counting the number of decimal places for multiplication is a common error. A zero may or may not count as a decimal place. There are two situations when it does not count:

1. When a zero is in front of the decimal point, it does not count: 0.103 counts as three decimal places for multiplication. The purpose of the zero in front is to emphasize the decimal point.

2. When a zero is at the end of the decimal, it does not count: 0.1030 counts as three decimal places for multiplication. The purpose of the zero at the end is to identify the accuracy and precision of the number. 0.1030 is more accurate and precise than 0.103.

Practice Problems: Give the number of decimal places in each number:

1. 1.52   2. 1.052   3. 0.52   4. 0.502   5. 0.5020

Answers: 5. 3   4. 3   3. 2   2. 3   1. 2

Study the next example, which combines subtraction, multiplication, and division of decimals.

EXAMPLE 1.17

Calculate:

$$\frac{0.5 \times 0.02}{0.11 - 0.06}$$

Solution Study the steps carefully. *The fraction line is like parentheses; you must first do the operations in the numerator and denominator before dividing.*

EXAMPLE 1.17 (Cont.)

Multiply the numbers in the numerator, counting the decimal places, and subtract the numbers in the denominator, lining up the decimal point:

$$\frac{0.5 \times 0.02}{0.11 - 0.06} = \frac{0.01}{0.05}$$

Then move the decimal point two places to the right in the numerator and denominator and divide:

$$\frac{0.010}{0.05} = \frac{1.0}{5} = 0.2$$

Percentages

A percentage (or percent) is a convenient way of writing a fraction whose denominator is 100. The numerator is written with the percent sign (%) which represents the denominator of 100. For example,

$$20\% = \frac{20}{100} = 0.20$$

$$150\% = \frac{150}{100} = 1.50$$

$$8\frac{1}{4}\% = \frac{8.25}{100} = 0.0825$$

Therefore:

> **Rule** To change from a percent to a decimal, move the decimal point two places to the left. To change from a decimal to a percent, move the decimal point two places to the right.

Study the next example, which shows how to change between percents, decimals, and fractions.

EXAMPLE 1.18

Express each fraction as a percent and a decimal:

$$\frac{1}{4}$$

$$\frac{3}{8}$$

Solution First, to express $\frac{1}{4}$ as a percent, first change the denominator to 100 by multiplying the numerator and denominator by 25. Then move the decimal point two places to the left to change to a decimal:

$$\frac{1\,(25)}{4\,(25)} = \frac{25}{100} = 25\% = 0.25$$

EXAMPLE 1.18 (Cont.)

Next, to express $\frac{3}{8}$ as a percent, the denominator cannot easily be changed to 100. First change the fraction to a decimal with two decimal places as follows. Add a decimal point and three zeroes to the numerator and divide 8 into 3.000. This will give you a decimal with three decimal places:

$$\frac{3}{8} = \frac{3.000}{8} = 0.375$$

Now move the decimal point to the right two places to express as a percent:

$$0.375 = 37.5\%$$

EXAMPLE 1.19

Express each decimal as a percent and a fraction:
$$0.80$$
$$0.015$$

Solution   First, to express 0.80 as a percent, move the decimal point two places to the right. Then write the fraction by putting the percent number over a denominator of 100 and reduce the fraction:

$$0.80 = 80\% = \frac{80}{100} = \frac{4}{5}$$

Second, to express 0.015 as a percent, move the decimal point two places to the right. Then write the fraction with a denominator of 100:

$$0.015 = 1.5\% = \frac{1.5}{100}$$

A fraction is not in simplest form if it contains a decimal in the numerator or denominator. Eliminate the decimal in the numerator by multiplying the numerator and denominator by 10; then reduce the fraction:

$$\frac{1.5\,(10)}{100\,(10)} = \frac{15}{1000} = \frac{3}{200}$$

EXAMPLE 1.20

Express each percentage as a decimal and a fraction:
$$7.5\%$$
$$110\%$$

Solution   To express 7.5% as a decimal, move the decimal point two places to the left:

$$7.5\% = 0.075$$

Write the fraction by putting the percent number 7.5 over 100. However, the fraction is not in simplest form until the decimal is eliminated. Multiply the numerator and denominator by 10; then reduce the fraction:

$$7.5\% = \frac{7.5}{100} = \frac{7.5\,(10)}{100\,(10)} = \frac{75}{1000} = \frac{3}{40}$$

EXAMPLE 1.20 (Cont.)

To express 110% as a decimal, move the decimal point two places to the left. Change the percent to a fraction with a denominator of 100 and reduce:

$$110\% = 1.10 = \frac{110}{100} = \frac{11}{10}$$

Note that 100% = 1 and a percent greater than 100 represents a number greater than one. For example, if a store marks up an item whose wholesale price is $20 by 200%, this means that the wholesale price is increased by two times or $40. The retail price is then

$$\$20 + (\$20)\,(200\%) = \$20 + (\$20)\,(2.00) = \$20 + \$40 = \$60.$$ ▪

Study the next example that closes the circuit and shows an electrical application of decimals and percents.

EXAMPLE 1.21

Close the Circuit

The current in a circuit is 1.5 A (amps). The voltage is increased by 12% while the resistance is kept constant. By Ohm's law, the current will increase by the same percentage. How much is the increase in the current, and what is the new current?

Solution   Multiply the original current by 12% to find the increase:

$$1.5(12\%) = 1.5(0.12) = 0.18 \ \text{A}$$

Then add this amount to the original current to find the new current. Change 1.5 to 1.50:

$$1.50 \ \text{A} + 0.18 \ \text{A} = 1.68 \ \text{A}$$

A 12% increase means the new current is 112% of the original current. Therefore, the new current can also be calculated by multiplying the original current by 112%:

$$1.5(112\%) = (1.5)(1.12) = 1.68 \ \text{A}$$ ▪

The next example illustrates a practical application of fractions, decimals, and percentages.

EXAMPLE 1.22

The list price of a computer is $900. A discount store sells it for $\frac{1}{3}$ off the list price. What is the discount price and the total cost including a tax of $7\frac{1}{2}\%$?

Solution   You can find the discount price by subtracting $\frac{1}{3}$ of the list price from itself:

$$\$900 - \tfrac{1}{3}(\$900) = \$900 - \$300 = \$600$$

Since $\frac{1}{3}$ off means there is $\frac{2}{3}$ left, you can also find the discount price by multiplying the list price by $\frac{2}{3}$:

$$\$900(\tfrac{2}{3}) = \$600$$

EXAMPLE 1.22 (Cont.)

The total cost, including a tax of $7\frac{1}{2}\%$, can be calculated by adding $7\frac{1}{2}\%$, or 7.5%, of the discount price to itself:

$$\$600 + 600(7.5\%) = 600 + 600(0.075) = 600 + 45 = \$645$$

More directly, you can multiply the discount price by 107.5%:

$$\$600(107.5\%) = 600(1.075) = \$645$$

It is often necessary to calculate the percent change in a value when you know the initial and final values. *Percent change* is given by the following formula:

$$\text{Percent change} = \frac{(\text{Difference in value}) \times 100\%}{\text{Initial value}} \qquad (1.7)$$

where the difference in value can be an increase or a decrease in value. If the final value is *greater* than the initial value, then the percent change is a percent increase. If the final value is *less* than the initial value, then the percent change is a percent decrease.

Study the next example, which closes the circuit and shows an electrical application involving the resistance of a wire.

EXAMPLE 1.23

Close the Circuit

The resistance of a copper wire increases when the temperature increases. A copper wire whose resistance is 2.0 Ω (ohms) increases to 2.1 Ω when heated. What is the percent increase in the resistance?

Solution Apply formula (1.7):

$$\text{Percent change} = \frac{(2.1 - 2.0)(100\%)}{2.0} = \frac{(0.1)(100\%)}{2.0} = \frac{10\%}{2.0} = 5\%$$

EXERCISE 1.3

In problems 1 through 16, *calculate each exercise by hand*. Use the calculator only to check the results.

1. $1.05 + 8.98 + 0.06$

2. $15.64 + 4.36 - 19.09$

3. 8.1×0.5

4. 0.031×0.20

5. $2.3 \times 1.5 \times 0.2$

6. $5.0 \times 0.5 \times 2.5$

7. $\dfrac{5.1}{0.17}$

8. $\dfrac{3.60}{0.030}$

9. $3.2 \times \dfrac{0.10}{4.0}$

10. $\dfrac{0.72}{3.0 \times 0.3}$

11. $(1.3 + 2.8) \times (1.6 + 1.4)$

12. $(0.4 + 0.1) \times (0.08 - 0.04)$

13. $\dfrac{1.2 \times 1.0 \times 0.03}{0.0050 + 0.0040}$

14. $\dfrac{0.02 \times (3.0 + 5.0)}{0.004}$

15. $\dfrac{0.5 \times 7.0 + 0.6 \times 7.0}{2.3 - 1.2}$

16. $\dfrac{8.0 \times 0.6 \times 0.04}{40 \times 30 \times 0.02}$

In problems 17 through 30, express each number as a fraction, decimal, and percent.

17. 20%

18. 86%

19. 0.17

20. 0.06

21. 5.6%

22. $8\frac{1}{2}\%$

23. 0.004

24. 0.025

25. $\dfrac{3}{4}$

26. $\dfrac{2}{5}$

27. $\dfrac{3}{20}$

28. $\dfrac{17}{1000}$

29. 1.50

30. $\dfrac{9}{8}$

In problems 31 through 40, *solve each applied problem by hand.* Use the calculator only to check your results.

31. Approximately 0.003% of seawater is salt. How many grams of salt are contained in 1,000,000 grams of seawater?

32. One discount technical supply store sells a voltmeter for one-third off its usual price of $59.40. Another discount store sells the same voltmeter for 25% off its usual price of $56.00.
   (a) How much is each voltmeter?
   (b) How much do you save by purchasing the less expensive voltmeter?

33. Your pocket computer costs $40.00 today. At a price increase of 5% a year,
   (a) How much will it cost to replace it one year from now?
   (b) How much will it cost to replace it two years from now?

34. A retailer pays the publisher 20% less than the list price for this text. Suppose the list price is $45:
   (a) How much does the retailer pay the publisher for the text?
   (b) What is the percent markup (increase) the retailer adds on to sell the book at the list price? (The answer is not 20%.) (See Example 1.23.)

35. The list price of a calculator is $50. A discount store sells it for 20% less than list.
   (a) What is the discount price?
   (b) What is the total cost including a tax of $7\frac{1}{2}\%$? (See Example 1.22.)

36. The price of an ammeter is $20.00. What is the total cost, including tax of $8\frac{1}{4}\%$? (Note: $8\frac{1}{4}\% = 0.0825$. See Example 1.22.)

37. Which rent increase costs less for two years for an apartment whose present rent is $1000 per month: one increase of 10% for the two years or two increases of 7%, one each year?

38. The earth's surface receives 317 Btu (British thermal units) per hour per square foot of solar energy on a clear day. How many Btu per hour are received by a solar panel whose total area is 8 square feet if 75% of it is covered with solar cells?

39. Arrange the following in order of increasing size:

$$1.19, \quad 110\%, \quad \frac{9}{8}, \quad 1\frac{1}{6}, \quad \frac{1120}{1000}$$

40. Arrange the following in order of increasing size:

$$0.03, \quad 0.50\%, \quad \frac{1}{40}, \quad \frac{0.5}{25}$$

Applications to Electronics   In problems 41 through 46 *solve each problem by hand.* Use the calculator to check your results.

41. Four batteries connected in series aiding have the following voltages: 13.01 V (volts), 12.52 V, 6.21 V, and 8.38 V. Series aiding means that the batteries are connected in such a way that *the total voltage equals the sum of the voltages.* Find the total voltage.

42. The current in a circuit is initially 1.10 mA (milliamps). The voltage is increased by 10%, causing the current to increase by the same percentage.
   (a) What is the increase in the current?
   (b) What is the new current? (See Example 1.21.)

43. The current in a circuit increases from 0.80 A to 0.84 A when the voltage is increased from 12.0 V to a higher value.
   (a) What is the percent increase in the current?
   (b) If the percent increase in the voltage is the same as that in the current, what is the new voltage? (See Example 1.21.)

44. An electronics company produces 200 microprocessors in one week. During the next week several employees are sick and the company only produces 150 microprocessors.
   (a) What is the percent decrease in production?
   (b) If 2% are defective, how many microprocessors are produced in the two weeks that are *not* defective?

45. The resistance of a wire increases from 0.5 Ω (ohms) to 0.6 Ω when heated. What is the percent increase in the resistance? (See Example 1.23.)

46. The current in a circuit decreases from 2.0 A to 1.8 A when the voltage is decreased. What is the percent decrease in the current? (See Example 1.23.)

☰ 1.4
PRECISION AND ACCURACY

The results predicted by electrical laws are not always easy to produce experimentally. In the laboratory, measurements taken with instruments and meters are subject to certain degrees of precision and accuracy. In addition, values of electrical and electronic components tend to vary under certain conditions. The numbers obtained are usually not exact. For example, suppose you read the scale on a voltmeter, when measuring the voltage of a battery, to be 1.5 V. This number, 1.5, is considered an *approximate* number. A more sensitive voltmeter with an expanded scale might enable you to read the voltage to be 1.53 V. However, this number, 1.53, is still considered to be an approximate number because a measurement can usually always be taken more accurately. Measurements are usually approximate. One of the only times a measurement is an exact number is when it is obtained by counting. For example, if you count the number of resistances in a circuit to be 3, then this number, 3, is considered an exact number.

   It is, therefore, necessary to understand what we mean by the precision and accuracy of numbers. However, before looking at precision and acccuracy, you must know the rules for rounding off a number. Consider the following fractions and their decimal equivalents:

$$\frac{1}{3} = 0.33333 \ldots \qquad \frac{2}{3} = 0.66666 \ldots$$

These decimals go on forever. To express them to two decimal places, you round off as follows:

$$0.3333 \ldots \approx 0.33 \qquad 0.6666 \ldots \approx 0.67$$

You either do not change the last digit, or you increase it by one, according to the following rules for rounding off:

1. If the digit to the right of the last rounded digit is *less than 5,* do not change the last rounded digit.
2. If the digit to the right of the last rounded digit is *5 or more than 5,* increase the last rounded digit by one.

Some further examples of rounding off numbers are

$$3.1416 \approx 3.142$$

$$57 \approx 60$$

$$1.034 \approx 1.03$$

$$2400 \approx 2000$$

Observe that a whole number such as 57, rounded off to one digit, becomes 60 and 2400 becomes 2000.

Precision and accuracy are two different ideas. The accuracy of an approximate number is based on what are called the *significant digits* of the number. Significant digits are defined as follows:

1. All nonzero digits are significant.
2. Any zeros that are *not* used as placeholders for the decimal point are significant.

For example, the number 0.123 has three significant digits. The zero is a placeholder for the decimal point and is not significant, whereas the number 10.23 has four significant digits. The zero is not a placeholder and counts as a significant digit.

▪ The *accuracy* of a number is measured by the number of significant digits the number contains.

▪ The number 10.23 is considered more accurate than 0.123 because it has more significant digits.

▪ The *precision* of a number is measured by the *decimal place of the last digit.*

For example, 0.123 is more precise than 10.23 since it contains three decimal places and 10.23 contains two decimal places. The number 0.123 is precise to the nearest thousandth whereas 10.23 is precise to the nearest hundredth.

The concept of accuracy and significant digits is used more often than precision and is important in experimental calculations. Study Table 1.1 which compares the precision and accuracy of six numbers.

TABLE 1.1   Precision and Accuracy of Six Numbers

Number	Significant Digits	Explanation	Comment
2313	Four	All nonzero digits are significant	Precise to nearest unit
30.2	Three	The zero is not a placeholder	Precise to the nearest tenth
18,000	Two	The zeros are assumed to be placeholders	Least precise
0.0703	Three	First two zeros are placeholders, last zero is not	Most precise
0.004	One	All zeros are placeholders	Least accurate
30.450	Five	Both zeros are not placeholders	Most accurate

Observe that the number 18,000 in Table 1.1 is shown as having only two significant digits. This number could have more significant digits; however, unless you have more information about the zeros at the end of an integer, they are assumed to be placeholders and not significant.

Computers and calculators can only work with a finite number of digits and must either round off or truncate numbers. *Truncate* means just dropping the digits to the right of the last rounded digit. For example, the number 2.0175 truncated to three digits is just 2.01. Most calculators can be set to round off to a certain number of decimal places using the ⌐FIX⌐ key. After pressing the ⌐FIX⌐ key, press the number key for the number of decimal places you want to round off to.

Note that, in general, *the results of calculations with experimental data cannot be more accurate than the data.* That is, if the data is accurate to three significant digits, the result can, at best, be accurate to three significant digits.

The following example serves to close the circuit between theory and practice. It shows how to apply the preceding mathematical ideas to a problem in electronics.

EXAMPLE 1.24

Close the Circuit

Figure 1–5 shows a series circuit containing three different resistances R_1, R_2, and R_3. A series circuit has only one current path, and the same current flows through each resistance. The current shown is electron flow from negative to positive. Electron flow is the current flow used throughout the text.

1. One student measures the voltage drops across R_1, R_2, and R_3 with a voltmeter to be $V_1 = 2.1$ V, $V_2 = 2.4$ V, and $V_3 = 1.85$ V. Find the total or applied voltage V_T to two significant digits.

EXAMPLE 1.24 (Cont.)

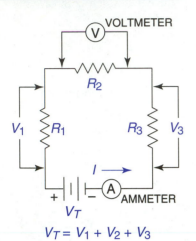

VT = V1 + V2 + V3

FIGURE 1−5   Series circuit for Example 1−24.

2. Three students measure the current with their ammeters to be 0.14 A, 0.12 A, and 0.11 A. Find the average current in the circuit to two significant digits.

Solution   1. In a series circuit, the sum of the voltage drops equals the total or applied voltage:

$$V_T = 2.1 \text{ V} + 2.4 \text{ V} + 1.85 \text{ V} = 6.35 \text{ V}$$

Rounding off to two digits, the total voltage is

$$V_T = 6.35 \approx 6.4 \text{ V}$$

2. The average current to two significant digits is

$$I = \frac{0.14 + 0.12 + 0.11}{3} = \frac{0.37}{3} = 0.12333\ldots \approx 0.12\,A$$

EXERCISE 1.4

In problems 1 through 10, tell how many significant digits are in each number.

1. 21.8
2. 30.2
3. 0.15
4. 0.030
5. 5.20

6. 10.01
7. 3210
8. 500
9. 6.060
10. 0.0050

In problems 11 through 18, round off each number to (a) two decimal places, (b) three significant digits.

11. 5.152
12. 3.100
13. 31.261
14. 0.4445

15. 321.872
16. 100.90
17. 28.999
18. 18.9625

In problems 19 and 20, calculate each result, using the calculator if necessary, and round off the answer.

19. The largest scientific building in the world is the Vehicle Assembly Building in Complex 39 at the John F. Kennedy Space Center, Cape Canaveral, Florida. It is 219 m (meters) long, 159 m wide and 160 m high. Assuming the building has a rectangular shape, find the volume it occupies in cubic meters to three significant digits. Note: Volume of a rectangular box = (length) × (width) × (height).

20. The fastest runner in baseball was Ernest Swanson, who took only 13.3 seconds to circle the bases, a distance of 360 ft, in 1932. What was his average speed in feet per second and in miles per hour to three significant digits? (Note: 1 ft/s = 0.6818 mi/h.)

Applications to Electronics In problems 21 through 28, calculate each result and round off the answer.

21. The screen of a computer monitor is measured to be 12 in. long and 8.5 in. wide. What is the area of the screen to two significant digits?

22. A microprocessor chip measured with a ruler is found to be 2.2 cm (centimeters) long and 1.8 cm wide. The thickness measured with a micrometer, which is a precision instrument for measuring small distances, is found to be 0.0115 cm. Find the volume of the chip to two significant digits. Note: Volume = (length) × (width) × (thickness).

23. Sophia precisely measures the resistance of a computer circuit at five equal intervals during a 1-h period of operation. Her readings are 1.28 Ω, 1.3 Ω, 1.25 Ω, 1.2 Ω, and 1.35 Ω. Find the average value of the resistance to (a) two significant digits and (b) three significant digits. Note: the result is only as accurate as the *least* accurate measurement.

24. The current I in the main power circuit of a house is measured by one ammeter as 11.2 A and an hour later by another ammeter as 7.3 A. During this time the voltage V has remained constant at 115 V. Given that power $P = V \times I$ or VI, find the *decrease* in power P in watts (W) to two significant digits.

25. Figure 1–6 shows two different resistances connected in series. The voltage drops across R_1 and R_2 are measured to be $V_1 = 5.3$ V and $V_2 = 3.4$ V. The current is measured to be $I = 0.14$ A. Ohm's law for resistance states

$$R = \frac{V}{I}$$

Find R_1 and R_2 to two significant digits by applying Ohm's law to each of the voltage drops.

26. In Figure 1–6, the resistances are measured separately with no current flow to be $R_1 = 68$ Ω and $R_2 = 55$ Ω. When connected in series to a voltage source, the current I is measured to be 0.15 A. Using Ohm's law for voltage,

$$V = I \times R \text{ or } IR$$

find the voltage drop across each resistance V_1 and V_2 to two significant digits.

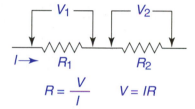

FIGURE 1–6 Ohm's law and series circuit for problems 25 and 26.

27. Truncate the following numbers to integers:
   (a) 3.6.
   (b) –22.3.
   (c) –1.25.
   (d) 0.98.
   (e) When can truncation result in a larger number?

28. The costs of three items are $12.82, $13.98 and $18.76. Suppose that a computer is programmed to truncate all numbers to integers before calculating.
   (a) What is the total cost of the three items as calculated by the computer?
   (b) If the numbers are not truncated, what is the difference between the total cost rounded off to the nearest integer and the total cost calculated by the computer?

≡ 1.5
HAND CALCULATOR OPERATIONS

We live in an age of electronic computation. Imagine a long division problem that might take you one minute to do by hand. The hand calculator takes a few hundredths of a second to do such a problem. One of the world's fastest computers, the CRAY-2, can do more than 200,000,000 of these problems in one second!

You need to master the calculator. It will allow you to spend less time doing calculations and more time learning important concepts. However, speed alone is of little value unless you can understand and apply concepts. For electricity and electronics, you need a scientific or engineering calculator. Both these types of calculators have an exponential key $\boxed{y^x}$ or $\boxed{x^y}$; trigonometric keys $\boxed{\sin}$, $\boxed{\cos}$, and $\boxed{\tan}$; and logarithmic keys $\boxed{\log}$ and $\boxed{\ln}$. The engineering calculator is very useful for electronics and has keys for metric notation: $\boxed{M}$ for mega, $\boxed{k}$ for kilo, $\boxed{m}$ for milli, and $\boxed{\mu}$ for micro. Graphing calculators are also available and are helpful with graphs of voltage and current waves in ac circuits, exponential graphs and other graphs of electronic phenomena.

This section is not designed to replace the instruction manual for your calculator. It is designed to be used with your manual. Study your instruction manual thoroughly to understand your particular calculator. Some of your keys and operations may be a little different from the examples given here. Some calculators have an execute $\boxed{\text{EXE}}$ key or $\boxed{\text{ENT}}$ key instead of an $\boxed{=}$ key such as the Hewlett Packard calculators that use reverse polish notation (RPN) or the graphics calculators produced by Casio, Sharp, and Texas Instruments. In this text, all calculator operations are shown with an $\boxed{=}$ key. The key strokes shown will work on most scientific calculators.

One of the critical things to keep in mind about your calculator is that *it cannot think!* You possess this unique ability. You must understand the problem, interpret the information, and key the information into the calculator correctly. Furthermore, and most importantly, you must be able to judge if the answer makes sense. If it does not, and this can happen often, you must understand the mathematical concepts well enough to troubleshoot for the error. You should estimate or approximate answers, whenever possible, before calculating, so you have some idea about the size of the answer. This not only provides a check of the results but helps you to understand the concepts better. These are some of the skills that this section and others in the text are designed to help you develop.

Arithmetic Operations and Memory

Scientific calculators are programmed to perform the basic arithmetic operations $\boxed{+}$, $\boxed{-}$, $\boxed{\times}$, and $\boxed{\div}$ according to the order of operations discussed in Section 1.1. This is called the *algebraic operating system*. See Examples 1.1 and 1.2 at the beginning of the chapter to test the order of operations on your calculator. If your calculator has a $\boxed{\text{MODE}}$ key, make sure it is set for normal calculation. The following examples show how to use the basic operations on the calculator.

EXAMPLE 1.25

Calculate:

$$\frac{5 \times 21 \times 12}{15 \times 7 \times 3}$$

Solution This is Example 1.3 in this chapter. One way to calculate this problem is to first multiply the numbers in the denominator and store the result in the memory. Then multiply the numbers in the numerator and divide by the memory:

$$15 \; \boxed{\times} \; 7 \; \boxed{\times} \; 3 \; \boxed{=} \; \boxed{M_{in}} \; 5 \; \boxed{\times} \; 21 \; \boxed{\times} \; 12 \; \boxed{\div} \; \boxed{MR} \; \boxed{=} \; \rightarrow 4$$
$$\qquad\qquad\downarrow\qquad\qquad\qquad\qquad\qquad\uparrow$$
$$\qquad\qquad 315 \qquad\qquad\qquad\qquad\qquad 315$$

Observe that you have to press the $\boxed{=}$ key after you multiply the numbers in the denominator so that the calculator will display the product 315 before you store it in the memory. The $\boxed{M_{in}}$ key stores the result 315. The $\boxed{MR}$ key recalls the result and enters it into the operations being performed. Some calculators use $\boxed{STO}$ or $\boxed{X \rightarrow M}$ to store a number and $\boxed{RCL}$ or $\boxed{RM}$ to recall a number from memory. Some calculators have more than one memory and it is necessary to key in the number of the memory after $\boxed{M_{in}}$ and $\boxed{MR}$, such as $\boxed{M_{in}}$ 07 and $\boxed{MR}$ 07.

Another way to calculate Example 1.25 is to use parentheses and group the numerator and denominator since the fraction line is like parentheses:

$$\boxed{(} \; 5 \; \boxed{\times} \; 21 \; \boxed{\times} \; 12 \; \boxed{)} \; \boxed{\div} \; \boxed{(} \; 15 \; \boxed{\times} \; 7 \; \boxed{\times} \; 3 \; \boxed{)} \; \boxed{=} \; \rightarrow 4$$

There are other ways to do this example on the calculator. You should try another way and see if you can get the correct answer.

◾

EXAMPLE 1.26

Calculate:

$$\frac{13}{8} - \frac{7}{5} \times \frac{15}{14} + 2 \div \frac{8}{15}$$

Solution This is Example 1.12 in this chapter. One direct calculator solution is

$$13 \; \boxed{\div} \; \boxed{8} \; \boxed{-} \; 7 \; \boxed{\div} \; 5 \; \boxed{\times} \; 15 \; \boxed{\div} \; 14 \; \boxed{+} \; 2 \; \boxed{\times} \; 15 \; \boxed{\div} \; 8 \; \rightarrow 3.875$$

Note that the last fraction must be inverted and the operation changed to multiplication or else parentheses must be used. You should try other ways to do this example and see if you can get the correct answer.

◾

Since you will make mistakes keying in numbers or operations, you need some way of checking that the calculator answer makes sense and is not too large or small to be correct. The next two examples show how to estimate the answer to provide a check on the calculator result.

EXAMPLE 1.27

Estimate the answer *by hand* and then calculate:

$$\frac{(64+57)\times 320}{840-50\times 8}$$

Solution   An estimate to the actual answer can readily be done *without the calculator* by rounding off each number to one significant digit. Use the rules for rounding off in Section 1.4; that is, if the second digit is 5 or more, increase the first digit by one. The example with the numbers rounded off to one digit is

$$\frac{(64+57)\times 320}{840-50\times 8}\approx \frac{(60+60)\times 300}{800-50\times 8}$$

Note that 64 and 57 rounded off to one digit both become 60.
   The estimation, which can be done quickly by hand (or mentally), is

$$\frac{(60+60)\times 300}{800-50\times 8}=\frac{120\times 300}{800-400}=\frac{36,000}{400}=90$$

The estimate gives you an idea of the size of the actual answer. The calculator solution using the original numbers and the parentheses keys is

  (64 + 57) × 320 ÷ (840 − 50 × 8) = → 88

The estimate of 90 is close to the calculator answer of 88, which indicates that 88 is the correct answer. Suppose your calculator answer was not close to the estimate. If it was 880 or 8.80, this would indicate an error somewhere; you then need to troubleshoot for the error. ■

EXAMPLE 1.28

Estimate the answer by hand and then calculate:

$$\frac{(2.06)(0.0340-0.0163)}{(3.12)(0.0560)}$$

Solution   Observe that parentheses are used in the example to indicate multiplication. For the estimate, the numbers rounded off to one digit are

$$\frac{(2.06)(0.0340-0.0163)}{(3.12)(0.0560)}\approx \frac{(2)(0.03-0.02)}{(3)(0.06)}$$

The estimation to three digits is

$$\frac{(2)(0.03-0.02)}{(3)(0.06)}=\frac{(2)(0.01)}{0.18}=\frac{0.02}{0.18}=\frac{2}{18}=\frac{1}{9}=0.111$$

   One calculator solution, using the = key to first compute the numerator and then dividing by each factor in the denominator, is, to three digits,

  2.06 × (0.0340 − 0.0163) = ÷ 3.12 ÷ 0.0560 = → 0.209

The estimate of 0.111 is close to the calculator answer of 0.209 verifying the result. ■

Lengthy calculations can be checked in stages by recording and estimating intermediate results. You can also check an operation by doing the inverse operation and seeing if you get back to the original number. In the following chapters, use your calculator to help you with calculations, but remember your calculator cannot replace mathematical understanding and reasoning. *You should always check an answer and ask yourself if it makes sense.*

EXERCISE 1.5

In problems 1 through 12, *estimate the answer first* by rounding off the numbers to one digit as shown in Examples 1.27 and 1.28. Then choose what you think is the correct answer from the four choices. Finally, do the problem on the calculator, and see if you are correct.

1. $(82 + 68 - 86) \div 32$ [2, 12, 20, 22]

2. $930 \times 81 \div 9 - 90$ [82, 820, 828, 8280]

3. $4.85 + 60.8 \div 3.20 - 2.75$ [0.211, 2.11, 21.1, 211]

4. $20 \times (1560 - 230) \div 28$ [9.5, 95, 650, 950]

5. $\dfrac{72 + 52 - 20}{176 - 34 + 66}$ [0.1, 0.5, 0.9, 1.5]

6. $\dfrac{252 \times (86 + 61)}{(36 - 29) \times 36}$ [14.7, 47, 147, 1470]

7. $\dfrac{159 \times 91 \times 76}{247 \times 53 \times 35}$ [0.24, 2.4, 8.2, 24]

8. $\dfrac{31 \times 176 \times 90}{186 \times 120 \times 110}$ [0.20, 1.2, 2.0, 12]

9. $\dfrac{(0.0068)(0.48 - 0.31)}{0.0017}$ [0.068, 0.68, 6.8, 68]

10. $\dfrac{(0.0112)(0.0211)}{(0.0400)(0.700)}$ [0.00844, 0.0844, 0.344, 0.844]

11. $\dfrac{(0.301 + 0.271)(0.0504)}{(1.26)(1.60)}$ [14.3, 1.43, 0.0143, 0.0413]

12. $\dfrac{3.03}{2.02} - \dfrac{27.6 - 8.40}{30.0}$ [86.0, 8.60, 0.860, 0.0860]

In problems 13 through 20, do the same as in problems 1 through 12, except round off the calculator answer to *three significant digits.*

13. $3.01 + 23.2 \div 0.715$ [35.4, 35.5, 355, 305]

14. $100.1 - 7.02 + 5.132 \times 6.02$ [124, 588, 58.8, 12.4]

15. $\dfrac{0.1014}{(1.485)(0.0360)}$ [0.190,   1.09,   1.90,   1.89]

16. $\dfrac{(10.0)(5.00)(100.0)}{(0.0300)}$ [16,700,   170,000,   167,000,   1,670,000]

17. $\dfrac{(6,710)(0.0780)(0.890)}{51.2+5.15}$ [0.0827,   0.827,   8.27,   82.7]

18. $\dfrac{(5.60)(20.6)}{(1.02)(1.03)}$ [110,   109.8,   10.0,   100]

19. $\dfrac{0.101+0.202}{52.1}-\dfrac{47.3}{15,000}$ [2.66,   0.266,   0.0266,   0.00266]

20. $\dfrac{333}{22.1+11.2}+\dfrac{5.99}{0.559}$ [2.07,   2.071,   20.7,   20.71]

In problems 21 through 26, use the calculator to solve each applied problem to *three significant digits*.

21. The area of the earth covered by water is 140,000,000 mi² (square miles) or 70.98% of the total surface. How many square miles is the entire surface of the earth?

22. The circumference of a circle is given by

$$C = 2 \times \pi \times r$$

where $\pi = 3.142$ to four digits and $r =$ radius. Find the circumference of a circle whose radius $= 2.37$ cm.

23. In 1994, out of a total of 337,069 students enrolled in technology programs, 15,742 were electrical technology students, 64,440 were electronics students, 20,113 were civil technology students, and 26,395 were mechanical technology students. What percentage of the total were enrolled in
    (a) Electrical technology?
    (b) Electronics?

24. The Social Security tax for employees in 1990 was 7.65% of gross earnings up to $51,300 with no tax on any amount above $51,300. Carmen Velez earned $54,850 in 1990. How much Social Security tax did she pay? Give answer to the nearest cent.

25. Your calculator cannot think, but it can display words. Calculate the following correctly, and discover one of the artistic neighborhoods in New York City by reading the display upside down:

$$\dfrac{82\times5-5}{16+24\times41}$$

26. Here's a good calculator trick:
    (a) Tell somebody to key in six 9s and divide by 7.
    (b) Pick a number from 1 to 6 and multiply.
    (c) Then have the person tell you the first digit in the display, and *you will tell them all the other digits.*
    Here's how it works: The number you get when you divide by 7 is a "cyclic" number of six digits. If you multiply by any number from 1 to 6, you get the same order of six digits starting with a different digit each time. Memorize the cyclic number, and you can do the trick.

Applications to Electronics   In problems 27 through 30 use the calculator to solve each applied problem to *three significant digits.*

27. Figure 1–7 shows a circuit containing two different resistances R_1 and R_2 in parallel. Each resistance provides a separate path for the current to flow and the voltage across each resistance is the same. The total resistance R_T of the parallel circuit is given by the following formula:

$$R_T = \frac{R_1 R_2}{R_1 + R_2}$$

Calculate R_T to the nearest ohm when $R_1 = 1200\ \Omega$ (1.2 kΩ) and $R_2 = 1500\ \Omega$ (1.5 kΩ). Note: $R_1 R_2$ means $R_1 \times R_2$.

28. In problem 27, the voltage of the source is given by Ohm's law:

$$V_S = I_T \left(\frac{R_1 R_2}{R_1 + R_2} \right)$$

where I_T = total current through both resistances. Calculate the voltage when $I_T = 0.025$ A (25 mA), $R_1 = 220\ \Omega$ and $R_2 = 330\ \Omega$.

29. Figure 1–8 shows a parallel circuit containing three different resistances R_1, R_2, and R_3 in parallel. Each resistance provides a separate path for the current to flow, and the voltage across each resistance is the same. If $R_1 = R_2$, the total resistance is given by

$$R_T = \frac{R_1 R_3}{R_1 + 2R_3}$$

Calculate R_T when $R_1 = R_2 = 15\ \Omega$ and $R_3 = 24\ \Omega$. Note: $R_1 R_3$ means $R_1 \times R_3$.

30. In problem 29, calculate R_T when $R_1 = R_2 = 2500\ \Omega$ (2.5 kΩ) and $R_3 = 5300\ \Omega$ (5.3 kΩ).

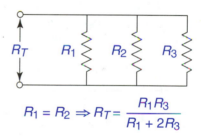

$$R_T = \frac{R_1 R_2}{R_1 + R_2}$$

FIGURE 1–7   Parallel circuit for problems 27 and 28.

$$R_1 = R_2 \Rightarrow R_T = \frac{R_1 R_3}{R_1 + 2R_3}$$

FIGURE 1–8   Parallel circuit for problems 29 and 30.

≡ CHAPTER HIGHLIGHTS

1.1 ARITHMETIC OPERATIONS

The *order of arithmetic operations* is

1. Perform operations in parentheses.
2. Do multiplication and division.
3. Do addition and subtraction.

1.2 FRACTIONS

To *multiply fractions,* divide out common factors first, and then multiply the numerators and the denominators:

$$\frac{A}{B} \times \frac{C}{D} = \frac{A \times C}{B \times D} \tag{1.4}$$

To *divide fractions,* invert the divisor and multiply:

$$\frac{A}{B} \div \frac{C}{D} = \frac{A}{B} \times \frac{D}{C} \qquad (1.5)$$

To *add fractions,* change each fraction to an equivalent fraction having the same lowest common denominator (LCD) by multiplying the top and bottom of each fraction by the necessary factor. The LCD must contain all the factors in each denominator. Then add the numerators over the LCD and reduce if possible:

$$\frac{A}{D} + \frac{B}{D} = \frac{A + B}{D} \qquad (1.6)$$

Study examples 1.9, 1.10, and 1.11.

1.3 DECIMALS AND PERCENTAGES

To *multiply decimals,* add the decimal places in each number to determine the number of decimal places in the answer.

To *divide decimals,* make the divisor a whole number by moving the decimal point to the right in the divisor *and* the dividend by as many places as there are in the divisor.

Percent means *hundredths.* To change a percent to a decimal, move the decimal point two places to the left. To change a decimal to a percent, move the decimal point two places to the right.

1.4 PRECISION AND ACCURACY

An *approximate number* is a number obtained by measurement or by rounding off a number.

Round off a number as follows:
1. If the digit to the right of the last rounded digit is less than 5, do not change the last rounded digit.
2. If the digit to the right of the last rounded digit is 5 or more, increase the last rounded digit by one.

Significant digits are
1. All nonzeros.
2. Zeros that are not placeholders for the decimal point.

Accuracy is measured by the number of significant digits.

Precision is measured by the decimal place of the last digit.

1.5 HAND CALCULATOR OPERATIONS

Most calculators perform operations according to the order of arithmetic operations stated in Section 1.1. Test your calculator by trying Examples 1.1 and 1.2.

Study the operations manual for your calculator to understand its key functions, particularly if it has a $\boxed{\text{MODE}}$ key. The keys $\boxed{\text{M}_{\text{in}}}$, $\boxed{\text{STO}}$, or $\boxed{\text{X}\rightarrow\text{M}}$ store a number in the memory. The keys $\boxed{\text{MR}}$, $\boxed{\text{RCL}}$, or $\boxed{\text{RM}}$ recall a number from memory.

Always check calculator results by rounding off all numbers to one digit and estimating the answer by hand. Study Examples 1.27 and 1.28.

▤ REVIEW QUESTIONS

In problems 1 through 14, calculate each exercise by hand. Use the calculator to check your results.

1. $8 \times 3 - 4 \div 2$

2. $(10 + 8) \times 3 \div 6$

3. $\dfrac{3 \times 15 \times 6}{2 \times 5}$

4. $\dfrac{5 + 4 \times (9 - 7)}{8 \times 3 + 2}$

5. $\dfrac{5}{6} \times \dfrac{12}{7} \div \dfrac{1}{21}$

6. $11 \times \dfrac{3}{22} \times \dfrac{4}{15}$

7. $\dfrac{4}{9} \times \dfrac{3}{5} + \dfrac{1}{2}$

8. $5 + \dfrac{7}{10} \div \dfrac{21}{20}$

9. $\dfrac{2}{3} + \dfrac{3}{4} - \dfrac{5}{6}$

10. $1 - \dfrac{3}{7} - \dfrac{5}{21}$

11. $3.2 + 2.1 \times 0.4$

12. $(0.10 - 0.09) \times (0.31 + 0.29)$

13. $\dfrac{1.2 + 3.0 \times 0.40}{0.02 \times 6.0}$

14. $\dfrac{0.03 \times 0.1}{6 \times 0.005}$

In problems 15 through 18, estimate each answer by hand. Then choose the correct answer, and check with the calculator.

15. $\dfrac{105 + 9 \times 25}{63 - 53}$ [3, 33, 60, 93]

16. $\dfrac{17 \times 2.0 \times 3.5}{2.5 \times 14}$ [3.4, 8.4, 34, 84]

17. $\dfrac{(0.08)(0.37 - 0.18)}{0.016}$ [0.19, 0.59, 0.95, 5.9]

18. $\dfrac{0.33}{5.5} - \dfrac{0.20}{28.3 + 71.7}$ [0.058, 0.098, 0.58, 0.98]

In problems 19 through 22, do the same as for problems 15 through 18, except round off the calculator answer to *three significant digits*.

19. $(3.70)\,(8.12) + 1.002$ [31, 31.0, 31.05, 25.0]

20. $1.15 - \dfrac{242}{856}$ [0.0867, 0.86, 0.867, 8.67]

21. $\dfrac{(39.03)(86.1)}{(10.0)(5.02)}$ [6.69, 66.9, 669, 6.70]

22. $\dfrac{0.707 + 0.808}{0.0102}$ [14.8, 14.9, 149, 148.5]

In problems 23 through 28, solve each applied problem.

23. For a 100-mi trip, a car travels the first 50 mi in 40 min and the last 50 mi in 1 h 10 min. What is the average speed for the entire trip?

24. A blue chip stock is selling at $99\frac{3}{8}$ at the opening of the day on the market. By noon the price has gone up $1\frac{5}{16}$. By the end of the day it closes down $\frac{3}{4}$ from its noon price. What is the closing price of the stock in fraction form?

25. The space for an office mail box is to be $6\frac{1}{2}$ ft long by 1 ft high. If it is to contain 17 equally spaced vertical mail slots and the partitions are $\frac{3}{8}$-in. thick, how wide will each slot be?

26. A school is given a grant of $60,000 for its computer laboratory. After spending 60% of the grant on new hardware, it spends $\frac{1}{3}$ of what is left on new furniture. After purchasing the furniture, 10% of the remainder is used to buy new software. How much is left of the grant?

27. A multimeter—which is an electrical meter for measuring voltage, current, and resistance—is advertised for $38. A discount store sells it at a 15% discount. What is the final price, including tax of 7%?

28. A lightning discharge from a cloud 2.5 mi high is measured at a speed of 550 mi/s (miles per second) for the downstroke and 10,000 mi/s for the powerful return stroke (ground to cloud). What is the total time for both strokes? Give your answer to two significant figures.

Applications to Electronics In problems 29 through 36, solve each applied problem.

29. Figure 1–9 shows a circuit containing two different resistances R_1 and R_2 in parallel. If the equivalent resistance of this parallel circuit $R_{eq} = 12\ \Omega$ and $R_1 = 30\ \Omega$, then R_2 is given by

$$R_2 = \frac{R_1 R_{eq}}{R_1 - R_{eq}}$$

Calculate the value of R_2. Note: $R_1 R_{eq}$ means $R_1 \times R_{eq}$.

30. In the circuit in Figure 1–9, if $R_2 = 3 \times R_1$, then R_{eq} is 25% less than R_1. Given $R_2 = 4.7\ k\Omega$ (kilohms), find R_{eq} under these conditions.

31. The current in a circuit is $I = 36$ mA (milliamps). The voltage decreases to $\frac{3}{4}$ of its value while the resistance remains constant. Ohm's law says that the current will decrease to the same fraction of its original value. Find the new value of the current.

32. The current in a circuit is 0.62 A. The resistance increases by half of its value while the voltage remains constant. Ohm's law says that the current will decrease to $\frac{2}{3}$ of its original value. Find the new value of the current.

33. Figure 1–10 shows a series-parallel circuit containing three different resistances R_1, R_2, and R_3. R_1 is connected in series to the two parallel resistances R_2 and R_3. The total current in the circuit passes through R_1 and is divided between R_2 and R_3. The equivalent resistance R_{eq} of this series-parallel circuit is given by

$$R_{eq} = R_1 + \frac{R_2 R_3}{R_2 + R_3}$$

Find R_{eq} when $R_1 = 30\ \Omega$, $R_2 = 50\ \Omega$, and $R_3 = 75\ \Omega$ to two significant figures. Note: $R_2 R_3$ means $R_2 \times R_3$.

34. In problem 33, find R_{eq} when $R_1 = 200\ \Omega$, $R_2 = 1000\ \Omega$, and $R_3 = 1200\ \Omega$ to three significant digits.

35. The current through a 50-Ω resistance is measured at five equal intervals to be 20 mA (milliamps), 21 mA, 20 mA, 21.5 mA, and 22 mA. Find the average current through the resistance to two significant digits.

36. The length of a wire is shortened from 15 cm to 10 cm causing its resistance to decrease by the same percentage.
    (a) What is the percent decrease in the length and the resistance?
    (b) If the resistance of the original wire was 2.5 Ω, what is the resistance of the shortened wire to two significant digits?

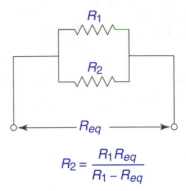

$$R_2 = \frac{R_1 R_{eq}}{R_1 - R_{eq}}$$

FIGURE 1–9 Parallel circuit for problems 29 and 30.

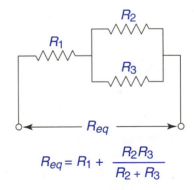

$$R_{eq} = R_1 + \frac{R_2 R_3}{R_2 + R_3}$$

FIGURE 1–10 Series-parallel circuit for problems 33 and 34.

2

Powers of Numbers

Courtesy of Hewlett Packard Company.

Engineer and technician use computer-based signal acquisition and analysis system to check for automobile vibrations.

In scientific and technical work, numbers are often expressed in terms of powers or exponents. This is especially important in electricity and electronics where powers of 10 and the metric system are used to express units and dimensions. This chapter studies powers of numbers, particularly powers of 10, and roots of numbers. It explains how they are used in engineering notation, scientific notation, and the metric system. The last section discusses how to calculate powers and roots on the calculator and how to solve some of the chapter examples using the calculator.

Chapter Objectives

In this chapter, you will learn:

- How to raise a number to a power and take the root of a number.
- How to simplify expressions with radical signs.
- The rules for working with powers of 10.
- How to express and work with numbers in engineering notation and scientific notation.
- The metric, or SI system, and how to perform conversions in the metric system.
- How to calculate powers and roots on the calculator and use the keys for engineering notation, scientific notation, and metric conversions.

2.1
POWERS AND ROOTS

Powers

A *power* or *exponent* defines repeated multiplication:

$$x^n = \underbrace{(x)\ (x)\ (x)\ \cdots\ (x)}_{n\ \text{times}} \qquad (n = \text{Positive whole number})$$

$$(2.1)$$

For example:

$$2^6 = (2)(2)(2)(2)(2)(2) = 64$$

$$6^3 = (6)(6)(6) = 216$$

On a calculator, raising to a power is done with the $\boxed{y^x}$ or $\boxed{x^y}$ key. The two examples just shown can be done on the calculator as follows:

$$2\ \boxed{x^y}\ 6\ \boxed{=}\ \to\ 64$$

$$6\ \boxed{x^y}\ 3\ \boxed{=}\ \to\ 216$$

Study the following examples, which show how to raise fractions and decimals to a power.

EXAMPLE 2.1

Evaluate:

$$\left(\frac{2}{5}\right)^4$$

Solution   Multiply the numerators and the denominators:

$$\left(\frac{2}{5}\right)^4 = \left(\frac{2}{5}\right)\left(\frac{2}{5}\right)\left(\frac{2}{5}\right)\left(\frac{2}{5}\right) = \frac{16}{625}$$

This example can be done on the calculator by using parentheses:

$$\boxed{(} \ 2 \ \boxed{\div} \ 5 \ \boxed{)} \ \boxed{y^x} \ 4 \ \boxed{=} \ \rightarrow 0.0256$$

▪

EXAMPLE 2.2

Evaluate:

$$(0.01)^2$$

Solution   Since there are two decimal places in 0.01, when you square the number (multiply it by itself), there will be four decimal places in the result:

$$(0.01)^2 = (0.01)(0.01) = 0.0001$$

To square a number on the calculator you can just press the square key:

$$0.01 \ \boxed{x^2} \ \rightarrow 0.0001$$

In the order of operations, *raising to a power (or taking a root) is done first*, before multiplication or division. For example, to evaluate:

$$\frac{(2)^5 (7)^2}{4^3}$$

you raise each number to the power first, then multiply and divide:

$$\frac{(2)^5 (7)^2}{4^3} = \frac{(32)(49)}{64} = \frac{49}{2} = 24.5$$

▪

The following examples illustrate further the order of operations.

EXAMPLE 2.3

Evaluate:

$$\left(\frac{3}{2}\right)^2 - 6\left(\frac{1}{2}\right)^3$$

EXAMPLE 2.3 (Cont.)

Solution First, raise the fractions to the powers. Second, do the multiplication, and third, subtract the fractions:

$$\left(\frac{3}{2}\right)^2 - 6\left(\frac{1}{2}\right)^3 = \frac{9}{4} - 6\left(\frac{1}{8}\right) = \frac{9}{4} - \frac{6}{8} = \frac{9}{4} - \frac{3}{4} = \frac{6}{4} = \frac{3}{2} \text{ or } 1.5$$

▪

EXAMPLE 2.4

Evaluate:

$$\frac{(0.1)^2 + (0.3)^3}{0.5}$$

Solution Raise to the powers first. Then, since the fraction line is like parentheses, you must do the addition in the numerator before you divide:

$$\frac{(0.1)^2 + (0.3)^3}{0.5} = \frac{0.01 + 0.027}{0.5} = \frac{0.037}{0.5} = \frac{0.37}{5} = 0.074$$

See Example 2.20 for the calculator solution of Example 2.4.

▪

Square Roots

The *square root* of a given number is a number that, when squared, exactly equals the given number. The square root is indicated with the root, or radical sign ($\sqrt{}$). For example,

$$\sqrt{64} = 8 \text{ because } 8^2 = 64$$

$$\sqrt{0.16} = 0.4 \text{ because } (0.4)^2 = 0.16$$

Taking a root is the *inverse* operation of raising to a power. This is the same relationship as that between subtraction, addition and/or between division and multiplication.

Numbers whose square roots are whole numbers, fractions, or finite decimals, such as 64 and 0.16 just illustrated, are called *perfect squares*. The whole numbers that are perfect squares are 4, 9, 16, 25, 36, 49, and so forth. You should memorize the whole number perfect squares up to 144. This will help you to do problems with squares and square roots and understand the concepts better.

Study the following examples, which further explain square roots.

EXAMPLE 2.5

Find the following square roots:

$$\sqrt{\frac{16}{49}}$$

$$\sqrt{1.44}$$

EXAMPLE 2.5 (Cont.)

Solution To find the square root of a fraction, find the square root of the numerator and the denominator separately:

$$\sqrt{\frac{16}{49}} = \frac{\sqrt{16}}{\sqrt{49}} \text{ or } = \frac{4}{7} \text{ because } \left(\frac{4}{7}\right)^2 = \frac{16}{49}$$

The square root of a decimal will have approximately half as many decimal places as the original number. Since 1.44 has two decimal places, its square root has one decimal place. The square root of 1.44 is a number a little larger than one because $1^2 = 1$ and $2^2 = 4$. Try $1.1^2 = 1.21$. Then try $1.2^2 = 1.44$ and you have the root. Therefore,

$$\sqrt{1.44} = 1.2$$

To find the square root on the calculator use the square root key:

$$1.44 \ \boxed{\sqrt{}} \ \rightarrow \ 1.2$$

Numbers that are not perfect squares, such as 2, 3, 5, and 7, have square roots that are infinite decimals, and it is necessary to round off the result:

$$\sqrt{2} = 1.414213562\ldots \approx 1.414$$

$$\sqrt{7} = 2.645751311\ldots \approx 2.646$$

The symbol "$\approx$" means approximately. That is, $\sqrt{2}$ is approximately 1.414, and $\sqrt{7}$ is approximately 2.646. Most square roots are not those of perfect squares, and because of this, we often calculate with them in the form of radicals.

For any positive number x, the definition of the square root or radical sign is as follows:

$$\sqrt{x^2} = \left(\sqrt{x}\right)^2 = x \tag{2.2}$$

The definition says that the operations of square and square root are inverse operations and tend to "cancel each other out." When a square root and a square appear together, the radical and the square disappear. For example,

$$\sqrt{9^2} = 9 \text{ and } \left(\sqrt{9}\right)^2 = 9$$

Two basic rules of calculation for the square roots of two positive numbers x and y are:

$$\sqrt{xy} = \left(\sqrt{x}\right)\left(\sqrt{y}\right) \tag{2.3}$$

$$\sqrt{\frac{x}{y}} = \frac{\sqrt{x}}{\sqrt{y}} \tag{2.4}$$

These rules help to simplify radical expressions. Products, or quotients, under a radical sign can be separated into products, or quotients, of separate radicals and vice versa. For example, using rule (2.3), you can separate radicals as follows:

$$\sqrt{12} = \sqrt{(4)(3)} = \left(\sqrt{4}\right)\left(\sqrt{3}\right) = 2\sqrt{3}$$

If the number under the radical sign (radicand) contains a factor that is a perfect square, such as the factor 4 above, it can be simplified by separating the perfect square. Note that the number in front of the radical means it is multiplied by the radical:

$$2\sqrt{3} = 2 \times \sqrt{3}$$

Also by rule (2.3) you can multiply under the radical as follows:

$$\left(\sqrt{0.2}\right)\left(\sqrt{1.8}\right) = \sqrt{(0.2)(1.8)} = \sqrt{0.36} = 0.6$$

Note carefully, however, that *you cannot add radicals that are different.*

EXAMPLE 2.6

Simplify:

$$\sqrt{75}$$

Solution   A radical can be simplified if it contains a factor that is a perfect square. The number 75 contains the perfect square 25. Therefore, you can apply rule (2.3):

$$\sqrt{75} = \left(\sqrt{25}\right)\left(\sqrt{3}\right) = 5\sqrt{3}$$

Using rule (2.4), you can separate a fraction under a radical as follows:

$$\sqrt{\frac{3}{100}} = \frac{\sqrt{3}}{\sqrt{100}} = \frac{\sqrt{3}}{10} \text{ or } 0.1\sqrt{3}$$

Also by rule (2.4), you can divide under one radical as follows:

$$\frac{\sqrt{28}}{\sqrt{7}} = \sqrt{\frac{28}{7}} = \sqrt{4} = 2$$

EXAMPLE 2.7

Simplify:

$$\sqrt{\frac{8}{9}}$$

Solution   Apply rule (2.4), and separate into two radicals. Then simplify each radical:

$$\sqrt{\frac{8}{9}} = \frac{\sqrt{8}}{\sqrt{9}} = \frac{\left(\sqrt{4}\right)\left(\sqrt{2}\right)}{3} = \frac{2\sqrt{2}}{3} \text{ or } \frac{2}{3}\sqrt{2}$$

Cube Roots

A cube root is the inverse of raising to the third power. A cube root is written using the radical sign with the *index* 3:

$$\sqrt[3]{}$$

The index of the root is the small number in the crook of the radical sign. For a square root, the index is 2, and it is understood. For example,

$$\sqrt[3]{1000} = 10 \text{ because } 10^3 = 1000$$

$$\sqrt[3]{0.064} = 0.4 \text{ because } (0.4)^3 = 0.064$$

The definition of a cube root for a number x is similar to that for a square root:

$$\sqrt[3]{x^3} = \left(\sqrt[3]{x}\right)^3 = x \tag{2.5}$$

Since raising to the third power is the inverse of taking a cube root, the two operations cancel each other. For example,

$$\sqrt[3]{8^3} = 8 \quad \text{and} \quad \left(\sqrt[3]{8}\right)^3 = 8$$

The cube root of a number can be found on the calculator by using the cube root key $\boxed{\sqrt[3]{}}$ or by applying the following rule that equates a cube root to a power:

$$\sqrt[3]{x} = x^{1/3} \tag{2.6}$$

For example, you can find the cube root of 10 on the calculator by raising to the $\frac{1}{3}$ power using the power key and the reciprocal key as follows:

$$10 \boxed{x^y} \ 3 \ \boxed{1/x} = \rightarrow 2.154$$

See Example 2.24 in Section 2.5 for further examples of finding cube roots.

EXERCISE 2.1

In problems 1 through 28 *evaluate each exercise by hand.* Use the calculator only to check your answers.

1. 3^4

2. 4^3

3. $\left(\frac{1}{2}\right)^2$

4. $\left(\frac{3}{5}\right)^4$

5. $(0.7)^3$

6. $(1.1)^2$

7. $\dfrac{4^4}{(8)^2(6)^2}$

8. $\dfrac{2^3 + 3^3}{5^2}$

9. $\left(\frac{1}{5}\right)^2 + \left(\frac{2}{5}\right)\left(\frac{1}{10}\right)^2$

10. $(6)\left(\frac{1}{4}\right)^2 - (4)\left(\frac{1}{6}\right)^2$

11. $\dfrac{(0.2)^3}{(0.5)^2 - (0.3)^2}$

12. $\dfrac{(0.7)^2 (0.3)^3}{(2.1)^2} - (0.1)^3$

13. $\dfrac{2^5}{8^2} - \left(\frac{3}{4}\right)^3$

14. $\left(\frac{1}{2} + \frac{1}{3}\right)^2 \left(\frac{1}{2} - \frac{1}{10}\right)^2$

15. $\sqrt{16}$

16. $\sqrt{121}$

17. $\sqrt{\dfrac{25}{4}}$

18. $\sqrt{\dfrac{9}{100}}$

19. $\sqrt{0.64}$

20. $\sqrt{0.36}$

21. $\sqrt{0.0081}$

22. $\sqrt{0.0144}$

23. $\sqrt[3]{27}$

24. $\sqrt[3]{125}$

25. $\sqrt[3]{\dfrac{1}{8}}$

26. $\sqrt[3]{\dfrac{27}{1000}}$

27. $\sqrt[3]{0.064}$

28. $\sqrt[3]{0.001}$

In problems 29 through 40, simplify each expression by applying the rules for radicals. Give the answers in radical form.

29. $\sqrt{8}$

30. $\sqrt{32}$

31. $2\sqrt{50}$

32. $3\sqrt{27}$

33. $\left(\sqrt{8}\right)\left(\sqrt{2}\right)+\left(\sqrt{10}\right)^2$

34. $\left(\sqrt{2}\right)\left(\sqrt{18}\right)-\left(\sqrt{12}\right)\left(\sqrt{3}\right)$

35. $\dfrac{\sqrt{12}}{\sqrt{3}}$

36. $\dfrac{\sqrt{3}}{\sqrt{75}}$

37. $\sqrt{\dfrac{7}{16}}$

38. $\sqrt{\dfrac{18}{25}}$

39. $\dfrac{\sqrt{0.09}}{\sqrt{9}}$

40. $\dfrac{\sqrt{0.4}}{\sqrt{0.04}}$

In problems 41 through 44, *solve each by hand.* Check with the calculator.

41. The volume of a cube $V = s^3$, where $s =$ side. Find the volume of a small cubic container in cubic centimeters when $s = 1.2$ cm.

42. At an inflation rate of 10% a year, the cost of a $10,000 car will increase to approximately $10{,}000(1.1)^3$ in 3 years. How much will this approximate cost be?

43. The area of a square circuit board is 10,000 mm² (square millimeters). How long is the side of the circuit board in mm and cm? Note: 10 mm = 1 cm.

44. The radius of a sphere is given approximately by

$$r=\sqrt[3]{\dfrac{V}{4.2}}$$

where $V =$ volume. Find r when V = 0.0042 ft³.

Applications to Electronics In problems 45 through 52 *solve each by hand.* Check your answers with the calculator.

45. Figure 2–1 shows a dc circuit containing a battery and a resistance R. The power in watts (W) of the circuit is given by

$$P = I^2 \times R = I^2R$$

Find P when $I = 0.50$ A and $R = 68$ Ω.

46. The power in watts of the circuit in Figure 2–1 is also given by $P = V_B{}^2 / R$. Find P when $V_B = 1.2$ V and $R = 12\ \Omega$.

47. Figure 2–2 shows a right triangle that relates the resistance R and the reactance X in an ac circuit. Reactance is a kind of resistance and is also measured in ohms. They combine together to form what is called the impedance Z according to the following formula:

$$Z = \sqrt{R^2 + X^2}$$

Find Z in ohms when $R = 16\ \Omega$ and $X = 12\ \Omega$.

48. In problem 47 find Z when $R = 51\ \Omega$ and $X = 68\ \Omega$.

49. Figure 2–3 shows a circuit containing two different resistances R_1 and R_2 in series. The current in the circuit is given by

$$I = \sqrt{\frac{P_T}{R_1 + R_2}}$$

where P_T = total power in watts. Find I when the power $P_T = 10$ W, $R_1 = 150\ \Omega$, and $R_2 = 100\ \Omega$.

50. The total voltage in the series circuit in Figure 2–3 is given by

$$V_T = \sqrt{(P_T) \times (R_1 + R_2)}$$

Find V_T when the power $P_T = 16$ W, $R_1 = 100\ \Omega$, and $R_2 = 44\ \Omega$.

51. In the BASIC computer language, an arrow (↑) or caret (∧) denotes an exponent, and SQR(x) denotes $\sqrt{x}$. For example, $\sqrt{49} - 2^3$ is written SQR(49) – 2∧3. Calculate the following expression written in the BASIC language:

$$4\wedge2/\text{SQR}(81) + 2/9$$

52. As in problem 51, calculate the following expression written in BASIC:

$$\text{SQR}(32/2) * 3\wedge2$$

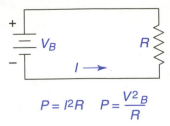

$$P = I^2 R \qquad P = \frac{V^2{}_B}{R}$$

FIGURE 2–1 Power in a dc circuit, problems 45 and 46.

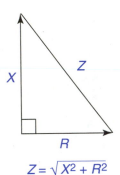

$$Z = \sqrt{X^2 + R^2}$$

FIGURE 2–2 Impedance triangle for problems 47 and 48.

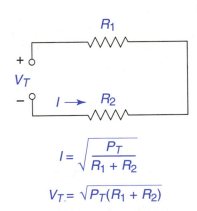

$$I = \sqrt{\frac{P_T}{R_1 + R_2}}$$

$$V_T = \sqrt{P_T(R_1 + R_2)}$$

FIGURE 2–3 Series circuit for problems 49 and 50.

Courtesy of New York Power Authority.

A technician checks electrical wiring in the Niagara Power Project.

≡ 2.2
POWERS OF 10

Our number system is the decimal system. It is based on 10 and the powers of 10. For example, when you write the number 546 you mean

$$546 = (5 \times 10^2) + (4 \times 10^1) + (6 \times 10^0)$$

where $10^2 = 100$, $10^1 = 10$, and $10^0 = 1$. Similiarly, when you write 8.72 you mean

$$8.72 = (8 \times 10^0) + (7 \times 10^{-1}) + (2 \times 10^{-2})$$

where $10^0 = 1$, $10^{-1} = 0.1$ or $\frac{1}{10}$, and $10^{-2} = 0.01$ or $\frac{1}{100}$.

Positive, zero, and negative powers of 10 are shown in Table 2-1. Note in the table that *the power of 10 determines the place of the decimal point.* For every increase of one in the power of 10, the decimal point moves one place to the right. For 0, 1, 2, 3, and so on, the power is just equal to the number of zeros. For example, $10^0 = 1$, $10^3 = 1,000$, and $10^5 = 100,000$. Similiarly, for every decrease of one in the power of 10, the decimal point moves one place to the left. Therefore, the negative powers of 10 are decimals between 0 and 1.

The *absolute value* of a number is the value *without the sign*. It is indicated by parallel lines. For example $|-9| = 9$ and $|+9| = 9$. Absolute value can be thought of as the positive value. For a negative power, the following two rules apply:

1. The absolute value of the power is equal to the number of decimal places. For example, $10^{-9} = 0.000000001$ and has $|-9| = 9$ decimal places. The number of zeros is eight, which is one less than the absolute value of the power.
2. For any positive number n,

$$10^{-n} = \frac{1}{10^n}$$

For example,

$$10^{-6} = \frac{1}{10^6} = \frac{1}{1,000,000} = 0.000001$$

TABLE 2.1   Powers of 10

Power	Decimal	Fraction or Whole Number
10^{-4}	0.0001	$\frac{1}{10,000}$
10^{-3}	0.001	$\frac{1}{1,000}$
10^{-2}	0.01	$\frac{1}{100}$
10^{-1}	0.1	$\frac{1}{10}$
10^{0}	1.0	1
10^{1}	10.0	10
10^{2}	100.0	100
10^{3}	1,000.0	1,000
10^{4}	10,000.0	10,000

Multiplication and Division

When multiplying with powers of 10, you add the exponents *algebraically*. There are three cases to consider:

1. When both powers are positive, add the exponents. For example,

$$10^3 \times 10^2 = 10^5$$

2. When both powers are negative, add their absolute (positive) values, but make the result negative. For example,

$$10^{-4} \times 10^{-1} = 10^{-(4+1)} = 10^{-5}$$

3. When one power is positive and the other negative, subtract the smaller absolute value from the larger absolute value, and use the sign of the larger value. For example,

$$10^6 \times 10^{-2} = 10^{6-2} = 10^4$$

$$10^{-6} \times 10^4 = 10^{-(6-4)} = 10^{-2}$$

The following is another example of multiplying with powers of 10.

EXAMPLE 2.8

Multiply:

$$10^3 \times 10^{-9} \times 10^0$$

Solution   First subtract 3 from 9 and make the result negative. Then add zero, which does not change the result:

$$10^3 \times 10^{-9} \times 10^0 = 10^{-(9-3)+0} = 10^{-6}$$

When you multiply by 10^0, you are multiplying by one, which does not change the number.

When dividing with powers of 10, you subtract the exponents algebraically: Change the sign of the power in the divisor, and then multiply by adding the exponents algebraically. For example, to divide

$$\frac{10^2}{10^6}$$

change the power of 6 to –6 and multiply adding the exponents algebraically:

$$10^2 \times 10^{-6} = 10^{-(6-2)} = 10^{-4}$$

The following is another example of dividing with powers of 10.

EXAMPLE 2.9

Divide:

$$\frac{10^{-3}}{10^3}$$

Solution   Change the power of 3 to –3, and multiply adding the exponents algebraically:

$$\frac{10^{-3}}{10^3} = 10^{-3} \times 10^{-3} = 10^{-(3+3)} = 10^{-6}$$

A very large or very small number is often expressed in terms of a number times a power of 10 as follows:

$$7,000,000 = 7 \times 1,000,000 = 7 \times 10^6$$

$$0.004 = 4 \times 0.001 = 4 \times 10^{-3}$$

When multiplying or dividing numbers with powers of 10, you calculate each part separately as follows:

1. Multiply or divide the numbers to obtain the number part of the answer.
2. Apply the rules for multiplying or dividing with powers of 10 to obtain the power of 10.

For example,

$$(2 \times 10^2)(6 \times 10^3) = (2)(6) \times 10^{2+3} = 12 \times 10^5$$

$$\frac{8 \times 10^{-3}}{4 \times 10^{-1}} = \frac{8}{4} \times (10^{-3})(10^1) = 2 \times 10^{-2}$$

Study the next example, which further shows how to multiply and divide with powers of 10.

EXAMPLE 2.10

Calculate:

$$\frac{(4 \times 10^{-6})(6 \times 10^3)}{(8 \times 10^{-3})}$$

Solution First multiply the numerator by multiplying the numbers 4 and 6 and applying the multiplication rule to the powers:

$$\frac{(4 \times 10^{-6})(6 \times 10^3)}{(8 \times 10^{-3})} = \frac{(4)(6) \times 10^{-(6-3)}}{8 \times 10^{-3}} = \frac{24 \times 10^{-3}}{8 \times 10^{-3}}$$

Then divide 24 by 8, and apply the division rule to the powers:

$$\frac{24 \times 10^{-3}}{8 \times 10^{-3}} = \frac{24}{8} \times 10^{-3} \times 10^3 = 3 \times 10^{(3-3)} = 3 \times 10^0 \text{ or } 3$$

Addition and Subtraction

Numbers times powers of 10 *cannot* be added or subtracted in power of 10 form *unless the powers are exactly the same*. If the powers are exactly the same, you add (or subtract) the numbers only. You do not change the power of 10:

$$(5 \times 10^6) + (7 \times 10^6) = (7+5) \times 10^6 = 12 \times 10^6$$

When the powers are different, the numbers must be changed to ordinary notation, or like powers, before adding or subtracting. For example, to subtract

$$(2 \times 10^{-3}) - (9 \times 10^{-4})$$

you can change to ordinary notation as follows:

$$(2 \times 10^{-3}) - (9 \times 10^{-4}) = (2 \times 0.001) - (9 \times 0.0001)$$
$$= 0.0020 - 0.0009 = 0.0011 \text{ or } 1.1 \times 10^{-3}$$

or you can change to like powers as follows:

$$(2 \times 10^{-3}) - (9 \times 10^{-4}) = (2 \times 10^{-3}) - (0.9 \times 10^{-3}) = 1.1 \times 10^{-3}$$

Note that 9×10^{-4} is changed to 0.9×10^{-3}. The power of 10 is *increased by one* from -4 to -3. To balance the increase in the power of 10, you move the decimal point in 9 one place to the left, which decreases it by a power of 10. The resulting number then has the same value as the original number.

> **Rule** If you move the decimal point to the left, decreasing the number, increase the power of 10 by the number of places you move the decimal point, and vice versa.

For example,

$$4 \times 10^5 = 0.4 \times 10^6 \text{ and } 7 \times 10^{-8} = 70 \times 10^{-9}$$

Raising to Powers and Taking Roots

To raise a number times a power of 10 to a higher power, raise the number to the higher power and *multiply* the powers:

$$(4 \times 10^3)^2 = 4^2 \times 10^{3 \times 2} = 16 \times 10^6$$

To take a root of a number times a power of 10, take the root of the number and *divide* the index of the root into the power of 10:

$$\sqrt[3]{27 \times 10^9} = \sqrt[3]{27} \times 10^{9/3} = 3 \times 10^3$$

The index of the root is the small number in the crook of the radical sign. For a square root, the index is 2, and it is understood.

Study the next example, which illustrates all the operations with the powers of 10.

EXAMPLE 2.11

Calculate:

$$\frac{\sqrt{9 \times 10^6}}{(2 \times 10^3)^3 (5 \times 10^{-3})}$$

Solution Apply the order of operations, and do the powers and roots first. Find the square root of the numerator and raise the number to the power in the denominator:

$$\frac{\sqrt{9 \times 10^6}}{(2 \times 10^3)^3(5 \times 10^{-3})} = \frac{\sqrt{9} \times 10^{6/2}}{(2^3 \times 10^{3 \times 3})(5 \times 10^{-3})} = \frac{3 \times 10^3}{(8 \times 10^9)(5 \times 10^{-3})}$$

EXAMPLE 2.11 (Cont.)

Then multiply the numbers in the denominator and divide:

$$\frac{3 \times 10^3}{(8 \times 10^9)(5 \times 10^{-3})} = \frac{3 \times 10^3}{40 \times 10^6} = \frac{3}{40} \times 10^{3-6}$$

$$= 0.075 \times 10^{-3} \text{ or } 75 \times 10^{-6}$$

Observe that the answer is written in more than one way with a power of 10 by applying the rule for moving the decimal point. The decimal point in 0.075 is moved three places to the right, which is an increase of 10^3 or 1000 times. At the same time, the power of 10 is decreased by three from −3 to −6, which balances the move. ▪

Study the next example, which closes the circuit and shows an application of powers of 10 to a problem involving Ohm's law.

EXAMPLE 2.12

Close the Circuit

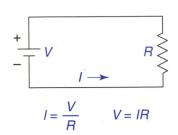

$$I = \frac{V}{R} \qquad V = IR$$

FIGURE 2–4   Ohm's law applied to a dc circuit for Example 2.12.

Figure 2–4 shows a dc circuit containing a resistance R connected to a voltage V.

1. Given Ohm's law $V = IR$, find V when the current $I = 4.2 \times 10^{-3}$ A (4.2 milliamps) and $R = 1.5 \times 10^3\ \Omega$ (1.5 kilohms). Note that IR means $I \times R$. In formulas, the multiplication sign is understood between two letters.

2. Given Ohm's law $I = \frac{V}{R}$, find I when $V = 9.3$ V and $R = 1.5 \times 10^3\ \Omega$.

Solution 1. To find V, substitute the given values for I and R in Ohm's law $V = IR$:

$$V = IR = (4.2 \times 10^{-3} \text{A})(1.5 \times 10^3\ \Omega) = (4.2)(1.5) \times 10^{(3-3)}$$
$$= 6.3 \times 10^0 \text{ V} = 6.3 \text{ V}$$

2. To find I, substitute the given values for V and R in Ohm's law $I = \frac{V}{R}$:

$$I = \frac{V}{R} = \frac{9.3V}{1.5 \times 10^3\ \Omega} = \frac{9.3 \times 10^0}{1.5 \times 10^3} = \frac{9.3}{1.5} \times 10^{(0-3)}$$

$$= 6.2 \times 10^{-3} \text{ A } (6.2 \text{ milliamps})$$
▪

EXERCISE 2.2

In problems 1 through 6, write each number as a power of 10.

1. 1000
2. 1,000,000
3. 0.000001
4. 0.001

5. $\dfrac{1}{1,000,000,000}$

6. $\dfrac{1}{100,000}$

In problems 7 through 12, write each power of 10 either as a whole number or as a decimal and a fraction.

7. 10^5

8. 10^8

9. 10^{-9}

10. 10^{-4}

11. $\dfrac{1}{10^3}$

12. $\dfrac{1}{10^6}$

In problems 13 through 50, *perform each calculation by hand,* and express the answer in terms of a power of 10. You can check your answers with the calculator.

13. $10^4 \times 10^2$

14. $10^3 \times 10^6$

15. $10^{-3} \times 10^{-2}$

16. $10^{-9} \times 10^{-1}$

17. $10^{-9} \times 10^3$

18. $10^{-3} \times 10^6$

19. $\dfrac{10^6}{10^3}$

20. $\dfrac{10^4}{10^9}$

21. $\dfrac{10^{-2}}{10^{-4}}$

22. $\dfrac{10^{-5}}{10^{-3}}$

23. $(2 \times 10^{-4})(6 \times 10^6)$

24. $(8 \times 10^5)(3 \times 10^{-4})$

25. $(3 \times 10^{-6})(2 \times 10^6)$

26. $(5 \times 10^{-9})(9 \times 10^5)$

27. $\dfrac{15 \times 10^{-6}}{5 \times 10^{-9}}$

28. $\dfrac{20 \times 10^{-6}}{4 \times 10^{-3}}$

29. $\dfrac{(2 \times 10^{-6})(5 \times 10^3)}{4 \times 10^2}$

30. $\dfrac{(3 \times 10^4)(6 \times 10^{-2})}{8 \times 10^{-3}}$

31. $\dfrac{6 \times 10^5}{(8 \times 10^2)(10^{-3})}$

32. $\dfrac{12 \times 10^{-4}}{(2 \times 10^{-3})(6 \times 10^{-6})}$

33. $(7 \times 10^3) + (9 \times 10^3)$

34. $(15 \times 10^{-6}) + (25 \times 10^{-6})$

35. $(18 \times 10^{-9}) - (11 \times 10^{-9})$

36. $(10 \times 10^{12}) - (5 \times 10^{12})$

37. $(5 \times 10^2)^3$

38. $(6 \times 10^4)^2$

39. $(2 \times 10^3)^4$

40. $(3 \times 10^4)^3$

41. $\sqrt{16 \times 10^6}$

42. $\sqrt{25 \times 10^4}$

43. $\sqrt[3]{8 \times 10^{12}}$

44. $\sqrt[3]{27 \times 10^3}$

45. $\dfrac{\sqrt{9 \times 10^8}}{(15 \times 10^{-3})(2 \times 10^2)}$

46. $\dfrac{\sqrt{16 \times 10^4}}{(4 \times 10^2)(5 \times 10^{-3})}$

47. $\dfrac{(3 \times 10^2)^2(2 \times 10^3)}{\sqrt{4 \times 10^{12}}}$

48. $\dfrac{(5 \times 10^{-4})(7 \times 10^2)^3}{\sqrt{25 \times 10^6}}$

49. $5 \times 10^3 + 6 \times 10^4$

50. $2 \times 10^{-3} + 9 \times 10^{-4}$

Applications to Electronics   In problems 51 through 56 *calculate each problem by hand.*

51. In Example 2.12, find V when $I = 2.5 \times 10^{-3}$ A and $R = 1.2 \times 10^3$ Ω.

52. In Example 2.12, find V when $I = 3.0 \times 10^{-6}$ A and $R = 3.3 \times 10^3$ Ω.

53. In Example 2.12, find I when $V = 6.0$ V and $R = 1.2 \times 10^3$ Ω.

54. In Example 2.12, find I when $V = 1.1 \times 10^3$ V and $R = 10 \times 10^3$ Ω.

55. In Example 2.12, given Ohm's law $R = \dfrac{V}{I}$, find R when $V = 5.4$ V and $I = 2.7 \times 10^{-3}$ A.

56. In problem 55, find R when $V = 7.5$ V and $I = 25 \times 10^{-6}$ A.

In problems 57 through 62, solve each applied problem to two significant digits.

57. Figure 2–5 shows a parallel circuit containing two different resistances R_1 and R_2. The power P_1 dissipated as heat in resistance R_1 is given by

$$P_1 = \frac{V_1^2}{R_1}$$

where V_1 is the voltage drop across R_1. Find P_1 in watts when $V_1 = 3.2 \times 10^3$ V and $R_1 = 1.6 \times 10^6$ Ω.

58. In the parallel circuit in Figure 2–5 the power dissipated as heat in resistance R_2 is given by

$$P_2 = I_2^2 R_2$$

where I_2 is the current through R_2. Find P_2 in watts when $I_2 = 2.2 \times 10^{-3}$ A and $R_2 = 1.2 \times 10^3$ Ω.

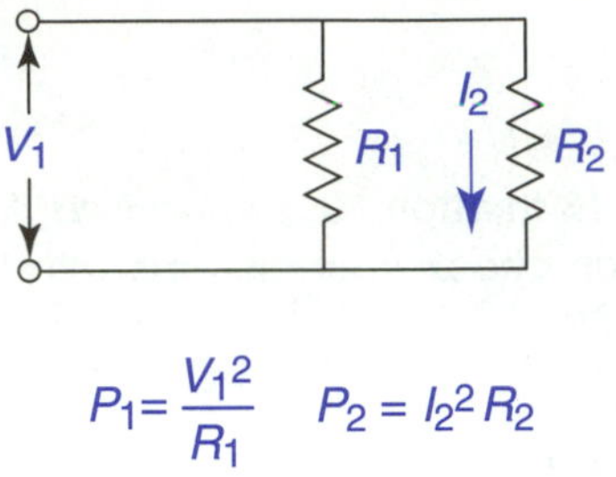

59. The total current I_T through the two resistances in series in Figure 2–6 is given by

$$I_T = \sqrt{\frac{P_T}{R_T}}$$

where the total resistance $R_T = R_1 + R_2$ and P_T is the total power dissipated in the resistances. Find I_T when $P_T = 24$ W, $R_1 = 5.6 \times 10^3$ Ω (5.6 kilohms), and $R_2 = 3.9 \times 10^3$ Ω (3.9 kilohms).

60. In the circuit in Figure 2–6, the total voltage drop across the two resistances in series is given by

$$V_T = \sqrt{P_T R_T}$$

where the total resistance $R_T = R_1 + R_2$ and P_T is the total power dissipated. Find V_T when $P_T = 24$ W, $R_1 = 5.6 \times 10^3$ Ω, and $R_2 = 3.9 \times 10^3$ Ω.

61. The resistance R of a wire is given by

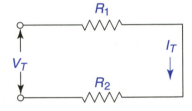

$$P_1 = \frac{V_1^2}{R_1} \qquad P_2 = I_2^2 R_2$$

FIGURE 2–5   Power in a parallel circuit for problems 57 and 58.

$$I_T = \sqrt{\frac{P_T}{R_T}} \qquad V_T = \sqrt{P_T R_T}$$

FIGURE 2–6   Current and voltage in a series circuit for problems 59 and 60.

$$R = \frac{\rho l}{A}$$

where ρ = resistivity, l = length and the cross-sectional area $A = \pi r^2$. Find R (in ohms) of a copper wire when $\rho = 1.75 \times 10^{-8}$ Ω–m, $l = 5$ m, and $r = 3 \times 10^{-3}$ m. Use $\pi = 3.14$ or the $\boxed{\pi}$ key on the calculator.

62. The coupling coefficient k of two coils in series is given by

$$k = \frac{L_M}{\sqrt{L_1 L_2}}$$

where L_M = mutual inductance, L_1 = inductance of the first coil, and L_2 = inductance of the second coil. Find k when $L_1 = 350 \times 10^{-6}$ H, $L_2 = 250 \times 10^{-6}$ H, and $L_M = 30 \times 10^{-6}$ H.

≡ **2.3**
SCIENTIFIC AND ENGINEERING NOTATION

Consider the following distance problem. The brightest star in the sky, Sirius, is 8.7 light-years away. This means it takes 8.7 years for light to travel from this star to the earth. If light travels at 186,000 mi/s (miles per second), how far away is Sirius? The distance to *three significant digits* is calculated as follows:

$$\left(186{,}000\ \frac{mi}{s}\right)\left(60\ \frac{s}{min}\right)\left(60\ \frac{min}{h}\right)\left(24\ \frac{h}{d}\right)\left(365\ \frac{d}{yr}\right)\left(8.7\ yr\right)$$
$$= 51{,}000{,}000{,}000{,}000\ \text{miles}$$

The answer is more than 50 trillion miles. Very large and very small numbers are used often in scientific work and particularly in electronics. These numbers are expressed using powers of 10 in scientific or engineering notation.

> **Scientific Notation**   A number between 1 and 10 times a power of 10.

The answer to the distance problem in scientific notation is

$$5.10 \times 10^{13}\ \text{miles}$$

When you do this problem on the calculator, it automatically switches to scientific notation when the result is too large (or too small) for ordinary notation. The power of 10 is shown on the right side of the display:

186000 ☒ 60 ☒ 60 ☒ 24 ☒ 365 ☒ 8.7 ☐ → 5.10315   13

> **Engineering Notation**   A number between 0.1 and 1000 times a power of 10 that *is divisible by three.*

The answer to the distance problem in engineering notation is

$$51.0 \times 10^{12}\ \text{miles}$$

Powers of 10 divisible by three correspond to units used in electronics and in the metric system. For example, 10^3 corresponds to kilo, 10^{-3} to milli, and 10^{-6} to micro. Table 2.3 in the next section shows these engineering prefixes.

To change ordinary notation to scientific notation, you move the decimal point to the right of *the first nonzero or significant digit.* This is called the zero position. Then count the number of places the decimal point has been moved. The number of places equals the power of 10: It is positive if the decimal point is moved to the left, negative if it is moved to the right.

For example,

$$5{,}320{,}000. \qquad = 5.32 \times 10^6$$
$$0.0702 \qquad = 7.02 \times 10^{-2}$$
$$9.68 \qquad = 9.68 \times 10^0$$
$$0.00000410 \quad = 4.10 \times 10^{-6}$$

The arrows show the zero position. Observe that when the decimal point is in the zero position it corresponds to 10^0 in scientific notation. For more information about significant digits, see Section 1.5 in chapter 1.

To change from ordinary notation to engineering notation, move the decimal point a multiple of three places: 3, 6, 9, etc., to change to a number between 0.1 and 1000. The number of places equals the power of 10: It is positive if moved to the left, negative if moved to the right.

For example,

$$0.0615 \quad\quad = 61.5 \times 10^{-3}$$

$$92{,}300. \quad\quad = 92.3 \times 10^3$$

$$0.00000544 = 5.44 \times 10^{-6}$$

$$249. \quad\quad\quad = 0.249 \times 10^3$$

Note that the last number, 249, can be expressed either way for engineering notation.

To change from scientific or engineering notation to ordinary notation, move the decimal point to the right if the exponent is positive, or to the left if it is negative, by the number of places indicated by the exponent.

For example,

$$3.40 \times 10^3 \quad = 3400$$
$$663 \times 10^{-3} \quad = 0.663$$
$$8.23 \times 10^{-1} = 0.823$$
$$0.768 \times 10^6 \quad = 768{,}000$$
$$55.2 \times 10^{-9} = 0.0000000552$$

The following example shows how to do calculations involving engineering notation or scientific notation.

EXAMPLE 2.13

Calculate and give the answer in engineering and scientific notation to three significant digits:

$$\frac{(328{,}000)(0.0850)}{0.000569}$$

Solution   Write the numbers in engineering (or scientific) notation:

$$\frac{(328 \times 10^3)(85.0 \times 10^{-3})}{0.569 \times 10^{-3}}$$

This problem can be done on the calculator, but first, as a check on the calculator, estimate the answer by rounding off to one digit:

$$\frac{(328 \times 10^3)(85.0 \times 10^{-3})}{0.569 \times 10^{-3}} \approx \frac{(300 \times 10^3)(90 \times 10^{-3})}{0.6 \times 10^{-3}}$$

Apply the rules for multiplying and dividing with powers of 10. Multiply out the numerator first, then divide:

$$\frac{(300)(90) \times 10^{3-3}}{0.6 \times 10^{-3}} = \frac{27{,}000 \times 10^0}{0.6 \times 10^{-3}} = \frac{270{,}000}{6} \times 10^{0+3} = 45{,}000 \times 10^3$$

EXAMPLE 2.13 (Cont.)

To change the estimate, which is in terms of a power of 10, to engineering notation, the decimal point is moved 3 places to the left, and the power of 10 is increased by 3 to balance the move:

$$45{,}000 \times 10^3 = 45 \times 10^6$$

Similarly, move the decimal point 4 places to the left for the estimate in scientific notation:

$$45{,}000 \times 10^3 = 4.5 \times 10^7$$

On the calculator, numbers can be entered with a power of 10 using the $\boxed{\text{EE}}$, $\boxed{\text{EEX}}$, or $\boxed{\text{EXP}}$ key. For example, 0.569×10^{-3} is entered as

$$0.569 \;\boxed{\text{EXP}}\; 3 \;\boxed{+/-}\; \rightarrow 0.569 \qquad -03$$

The factor of 10 does not appear on the display but is understood. Only the significant digits and the exponent are shown.

The calculator solution of Example 2.13 is then

$$328 \;\boxed{\text{EXP}}\; 3 \;\boxed{\times}\; 85 \;\boxed{\text{EXP}}\; 3 \;\boxed{+/-}\; \boxed{\div}\; 0.569 \;\boxed{\text{EXP}}\; 3 \;\boxed{+/-}\; \boxed{=}\; \rightarrow 4.90 \quad 07$$

Depending on how you set your calculator, this result might be in ordinary notation or scientific notation. The estimate of 4.5×10^7 agrees closely with the answer, thereby verifying the calculation.

The answer in engineering notation is

$$4.90 \times 10^7 = 49.0 \times 10^6$$

If you have an engineering calculator, you can do this problem using separate keys for engineering notation. The next section on the metric system illustrates further the use of engineering notation on the calculator.

▪

Study the next example, which closes the circuit and shows an application of engineering notation to a formula in electronics.

EXAMPLE 2.14

Close the Circuit

Figure 2–7 shows a circuit which contains a resistance, an inductance, and a capacitance in series. It is called an *RLC* circuit. The frequency f in hertz (Hz) of the circuit is given by

$$f = \frac{1}{2\pi \sqrt{LC}}$$

Calculate f in engineering notation when the inductance $L = 3.5 \times 10^{-3}$ H (henrys) and the capacitance $C = 150 \times 10^{-9}$ F (farads).

Solution Substitute the given values for L and C into the formula, and apply the rules for multiplying with powers of 10:

$$f = \frac{1}{2\pi\sqrt{LC}} = \frac{1}{2\pi\sqrt{(3.5 \times 10^{-3})(150 \times 10^{-9})}} = \frac{1}{2\pi\sqrt{525 \times 10^{-12}}}$$

EXAMPLE 2.14 (Cont.)

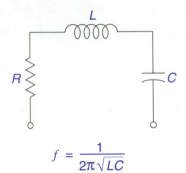

$$f = \frac{1}{2\pi\sqrt{LC}}$$

FIGURE 2–7 Frequency of an RLC circuit for Example 2.14.

Divide the index of the root into the power of 10, and use $\pi = 3.14$ or the $\boxed{\pi}$ key on the calculator. The result, accurate to three significant digits, is

$$f = \frac{1}{2\pi\left(\sqrt{525} \times 10^{-12/2}\right)} = \frac{1}{2\pi\left(22.9 \times 10^{-6}\right)} = 6950\,\text{Hz}$$

$$= 6.95 \times 10^3 \text{ Hz } (6.95 \text{ kHz})$$

One way this problem can be done on the calculator is as follows:

$$3.5 \boxed{\text{EXP}} \; 3 \boxed{+/-} \boxed{\times} \; 150 \boxed{\text{EXP}} \; 9 \boxed{+/-}$$

$$\boxed{=} \boxed{\sqrt{}} \boxed{\times} \; 2 \boxed{\times} \boxed{\pi} \boxed{=} \boxed{1/x} \rightarrow 6950$$

The size of numbers written with powers of 10 can be misleading. The increase from 10^2 to 10^3 is only 900, but the increase from 10^6 to 10^7 is 9,000,000! Figure 2–8 gives you an idea of how powers of 10 relate to some of the world's important (and not so important) scientific phenomena. For example, if you could add all the words ever spoken since people first started babbling, the number would still be less than 10^{18}!

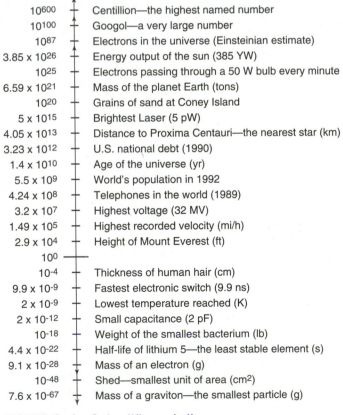

10^{600}	Centillion—the highest named number
10^{100}	Googol—a very large number
10^{87}	Electrons in the universe (Einsteinian estimate)
3.85×10^{26}	Energy output of the sun (385 YW)
10^{25}	Electrons passing through a 50 W bulb every minute
6.59×10^{21}	Mass of the planet Earth (tons)
10^{20}	Grains of sand at Coney Island
5×10^{15}	Brightest Laser (5 pW)
4.05×10^{13}	Distance to Proxima Centauri—the nearest star (km)
3.23×10^{12}	U.S. national debt (1990)
1.4×10^{10}	Age of the universe (yr)
5.5×10^9	World's population in 1992
4.24×10^8	Telephones in the world (1989)
3.2×10^7	Highest voltage (32 MV)
1.49×10^5	Highest recorded velocity (mi/h)
2.9×10^4	Height of Mount Everest (ft)
10^0	
10^{-4}	Thickness of human hair (cm)
9.9×10^{-9}	Fastest electronic switch (9.9 ns)
2×10^{-9}	Lowest temperature reached (K)
2×10^{-12}	Small capacitance (2 pF)
10^{-18}	Weight of the smallest bacterium (lb)
4.4×10^{-22}	Half-life of lithium 5—the least stable element (s)
9.1×10^{-28}	Mass of an electron (g)
10^{-48}	Shed—smallest unit of area (cm²)
7.6×10^{-67}	Mass of a graviton—the smallest particle (g)

FIGURE 2–8 Scientific notation.

EXERCISE 2.3

In problems 1 through 12, change each number to (a) scientific notation and (b) engineering notation.

1. 42,600,000,000
2. 11,700,000
3. 0.0000000930
4. 0.0000301
5. 2.35
6. 0.00000117

7. 62,200
8. 7,760
9. 564×10^5
10. 3.36×10^{-5}
11. 11.4×10^{-8}
12. 0.0528×10^7

In problems 13 through 22, change each number to ordinary notation.

13. 2.64×10^{10}
14. 3.90×10^{-6}
15. 93.1×10^{-12}
16. 0.112×10^9
17. 32.0×10^{-6}

18. 8.71×10^0
19. 4.44×10^6
20. 1.25×10^3
21. 1.01×10^0
22. 0.550×10^{-3}

In problems 23 through 30, estimate the result by rounding off to one digit. Then choose the correct answer, and check with the calculator. *Express answers to three significant digits.* (See Example 2.13.)

23. $(0.00000000101)(9,330,000)$ $[9.42 \times 10^{-4}, \quad 9.42 \times 10^{-3}, \quad 9.42 \times 10^{-2}, \quad 9.42 \times 10^{-1}]$

24. $(32,200)(86,500)(2830)$ $[7.88 \times 10^{10}, \quad 78.8 \times 10^9, \quad 78.8 \times 10^{12}, \quad 7.88 \times 10^{12}]$

25. $\dfrac{8.89 \times 10^{12}}{9.89 \times 10^6}$ $[0.899 \times 10^6, \quad 8.99 \times 10^6, \quad 89.9 \times 10^6, \quad 899 \times 10^6]$

26. $\dfrac{3.03 \times 10^{-3}}{6.06 \times 10^3}$ $[0.500 \times 10^{-6}, \quad 5.00 \times 10^{-6}, \quad 50.0 \times 10^{-6}, \quad 500 \times 10^{-6}]$

27. $\dfrac{0.0000000656}{0.00000123}$ $[5.33 \times 10^{-1}, \quad 53.3 \times 10^{-2}, \quad 533 \times 10^{-3}, \quad 53.3 \times 10^{-3}]$

28. $\dfrac{(0.000210)(0.00349)}{451}$ $[163 \times 10^{-9}, \quad 16.3 \times 10^{-9}, \quad 1.63 \times 10^{-9}, \quad 0.163 \times 10^{-9}]$

29. $\dfrac{1.26 \times 10^{12}}{(35.0 \times 10^{-3})(0.900 \times 10^6)}$ $[4.00 \times 10^6, \quad 40.0 \times 10^6, \quad 0.400 \times 10^6, \quad 400 \times 10^6]$

30. $\dfrac{(116 \times 10^3)(0.203 \times 10^6)}{94.1 \times 10^{-3}}$ $[0.250 \times 10^{12}, \quad 2.50 \times 10^{12}, \quad 25.0 \times 10^{12}, \quad 250 \times 10^{12}]$

In problems 31 through 36, solve each applied problem to *three significant digits.*

31. In 1989, General Motors made one of the greatest profits ever achieved by an industrial company. Worldwide sales totaled $136,975,000,000, with its assets valued at $180,236,500,000. What percentage of the assets do the total sales represent?

32. Parker Bros. Inc., manufacturer of the board game *Monopoly,* printed $18,500,000,000,000 of toy money in 1990 for all its games, which is more than all the real money in circulation in the world. If all this "money" were distributed equally among the world's population, estimated at 5.3 billion (5,300,000,000) in 1990, how much would each person get?

33. The shortest blip of light produced at the AT&T laboratories in New Jersey lasts 8×10^{-15} seconds. If the speed of light is 300×10^6 meters per second, how many meters does light travel in that time? Give the answer in engineering notation.

34. The lowest temperature ever achieved on earth was $T = 2.00 \times 10^{-9}$ Kelvin by Professor Lounasmaa working with a team of scientists in Finland in 1989. The Kelvin scale starts at absolute zero (−273.15°C or −459.67°F). At absolute zero, a gas theoretically shows no pressure. Absolute temperatures are defined in terms of the ratio:

$$\frac{273.15}{T}$$

Compute this ratio for the lowest temperature ever achieved on earth. Give the answer in engineering notation.

35. If there are 6.02×10^{23} molecules in 32 g (grams) of oxygen, how many grams does one molecule weigh? Give the answer in scientific notation.

36. The most massive living thing on earth is the giant Sequoia named the General Sherman in Sequoia National Park, California. Its weight is estimated at 5.51×10^6 lbs. The seed of such a tree weighs only 1.67×10^{-4} oz. Calculate the weight ratio of the mature tree compared with its seed. Give the answer in scientific notation. Note: 1 lb = 16 oz.

Applications to Electronics In problems 37 through 44, solve each applied problem. Give the answer in engineering notation to *three significant digits*.

37. The world's most expensive pipeline is the Alaska pipeline, which is built to carry 2,000,000 barrels a day of crude oil. If 6.65×10^6 barrels of oil (1 million tons) can generate 4.00×10^9 kWh (kilowatt-hours) of electricity, how many kilowatt-hours could be generated in one year from Alaska pipeline oil?

38. One of the world's fastest computers, the CRAY-2, can do a simple addition in 270×10^{-12} second (270 picosecond). How many additions can the CRAY-2 do in 1 minute?

39. Figure 2–9 shows three different resistances R_1, R_2, and R_3 connected in series. The total resistance R_T of the series circuit is given by

$$R_T = R_1 + R_2 + R_3$$

Find R_T when $R_1 = 3.30 \times 10^3 \ \Omega$ (3.30 kΩ), $R_2 = 910 \ \Omega$, and $R_3 = 1.20 \times 10^3 \ \Omega$ (1.20 kΩ). (Note: You cannot add powers of 10 unless they are the same.)

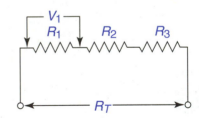

$$R_T = R_1 + R_2 + R_3$$

FIGURE 2–9 Series circuit for problems 39 and 40.

40. In the series circuit in Figure 2–9, $R_1 = 3.30 \times 10^3 \ \Omega$ (3.30 kΩ), and the voltage drop across R_1 is $V_1 = 350 \times 10^{-3}$ V (350 mV). Using Ohm's law for current: $I = \dfrac{V}{R}$, find the current through R_1.

41. Figure 2–10 shows two different capacitances C_1 and C_2 connected in series. The total capacitance is given by

$$C_T = \frac{C_1 C_2}{C_1 + C_2}$$

Find C_T when $C_1 = 2.20 \times 10^{-9}$ F (farads) and $C_2 = 3.30 \times 10^{-9}$ F.

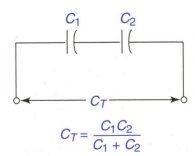

$$C_T = \frac{C_1 C_2}{C_1 + C_2}$$

FIGURE 2–10 Capacitances in series for problems 41 and 42.

42. In problem 41, find C_T when $C_1 = 8.50 \times 10^{-6}$ F (8.50 μF) and $C_2 = 0.0000550$ F.

43. In Example 2.14 calculate f when the inductance $L = 15.0 \times 10^{-3}$ H (15.0 mH) and the capacitance $C = 200 \times 10^{-9}$ F (200 nf).

44. In Example 2.14 calculate f when $L = 750 \times 10^{-3}$ H (750 mH) and $C = 500 \times 10^{-12}$ F (500 pf).

2.4
METRIC SYSTEM (SI)

The metric system is used throughout electricity and electronics. The United States is moving "inch by 2.54 centimeters" toward using the metric system completely. In 1960 the metric system, which was first established in France in 1790, became the more complete International System, abbreviated SI, which stands for "Systeme International." Since then, almost every country has converted to the SI system except the United States. The U.S. customary system, which evolved from the old English system, is gradually being replaced by the SI system. Almost all technical and scientific work uses SI units.

The metric system is based on powers of 10 as is our decimal system. The U.S. customary system is based on numbers such as 12 and 16, which were the basis of the old English system. In part, these numbers were used because they had several divisors and were easy to work with as fractions. This is no longer important in the age of electronic computation.

Important SI units, their symbols, and what they measure are shown in Table 2.2. Study this table. There are seven base units in the SI system. The other units are called derived units. For example, the base unit for electricity is the ampere, named after Andre Marie Ampere.

TABLE 2.2 SI Base Units and Other Important SI Units

	Quantity	Unit	Symbol
Base Units	Length	meter	m
	Time	second	s
	Mass	kilogram	kg
	Electric current	ampere	A
	Thermodynamic temperature	kelvin	K
	Molecular subtance	mole	mol
	Light intensity	candela	cd
	Electric charge	coulomb	C
	Electric voltage	volt	V
	Electric resistance	ohm	Ω
	Electric power	watt	W
	Electric conductance	siemen	S
	Electric capacitance	farad	F
	Electric inductance	henry	H
	Force	newton	N
	Energy	joule	J
	Frequency	hertz	Hz
	Temperature	Celsius	°C
	Pressure	kilopascal	kPa
	Angle	radian	rad
	Capacity	liter	L
	Area	square meter	m^2

> **Ampere** The constant current in two parallel conductors of infinite length and small cross section placed 1 meter apart in a vacuum that will produce a force between them of 2×10^{-7} newtons per meter.

The other electrical units, such as coulomb, volt, ohm, and watt, are derived units. For example, the unit of charge, the coulomb, is defined as follows.

> **Coulomb** The quantity of electric charge moved in one second by a current of one ampere: 1 coulomb = 1 C = 1 ampere × 1 second = 1 A · s

Concerning temperature, kelvin (K) is the base unit for thermodynamic or absolute temperature, but degrees Celsius is more commonly used. The conversion formula is

$$K = C + 273.15°$$

For units of time, s is used for seconds, min for minutes, h for hours, and d for days. The unit of energy, the joule, is used for electrical energy, mechanical energy, and heat energy.

Table 2.3 shows common prefixes based on powers of 10 which are used to denote quantities of metric units. For example,

$$
\begin{aligned}
1 \;\; &km \;\; (kilometer) &&= 1 \times 10^{3} \; m \; \text{ or } \; 1000 \; m \\
1 \;\; &\mu F \;\; (microfarad) &&= 1 \times 10^{-6} \; F \; \text{ or } \; 0.000001 \; F \\
1 \;\; &mA \;\; (millamp) &&= 1 \times 10^{-3} \; A \; \text{ or } \; 0.001 \; A \\
1 \;\; &MV \;\; (megavolt) &&= 1 \times 10^{6} \; V \; \text{ or } \; 1,000,000 \; V
\end{aligned}
$$

TABLE 2.3 Common SI Prefixes

Power of 10	Prefix	Symbol	Factor
10^{-12}	pico	p	trillionth (0.000000000001)
10^{-9}	nano	n	billionth (0.000000001)
10^{-6}	micro	μ	millionth (0.000001)
10^{-3}	milli	m	thousandth (0.001)
10^{-2}	centi	c	hundredth (0.01)
10^{3}	kilo	k	thousand (1000)
10^{6}	mega	M	million (1,000,000)
10^{9}	giga	G	billion (1,000,000,000)

Changes within the metric system are much easier to perform than within the U.S. system because of the powers of 10. Most changes can be done by just moving the decimal point 3 places to the right or left. For example,

$$
\begin{aligned}
1.28 \;\; kW &= 1280 \; W \\
505 \;\; \mu A &= 0.505 \; mA
\end{aligned}
$$

The procedure for changing these units is based on multiplying by a conversion factor equal to 1 or a unit ratio. To change 1.28 kilowatts to watts, you set up the fraction that equates kilowatts and watts:

$$\frac{1000 \text{ W}}{1 \text{ kW}}$$

Make the unit you are changing to, watts, the numerator. This fraction represents a unit ratio and is equal to 1:

$$\frac{1000 \text{ W}}{1 \text{ kW}} = \frac{1 \text{ kW}}{1 \text{ kW}} = 1$$

You can then convert 1.28 kW to watts by multiplying by this unit ratio, which does not change the value of a quantity:

$$1.28 \text{ kW} \left(\frac{1000 \text{ W}}{1 \text{ kW}} \right) = 1.28 \times 10^3 \text{ W} = 1280 \text{ W}$$

You can divide out units the same way as numbers. This helps to determine the correct units for the answer. Observe that the change of units is equivalent to *moving the decimal point 3 places to the right*.

To change 505 microamps to milliamps, consider that 1000 mA = 1 A = 1,000,000 μA. Therefore, 1 mA = 1000 μA, and you can multiply by the unit ratio:

$$\frac{1 \text{ mA}}{1000 \text{ μA}} = 1$$

Then,

$$505 \text{ μA} \left(\frac{1 \text{ mA}}{1000 \text{ μA}} \right) = 505 \times 10^{-3} \text{ mA} = 0.505 \text{ mA}$$

Observe that the change of units can be done by *moving the decimal point 3 places to the left*.

These two examples lead to the following direct procedure when you are changing units in the SI system.

The number of places you move the decimal point is usually 3 for most changes.

> **Changing Units**   Move the decimal point to the left when changing to a larger unit and to the right when changing to a smaller unit. The number of places is equal to the difference in the powers of 10.

Study the following example, which shows how to apply this procedure.

EXAMPLE 2.15

Change each of the following units:

1. 0.085 M Ω to kilohms
2. 280 pF to nanofarads

EXAMPLE 2.15 (Cont.)

Solution 1. Kilohms (10^3 Ω) is a smaller unit than megohms (10^6 Ω). The difference in the powers of 10 is 3. Therefore, move the decimal point 3 places to the right since you are changing to a smaller unit:

$$0.085 \ \text{M}Ω = 85 \ \text{k}Ω$$

2. Nanofarad (10^{-9}) is a larger unit than picofarad (10^{-12}). The number −9 is larger than −12. The difference in the powers of 10 is 3. Therefore, move the decimal point 3 places to the left since you are changing to a larger unit:

$$280 \ \text{pF} = 0.280 \ \text{nF}$$

▪

The next example shows some less common changes between units where the decimal point is moved a number of places other than 3.

EXAMPLE 2.16

Change each of the following units:

1. 670,000 mV to kilovolts
2. 0.034 m^2 to square centimeters

Solution 1. Kilovolts (10^3) is a larger unit than millivolts (10^{-3}). The difference in the powers of 10 is 6. Therefore, move the decimal point to the left 6 places since you are changing to a larger unit:

$$670,000 \ \text{mV} = 0.670 \ \text{kV}$$

2. To change square meters to square centimeters, note that, since 1 m = 100 cm, it follows that

$$(1 \ \text{m})^2 = (100 \ \text{cm})^2$$
$$\text{or} \ 1 \ \text{m}^2 = 10^4 \ \text{cm}^2$$

That is, 1 square meter = 10,000 square centimeters. Square centimeter is a smaller unit than square meter, and the difference in the powers of 10 is 4. Therefore, move the decimal point 4 places to the right:

$$0.034 \ \text{m}^2 = 340 \ \text{cm}^2$$

▪

The next example closes the circuit and shows an application of how to change units in an electrical problem.

EXAMPLE 2.17

Close the Circuit

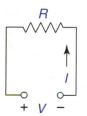

$V = IR$

FIGURE 2–11
Voltage in a
dc circuit for
Example 2.17.

Figure 2–11 shows a basic dc circuit containing a resistance connected to a voltage source. The voltage in volts is given by Ohm's law $V = IR$, where I = current in *amps* and R = resistance in *ohms*. Find V when $R = 11.0$ kΩ and $I = 5.50$ mA.

Solution The units in the formula must be in ohms and amps. Change milliamps to amps and kilohms to ohms before using the formula. Apply the procedure given earlier.

Since amps is a larger unit than milliamps, move the decimal point 3 places to the left:

$$5.50 \text{ mA} = 0.00550 \text{ A}.$$

Since ohms is a smaller unit than kilohms, move the decimal point 3 places to the right:

$$11.0 \text{ kΩ} = 11,000 \text{ Ω}$$

Then, the voltage to three significant digits is:

$$V = IR = (0.00550 \text{ A})(11,000 \text{ Ω}) = 60.5 \text{ V}$$

When changing units to a base unit (volts, amps, ohms, watts), you can just apply the power of 10 that corresponds to the prefix:

$$5.50 \text{ mA} = 5.50 \times 10^{-3} \text{ A} = 0.00550 \text{ A}$$
$$11.0 \text{ kΩ} = 11.0 \times 10^3 \text{ Ω} = 11,000 \text{ Ω}$$

Example 2.17 can then also be done with powers of 10:

$$V = IR = (5.50 \times 10^{-3} \text{ A})(11.0 \times 10^3 \text{ Ω}) = 60.5 \times 10^0 \text{ V} = 60.5 \text{ V}$$

Example 2.17 can be done on an electrical engineering calculator that has special units keys without converting the units. You enter the units after entering each number. Depending on your calculator, you may have to first set the mode to engineering and then press the [Shift] key or a special units key to enter the units after each number:

$$5.5 \boxed{\text{Shift}} \boxed{\text{m}} \boxed{\times} 11 \boxed{\text{Shift}} \boxed{\text{k}} \boxed{=} \rightarrow 60.5 \text{ V}$$

ERROR BOX

A common error to watch out for is moving the decimal point the wrong way when changing units in the metric system. Check if your answer makes sense as follows. If you are changing to a *smaller* unit, then you should be *increasing* the number of units. If you are changing to a *larger* unit, then you should be *decreasing* the number of units. See if you can do the practice problems by applying this reasoning.

Practice Problems: Change each of the following units:

1. 150 mA to amps
2. 0.15 mA to microamps
3. 1.5 MΩ to kilohms
4. 1500 Ω to kilohms
5. 0.015 V to millivolts
6. 15 pF to nanofarads

Answers:
1. 0.15 A
2. 150 μA
3. 1500 kΩ
4. 1.5 kΩ
5. 15 mV
6. 0.015 nF

The U.S. System and the International System _____

Some of the major differences and conversion factors between units of the U.S. customary system and the International System are shown in Table 2.4. The U.S. base units for length and mass are different than the SI units, but the other five base units are the same (see Table 2.2). In both systems all electrical units are the same. However, the units for work energy, heat energy, and mechanical power are different in the U.S. system.

"Thinking metric," observe in Table 2.4 that

TABLE 2.4   The U.S. Customary System and SI Conversions

Quantity	U.S. Unit	Symbol	SI Conversion Factor
Length	foot	ft	1 m = 3.281 ft
Mass	pound	lb	1 kg = 2.205 lb
Force	pound force	lb_f	1 N = 0.2248 lb_f
Temperature	degree Fahrenheit	°F	°C = (5/9) (°F − 32°)
Area	square foot	ft^2	1 m^2 = 10.76 ft^2
Velocity	feet per second	ft/s	1 m/s = 3.281 ft/s
Pressure	pounds per square foot	lb/ft^2	1 kPa = 20.89 lb/ft^2
Mechanical power	horsepower	hp	1 W = 0.001341 hp
Work energy	foot-pound	ft·lb	1 J = 0.7376 ft·lb
Heat energy	British thermal unit	Btu	1 J = 0.0009485 Btu

1. A meter is a little longer than a yard (3 feet).
2. A kilogram is a little heavier than 2 pounds.
3. A newton is about a quarter of a pound force.

To convert units between systems, you set up a unit ratio using the conversion factor and multiply by this ratio. For example, to convert 28.0 feet to meters, obtain the conversion factor from Table 2.4, and set up the unit ratio:

$$\frac{1\text{ m}}{3.281\text{ ft}} = 1$$

Then multiply by this ratio, which does not affect the value of a quantity, and divide out units to obtain the correct units:

$$28.0\,\cancel{\text{ft}} \left(\frac{1\text{ m}}{3.281\,\cancel{\text{ft}}} \right) = 8.53\text{ m}$$

All conversions are rounded off to *three significant digits.*

EXAMPLE 2.18

Convert 25.0 J (joules) to foot-pounds.

Solution   From Table 2.4, the unit ratio conversion factor from joules to foot-pounds is

$$\frac{0.7376 \, ft \cdot lb}{1 \, J}$$

Multiply by this ratio, dividing out the units, to obtain the correct units:

$$25.0 \, J \left(\frac{0.7376 \, ft \cdot lb}{1 \, J} \right) = 18.4 \, ft \cdot lb$$

EXAMPLE 2.19

Convert 37.0 kW to horsepower.

Solution   First change to watts:

$$37.0 \ kW = 37,000 \ W$$

Then change to horsepower. Multiply by the unit ratio conversion factor for watts to horsepower:

$$37,000 \, W \left(\frac{0.001341 \, hp}{1 \, W} \right) = 49.6 \, hp$$

The one exception to this procedure is as follows: To change from Fahrenheit to Celsius temperature, use the formula shown in Table 2.4 and substitute. For example, 50°F in Celsius is, to the nearest degree:

$$°C = (5/9)(50°F - 32°) = (5/9)(18°) = 10°C$$

EXERCISE 2.4

In problems 1 through 20, change each of the following units. *Express answers to three significant digits.*

1. 2.30 kV to volts
2. 0.560 kW to watts
3. 310 mA to amps
4. 1.50 kΩ to ohms
5. 5.20 W to milliwatts
6. 0.780 C to millicoulombs
7. 860 μA to milliamps
8. 7400 kV to megavolts
9. 0.130 nF to picofarads
10. 1.20 MΩ to kilohms

11. 3880 kHz to megahertz
12. 220 μs to milliseconds
13. 4100 mm to meters
14. 230 g to kilograms
15. 0.0600 m² to square centimeters
16. 55.0 mm² to square centimeters
17. 0.00350 A to microamps
18. 900,000 W to megawatts
19. 100°C to degrees kelvin
20. 3.14 rad to degrees (1 rad = 57.3°)

In problems 21 through 32, convert each of the following units. *Express answers to three significant digits.* Express 31 and 32 to the nearest degree.

21. 120 lb to kilograms
22. 15.0 kg to pounds
23. 135 ft to meters

24. 5.30 in. to centimeters
25. 135 W to horsepower
26. 0.250 hp to watts

27. 6.40 J to foot-pounds

28. 2.50 Btu to joules

29. 35.0 m/s to feet per second

30. 240 ft² to square meters

31. 95°F to Celsius

32. 32°F to Celsius

In problems 33 through 38, solve each conversion problem *to three significant digits.*

33. The total mass of a large computer system is 295 kg. How many pounds is this?

34. The formula for work is $W = Fd$, where F = force and d = distance. When F is in newtons and d in meters, W is in joules. Find W in joules when $F = 98.0$ N and $d = 55.0$ cm.

35. One of the world's thinnest calculators is manufactured by Sharp Electronics Corporation. It is 1.60 mm thick. How many inches thick is this calculator?

36. The tallest office building in the world is the Sears Tower in Chicago, which was completed in 1974. It has 110 stories and rises to 1454 ft. Its total ground area is 4,400,000 ft².
(a) What is the building's height in meters?
(b) What is the total ground area in square meters?

37. If one nautical mile (nmi) = 1.15 statute (land) miles, how many kilometers equal 1 nmi? Note: 5,280 ft = 1 mi.

38. Einstein's formula for atomic energy is $E = mc^2$, where m = mass in kilograms and c = speed of light = 3.00×10^8 m/s. Calculate how many joules of atomic energy are contained in 1 mg of mass.

Applications to Electronics   In problems 39 through 48, solve each applied problem *to three significant digits.*

39. Using Ohm's law for voltage $V = IR$, find V in volts when $R = 5.10$ kΩ and $I = 25.0$ μA. (See Example 2.17.)

40. Using Ohm's law for current $I = \dfrac{V}{R}$, find I in amps when $V = 240$ V and $R = 1.50$ kΩ.

41. Using Ohm's law for resistance $R = \dfrac{V}{I}$, find R when $V = 3.10$ V and $I = 500$ μA.

42. Using the formula for power $P = VI$, find P when $V = 500$ mV and $I = 50.0$ mA.

43. The resistance of a no. 8 copper wire whose cross section is 8.37 mm² is 2.06 Ω per kilometer.
(a) What is the resistance in ohms per 1000 ft?
(b) What is the cross section in square inches?

44. The wave length λ of a radio wave in meters is given by

$$\lambda = \frac{v}{f}$$

where the velocity of the wave $v = 3.00 \times 10^8$ m/s and f = the frequency of the wave in hertz. Find the wavelength of a radio wave whose frequency is 1010 kHz.

45. The surface area of a microprocessor is 1.90 cm². How many square inches is this area?

46. The solar energy falling on one square centimeter of the earth is 1.93 calories per minute. If 1 kW = 239 calories per second, how many kilowatts per square meter fall on the earth?

47. Figure 2–12 shows a circuit containing two different resistances R_1 and R_2 connected in series. The circuit contains an ammeter in line with the current and a voltmeter connected in parallel across R_1. If the ammeter reads 3.80 mA and the voltmeter reads 5.60 V, using Ohm's law for resistance $R = \dfrac{V}{I}$, find R_1 in ohms and kilohms.

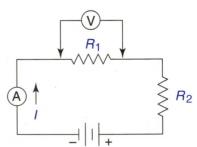

FIGURE 2–12 Series circuit for problems 47 and 48.

48. In Figure 2–12, $R_2 = 180 \ \Omega$, and the ammeter reads 28.0 mA. Using the formula $P = I^2R$, find the power dissipated across R_2 in watts and milliwatts.

☰ 2.5

HAND CALCULATOR OPERATIONS

Powers and Roots

Powers or exponents are done on the calculator with the power key $\boxed{x^y}$ or $\boxed{y^x}$. For example, 3^5 is calculated

$$3 \ \boxed{y^x} \ 5 \ \boxed{=} \ \rightarrow \ 243$$

To square a number, press $\boxed{x^2}$:

$$0.7 \ \boxed{x^2} \ \rightarrow \ 0.49$$

Scientific calculators are programmed to perform the order of operations correctly. Raising to a power or taking a root is done before multiplying or dividing.

EXAMPLE 2.20

Calculate:

$$\frac{(0.1)^2 + (0.3)^3}{0.5}$$

Solution   This is Example 2.4 in this chapter. One calculator solution is

$$0.1 \ \boxed{x^2} \ \boxed{+} \ 0.3 \ \boxed{y^x} \ 3 \ \boxed{=} \ \boxed{\div} \ 0.5 \ \boxed{=} \ \rightarrow \ 0.074$$

It is necessary to press the $\boxed{=}$ key after putting in the operations in the numerator. This computes the numerator before dividing. Remember that the fraction line acts like parentheses around the numerator and the denominator.

Another solution that uses parentheses corresponding to the fraction line is

$$\boxed{(} \ 0.1 \ \boxed{x^2} \ \boxed{+} \ 0.3 \ \boxed{y^x} \ 3 \ \boxed{)} \ \boxed{\div} \ 0.5 \ \boxed{=} \ \rightarrow \ 0.074$$

EXAMPLE 2.21

Estimate and then find the solution on the calculator:

$$\left(\frac{1}{0.25}\right)^2 + \frac{(4.5)^3}{(2.7)^2}$$

Solution   Estimate the answer by rounding off to one digit:

$$\left(\frac{1}{0.25}\right)^2 + \frac{(4.5)^3}{(2.7)^2} \approx \left(\frac{1}{0.3}\right)^2 + \frac{(5)^3}{(3)^2} = \frac{1}{0.09} + \frac{125}{9} = \frac{100}{9} + \frac{125}{9} = \frac{225}{9} = 25$$

Note that the fraction $\frac{1}{0.09}$ is changed to $\frac{100}{9}$ before adding.

The calculator solution can be done using the reciprocal key $\boxed{1/x}$:

$$0.25 \ \boxed{1/x} \ \boxed{x^2} \ \boxed{+} \ 4.5 \ \boxed{y^x} \ 3 \ \boxed{\div} \ 2.7 \ \boxed{x^2} \ \boxed{=} \ \rightarrow \ 28.5$$

The estimate of 25 agrees closely with the calculator answer of 28.5 and verifies the calculation.

Square roots are done on the calculator by pressing the square root key $\boxed{\sqrt{x}}$:

$$36 \ \boxed{\sqrt{x}} \ \rightarrow \ 6$$

EXAMPLE 2.22

Estimate and then compute with the calculator:

$$\sqrt{\frac{(16.5)^2(2.20)}{(6.82)(3.10)}}$$

Solution   Estimate by first rounding off to one digit and performing the operations under the radical:

$$\sqrt{\frac{(16.5)^2(2.20)}{(6.82)(3.10)}} \approx \sqrt{\frac{(20)^2(2)}{(7)(3)}} = \sqrt{\frac{800}{21}}$$

The estimate should be done as simply and quickly as possible *by hand* to provide a check on the calculator. To speed up the process, you can continue to round off to one digit as you estimate.

Change the 21 above to 20. Then *change the square root to the nearest perfect square* so you can do it mentally:

$$\sqrt{\frac{800}{21}} \approx \sqrt{\frac{800}{20}} = \sqrt{40} \approx \sqrt{36} = 6$$

There is little loss in accuracy by continuing to round off and the estimate will still be reliable.

One calculator solution to three significant digits is

$$16.5 \ \boxed{x^2} \ \boxed{\times} \ 2.20 \ \boxed{\div} \ 6.82 \ \boxed{\div} \ 3.10 \ \boxed{=} \ \boxed{\sqrt{x}} \ \rightarrow \ 5.32$$

Observe that you can divide by each factor in the denominator separately. The estimate of 6 is close to the calculator answer 5.32 and verifies the calculation.

EXAMPLE 2.23

Estimate and then calculate:
$$\sqrt{0.472} + \sqrt{0.0789}$$

Solution   To estimate the answer, change each decimal to the nearest perfect square. Note that the nearest perfect square to 0.472 is 0.49 and the nearest perfect square to 0.0789 is 0.09 as follows:

$$\sqrt{0.49} + \sqrt{0.09} = 0.7 + 0.3 = 1.0$$

The calculator solution is

$$0.472 \ \boxed{\sqrt{x}} \ \boxed{+} \ 0.0789 \ \boxed{\sqrt{x}} \ \boxed{=} \ \rightarrow \ 0.968$$

The estimate of 1 is close to the calculator answer 0.968.

EXAMPLE 2.24

Calculate:

$$\sqrt[3]{4.86}$$

Solution Cube roots can be done in one or more ways on your calculator, depending on which keys you have:

1. 4.86 [INV] [y^x] 3 [=] → 1.69 (Some calculators use [2nd F] or [SHIFT] instead of [INV].)
2. 4.86 [$\sqrt[x]{y}$] 3 [=] → 1.69
3. 4.86 [$\sqrt[3]{}$] → 1.69
4. 4.86 [$x^{\frac{1}{y}}$] 3 [=] → 1.69
5. 4.86 [y^x] 3 [1/x] → 1.69

You can estimate a cube root by changing it to the closest perfect cube root:

$$\sqrt[3]{4.86} \approx \sqrt[3]{8} = 2$$

Scientific Notation

To enter a number in scientific notation on the calculator:

1. Enter the significant digits with the decimal point to the right of the first digit.
2. Press [EE], [EEX], or [EXP].
3. Enter the exponent.

For example, to enter 9.78×10^{-3}, press

$$9.78 \ [\text{EXP}] \ 3 \ [+/-] \rightarrow 9.78 \qquad -03$$

The exponent appears to the right on the display. You do not enter the number 10; it is understood. Note that the [+/-] key changes the sign on the display.

When a calculation in ordinary notation produces a result with too many places for the display, the calculator automatically switches to scientific notation.

Engineering Notation

You can use the same keys for scientific notation to enter a number in engineering notation. However, if you have an engineering calculator with electrical units, you do not need to enter the exponent. When you press the units key after entering a number, the correct power of 10 corresponding to the units is entered. Depending on your calculator, you may first have to set the mode to engineering and then press the [Shift] key or a special units key before entering the units. Study the following example that closes the circuit and shows two ways to do an electrical problem with engineering notation.

EXAMPLE 2.25

Close the Circuit

Given a resistance $R = 1.50$ kΩ with a voltage drop across the resistance $V = 220$ mV, find the current through the resistance.

Solution   The current is given by Ohm's law:

$$I = \frac{V}{R}$$

One solution using the [EXP] key is as follows: Enter the number and then the power of 10:

220 [EXP] 3 [+/-] [÷] 1.5 [EXP] 3 [=] → 0.000147 or 1.47 −04

The result will be in ordinary or scientific notation, depending on how you set your calculator. You need to change the answer to engineering notation:

$$I = 1.47 \times 10^{-4} \text{ A} = 147 \times 10^{-6} \text{ A} = 1.47 \text{ } \mu\text{A}$$

Using an engineering calculator, you enter each number, and then press the corresponding units key:

220 [Shift] [m] [÷] 1.5 [Shift] [k] [=] → 147 μ

The calculator gives the answer with the correct units: microamps.

■

EXERCISE 2.5

In problems 1 through 28, *estimate the answer first* and choose what you think is the correct answer from the four choices. Then check with the calculator rounding off to three significant digits.

1. $(3.20)(4.26)^2$ [5.81, 58.1, 88.0, 581]

2. $(0.913)^2(1.02)^3$ [0.185, 0.885, 8.85, 88.5]

3. $\left(\dfrac{0.112}{0.518}\right)^3$ [0.0101, 0.101, 0.110, 1.10]

4. $\dfrac{(9.87)^3}{(2.03)^4}$ [0.566, 5.66, 56.6, 566]

5. $\dfrac{(1.25)^4 - (0.831)^4}{3.00}$ [0.065, 0.655, 0.950, 0.955]

6. $\left(\dfrac{1}{8.51}\right)^2 + \left(\dfrac{1}{6.66}\right)^2$ [0.0364, 0.364, 3.64, 6.43]

7. $100\left(1.00 + \dfrac{0.180}{4.00}\right)^4$ [3.91, 11.9, 39.1, 119]

8. $13.5(0.281 + 0.591)^3$ [0.895, 6.63, 8.95, 66.3]

9. $\sqrt{0.0588}$ [0.242, 0.842, 2.42, 8.42]

10. $\sqrt[3]{25.6}$ [1.95, 2.95, 4.95, 6.95]

11. $\left(\sqrt{19.0}\right)\left(\sqrt{171}\right)$ [5.70, 27.0, 57.0, 570)

12. $(0.789)^2\left(\sqrt{0.105}\right)$ [0.202, 0.388, 2.02, 3.88]

13. $\sqrt{93.2} + \sqrt{29.1}$ [1.50,   15.0,   150,   650]

14. $\sqrt[3]{8.31} - \sqrt[3]{1.28}$ [0.0940,   0.940,   9.40,   94.0]

15. $\dfrac{\sqrt{600}}{\sqrt{5.88}}$ [1.01,   10.1,   50.1,   101]

16. $\dfrac{\sqrt{(0.130)(22.4)}}{\sqrt{7.92}}$ [0.606,   6.06,   9.09,   60.6]

17. $\sqrt{(47)^2 + (33)^2}$ [1.74,   5.74,   57.4,   75.4]

18. $\sqrt{(3.15)^2 - (2.26)^2}$ [0.219,   1.29,   2.19,   12.9]

19. $\dfrac{201}{\sqrt[3]{(3.14)^2}}$ [3.97,   9.37,   39.7,   93.7]

20. $\sqrt{\dfrac{(3.01)(5.05)^2}{10.2}}$ [1.47,   1.74,   2.74,   7.74]

21. $(58.6 \times 10^3)(9.19 \times 10^{-3})$ [5.39,   53.9,   539,   5390]

22. $(86.6 \times 10^3)(94.1 \times 10^3)(5.31 \times 10^3)$ [0.433 \times 10^{12},   4.33 \times 10^{12},   43.3 \times 10^{12},   433 \times 10^{12}]

23. $\dfrac{55 \times 10^9}{0.220 \times 10^3}$ [2.50 \times 10^6,   25.0 \times 10^6,   25.0 \times 10^9,   0.250 \times 10^9]

24. $\dfrac{6.67 \times 10^3}{2.11 \times 10^{-6}}$ [31.6 \times 10^6,   316 \times 10^6,   3.16 \times 10^9,   31.6 \times 10^9]

25. $\dfrac{10.1 \times 10^6}{(71.1 \times 10^6)(0.813)^2}$ [0.215 \times 10^0,   21.5 \times 10^{-3},   2.15 \times 10^0,   2.15 \times 10^{-3}]

26. $\dfrac{(2.50 \times 10^{-3})^2}{(0.244)^3}$ [4.30 \times 10^{-3},   43.0 \times 10^{-3},   43.0 \times 10^{-6},   430 \times 10^{-6}]

27. $\sqrt{4.32 \times 10^{12}}$ [2.08 \times 10^{12},   2.08 \times 10^6,   20.8 \times 10^{12},   20.8 \times 10^6]

28. $\sqrt{0.0893 \times 10^6}$ [0.299 \times 10^3,   0.299 \times 10^0,   2.99 \times 10^3,   2.99 \times 10^0]

In problems 29 through 32, solve each applied problem *to three significant digits*.

29. The heat energy radiated by a certain black body is given by

$$E = \sigma(T^4 - T_0^4)$$

where $\sigma = 48.8 \times 10^{-9}$, $T = 291$ K (kelvin), and $T_0 = 273$ K. Calculate E.

30. The formula for the volume of a sphere is

$$V = \frac{4}{3}\pi r^3$$

where r = radius. Find V when $r = 3.28$ cm. Use $\pi = 3.142$ or the $\boxed{\pi}$ key on the calculator.

31. From problem 30, the radius of a sphere is given by

$$r = \sqrt[3]{\frac{3V}{4\pi}}$$

where V = volume. Find r when $V = 1.13 \times 10^3$ cm^3.

32. The velocity of sound in meters per second is given by

$$v = v_0 \sqrt{1 + \frac{t}{273}}$$

where v_0 = velocity of sound at 0°C and t = temperature in degrees Celsius. Find v when v_0 = 332 m/s and t = 15.0°C.

Applications to Electronics   In problems 33 through 38 solve each applied problem *to three significant digits.*

33. The current through a resistance R = 1.20 kΩ is I = 3.50 mA. Using Ohm's law for voltage $V = IR$, find the voltage drop across the resistance.

34. The current through a resistance R is I = 220 μA. If the voltage drop across the resistance is V = 53.0 mV, use Ohm's law for resistance $R = \frac{V}{I}$ to find the value of R.

35. The current in a circuit I = 42.5 mA, and the power P = 1.35 W. Using the power formula for resistance $R = \frac{P}{I^2}$, find the resistance of the circuit.

36. The voltage in a circuit V = 125 V and the resistance R = 1.60 kΩ. Using the formula for power $P = \frac{V^2}{R}$, find the power in the circuit.

37. The impedance Z in ohms of an ac circuit is given by

$$Z = \sqrt{R^2 + X^2}$$

Find Z when R = 14.7 kΩ and X = 26.5 kΩ.

38. The resultant voltage V_T of two voltages V_R and V_L 90° out of phase is given by

$$V_t = \sqrt{V_R{}^2 + V_L{}^2}$$

Find V_T when V_R = 140 V and V_L = 120 V.

≡ CHAPTER HIGHLIGHTS

2.1 POWERS AND ROOTS

A power defines repeated multiplication:

$$x^n = \underbrace{(x)(x)(x) \cdots (x)}_{n \text{ times}} \quad (n = \text{positive whole number}) \tag{2.1}$$

The definition of a square root is

$$\sqrt{x^2} = \left(\sqrt{x}\right)^2 = \left(\sqrt{x}\right)\left(\sqrt{x}\right) = x \tag{2.2}$$

Square and square root are inverse operations and cancel each other. Two basic rules for square roots are for x and y positive numbers

$$\sqrt{xy} = \left(\sqrt{x}\right)\left(\sqrt{y}\right) \tag{2.3}$$

$$\sqrt{\frac{x}{y}} = \frac{\sqrt{x}}{\sqrt{y}} \tag{2.4}$$

The cube root of a number can be found on a calculator by applying the following rule:

$$\sqrt[3]{x} = x^{1/3} \tag{2.6}$$

2.2 POWERS OF 10

For n = positive whole number, 10^n is equal to 1 followed by n zeros, and $10^{-n} = \frac{1}{10^n}$ or 1 preceded by n decimal places.

To multiply powers of 10, add the exponents algebraically as follows:

1.   When both powers are positive, add them.
2.   When both powers are negative, add the absolute (positive) values and make the result negative.
3.   When one power is positive and one is negative, subtract the smaller absolute value from the larger, and use the sign of the larger.

To divide powers of 10, change the sign of the power in the divisor, and apply the rules for multiplying. To raise a power of 10 to a higher power, multiply the exponents. To take a root of a power of 10, divide the root index into the power.

You can add coefficients of powers of 10 *only* if the exponents are the same.

2.3 SCIENTIFIC AND ENGINEERING NOTATION

To express a number in scientific notation, move the decimal point to the right of the first significant digit. Count the number of places the decimal point was moved. The number of places equals the power of 10: it is positive if moved to the left, negative if moved to the right.

To express a number in engineering notation, move the decimal point a multiple of three places to the right, or left, to change to a number between 0.1 and 1000. The number of places equals the power of 10: it is positive if moved to the left, negative if moved to the right.

To change to ordinary notation, move the decimal point the number of places equal to the power of 10: positive to the right, negative to the left.

2.4 METRIC SYSTEM (SI)

See Tables 2.2 and 2.3 for metric units and prefixes. Changes within the metric system are done by moving the decimal point to the left if changing to a larger unit and vice versa. The number of places is the difference in the powers of 10 and is equal to 3 for most changes.

Conversions between the U.S. system and metric system are done by multiplying by the unit ratio factor in Table 2.4. (See Examples 2.18 and 2.19.)

2.5 HAND CALCULATOR OPERATIONS

To raise to a power use $\boxed{y^x}$ or $\boxed{x^y}$. To square a number use $\boxed{x^2}$. To take the square root use $\boxed{\sqrt{\ }}$. To take the cube root use $\boxed{x^{\frac{1}{y}}}$, $\boxed{\text{INV}}$ $\boxed{y^x}$, $\boxed{\sqrt[x]{y}}$, $\boxed{\sqrt[3]{\ }}$, or a similar key. (See Example 2.24.)

To enter a number in scientific or engineering notation, use $\boxed{\text{EE}}$, $\boxed{\text{EEX}}$, or $\boxed{\text{EXP}}$ to enter the power of 10. To use an engineering calculator, enter the units after the number using the units key.

Estimate calculator results by rounding numbers to one digit and roots to the closest perfect root and computing by hand or mentally.

≡ REVIEW QUESTIONS

In problems 1 through 8, *calculate each by hand.* Check with the calculator.

1. $\dfrac{6^2 + 8^2}{5^2}$

2. $(6)^2 \left(\dfrac{1}{3}\right)^3 - (18)\left(\dfrac{1}{9}\right)^2$

3. $\sqrt{81}$

4. $\sqrt{\dfrac{25}{49}}$

5. $\sqrt{6.25}$

6. $\sqrt[3]{0.008}$

7. $\dfrac{(2 \times 10^{-3})(0.5 \times 10^6)}{(4 \times 10^{-3})}$

8. $\dfrac{\sqrt{64 \times 10^6}}{(6 \times 10^3)(4 \times 10^6)}$

In problems 9 through 12, simplify each expression.

9. $\sqrt{75}$

10. $\dfrac{\sqrt{24}}{\sqrt{8}}$

11. $\sqrt{\dfrac{0.48}{12}}$

12. $\left(\sqrt{20}\right)^2 - \left(\sqrt{5}\right)\left(\sqrt{45}\right)$

In problems 13 through 24, *estimate the answer first,* and choose what you think is the correct answer from the four choices. Then check with the calculator rounding off to three significant digits.

13. $(9.38)^2\left(\sqrt{9.91}\right)$ [1.77,   2.77,   27.7,   277]

14. $\left(\sqrt[3]{28.3}\right)(0.101)$ [0.308,   0.808,   3.08,   30.8]

15. $\dfrac{\sqrt{66.6}}{\left(\sqrt{50}\right)\left(\sqrt{72}\right)}$ [0.0136,   0.136,   0.936,   1.36]

16. $\sqrt{\dfrac{(1.21)(8.13)}{(0.331)^2}}$ [0.948,   1.98,   9.48,   19.8]

17. $(9.21 \times 10^6)(3.58 \times 10^{-3})$ [33.0×10^3, 3.30×10^3, 330×10^3, 330×10^6]

18. $(5.21 \times 10^3)(8.11 \times 10^{-6})$ [4.23×10^{-3}, 42.3×10^{-3}, 423×10^{-3}, 0.423×10^3]

19. $(0.325 \times 10^6)(6.31 \times 10^{-9})$ [0.205×10^{-3}, 2.05×10^{-3}, 20.5×10^{-3}, 205×10^{-3}]

20. $(28,100)(32,500)(100,000)$ [9.13×10^{12}, 91.3×10^{12}, 913×10^{12}, 913×10^{15}]

21. $\dfrac{101,000,000}{505,000,000,000}$ [20×10^{-3}, 200×10^{-3}, 20×10^{-6}, 200×10^{-6}]

22. $\dfrac{3.33 \times 10^6}{7.89 \times 10^{-3}}$ [422×10^9, 42.2×10^9, 4.22×10^9, 0.422×10^9]

23. $\dfrac{\sqrt{2.86 \times 10^6}}{0.532 \times 10^3}$ [0.317,   31.7,   0.318,   3.18]

24. $\left(\sqrt{9.33 \times 10^2}\right)(8.93 \times 10^2)^3$ [218×10^6, 21.8×10^6, 21.8×10^9, 2.18×10^9]

In problems 25 through 32, convert each of the following units. *Round answers to three significant digits.*

25. 0.510 ms to microseconds

26. 120 V to kilovolts

27. 1500 kΩ to megohms

28. 150 μA to milliamps

29. 2,300,000 W to megawatts

30. 700 pF to nanofarads

31. 2.50 ft to meters

32. 3.50 hp to watts

In problems 33 through 37 solve each applied problem *to three significant digits.*

33. The mean radius of the earth is 6,370,000 m. What is the radius in kilometers?

34. A "22-caliber" rifle means the bore is 0.22 inch. What is the caliber of a rifle whose bore is 7.62 mm to two digits?

35. A power plant is rated at 150 MW. How many horsepower is this equivalent to?

36. The fastest runner in baseball was Ernest Swanson, who took only 13.30 s to circle the bases, a distance of 360 ft, in 1932. What was his average speed in kilometers per hour?

37. Under certain conditions, when the driver jams on the brakes of an automobile weighing 4000 lb and traveling 60 mi/h, the car will skid a distance in feet given by

$$d = \dfrac{4000(13.9)^2}{1200(1.22)(9.81)}$$

Calculate the distance.

38. The world produced more than 6.82×10^{12} kWh of electricity in 1978. If 1 kWh is equivalent to 7.60×10^{-2} gal of oil, how many gallons of oil would it take to produce that much energy? Give the answer in scientific notation.

Applications to Electronics   In problems 39 through 48 solve each applied problem *to three significant digits*.

39. An ammeter measures the current though a resistance $R = 4.70$ kΩ to be $I = 12.0$ mA. Using Ohm's law for voltage $V = IR$, find the voltage drop across the resistance.

40. In problem 39, using the power formula $P = I^2R$, find the power dissipated in the resistance.

41. The power P dissipated in a resistance $R = 750$ Ω is 90.0 mW. Using the power formula for current $I = \sqrt{\frac{P}{R}}$, find the current I.

42. Figure 2–13 shows a circuit containing two different resistances R_1 and R_2 connected in parallel. The equivalent resistance of the circuit is given by

$$R_{eq} = \frac{R_1 R_2}{R_1 + R_2}$$

If $R_1 = 820$ Ω and $R_2 = 1.20$ kΩ, find R_{eq}.

43. The impedance Z of the *RL* (resistance-inductance) circuit in Figure 2–14 is given by

$$Z = \sqrt{R^2 + X_L{}^2}$$

where X_L = inductive reactance. Find Z when $R = 620$ Ω and $X_L = 250$ Ω.

44. In the *RL* circuit in Figure 2–14, find Z when $R = 910$ kΩ and $X_L = 1.50$ MΩ. Note: Units must be the same in the formula.

45. Figure 2–15 shows two inductances in parallel. The total inductance is given by

$$L_T = \frac{L_1 L_2}{L_1 + L_2}$$

Find L_T when $L_1 = 50.0$ mH and $L_2 = 100$ mH.

46. Find L_T in Figure 2–15 when $L_1 = 500$ μH and $L_2 = 1.50$ mH. Note: Units must be the same in the formula.

47. The inductance L of a coil is given by

$$L = \frac{n^2 \mu A}{l}$$

Calculate L in henrys when $n = 200$ turns, $\mu = 1.26 \times 10^{-6}$ N/A^2, A $= 3.00 \times 10^{-4}$ m^2, and $l = 0.300$ m.

48. Calculate the following written in BASIC:

3.82^3*SQR(83.8)/9.31

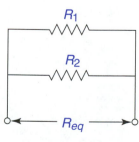

$$R_{eq} = \frac{R_1 R_2}{R_1 + R_2}$$

FIGURE 2–13   Parallel circuit for problem 42.

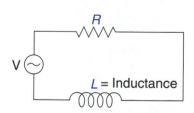

X_L = Inductive reactance

$$Z = \sqrt{R^2 + X_L{}^2}$$

FIGURE 2–14   Impedance Z in an *RL* circuit for problems 43 and 44.

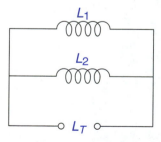

$$L_T = \frac{L_1 L_2}{L_1 + L_2}$$

FIGURE 2–15 Inductances in parallel for problems 45 and 46.

Basic Algebra

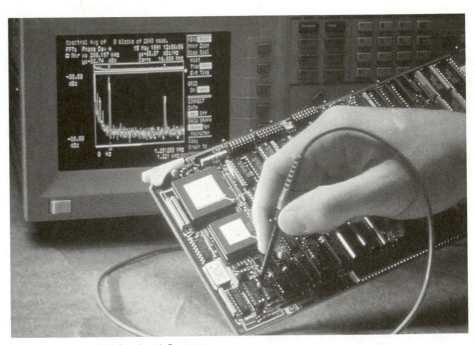

Courtesy of Hewlett Packard Company.

Designers of electronic and electromechanical products determine sources of phase noise with a modulation domain analyzer.

This chapter presents the fundamentals of algebra necessary for working with the basic formulas and equations of electricity and electronics. It also provides essential background for a better understanding of the important concepts and the theory of electricity. Signed numbers and algebraic terms are explained first, followed by the rules for working with positive and negative exponents. How to multiply binomials and factor polynomials is then studied, which leads to algebraic fractions. Multiplication and addition of algebraic fractions are presented in the last two sections. You must have a good understanding of how to add fractions in arithmetic to be able to add algebraic fractions. The last section reviews addition of fractions in arithmetic and shows many examples that clearly explain how to add algebraic fractions.

Chapter Objectives

In this chapter, you will learn:

- How to add, subtract, multiply, and divide signed numbers.
- How to combine algebraic terms.
- The rules for working with exponents.
- The meaning of negative and zero exponents.
- How to multiply binomials and polynomials.
- How to factor polynomials.
- How to reduce algebraic fractions.
- How to multiply and divide algebraic fractions.
- How to add algebraic fractions.

3.1
SIGNED NUMBERS AND ALGEBRAIC TERMS

Signed Numbers

One of the basic features of algebra is the use of positive and negative signed numbers. The signed numbers and zero can be represented as a number line, where the numbers increase from left to right:

$$\dots\ -3,\ -2,\ -1,\ 0,\ +1,\ +2,\ +3,\ \dots$$

Observe that -2 is mathematically a larger number than -3 because the numbers increase from left to right. However, the sign is also used to indicate direction in electricity and electronics. For example, a voltage of zero usually indicates ground potential, with *positive* meaning a voltage above ground and *negative* a voltage below ground. A voltage of $+5$ V has the same *magnitude* or *absolute value* as a voltage of -5 V but with a different reference point.

For current, positive indicates current in one direction, and negative indicates current in the opposite direction. A current of $+2$ mA has the same magnitude as a current of –2 mA, but it flows in the opposite direction. Positive and negative numbers are also used for angles where positive represents counterclockwise rotation and negative represents clockwise rotation. In Chapter 2, positive and negative numbers are used for powers of 10: $10^0 = 1$, $10^{-1} = 0.1$, $10^{-2} = 0.01$.

The positive and negative whole numbers and zero are called the *integers*. The *absolute value* of an integer is the value of the number without the sign, and it represents the magnitude of the number. Absolute value is equal to the positive value and is indicated with vertical lines:

$$|\text{-}6| = 6$$
$$|\text{+}9| = 9$$

The rules for adding, subtracting, multiplying and dividing signed numbers are given below.

Adding Like Signs	To add two signed numbers with like signs, add the numbers and apply the common sign. (3.1)

For example,

$$(-8) + (-7) = -(8 + 7) = -15$$

$$5 + (+2) = +7 \text{ or } 7$$

Note that when there is no sign written it is understood to be positive: 5 means +5.

Adding Unlike Signs	To add two signed numbers with unlike signs, subtract the smaller absolute value from the larger absolute value, and apply the sign of the larger absolute value. (3.2)

For example,

$$(-9) + (+6) = -(9 - 6) = -3$$

$$13 + (-5) = +(13 - 5) = +8$$

To add several signed numbers, you add the like signs first, and then take the difference of the positive and negative numbers:

$$6.6 + (-2.3) + (-8.8) + 1.2 = (6.6 + 1.2) - (2.3 + 8.8) = 7.8 - 11.1 = -3.3$$

Subtraction in algebra is just addition with the sign changed.

Subtracting Signed Numbers	To subtract two signed numbers, first change the sign of the number that follows the negative sign. Then add the numbers applying the rules (3.1) and (3.2). (3.3)

For example,

$$(10) - (-3) = 10 + (+3) = 10 + 3 = 13$$
$$(-14) - (+6) = -14 + (-6) = -14 - 6 = -20$$
$$(12) - (7) = 12 - 7 = 5$$

Observe carefully that a minus sign just has the effect of changing the sign that follows. A minus followed by a minus equals a plus. A minus followed by a plus equals a minus. Study the following example, which applies these ideas.

EXAMPLE 3.1

Perform the operations:

$$3.2 + (-4.1) - (-5.5) - (1.2)$$

Solution This example contains addition and subtraction of signed numbers. However, since subtraction is a type of addition, it is helpful to think of addition and subtraction of signed numbers as just an *algebraic combination* of signed numbers. Apply these two ideas:

1. Any sign that follows a plus sign you leave the same,

2. Any sign that follows a minus sign you change.

The parentheses in the example can then be removed and it becomes

$$3.2 + (-4.1) - (-5.5) - (1.2) = 3.2 - 4.1 + 5.5 - 1.2$$

Then combine the positive and negative numbers and take the difference of the two:

$$3.2 + 5.5 - 4.1 - 1.2 = 8.7 - 5.3 = 3.4$$

■

Multiplication or division of signed numbers follows a straightforward rule.

Multiplying or Dividing Signed Numbers	To multiply or divide two signed numbers, multiply or divide the absolute values. Then make the result positive if the signs are the same and negative if the signs are different. (3.4)

For example,

$$(-5)(-3) = +15$$

$$(3)(-5)(2) = -15(2) = -30$$

$$(5)(-3) \div 15 = -15 \div 15 = -1$$

Multiplication of numbers in algebra is indicated by parentheses or a dot: $(-5)(3) = -5 \cdot 3$. The multiplication sign is not used in algebra because it can be confused with the letter x.

The order of operations in arithmetic also applies in algebra:

1. Perform operations in parentheses.
2. Do multiplication or division.
3. Do addition or subtraction.

Study the next two examples, which show a combination of operations in algebra.

EXAMPLE 3.2

Perform the operations:

$$-0.5(-7+3) + \frac{8(-3)}{6}$$

Solution   Perform the operation in parentheses first; then multiply and divide. Next, add or subtract:

$$-0.5\,(-7+3) + \frac{8(-3)}{6} = -0.5\,(-4) + \frac{8(-3)}{6} = 2 + \frac{-24}{6} = 2 + (-4) = -2$$

■

EXAMPLE 3.3

Perform the operations:

$$\frac{-8(3) - 4(2)}{2 + 3(-1)}$$

Solution   Remember, *a fraction line is like parentheses*. Do the operations in the numerator and denominator first, then divide:

$$\frac{-8(3) - 4(2)}{2 + 3(-1)} = \frac{-24 - 8}{2 - 3} = \frac{-32}{-1} = 32$$

■

Combining Terms

An *algebraic term* is a combination of letters, numbers, or both, joined by the operations of multiplication or division.

Examples of six different algebraic terms are as follows:

$$4x^2$$

$$10b$$

$$-3I^2R_1$$

$$\frac{V_T}{R_T}$$

$$-1.8C$$

$$2\pi r$$

Each algebraic term has two parts: the *coefficient*, which is the constant number in front, and the *literal part:*

Coefficient	**Literal part**
4	x^2
-3	I^2R_1
2π	r

Note that π is part of the coefficient because it is a constant number, $2\pi \approx 6.28$. The literal part consists of letters that can take on different values and are called *variables*. The subscript 1 used in the above term for R_1 means a specific value of the variable R. R_1 is a different variable than R_2 or R_T or just R.

When the coefficient is 1, it is understood and so not written:

$$\frac{V_T}{R_T} = (1)\frac{V_T}{R_T}$$
$$-\pi d = -(1)\pi d$$

Observe that the minus sign represents a coefficient of -1.

Like terms are terms that have *exactly the same literal part.* For example, pairs of like terms are

$$4x^2 \text{ and } -10x^2$$
$$-3I^2R_1 \text{ and } 5I^2R_1$$

Like terms *can* be combined or added together.

Unlike terms do not have the same literal part. For example, pairs of unlike terms are

$$4x^2 \text{ and } 3x$$
$$-3I^2R_2 \text{ and } 5I^2R_1$$

Here, $-3I^2R_2$ and $5I^2R_1$ are unlike terms because the subscripts are different. Unlike terms *cannot* be combined.

> **Combining Like Terms**   To combine like terms, add the coefficients only. Do not change the literal part including the exponents. $\qquad$ (3.5)

For example,

$$-5y + 3y = (-5 + 3)y = -2y$$
$$6I^2R_1 - 3I^2R_1 + 2I^2R_1 = (6 - 3 + 2)I^2R_1 = (8 - 3)I^2R_1 = 5I^2R_1$$

Study the next two examples, which further illustrate these ideas.

EXAMPLE 3.4

Perform the operations:

$$5V_x + V_y - (4V_y - 3V_x)$$

Solution   Perform the operation on the parentheses first. A minus sign in front of a parenthesis is like multiplying every term within the parentheses by -1. *It changes the sign of every term in the parentheses:*

$$5V_x + V_y - (4V_y - 3V_x) = 5V_x + V_y - (4V_y) - (-3V_x) = 5V_x + V_y - 4V_y + 3V_x$$

EXAMPLE 3.4 (Cont.)

Now note that V_x is different than V_y and you can only combine like terms. Therefore,

$$5V_x + V_y - 4V_y + 3V_x = (5 + 3)V_x + (1 - 4)V_y = 8V_x - 3V_y$$

■

Sometimes you can have a set of parentheses inside of another parentheses or grouping. Parentheses are used first, then brackets [ ], then braces { }.

EXAMPLE 3.5

Perform the operations:

$$2V - [(2V - 2IR) - (V - IR)]$$

Solution   To evaluate such an expression, you work from the inside out, paying careful attention to the negative signs in front of the parentheses and the brackets. First remove parentheses:

$$2V - [(2V - 2IR) - (V - IR)] = 2V - [2V - 2IR - V + IR]$$

Observe that a plus sign in front of the parentheses does not change any signs, and you can just remove the parentheses:

$$(2V - 2IR) = 2V - 2IR$$

Now combine terms in the brackets:

$$2V - [2V - 2IR - V + IR] = 2V - [V - IR]$$

Then remove the brackets:

$$2V - V + IR$$

Combine terms again:

$$V + IR$$

■

EXERCISE 3.1

In problems 1 through 42, perform the indicated operations *without the calculator*.

1. $2 - 5$

2. $-2 - 8$

3. $3 - (-1)$

4. $15 - (+10)$

5. $8 - 6(-2) + 1$

6. $2.2 - 1.5 - (-3.1)$

7. $-8 \cdot 5.5$

8. $(-6)\left(-\dfrac{1}{2}\right)$

9. $\dfrac{63}{-9}$

10. $\dfrac{-10}{-2}$

11. $(-2)(0.1)(-0.3)$

12. $(-5)(-5)(-3)$

13. $\dfrac{6(-3)}{9}$

14. $\dfrac{1.21}{10(-1.1)}$

15. $-2 \cdot 6 - 3 \cdot 8$

16. $\dfrac{3 - 8}{-1(-2)}$

17. $\dfrac{5(-6)(-8)}{2(-5) + 5(3)}$

18. $\dfrac{(0.3)(-6)}{3.3 - 2.7}$

19. $-3\left(-\dfrac{1}{5}\right) + \dfrac{4(-2)}{10}$

20. $\left(\dfrac{1}{2}\right)\left(-\dfrac{1}{3}\right) - \left(\dfrac{2}{3}\right)\left(-\dfrac{1}{4}\right)$

21. $\dfrac{1.2 - 3.3}{6(0.7)}$

22. $\dfrac{10(-1.1)}{2.8 - 0.6}$

23. $5\left(\dfrac{-3}{10}\right) + 4\left(\dfrac{7}{-12}\right)$

24. $-6 - \dfrac{-16}{2} + 3\left(-\dfrac{1}{6}\right)$

25. $2(-0.5) + (0.6)\left(\dfrac{5}{6}\right)$

26. $\dfrac{8}{2(-0.2)} + \left(\dfrac{1}{2}\right)(-8)(0.5)$

27. $2x - 10x + 3y$

28. $11q + 2p - 6p$

29. $5.3V_1 - 2.8V_2 - 6.8V_1 + 0.5V_2$

30. $8.2C_1 + 4.3C_2 - 1.3C_2 - 2.4C_1$

31. $IR^2 - I^2R + 3I^2R - IR^2$

32. $2\lambda f^2 + 3\lambda f - 6\lambda^2 f - \lambda f^2$

33. $(by - 4) - (2by - 3)$

34. $5V - (-2V + 4I) + 3I$

35. $(8X_C^2 + 2X_C + 1) - (3X_C^2 + 3X_C + 8)$

36. $(2u + 3w - 1) - (-3u + 4w - 2)$

37. $\dfrac{2R_1}{3} - \dfrac{R_2}{2} - \left(\dfrac{R_1}{3} - \dfrac{R_2}{2}\right)$

38. $\dfrac{-8.6I_1}{2} - (3.5I_2 - 2.1I_1) + (6.8I_1 - I_2)$

39. $3V_0 - [(V_1 - V_0) - (V_0 - V_1)]$

40. $5 + [r\theta - (r\theta - 1)]$

41. $[n^2 + (2n + 1)] - (n - 1)$

42. $4 - [(f - g) - (f + g - 3)]$

In applied problems 43 through 48 *do all calculations by hand.*

43. Mount Everest, the highest mountain on earth, is 8848 m or 29,028 ft above sea level. The Mariana trench, the greatest sea depth, is 11,034 m or 36,201 ft below sea level. Express these distances as signed numbers, and calculate the difference between the two numbers in meters and feet.

44. What is the temperature difference between the melting point of petroleum, $-70.5°C$, and the boiling point, $250°C$?

45. A ship travels north (0°) for 1 hour, turns clockwise 45° and travels northeast for 1 hour, then turns counterclockwise 90° and travels northwest for 1 hour. Using positive for counterclockwise, express these angles as signed numbers, and combine them to calculate the net degrees the ship has turned.

46. A football team is pushed back 8 yd on a poor play. The team then completes a pass gaining 14 yd, but incurs a penalty of 5 yd. Express these changes as signed numbers, and combine them to determine the team's gain or loss.

47. Simplify the following expression from a problem involving the volume of a solid:

$$(\pi r^2 h + 2\pi r) - (4\pi r - 2\pi r^2 h)$$

48. Two forces are acting on a body in the same direction. If one of the forces $F_1 = 3d^2 - 2d + 1$ and the other force $F_2 = 4d^2 + 3d - 5$, what is the resultant force on the body?

Applications to Electronics In applied problems 49 through 56 *do all calculations by hand.*

49. The temperature of a light filament changes from 175°C to 17°C when the power is turned off. What signed number represents this change?

50. The voltage in a circuit with respect to ground changes from 2.5 V to –0.5 V when the ground connection is changed. What signed number represents this change?

51. Flowing into a junction in a circuit are the two currents $I_1 = 1.3$ A and $I_2 = 0.8$ A. Flowing out of the same junction are the two currents $I_3 = 1.7$ A and $I_4 = 0.4$ A. Expressing the currents flowing in as positive and the currents out as negative, show that the algebraic sum of the currents at the junction is zero. This is Kirchhoff's current law.

52. Going clockwise around a closed path in a circuit the voltage *drops* are 10.3 V, 8.2 V, and 7.0 V. The voltage *increases* are 12.6 V and 12.9 V. Expressing the drops as negative and the increases as positive, show that the algebraic sum of the voltages is zero. This is Kirchhoff's voltage law.

53. In Figure 3–1 the voltage drop across one resistance is given by $V_1 = 1.5IR - 0.4$ and the voltage drop across a second resistance by $V_2 = 2.3IR + 0.8$. What is the total voltage drop V_T across both resistances?

54. The power in one circuit is given by $8.34I^2R + 9.89$, and the power in a lower rated circuit by $5.76I^2R - 7.67$. What is the expression for the difference in power between the circuits?

55. The work output of one computer is given by $\frac{1}{3}W - 5$ and a second computer by $\frac{1}{4}W - 7$. How much more work, in terms of W, is done by the computer with the greater output?

56. The total heat produced by two generators is given by $7Q^4 + \frac{1}{3}Q^2$. If the heat produced by one generator is given by $5Q^4 + \frac{1}{6}Q^2$, what is the expression for the heat produced by the other generator?

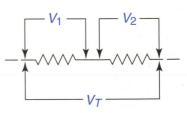

$$V_1 = 1.5IR - 0.4$$

$$V_2 = 2.3IR + 0.8$$

$$V_T = V_1 + V_2$$

FIGURE 3–1 Voltage drops for problem 53.

≡ 3.2
RULES FOR EXPONENTS

As shown in Section 2.1, an exponent defines repeated multiplication:

$$x^n = \underbrace{(x)(x)(x) \cdots (x)}_{n \text{ times}} \qquad (n = \text{positive integer})$$

The number x is called the *base*. When performing operations on numbers with exponents, there are certain rules that simplify the calculations.

Five Basic Rules

| **Addition Rule** | $(x^m)(x^n) = x^{m+n}$ | (3.6) |

For example,

$$(x^3)(x^2) = x^{3+2} = x^5$$

$$(2^4)(2^2) = 2^6 \text{ or } 64$$

Note that you do not multiply the 2s, which are the base. You just add the exponents.

| **Subtraction Rule** | $\dfrac{x^m}{x^n} = x^{m-n}$ | (3.7) |

For example,

$$\frac{P^5}{P} = P^{5-1} = P^4$$

Observe that if there is no exponent, it is understood to be 1: $P = P^1$.

| **Multiplication Rule** | $(x^m)^n = x^{mn}$ | (3.8) |

For example,

$$(I^2)^2 = I^{2 \cdot 2} = I^4$$
$$3(y^2)^3 = 3(y^{2 \cdot 3}) = 3y^6$$

| **Product Rule** | $(xy)^n = x^n y^n$ | |
| | $\left(\dfrac{x}{y}\right)^n = \dfrac{x^n}{y^n}$ | (3.9) |

For example,

$$(A^2 B)^3 = A^{2 \cdot 3} B^{1 \cdot 3} = A^6 B^3$$

$$\left(\frac{2}{R^2}\right)^2 = \frac{2^2}{R^{2 \cdot 2}} = \frac{4}{R^4}$$

The last rule applies to radicals.

| **Division Rule** | $\sqrt[n]{x^m} = x^{\frac{m}{n}}$ | (3.10) |

For example,

$$\sqrt[3]{V^6} = V^{\frac{6}{3}} = V^2$$

Study the following example, which shows how to apply several of the rules.

EXAMPLE 3.6

Evaluate:
$$\frac{10X^3 \cdot (X^2)^2}{10X^2}$$

Solution Perform the operations in the numerator first. In the order of operations, *raising to a power and taking a root are done before multiplication or division*:

$$\frac{10X^3 \cdot (X^2)^2}{10X^2} = \frac{10X^3 \cdot X^4}{10X^2} = \frac{10X^7}{10X^2}$$

Note that when you multiply $(10X^3)(X^4)$ you add the exponents and carry along the factor of 10. Now apply the division rule:

$$\frac{10X^7}{10X^2} = X^5$$

When $n = m$ in the subtraction rule, you obtain a zero exponent:

$$\frac{x^3}{x^3} = x^{3-3} = x^0$$

Since any number divided by itself is 1, the zero exponent is defined as 1 for *any* number except zero:

$$x^0 = 1 \quad (x \neq 0) \tag{3.11}$$

For example,
$$10^0 = 1$$
$$(-4)^0 = 1$$
$$W^0 = 1$$
$$(5.3I)^0 = 1$$

When $m < n$ (m less than n) in the subtraction rule, you obtain a negative exponent:

$$\frac{x^2}{x^4} = x^{2-4} = x^{-2}$$

This fraction can also be evaluated by dividing out factors:

$$\frac{x^2}{x^4} = \frac{\cancel{x} \cdot \cancel{x}}{\cancel{x} \cdot \cancel{x} \cdot x \cdot x} = \frac{1}{x^2}$$

Then,
$$x^{-2} = \frac{1}{x^2}$$

Therefore, a negative exponent is defined as a reciprocal:

$$x^{-n} = \frac{1}{x^n} \text{ and } \left(\frac{x}{y}\right)^{-n} = \left(\frac{y}{x}\right)^n \quad (x \text{ or } y \neq 0) \tag{3.12}$$

That is, a negative power means *invert and raise to the positive power*. For example,

$$4^{-3} = \frac{1}{4^3} = \frac{1}{64}$$

$$2^{-1} = \frac{1}{2}$$

Note that the –1 power just means reciprocal:

$$x^{-1} = \frac{1}{x}$$

When a fraction is raised to a negative exponent, you invert the fraction and change the exponent to positive:

$$\left(-\frac{2}{5}\right)^{-2} = \left(-\frac{5}{2}\right)^{2} = \frac{25}{4}$$

To raise a number to a negative exponent on the calculator, you can use the $\boxed{+/-}$ key or the reciprocal key $\boxed{1/x}$. For example, three ways to calculate 5^{-3} are

$$5 \;\boxed{y^x}\; 3 \;\boxed{+/-}\; \boxed{=} \;\rightarrow 0.008$$
$$5 \;\boxed{1/x}\; \boxed{y^x}\; 3 \;\boxed{=} \;\rightarrow 0.008$$
$$5 \;\boxed{y^x}\; 3 \;\boxed{=}\; \boxed{1/x} \;\rightarrow 0.008$$

All the rules for exponents [(3.6) to (3.10)] also apply to zero and negative exponents. For example, by the addition rule:

$$(a^{-2})\,(a^{3}) = a^{-2+3} = a^{+1} = a$$

and by the multiplication rule:

$$(R_1{}^{-2})^3 = R_1{}^{(-2)(3)} = R_1{}^{-6}$$

Study the next two examples that apply several of the rules to zero and negative exponents.

EXAMPLE 3.7

Evaluate:

$$\left(\frac{10^{-3}}{10^{-2}}\right)^2$$

Solution Apply the division rule to the fraction and subtract the exponents:

$$\left(\frac{10^{-3}}{10^{-2}}\right)^2 = (10^{-3-(-2)})^2 = (10^{-3+2})^2 = (10^{-1})^2$$

Now apply the multiplication rule:

$$(10^{-1})^2 = 10^{(-1)(2)} = 10^{-2} \text{ or } 0.01$$

The rules for exponents include the rules for powers of 10 from Section 2.2 as a special case when 10 is the base.

EXAMPLE 3.8

Evaluate:

$$(-1.1T^2)^2\,(T^{-1})(T^0)$$

Solution Apply the product rule, multiplication rule, and the addition rule:

$$(-1.1T^2)^2\,(T^{-1})(T^0) = (-1.1)^2\,(T^4)(T^{-1})(T^0) = 1.21T^{4-1+0} = 1.21T^3$$

Note that when you evaluate $(-1.1T^2)^2$, you apply the exponent to the coefficient and to T^2. This is an application of product rule (3.9). You can also replace T^0 by 1 in the first step.

Multiplying and Dividing Terms

To multiply or divide algebraic terms with exponents, you multiply or divide the coefficients separately and apply the rules for exponents to the literal parts. For example,

$$(-4v^2)(3v^{-3}) = (-4)(3)v^{2-3} = -12v^{-1} \text{ or } \frac{-12}{v}$$

$$\frac{15A^3}{5A^2} = \left(\frac{15}{5}\right)A^{3-2} = 3A$$

The next two examples further show how to multiply and divide algebraic terms with exponents.

EXAMPLE 3.9

Multiply:

$$5x(x^2 + 2x - 1)$$

Solution Each term in the parentheses must be multiplied by the factor on the outside:

$$5x(x^2 + 2x - 1) = 5x(x^2) + 5x(2x) - 5x(1) = 5x^3 + 10x^2 - 5x$$

▪

EXAMPLE 3.10

Divide:

$$\frac{\alpha t^3 - \alpha^2 t^2 - \alpha^3 t}{\alpha t}$$

Solution The fraction line is like parentheses, and each term in the numerator must be divided by the denominator:

$$\frac{\alpha t^3 - \alpha^2 t^2 - \alpha^3 t}{\alpha t} = \frac{\alpha t^3}{\alpha t} - \frac{\alpha^2 t^2}{\alpha t} - \frac{\alpha^3 t}{\alpha t} = t^2 - \alpha t - \alpha^2$$

▪

Study the following example that closes the circuit and shows an application to electronics, involving a calculation using negative powers of 10:

EXAMPLE 3.11

Close the Circuit

Figure 3–2 shows a series circuit containing a resistance, an inductance, and a capacitance in series. It is called an *RLC* series circuit. The formula for the resonant frequency of the circuit is given by

$$f_r = \frac{1}{2\pi \sqrt{LC}}$$

where L is in henrys, C is in farads, and f_r is in hertz. The resonant frequency is the frequency that produces maximum power in the circuit. Calculate f_r when $L = 125 \ \mu\text{H}$ and $C = 245 \ \text{pF}$.

EXAMPLE 3.11 (Cont.)

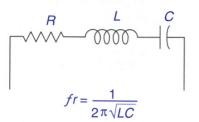

$$fr = \frac{1}{2\pi\sqrt{LC}}$$

FIGURE 3–2 Resonant frequency for example 3.11.

Solution First express L in henrys and C in farads using powers of 10:

$$L = 125 \ \mu H = 125 \times 10^{-6} \ H$$
$$C = 245 \ pF = 245 \times 10^{-12} \ F$$

Substitute in the formula using $2\pi = 6.28$, and apply the addition rule to the powers of 10 under the radical:

$$f_r = \frac{1}{6.28\sqrt{LC}} = \frac{1}{6.28\sqrt{(125\times 10^{-6})(245\times 10^{-12})}}$$

$$= \frac{1}{6.28\sqrt{(125)(245)\times 10^{-6-12}}} = \frac{1}{6.28\sqrt{30,625\times 10^{-18}}}$$

Observe that when working with powers of 10, the number is treated like the coefficient in an algebraic term, and the power of 10 is treated like the literal part of the term.

Take the square root, applying the division rule to the power of 10 in the radical. Divide the exponent by the index 2:

$$f_r = \frac{1}{6.28\sqrt{30,625\times 10^{-18}}} = \frac{1}{6.28\,(175\times 10^{-9})} = \frac{1}{1099.6\times 10^{-9}}$$

The answer to 3 significant digits is then

$$f_r = \frac{1}{1099.6\times 10^{-9}} = 909,000 \ Hz = 909 \ kHz$$

This problem can also be done completely on the calculator. One possible way, starting with the denominator and the square root and using the $\boxed{\pi}$ key, is

125 $\boxed{EXP}$ 6 $\boxed{+/-}$ $\boxed{\times}$ 245 $\boxed{EXP}$ 12 $\boxed{+/-}$
$\boxed{=}$ $\boxed{\sqrt{}}$ $\boxed{\times}$ 2 $\boxed{\times}$ $\boxed{\pi}$ $\boxed{=}$ $\boxed{1/x}$ $\rightarrow$ 909,000

▪

EXERCISE 3.2

In problems 1 through 36, perform the operations *without the calculator* by applying the rules for exponents.

1. $(x^4)(x^2)$

2. $(3^2)(3^2)$

3. $\dfrac{R^7}{R^5}$

4. $\dfrac{n^5}{n^3 \cdot n}$

5. $(3pv^2)^3$

6. $(-2c^2d)^4$

7. $(10^6 \cdot 10^0)^2$

8. $\dfrac{10^2}{10 \cdot 10^3}$

9. $\left(\dfrac{x}{3b^2}\right)^2$

10. $\left(\dfrac{0.1w^2}{p}\right)^3$

11. $\sqrt{I^4}$

12. $\sqrt[3]{\mu^6}$

13. 5^{-2}

14. $(-3)^{-3}$

15. $\left(\dfrac{2}{9}\right)^{-1}$

16. $\dfrac{1}{7^{-2}}$

17. $10^0 \cdot 10^{-2} \cdot 10^4$

18. $\left(\dfrac{10^4}{10^2}\right)^{-2}$

19. $(-3T)^2(T^{-2})^2$

20. $\dfrac{-v^3}{(-2v)^2}$

21. $\sqrt[3]{Y^{-3}}$

22. $\sqrt{V^4R^{-2}}$

23. $(2Q^3)(-7Q^2)$

24. $(5P_1^2)(3P_1)^2$

25. $\dfrac{18y^6}{6y^3}$

26. $\dfrac{2.8\omega L^2}{1.4\omega L}$

27. $2m(m^2 - m + 3)$

28. $-3p(2p^2 + 3p - 1)$

29. $\dfrac{6a^2b - 3ab^2}{3ab}$

30. $\dfrac{10k^3 - 5k^2 + 15k}{-5k}$

31. $\dfrac{6.8\pi f_r{}^2 - 4.4f_r}{2.0f_r}$

32. $\dfrac{C_1^2 + C_1C^2}{C_1}$

33. $(5 \times 10^{-3})^{-2}$

34. $(3 \times 10^{-6})^2$

35. $\sqrt[3]{8 \times 10^{-6}}$

36. $\sqrt{16 \times 10^{-12}}$

In problems 37 through 42, perform each calculation. Round answer to three significant digits in engineering or scientific notation.

37. $\left(\sqrt{5.31 \times 10^{-4}}\right)(3.15 \times 10^{-2})^3$

38. $\left(\sqrt[3]{9.22 \times 10^{-3}}\right)(8.77 \times 10^{-3})^{-2}$

39. $\sqrt{23 \times 10^{-6}}\,\sqrt{18 \times 10^{-2}}$

40. $\dfrac{\sqrt{310 \times 10^{-12}}}{\sqrt{550 \times 10^{-6}}}$

41. $\dfrac{\sqrt[3]{8.37 \times 10^{-6}}}{(3.12 \times 10^{-3})^{-2}}$

42. $\dfrac{(2.23 \times 10^{-3})^2}{\sqrt{6.16 \times 10^{-6}}}$

In problems 43 through 44, solve each applied problem.

43. The efficiency of a turbojet engine is given by the following expression:

$$Eff = 1 - \left(\frac{P_1}{P_2}\right)^{\frac{\alpha}{1-\alpha}}$$

Where $\dfrac{P_1}{P_2}$ is the compression ratio. Simplify this expression for $\alpha = 1.5$ using only positive exponents.

44. In studying the vibration of a machine on spring mounts, the following expression arises:

$$\frac{e^{-\beta(n+1)t}}{e^{-\beta nt}}$$

Simplify this expression, using only positive exponents.

Applications to Electronics In problems 45 through 56 solve each applied problem.

45. The power in one circuit is given by

$$P_1 = \sqrt{\frac{9V^4}{16}}$$

The power in a second circuit is given by

$$P_2 = \sqrt{\frac{25V^4}{4}}$$

Write the expression for the total power $P_T = P_1 + P_2$, and simplify the expression.

46. The force of a strong electric charge is given by

$$\frac{qr^2}{r}$$

The force of a weaker charge is given by

$$\frac{rq^2 - 10q^3}{q}$$

Write the expression that represents the difference in the charges and simplify the expression.

47. Figure 3–3 shows a circuit containing two resistances connected in parallel. The total resistance of the circuit can be expressed in terms of negative exponents as

$$R_T = \left[(R_1)^{-1} + (R_2)^{-1}\right]^{-1}$$

Find R_T when $R_1 = 85\ \Omega$ and $R_2 = 68\ \Omega$.

48. In problem 47, find R_T when $R_1 = 1.50\ k\Omega$ and $R_2 = 750\ \Omega$.

49. The resistance of a wire in ohms (Ω) is given by

$$R = \frac{\rho l}{A}$$

Given a copper wire where the resistivity $\rho = 17.5 \times 10^{-9}\ \Omega - m$, the length $l = 5.00$ m, and the cross sectional area $A = \pi r^2$. Find R to three significant digits when the radius $r = 4.00$ mm. (Note: The units for r must agree with l and ρ.)

50. In problem 49, find R if the radius decreases by half to $r = 2.00$ mm and the other values do not change.

51. In Example 3.11 find f_r when $L = 150\ \mu H$ and $C = 0.175$ nF.

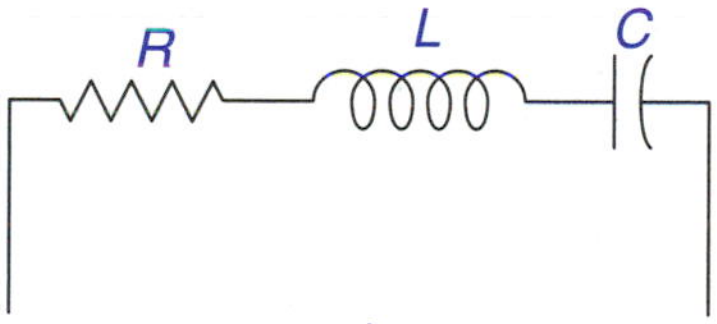

52. In Example 3.11, find f_r when $L = 2.50$ mH and $C = 400$ pF.

53. Figure 3– 4 shows an *RLC* series circuit. The capacitance C is given by

$$C = \frac{1}{4\pi^2 f_r^2 L}$$

where f_r is the resonant frequency. Find C when $f_r = 1020$ kHz and $L = 220\ \mu H$. (See Example 3.11.)

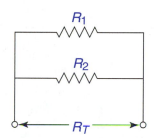

$$R_T = [(R_1)^{-1} + (R_2)^{-1}]^{-1}$$

FIGURE 3–3 Two resistances in parallel for problems 47 and 48.

$$C = \frac{1}{4\pi^2 fr^2 L}$$

$$L = \frac{1}{4\pi^2 fr^2 C}$$

FIGURE 3–4 *RLC* series circuit for problems 53 and 54.

54. In the *RLC* series circuit in Figure 3–4 the inductance *L* is given by

$$L = \frac{1}{4\pi^2 f_r^2 C}$$

where f_r is the resonant frequency. Find *L* when $f_r = 880$ kHz and $C = 5.50$ nF.

55. Multiply the following BASIC expression, and express the result in BASIC:

$$2*X*(X\wedge2 - 3*X + 5)$$

56. Divide the following BASIC expression, and express the result in BASIC:

$$(2*X\wedge3 + 4*X\wedge2 - 6*X)/2*X$$

≡ 3.3
POLYNOMIALS AND FACTORING

Multiplying Polynomials

A *polynomial* is an algebraic expression containing two or more terms. The most common polynomial is a *binomial* which contains only two terms. For example, $x + 2$ and $3x - 1$ are binomials. To multiply two binomials, a useful step-by-step method is the *FOIL* method. The acronym FOIL represents the order in which the numbers are multiplied: First, Outside, Inside, and Last. For example, to multiply the two binomials $(x - 2)(3x - 1)$, you multiply the four possible products as follows:

$$(x + 2)(3x - 1) = (x)(3x) + (x)(-1) + (2)(3x) + (2)(-1)$$

$$= 3x^2 - x + 6x - 2 = 3x^2 + 5x - 2$$

Here, F is the product of the *first* two terms, O is the product of the two *outside* terms, I is the product of the two *inside* terms, and L is the product of the two *last* terms. Note that the outside and inside products combine to give a middle term:

$$-x + 6x = 5x.$$

This happens when the binomials contain similar terms, which is the most common case.

The FOIL method is an application of the distributive law (1-3):

$$a(b + c) = ab + ac$$

Suppose $a = (x + 2)$, $b = 3x$ and $c = -1$, then you have

$$(x + 2)(3x - 1) = (x + 2)(3x) + (x + 2)(-1)$$

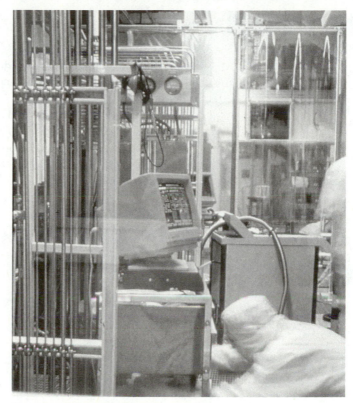

Courtesy of Intel Corporation.

Technicians use microprocessor-based systems to install
new equipment in a computer chip manufacturing plant.

The products can be commuted (switched around) and the distributive law applied twice again to yield:

$$(3x)(x + 2) + (-1)(x + 2) = (3x)(x) + (3x)(2) + (-1)(x) + (-1)(2)$$

Looking carefully at this example, you can see that these are the same four products as the FOIL method.

Two more examples of the FOIL method are shown below.

$$\begin{matrix} \mathsf{F} & \mathsf{O} & \mathsf{I} & \mathsf{L} \end{matrix}$$

$$(4N - 3)(3N + 2) = 12N^2 + 8N - 9N - 6 = 12N^2 - N - 6$$

$$\begin{matrix} \mathsf{F} & \mathsf{O} & \mathsf{I} & \mathsf{L} \end{matrix}$$

$$(2V_1 - 3V_2)(3V_1 - 2V_2) = 6V_1{}^2 - 4V_1 V_2 - 9V_1 V_2 + 6V_2{}^2 = 6V_1{}^2 - 13V_1 V_2 + 6V_2{}^2$$

In each of these cases, the outside and inside products combine to form a middle term, resulting in a polynomial of three terms called a *trinomial*.

Study the next two examples, which illustrate special binomial products.

EXAMPLE 3.12

Multiply:

$$(A + B)(A - B)$$

Solution Note that the binomials differ only by the middle sign. When you multiply this combination, the outside and inside products cancel, and there is no middle term:

$$(A + B)(A - B) = A^2 - AB + AB - B^2 = A^2 - B^2$$

The result is called the *difference of two squares* and is an important product used to simplify more complex expressions. This type of product can be done mentally. Only the first and last terms need to be squared and a minus sign placed in between. For example,

$$(3P + 2I)(3P - 2I) = (3P)^2 - (2I)^2 = 9P^2 - 4I^2$$

■

EXAMPLE 3.13

Multiply:

$$(A + B)(A + B) = (A + B)^2$$

Solution This product is the square of a binomial. The outside and the inside products are the same, resulting in a middle term that is twice the product of the two terms in the binomial:

$$(A + B)(A + B) = (A + B)^2 = A^2 + AB + AB + B^2 = A^2 + 2AB + B^2$$

■

The next example shows how to multiply a binomial by a trinomial.

EXAMPLE 3.14

Multiply:

$$(Z - 2)(Z^2 + 3Z + 1)$$

Solution To multiply a binomial by a trinomial, you can use a method similar to long multiplication in arithmetic. Set up the binomial below the trinomial, and multiply all the possible six products lining up the like terms:

$$Z^2 + 3Z + 1$$
$$Z - 2$$

Multiply Z by $(Z^2 + 3Z + 1)$: $Z^3 + 3Z^2 + Z$

Multiply -2 by $(Z^2 + 3Z + 1)$
  lining up similar terms: $-2Z^2 - 6Z - 2$

Add similar terms: $Z^3 + Z^2 \quad - 5Z - 2$

■

Factoring Polynomials

To factor means to separate a number or algebraic expression into its divisors. Factoring is the reverse of multiplication and results in a simplification of the expression. The first step in factoring a polynomial is to separate any common monomial (single term) factors from each term. For example, in the polynomial

$$30ax^2 - 42a^2x$$

each term has the common factor of $6ax$ since this factor can be divided evenly into each term. It can then be separated as follows:

$$30ax^2 - 42a^2x = 6ax(5x - 7a)$$

The terms in the parentheses are obtained by dividing $6ax$ into $30ax^2$ and $-42a^2x$. This is an application of the distributive law (1.3). If you now multiply out the parentheses, you will get the original polynomial. When you separate a factor, look for the highest common factor. This simplifies the expression as much as possible. For example, you could factor the original expression as

$$30ax^2 - 42a^2x = 3x(10ax - 14a^2)$$

However, it is not factored completely. The factors 2 and a can still be separated.

The following is another example of separating a common factor.

EXAMPLE 3.15

Factor:

$$20P_T{}^4 - 15P_T{}^3 + 10P_T{}^2$$

Solution First look for the highest number that divides into all the coefficients, which is 5. Then find the highest power of the variable or variables that is common to each term, which is $P_T{}^2$. Then divide the common factor $5P_T{}^2$ into each term to get the terms in the other factor:

$$20P_T{}^4 - 15P_T{}^3 + 10P_T{}^2 = 5P_T{}^2(4P_T{}^2 - 3P_T + 2)$$

Binomial Factors

Factoring polynomials into two binomials is the reverse of the FOIL method and is generally done by trying different pairs of binomials until the correct combination is obtained. One of the simpler types of factoring into two binomials is the difference of two squares illustrated in Example 3.12:

$$A^2 - B^2 = (A + B)(A - B)$$

A binomial consisting of two perfect squares *separated by a minus sign* can be readily factored into the sum and difference of the square roots as follows:

$$4a^2 - 49 = (2a + 7)(2a - 7)$$

$$25R_1{}^2 - 36R_2{}^2 = (5R_1 + 6R_2)(5R_1 - 6R_2)$$

EXAMPLE 3.16

Factor completely:

$$3\pi r^2 - 3\pi h^2$$

Solution   If there is a common factor, it should be separated first. Separate 3π and you will see that the other factor is the difference of two squares. You can then factor further:

$$3\pi r^2 - 3\pi h^2 = 3\pi(r^2 - h^2) = 3\pi(r + h)(r - h)$$

The common factor must be carried along, and the answer contains three factors.

To factor any trinomial into two binomials, you must know the FOIL method very well to work the process backward. In addition, you must understand the basic sign relations of binomial products. Study these four combinations:

$$x^2 + 5x + 6 = (x + 2)(x + 3)$$
$$x^2 - 5x + 6 = (x - 2)(x - 3)$$

Like signs

$$x^2 + 5x - 6 = (x + 6)(x - 1)$$
$$x^2 - 5x - 6 = (x - 6)(x + 1)$$

Unlike signs

Each of the four trinomials has the same terms except for the signs. This makes the factors different for each. When the sign of the last term is positive, the binomials have like signs, which are the same as the middle term. When the last term is negative, the binomials have unlike signs. You should always check that the factors are correct by multiplying them back with the FOIL method. Study the next three examples, which explain the step-by-step process of trinomial factoring.

EXAMPLE 3.17

Factor completely:

$$L^2 + 4L - 12$$

Solution   Consider the signs first. The last term is negative, which means the binomial factors have unlike signs. Now consider the factors of the first term. The only factors of L^2 are $L \cdot L$. The first term of each binomial is then L, and the binomials must look like:

$$(L + \quad)(L - \quad)$$

The possible factors of 12 are $1 \cdot 12$, $2 \cdot 6$, and $3 \cdot 4$. It is now necessary to try different possible pairs of binomials until the correct pair is found:

$$(L + 1)(L - 12) = L^2 - 11L - 12$$
$$(L + 2)(L - 6) = L^2 - 4L - 12$$
$$(L + 6)(L - 2) = L^2 + 4L - 12 \quad \textbf{Correct}$$

As you try different products you see certain things to look for. For example, the middle term, $4L$, is positive; therefore, the larger factor 6 should be positive.

EXAMPLE 3.18

Factor completely:

$$6y^2 - 19y + 8$$

Solution   The last term is positive, and the middle term is negative. The binomial factors therefore have the form:

$$(\; - \;)(\; - \;)$$

The possible factors of the first term $6y^2$ are $6y \cdot y$ and $2y \cdot 3y$. The possible factors of the last term 8 are $1 \cdot 8$ and $2 \cdot 4$. You must consider all the possible combinations of factors. There are eight possible combinations, and you need to try them one by one with the FOIL method until you get the correct factors. After trying some, you will find the correct pair of binomials whose outside and inside products combine to give the middle term, $-19y$:

$$6y^2 - 19y + 8 = (2y - 1)(3y - 8)$$

There is no quick way to do this problem. However, the more factoring you do, the more you begin to see certain combinations and do some of the products mentally, which speeds up the process.

EXAMPLE 3.19

Factor completely:

$$6\alpha T^2 - 9\alpha T - 15\alpha$$

Solution   Always look for a common factor first. In this trinomial, there is the common factor of 3α, which can be separated:

$$3\alpha(2T^2 - 3T - 5)$$

Then factor the trinomial into two binomials. The last sign is negative, so the binomials have unlike signs. After considering the factors of $2T^2$ and 5 and trying different combinations, the answer turns out to be

$$6\alpha^2 - 9\alpha T - 15\alpha = 3\alpha(2T - 5)(T + 1)$$

EXERCISE 3.3

In problems 1 through 20, multiply out each of the products.

1. $(x + 2)(x + 4)$

2. $(y - 6)(2y + 4)$

3. $(2P + 4)(2P - 5)$

4. $(5f - g)(4f - g)$

5. $(2I - 3R)(3I - 2R)$

6. $(4A + 7B)(A - 5B)$

7. $(V_1 + 3)(V_1 - 3)$

8. $(2R_2 + 1)(2R_2 - 1)$

9. $2(5L + 2C)(5L - 2C)$

10. $3(\alpha + 2\beta)(\alpha - 2\beta)$

11. $(t - 4)^2$

12. $(2\theta - 3)^2$

13. $(xy + 1)^2$

14. $(k^2 + 1)^2$

15. $(A - B)(X + Y)$

16. $(r + h)(r + 2)$

17. $(E + 1)(E^2 - 2E + 1)$

18. $(c + d)(c^2 - cd + d^2)$

19. $(G + 2)(G^2 - G + 3)$

20. $(Q + 1)(2Q^2 + Q - 4)$

In problems 21 through 28, factor each expression by separating the highest common monomial factor.

21. $6Y^2 + 3Y$

22. $10\omega L - 20\omega L^2$

23. $4I^2R_1 + 2I^2R_2$

24. $pq^2 - p^2q$

25. $10s^2 + 20s - 15$

26. $2m^3 - 4m^2 + 6m$

27. $x^2yz + xy^2z + xyz^2$

28. $36cd^3 + 16c^2d^2 + 12c^3d$

In problems 29 through 54, factor each expression completely.

29. $a^2 - 4b^2$

30. $16x^2 - 25y^2$

31. $50m^2 - 200$

32. $3h^2 - 27$

33. $X_C^2 - 0.16$

34. $0.04 - f_r^2$

35. $Z^2 + 6Z + 5$

36. $\delta^2 + 3\delta + 2$

37. $v^2 - 4v + 3$

38. $E^2 - 8E + 12$

39. $C_1^2 - 6C_1 + 9$

40. $V_T^2 + 16V_T + 64$

41. $3G^2 + 4G + 1$

42. $6f^2 - 7f - 3$

43. $5x^2 + xy - 6y^2$

44. $8a^2 - 2ab - 15b^2$

45. $8R_T^2 + 28R_T - 16$

46. $18t^2 + 60t + 50$

47. $10PV^2 + 15PV - 25P$

48. $2gy^2 + 12gy + 10g$

49. $s_R^2 - 0.6s_R + 0.09$

50. $V_0^2 + V_0 + 0.25$

51. $L^4 + 2L^2 + 1$

52. $I^4 - 3I^2 + 2$

53. $T^4 - 10^4$

54. $5R^4 - 80$

In problems 55 through 58, solve each applied problem.

55. The length of a rectangular circuit component is given by $(n + 3)$ and the width by $(n - 2)$. Write the expression for the area of the component, and multiply out the expression.

56. Given that w = width of a rectangular computer cabinet, $w + 3$ = length and $w - 2$ = height. Write the expression for the volume of the cabinet, and multiply out this expression. [Note: Volume = (length × width × height).]

57. The height s of a rocket is given by $s = 4t^2 + 28t + 48$, where t = time. Factor this expression.

58. The energy E radiated by a blackbody is given by

$$E = KT^4 - KT_0^4$$

where K = constant, T = absolute temperature of the body, and T_0 = absolute temperature of the surroundings. Factor this expression.

Applications to Electronics   In problems 59–66 solve each applied problem.

59. The current in a circuit is given by $I = 2t - 3$ and the voltage by $V = 3t + 2$, where t = time. Using the power law $P = VI$, write the expression for the power in the circuit, and multiply out the expression.

60. The relationship between the voltage V and the temperature T of a transmission line is given by

$$V^2 = 350R(1 + 0.003T)(T - 24)$$

where R = resistance of the line. Multiply out this expression.

61. Figure 3–5 shows a circuit containing a resistance R in series with a capacitance C (series RC circuit). The current at a certain time after the switch is closed is given by

$$I = \frac{Ve^{-1}}{R} - \frac{V_0 e^{-1}}{R}$$

where the constant $e \approx 2.718$, $V =$ applied voltage, and $V_0 =$ capacitor voltage when the switch is closed. Factor this expression for I.

62. Figure 3–6 shows a circuit containing a resistance R in series with an inductance L (series RL circuit). The current at a certain time after the switch is closed is given by

$$I = \frac{V}{R} - \frac{Ve^{-1}}{R}$$

where the constant $e \approx 2.718$ and $V =$ applied voltage. Factor this expression for I.

63. The electric power in one circuit is given by

$$\frac{0.50V^2}{R_1}$$

and in a second circuit by

$$\frac{0.25V^2}{R_2}$$

Factor the expression that represents the total power consumed in both circuits.

64. Figure 3–7 shows a series-parallel circuit that contains a resistance R_1 connected in series to two parallel resistances R_2 and R_3. The total current I_1 through R_1 divides into I_2 through R_2 and I_3 through R_3. The power dissipated by R_1 is $P_1 = I_1^2 R_1$. If $R_2 = R_1$, the power dissipated by R_2 is given by $P_2 = I_2^2 R_1$. If I_1 is greater than I_2, factor the expression that represents the difference in power dissipated by the two resistances, $P_1 - P_2$.

65. A variable with a subscript such as V_1 is written in BASIC and other computer languages as V1. Write this BASIC expression in algebraic notation, and factor the expression:

$$\text{V1\^{}2/R } - \text{ V2\^{}2/R}$$

66. As in problem 59 write this BASIC expression in algebraic notation, and factor the expression:

$$\text{K\^{}2}*\text{T\^{}2 } - \text{ 2}*\text{K}*\text{T}*\text{T1 } + \text{ T1\^{}2}$$

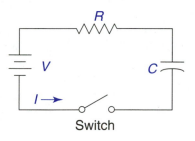

$$I = \frac{Ve^{-1}}{R} - \frac{V_0 e^{-1}}{R}$$

FIGURE 3–5 *RC* circuit for problem 61.

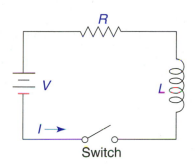

$$I = \frac{V}{R} - \frac{Ve^{-1}}{R}$$

FIGURE 3–6 *RL* circuit for problem 62.

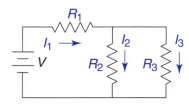

$$P_1 = I_1^2 R_1 \qquad P_2 = I_2^2 R_2$$

FIGURE 3–7 Power in a series-parallel circuit for problem 64.

☰ 3.4

MULTIPLICATION AND DIVISION OF FRACTIONS

Reducing Fractions

When working with algebraic fractions, factors play an important role. The more you know how to factor expressions, the easier it will be to multiply and add fractions.

To reduce an algebraic fraction, look for a common factor in the numerator and denominator that can be divided out. For example, the following fraction has a common factor of $4x$ that can be divided out and the fraction reduced:

$$\frac{12x^2}{28x} = \frac{(4x)(3x)}{(4x)(7)} = \frac{3x}{7}$$

When the numerator or denominator has more than one term, a common factor must first be separated from each term before dividing, such as the factor $5y$ in the following:

$$\frac{15y^2 + 5y}{5y} = \frac{5y(3y + 1)}{5y} = 3y + 1$$

If you do not separate the factor first, you may make the common error of dividing one term in a fraction without dividing the other. For example, you may divide only the $5y$ in the top with the $5y$ in the bottom without dividing the $15y^2$. This would produce the wrong answer $15y^2 + 1$. The fraction line is like parentheses, and every term in the numerator must be divided by the denominator. Remember: *you can only divide out factors, not terms.*

Study the next example, which illustrates this process further.

EXAMPLE 3.20

Simplify the fraction:

$$\frac{12N^2 + 4N}{2N^2 - 2N}$$

Solution   Before you divide you must factor the numerator and the denominator completely. Separate any common factors in the numerator and the denominator and then divide:

$$\frac{12N^2 + 4N}{2N^2 - 2N} = \frac{\overset{2}{4N}(3N + 1)}{2N(N - 1)} = \frac{2(3N + 1)}{N - 1}$$

Look again at this example. You may have been tempted to divide $2N^2$ into $12N^2$ and $-2N$ into $4N$. The result would then be $6 - 4 = 2$ which is very different from the correct answer. Remember you can only divide out factors.

▪

A fraction may also have the same binomial factor in the numerator and the denominator that can be divided out, such as

$$\frac{R^2 - 4T^2}{R - 2T} = \frac{(R + 2T)(R - 2T)}{R - 2T} = \frac{R + 2T}{1} = R + 2T$$

EXAMPLE 3.21

Simplify the fraction:

$$\frac{a^2 + 2ab + b^2}{a^2 - ab - 2b^2}$$

Solution Again you must factor before you can simplify. You cannot divide the a^2 in the numerator with the a^2 in the denominator because they are terms, not factors. Factor the numerator and denominator completely, and then divide out the common binomial factor:

$$\frac{a^2 + 2ab + b^2}{a^2 - ab - 2b^2} = \frac{(a+b)(a+b)}{(a+b)(a-2b)} = \frac{a+b}{a-2b}$$

You will probably make the mistake of dividing out terms that are not factors at least once in your life. Try not to make it more than once! When you cross out factors, you are doing division, which is the inverse of multiplication. Therefore, before you cross out anything, make sure it is separated from the rest of the expression by multiplication, not by plus and minus signs, and is a factor. See the Error Box in this section to better understand this idea.

EXAMPLE 3.22

Simplify the fraction:

$$\frac{P^2 - V^2}{P^2 + V^2}$$

Solution The numerator can be factored but *not* the denominator:

$$\frac{(P - V)(P + V)}{P^2 + V^2}$$

Resist any temptation to cross out the Ps or the Vs. The sum of two squares cannot be factored, and this fraction cannot be reduced.

Multiplying and Dividing Fractions

The rules for *multiplying and dividing* fractions are the same as those in Chapter 1 for arithmetic:

$$\frac{A}{B} \cdot \frac{C}{D} = \frac{AC}{BD} \qquad \text{(Multiply across the top and bottom)}$$

$$\frac{A}{B} \div \frac{C}{D} = \frac{A}{B} \cdot \frac{D}{C} = \frac{AD}{BC} \qquad \text{(Invert and multiply)}$$

As in arithmetic, you can first divide out common factors in any numerator and denominator before multiplying as follows:

$$\frac{12y^3}{7x^2} \cdot \frac{21x}{24y^2} = \frac{\overset{y}{\cancel{12y^3}}}{\underset{x}{\cancel{7x^2}}} \cdot \frac{\overset{3}{\cancel{21x}}}{\underset{2}{\cancel{24y^2}}} = \frac{3y}{2x}$$

Study the following examples, which show further illustrations of multiplying and dividing fractions.

EXAMPLE 3.23

Multiply the fractions:

$$\frac{I_1^2 - I_2^2}{4I_1} \cdot \frac{8I_1^2}{2I_1 + 2I_2}$$

Solution Factor each numerator and denominator completely before dividing out any common factors:

$$\frac{I_1^2 - I_2^2}{4I_1} \cdot \frac{8I_1^2}{2I_1 + 2I_2} = \frac{\overset{1}{\cancel{(I_1+I_2)}}(I_1 - I_2)}{\cancel{4I_1}} \cdot \frac{\overset{I_1}{\cancel{8I_1^2}}}{2\cancel{(I_1+I_2)}} = \frac{I_1(I_1 - I_2)}{1} = I_1(I_1 - I_2)$$

The next example closes the circuit and shows an application of algebraic fractions to a formula in electricity.

EXAMPLE 3.24

Close the Circuit

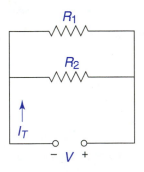

$$R_T = \frac{R_1 R_2}{R_1 + R_2}$$

FIGURE 3–8 Two parallel resistances for example 3.24.

Figure 3–8 shows a circuit containing two different resistances R_1 and R_2 in parallel connected to a voltage source V. The formula for the total resistance is given by:

$$R_T = \frac{R_1 R_2}{R_1 + R_2}$$

By Ohm's law, the voltage $V = I_T R_T$, where I_T is the total current. If $R_2 = 2R_1$, express V in terms of R_1.

Solution Substitute $2R_1$ for R_2 in the expression for R_T, and reduce the fraction:

$$R_T = \frac{R_1 R_2}{R_1 + R_2} = \frac{R_1(2R_1)}{R_1 + (2R_1)} = \frac{2R_1^2}{3R_1} = \frac{2R_1}{3}$$

Then substitute this expression for R_T in $V = I_T R_T$ to produce the desired expression:

$$V = I_T\left(\frac{2R_1}{3}\right) = \frac{2I_T R_1}{3}$$

EXAMPLE 3.25

Divide the fractions:

$$\frac{mv^2 + 4mv + 4m}{3v} \div \frac{2v^2 + v - 6}{6v - 9}$$

Solution   Invert the second fraction, factor as much as possible, and divide out common factors:

$$\frac{mv^2 + 4mv + 4m}{3v} \div \frac{2v^2 + v - 6}{6v - 9} = \frac{m(v^2 + 4v + 4)}{3v} \cdot \frac{6v - 9}{2v^2 + v - 6}$$

$$= \frac{m(v + 2)^2}{3v} \cdot \frac{3(2v - 3)}{(2v - 3)(v + 2)} = \frac{m(v + 2)}{v}$$

ERROR BOX

A very common error to watch out for is dividing out terms that are not factors. This is such a common error that it is necessary to emphasize it again. Suppose you had to simplify the following fraction:

$$\frac{3 + 5}{3 - 1}$$

You would never think to incorrectly cross out the 3s because you can simply calculate the result:

$$\frac{3 + 5}{3 - 1} = \frac{8}{2} = 4$$

But if you had to simplify the following fraction in algebra:

$$\frac{a + b}{a - c}$$

You might be tempted to cross out the as because there is nothing that can be done to simplify this fraction. However, when there is more than one term in the numerator or denominator, of a fraction do three things before you cross out any expression:

1. Separate all common factors from every term in the numerator *and* the denominator.

2. Factor further as much as possible into binomials.

3. Ask yourself: Is the expression a factor of *every* term in the numerator *and* the denominator?

Practice Problems: See if you can correctly reduce the following fractions:

1. $\dfrac{I^2 - 1}{I + 1}$ 2. $\dfrac{I^2 + 1}{I + 1}$ 3. $\dfrac{I^2R + V}{IR + V}$ 4. $\dfrac{I^2R + IV}{IR + V}$

Answers: 1. $I - 1$ 2. Not reducible   3. Not reducible   4. I

EXAMPLE 3.26

Multiply:

$$\frac{\beta + t}{3\beta^2 + 5\beta t + 2t^2} \cdot (3\beta + 2t)$$

Solution   Put $3\beta + 2t$ over 1, factor and divide common factors:

$$\frac{\beta + t}{3\beta^2 + 5\beta t + 2t^2} \cdot \frac{3\beta + 2t}{1} = \frac{\cancel{\beta + t}}{\cancel{(3\beta + 2t)}\cancel{(\beta + t)}} \cdot \frac{\cancel{3\beta + 2t}}{1} = \frac{1}{1} = 1$$

Note that when all the factors divide out, the answer is 1, not 0.

EXERCISE 3.4

In problems 1 through 18, reduce each fraction to lowest terms.

1. $\dfrac{210\,y^3}{270\,y}$

2. $\dfrac{66b^2}{165b^3}$

3. $\dfrac{AC}{AB + AD}$

4. $\dfrac{QX - QY}{TX - TY}$

5. $\dfrac{2R_1^2 - 2R_2^2}{10R_1 + 10R_2}$

6. $\dfrac{5.1aV_1 + 5.1aV_2}{1.7V_1 + 1.7V_2}$

7. $\dfrac{\alpha^2 + 6\alpha t + 9t^2}{3\alpha^2 + 8\alpha t - 3t^2}$

8. $\dfrac{4C^2 - 6CV - 4V^2}{4C - 8V}$

9. $\dfrac{2.5I_1^2 + 2.5I_2^2}{6.6I_1^2 - 6.6I_2^2}$

10. $\dfrac{m^2 + 8mn + 16n^2}{m^2 + 4mn + 4n^2}$

11. $\dfrac{7X_L^2 + 13X_L - 2}{9X_L^2 + 14X_L - 8}$

12. $\dfrac{6r^2 - 17r - 3}{11r^2 - 29r - 12}$

13. $\dfrac{6V + 15R}{6V^2 + 9VR - 15R^2}$

14. $\dfrac{16E_1^2 - 16E_2^2}{E_1^2 - E_1E_2 - 2E_2^2}$

15. $\dfrac{2(f_c - 1) - (f_c - 3)}{f_c^2 - 1}$

16. $\dfrac{3(t - 2) - 2(t - 4)}{t^2 + 4t + 4}$

17. $\dfrac{(3Q - 1)(Q + 1) - (Q + 1)(2Q - 1)}{Q^2 + 3Q + 2}$

18. $\dfrac{(3P_1 + P_2)(2P_1 - P_2) - P_2(P_2 - 5P_1)}{P_1^2 - P_2^2}$

In problems 19 through 40, perform the indicated operations. Divide out common factors first.

19. $\dfrac{9x}{2} \cdot \dfrac{4}{3x^2}$

20. $\dfrac{5b^2y}{8by^3} \cdot \dfrac{4y}{10b}$

21. $\dfrac{0.09V}{0.6I} \div \dfrac{0.3V}{I}$

22. $\dfrac{5.6}{0.80Z} \div \dfrac{1.4}{1.0Z^2}$

23. $\dfrac{21u - 7v}{9u^2 - v^2} \cdot \dfrac{3u + v}{uv}$

24. $\dfrac{8g^2 - 8h^2}{16h} \cdot \dfrac{5g}{7g + 7h}$

25. $\dfrac{s^2 - 2st + t^2}{s} \div (s - t)$

26. $\dfrac{x + i}{x - i} \cdot (x^2 - i^2)$

27. $\dfrac{I^2 + 5I + 4}{2I + 2} \cdot \dfrac{4I + 4}{3I + 12}$

28. $\dfrac{6V^2 - 13V + 6}{4V + 1} \cdot \dfrac{16V^2 - 1}{6V - 4}$

29. $(2D^2 - 5D + 2) \div \dfrac{D^2 - 4}{5D + 10}$

30. $\dfrac{4m^2 - 12mn + 9n^2}{2n} \div (2m - 3n)$

31. $\dfrac{V_0^2 - V_0}{V_0^2} \cdot \dfrac{V_0 + 1}{V_0 - 1}$

32. $\dfrac{R_1 - R_2}{R_1} \cdot \dfrac{R_2}{R_1^2 - R_2^2}$

33. $\dfrac{(9 \times 10^6)(K^2 - 1)}{K} \cdot \dfrac{2K}{(18 \times 10^3)(K + 1)}$

34. $\dfrac{(5 \times 10^{-3})N^2}{2N + 1} \div \dfrac{(15 \times 10^{-6})N}{4N + 2}$

35. $\dfrac{3A^2 - 12AB + 12B^2}{2A^2 + 2A - 12} \div \dfrac{6A^2 - 24B^2}{8A + 24}$

36. $\dfrac{3W^2 - 27}{6W^2 - 6} \cdot \dfrac{6W^2 - 12W + 6}{15W + 45}$

37. $\dfrac{3.5x}{2.5y} \cdot \dfrac{0.5x^2y}{2.1xy^2} \div \dfrac{0.1x^2}{0.3y^2}$

38. $\dfrac{1.2c^2d}{0.4cd^2} \div \left(\dfrac{0.6c}{0.3d} \cdot \dfrac{1.8c}{1.2d} \right)$

39. $\dfrac{3I_T - 3}{2I_T^2 - I_T - 1} \cdot \dfrac{2I_T + 1}{I_T + 1} \cdot \dfrac{2I_T^2 + 3I_T + 1}{6I_T + 3}$

40. $\dfrac{2C + L}{2C^2 + 3CL + L^2} \cdot \dfrac{2C - L}{4C + 8L} \div \dfrac{1}{4C + 4L}$

In problems 41 through 44, solve each applied problem.

41. The input force of a machine is given by $1.5F_x - 1.5F_y$ and the output force by $3F_x^2 - 3F_y^2$. The mechanical advantage is defined as

$$MA = \frac{\text{Output force}}{\text{Input force}}$$

Write and simplify the expression for the mechanical advantage.

42. The height of a VDT (video display terminal) is given by

$$6l + 2w$$

The volume is given by

$$24l^3 - 16l^2w - 8lw^2$$

Find and simplify the expression for the area of the base of the terminal. (Note: Base area = volume/height.)

43. In calculating the center of gravity of a body, the expression arises:

$$\left(\frac{4\pi r^3}{3} \right)\left(\frac{2}{\pi r^2} \right) \div 2\pi$$

Simplify this expression.

44. The dimensions of a rectangular computer cabinet are given by

$$h = \text{height}, \quad \frac{h}{h + 2} = \text{width}, \quad \text{and} \quad \frac{3h + 6}{h} = \text{length}.$$

Write and simplify the expression for the volume of the computer. (Note: Volume = length · width · height.)

Applications to Electronics   In problems 45 through 50 solve each applied problem.

45. Given Ohm's law $I = \frac{V}{R}$ and the power formula $P = VI$, show that $P = \frac{V^2}{R}$ by substituting the expression for I given by Ohm's law for I in the power formula.

46. Given Ohm's law $R = \frac{V}{I}$ and the power formula $V = \frac{P}{I}$, show that $R = \frac{P}{I^2}$ by substituting the expression for V given by the power formula for V in Ohm's law.

47. In Example 3.24 express V in terms of R_1 if $R_2 = 3R_1$.

48. A computer processes information based on the formula

$$\frac{I^2 - 4}{2n + 6}$$

If $n = I - 5$, simplify the formula in terms of I.

49. In a circuit, the resistance as a function of time t is given by

$$R = 5t^2 + 10t + 5$$

and the current is given by

$$I = \frac{t}{t + 1}$$

Express the voltage $V = IR$ as a function of t, and simplify the expression.

50. In problem 49, express the power $P = I^2R$ as a function of t, and simplify the expression.

≡ 3.5
ADDITION OF FRACTIONS

Addition of algebraic fractions requires a clear understanding of the process of adding fractions in arithmetic and a good grasp of factoring. Therefore, we first review how you add arithmetic fractions and emphasize the steps that are important in adding algebraic fractions. Consider the following addition problem. How do you add

$$\frac{5}{24} + \frac{7}{45}$$

You look for the lowest common denominator (LCD) that both denominators divide into evenly. This can be done by trying multiples of the larger denominator: 45, 90, 135, and so on, until you find a multiple that both denominators divide into. Another method, which was introduced in Chapter 1, is to factor each denominator into its smallest factors or primes and list them as follows to obtain the LCD:

$$24 = 2 \cdot 2 \cdot 2 \cdot 3 \qquad\qquad = 2^3 \cdot 3$$
$$45 = \phantom{2 \cdot 2 \cdot 2 \cdot{}} 3 \cdot 3 \cdot 5 \quad = 3^2 \cdot 5$$

$$\text{LCD} = 2 \cdot 2 \cdot 2 \cdot 3 \cdot 3 \cdot 5$$
$$\text{LCD} = 2^3 \cdot 3^2 \cdot 5 = 360$$

Note that *the LCD contains the highest power of each prime factor.* Both the denominators divide evenly into the LCD since it contains all the prime factors of each:

$$\frac{360}{24} = \frac{\cancel{2} \cdot \cancel{2} \cdot \cancel{2} \cdot \cancel{3} \cdot 3 \cdot 5}{\cancel{2} \cdot \cancel{2} \cdot \cancel{2} \cdot \cancel{3}} = 15$$

$$\frac{360}{45} = \frac{2 \cdot 2 \cdot 2 \cdot \cancel{3} \cdot \cancel{3} \cdot \cancel{5}}{\cancel{3} \cdot \cancel{3} \cdot \cancel{5}} = 8$$

The method of prime factors is the method used to find the LCD for algebraic fractions. Once you find the LCD, you then apply the *fundamental rule of fractions:*

Rule of Fractions   Multiplying or dividing the numerator and the denominator by the same quantity (other than zero) does not change the value of a fraction.

Multiply the top and bottom of each fraction by the factor obtained when its denominator is divided into the LCD as was just shown. The first fraction is multiplied top and bottom by 15 and the second fraction top and bottom by 8. This changes the denominator of each fraction into the LCD, so you can add the fractions:

$$\frac{5\,(15)}{24\,(15)} + \frac{7\,(8)}{45\,(8)} = \frac{75}{360} + \frac{56}{360} = \frac{75+56}{360} = \frac{131}{360}$$

Study the following example, which shows you how to apply the method of prime factors to add algebraic fractions. Consider the following addition problem:

$$\frac{3}{6x^2y} + \frac{7}{18xy}$$

The prime factors of each denominator are

$$6x^2y = 2 \cdot 3 \cdot x \cdot x \cdot y = 2 \cdot 3 \cdot x^2 \cdot y$$

$$18xy = 2 \cdot 3 \cdot 3 \cdot x \cdot y = 2 \cdot 3^2 \cdot x \cdot y$$

Take the *highest power* of each prime factor to make up the LCD:

$$\text{LCD} = 2 \cdot 3^2 \cdot x^2 \cdot y = 18x^2y$$

For each fraction, divide its denominator into the LCD, and multiply the numerator and denominator by the result. The first fraction is multiplied by 3 and the second fraction by x. This changes the denominator of each fraction to the LCD, so you can add the numerators over the same denominator:

$$\frac{3\,(3)}{6x^2y(3)} + \frac{7\,(x)}{18xy\,(x)} = \frac{9}{18x^2y} + \frac{7x}{18x^2y} = \frac{9+7x}{18x^2y}$$

This example can be done in a more convenient way by adding and multiplying in one step over the LCD:

$$\frac{3}{6x^2y} + \frac{7}{18xy} = \frac{3(3) + 7(x)}{18x^2y} = \frac{9 + 7x}{18x^2y}$$

Study the following examples, which show how to add many of the important types of fractions you need to know.

EXAMPLE 3.27

Add the fractions:

$$\frac{5}{G} + \frac{G+1}{G-1}$$

Solution   When the denominator contains more than one term, you must carefully identify prime factors. The denominator $(G - 1)$ represents only *one* prime factor. The LCD therefore contains the two prime factors: G and $(G - 1)$. Change each fraction to the LCD. Multiply the first fraction numerator and denominator by $(G - 1)$. Multiply the second fraction numerator and denominator by G. Show the factors in the numerators over the LCD as follows:

$$\frac{5}{G} + \frac{G+1}{G-1} = \frac{5(G-1)}{G(G-1)} + \frac{(G+1)(G)}{(G-1)(G)} = \frac{5(G-1) + (G+1)(G)}{(G)(G-1)}$$

Note that *the fraction line has the same effect as parentheses*. To avoid an error when there is more than one term in the numerator, enclose it in parentheses, as $(G + 1)$ and $(G - 1)$ have been enclosed. Now multiply the terms in the numerator and simplify:

$$\frac{5(G-1) + (G+1)(G)}{(G)(G+1)} = \frac{5G - 5 + G^2 + G}{(G)(G-1)} = \frac{G^2 + 6G - 5}{(G)(G-1)}$$

EXAMPLE 3.28

Add the fractions:

$$\frac{2}{T+1} - \frac{T-3}{T^2-1}$$

Solution   Factor the second denominator into two prime factors to find the LCD $= (T + 1)(T - 1)$. Multiply the first fraction by $(T - 1)$. The second fraction does not need to be changed:

$$\frac{2}{T+1} - \frac{T-3}{T^2-1} = \frac{2(T-1) - (T-3)}{\underbrace{(T+1)(T-1)}_{\text{LCD}}}$$

Note that even though the second fraction does not need to be multiplied by a factor, *every term in the numerator must be multiplied by the minus sign* because the fraction line is like a parentheses.

EXAMPLE 3.28 (Cont.)

Multiply out the numerator, simplify, and reduce the fraction:

$$\frac{2T - 2 - T + 3}{(T+1)(T-1)} = \frac{T+1}{(T+1)(T-1)} = \frac{\cancel{T+1}}{\cancel{(T+1)}(T-1)} = \frac{1}{T-1}$$

Observe that answers should be reduced whenever possible.

■

Study the following example that closes the circuit and shows how to simplify a formula for the current in an electric circuit by adding fractions.

EXAMPLE 3.29

Close the Circuit

Figure 3–9 shows a series-parallel circuit containing a resistance R_1 in parallel with two series resistances R_2 and R_3. The voltage across $R_1 = 6V$, which is the *same* as the combined voltage across R_2 and R_3. The total current I_T in the circuit is given by Ohm's law applied to each parallel branch:

$$I_T = \frac{6}{R_1} + \frac{6}{R_2 + R_3}$$

If $R_3 = R_1$, simplify the expression for I_T by adding the fractions.

Solution   Let $R_3 = R_1$ in the formula, which then becomes

$$I_T = \frac{6}{R_1} + \frac{6}{R_2 + R_1}$$

Then the LCD $= R_1(R_2 + R_1)$ or $R_1(R_1 + R_2)$. The first fraction needs to be multiplied by $(R_1 + R_2)$ and the second fraction by R_1:

$$I_T = \frac{6(R_1 + R_2) + 6(R_1)}{(R_1)(R_1 + R_2)} = \frac{6R_1 + 6R_2 + 6R_1}{R_1(R_1 + R_2)}$$

Combine terms and simplify:

$$I_T = \frac{12R_1 + 6R_2}{R_1(R_1 + R_2)} \text{ or } \frac{6(2R_1 + R_2)}{R_1(R_1 + R_2)}$$

Observe that the numerator of the answer can be factored as $6(2R_1 + R_2)$, but the fraction cannot be reduced because the factors in the numerator and the denominator are different.

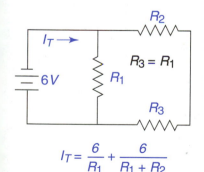

$$I_T = \frac{6}{R_1} + \frac{6}{R_1 + R_2}$$

FIGURE 3–9 Series-parallel circuit for example 3.29.

■

The last example requires a good grasp of algebra. Study it carefully, and you will reinforce your understanding of how to add fractions.

EXAMPLE 3.30

Add the fractions:

$$\frac{A^2 + B^2}{A^2 - B^2} - \frac{A + B}{2A - 2B}$$

Solution Factor both denominators completely:

$$\frac{A^2 + B^2}{(A + B)(A - B)} - \frac{A + B}{2(A - B)}$$

The LCD is then $2(A + B)(A - B)$. Multiply the first fraction by 2 and the second fraction by $(A + B)$:

$$\frac{A^2 + B^2}{(A + B)(A - B)} - \frac{A + B}{2(A - B)} = \frac{(2)(A^2 + B^2) - (A + B)(A + B)}{2(A + B)(A - B)}$$

Multiply out the numerator and simplify. Pay careful attention to the minus sign in front of the second numerator. It must be multiplied by every term in the product $(A + B)(A + B)$. First multiply the binomials $(A + B)(A + B)$, then multiply each of the three terms in the product by -1:

$$\frac{2A^2 + 2B^2 - (A^2 + 2AB + B^2)}{2(A + B)(A - B)} = \frac{2A^2 + 2B^2 - A^2 - 2AB - B^2}{2(A + B)(A - B)} = \frac{A^2 - 2AB + B^2}{2(A + B)(A - B)}$$

The answer can be reduced to a simpler fraction by factoring the numerator as follows:

$$= \frac{(A - B)(A - B)}{2(A + B)(A - B)} = \frac{A - B}{2(A + B)}$$

EXERCISE 3.5

In problems 1 through 36, add the fractions and simplify.

1. $\dfrac{3}{8} + \dfrac{7}{12}$

2. $\dfrac{4}{9} + \dfrac{11}{15}$

3. $\dfrac{7x}{18} + \dfrac{13x}{30}$

4. $\dfrac{11y}{90} - \dfrac{7y}{60}$

5. $\dfrac{X_L}{36} + \dfrac{7X_C}{30}$

6. $\dfrac{5\theta}{21} - \dfrac{\varphi}{15}$

7. $\dfrac{3I_1}{I_2} + \dfrac{2I_2}{I_1}$

8. $\dfrac{t}{s} + \dfrac{3t}{2s}$

9. $\dfrac{8.5 \times 10^{-3}}{4} + \dfrac{2.2 \times 10^{-2}}{16}$

10. $\dfrac{2.8 \times 10^3}{24} - \dfrac{4.0 \times 10^2}{6}$

11. $\dfrac{5}{2 \times 10^3} - \dfrac{20}{4 \times 10^4}$

12. $\dfrac{3}{2 \times 10^{-6}} + \dfrac{16}{4 \times 10^{-5}}$

13. $\dfrac{6}{5r\omega^2} - \dfrac{8}{15r\omega}$

14. $\dfrac{x}{at^2} + \dfrac{y}{a^2t}$

15. $1 + \dfrac{3c}{d} - \dfrac{5c^2}{2d^2}$

16. $\dfrac{3w}{x} + \dfrac{x}{w} - 2$

17. $\dfrac{1}{V_R - 2} - \dfrac{4}{V_R^2 - 4}$

18. $\dfrac{\pi}{r+1} - \dfrac{\pi}{r-1}$

19. $\dfrac{2.3}{I} + \dfrac{5.8}{I + kI}$

20. $\dfrac{0.31}{P - nP} - \dfrac{0.46}{P}$

21. $\dfrac{4}{3} - \dfrac{D+1}{D} + \dfrac{2D+6}{6D}$

22. $\dfrac{X-1}{X} + \dfrac{X+2}{2X} - \dfrac{1}{2}$

23. $\dfrac{1}{2V} - \dfrac{7}{6V} + \dfrac{4V-1}{6V^2 - 6V}$

24. $\dfrac{8R-2}{5R+15} + \dfrac{1}{R+3} - \dfrac{11}{10}$

25. $\dfrac{3t-5}{t^2-1} - \dfrac{4}{t+1}$

26. $\dfrac{2m^2}{k^2 - m^2} - \dfrac{k}{k+m}$

27. $\dfrac{X^2 + Y^2}{X^2 - Y^2} - \dfrac{X+Y}{2X - 2Y}$

28. $\dfrac{V_x + V_y}{V_x - V_y} + \dfrac{V_x V_y}{V_x^2 - V_y^2}$

29. $\dfrac{m_0 - 1}{3m_0^2 + 4m_0 + 1} + \dfrac{m_0}{3m_0 + 3}$

30. $\dfrac{2b+2}{b^2 + 6b + 9} + \dfrac{3b}{2b+6}$

31. $\dfrac{3}{I^2 + 3IR + 2R^2} - \dfrac{5}{I^2 - R^2}$

32. $\dfrac{x}{x^2 - y^2} - \dfrac{y}{x^2 + 2xy + y^2}$

33. $\dfrac{3}{I_1 + I_2} - \dfrac{1}{I_1} + \dfrac{7}{I_2}$

34. $\dfrac{5}{V_1} - \dfrac{1}{V_2 - V_1} + \dfrac{8}{V_2}$

35. $\dfrac{3}{C^2 - 1} + \dfrac{4C}{(C-1)^2} + \dfrac{2}{C+1}$

36. $\dfrac{12T}{T-4} + \dfrac{14T}{T+4} - \dfrac{15}{T^2 - 16}$

In problems 37 through 40, perform the operations and simplify.

37. $\dfrac{i}{i^2 - 1} \cdot \dfrac{i-1}{i+1} + \dfrac{1}{(i+1)^2}$

38. $\dfrac{w}{w-1} - \dfrac{w^2 - 1}{2} \cdot \dfrac{w}{(w-1)^2}$

39. $\dfrac{\dfrac{1}{\alpha} - \alpha}{\dfrac{1}{\alpha} + 1}$

40. $\dfrac{\dfrac{x^2}{y} - y}{\dfrac{x}{y} + 1}$

In problems 41 through 44, solve each applied problem.

41. The average rate for a round trip whose one way distance is D is given by

$$R_{avg} = \dfrac{2D}{\dfrac{D}{r_1} + \dfrac{D}{r_2}}$$

where r_1 = average rate going and r_2 = average rate returning.

(a) Simplify the complex fraction formula for R_{avg} by adding the fractions in the denominator and dividing.

(b) If you average 20 mi/h one way on a trip and 30 mi/h returning, what is your average rate for the entire trip? Use the formula from (a). (Note: It is not 25 mi/h.)

42. Figure 3–10 shows an angle θ (theta) in a circle of radius r on an XY coordinate graph. The trigonometric functions of the angle are defined as

$$\sin\ \theta = y/r$$
$$\cos\ \theta = x/r$$

where (x,y) is the point on the circle where the side of the angle cuts the circle. Using the Pythagorean theorem $x^2 + y^2 = r^2$, show that:

$$(\sin\ \theta)^2 + (\cos\ \theta)^2 = 1$$

43. The following formula comes from the theory of lenses in a microscope:

$$\frac{1}{F} = \frac{1}{f_1} + \frac{1}{f_2} - \frac{d}{f_1 f_2}$$

Find the expression for F by adding the fractions and inverting.

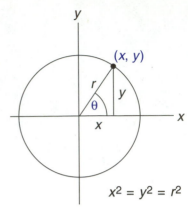

$$x^2 = y^2 = r^2$$

$$\sin\theta = \frac{y}{r} \qquad \cos\theta = \frac{x}{r}$$

FIGURE 3–10 Trigonometric functions for problem 42.

44. In working with the deflection δ of a beam, the following formula arises:

$$\delta = \frac{fl^4}{384EI} + \frac{Fl^3}{48\,EI}$$

Simplify the expression for δ by adding the fractions.

Applications to Electronics   In problems 45 through 50, solve each applied problem.

45. Figure 3–11 shows a circuit containing three resistances R_1, R_2, and R_3 in parallel connected to a voltage source V. The total resistance of this circuit is given by

$$\frac{1}{R_T} = \frac{1}{R_1} + \frac{1}{R_2} + \frac{1}{R_3}$$

Find the expression for R_T by adding the three fractions and then inverting.

46. In the circuit in Figure 3–11, if I_a is the current through R_1 and I_b is the current through R_2 and R_3, the total current from the source can be expressed by

$$I_T = I_a + \frac{I_b R_2}{R_2 + R_3} + \frac{I_b R_3}{R_2 + R_3}$$

Show that $I_T = I_a + I_b$ by adding the fractions and simplifying the expression for I_T.

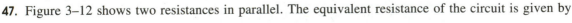

$$\frac{1}{R_T} = \frac{1}{R_1} + \frac{1}{R_2} + \frac{1}{R_3}$$

$$I_T = I_a + I_b$$

FIGURE 3–11 Parallel resistances for problems 45 and 46.

47. Figure 3–12 shows two resistances in parallel. The equivalent resistance of the circuit is given by

$$R_{eq} = \frac{R_1 R_2}{R_1 + R_2}$$

Each resistance can be expressed in terms of its conductance as

$$R_1 = \frac{1}{G_1}$$

$$R_2 = \frac{1}{G_2}$$

and the equivalent conductance of the circuit is given by

$$G_{eq} = \frac{1}{R_{eq}}$$

Show that the equivalent conductance of the circuit can be written

$$G_{eq} = G_1 + G_2$$

Substitute in the formula for R_{eq}, simplify, and then invert.

48. Figure 3–13 shows two capacitances in series. The total capacitance is given by

$$C_T = \frac{1}{\dfrac{1}{C_1} + \dfrac{1}{C_2}}$$

Simplify this expression for C_T.

49. Figure 3–14 shows a series-parallel circuit containing two series resistances R_1 and $R_2 = 100\ \Omega$ in parallel with R_3. When $R_3 = R_1$, the total current is given by

$$I_T = \frac{V}{R_1 + 100} + \frac{V}{R_1}$$

Simplify this expression for I_T by adding the fractions.

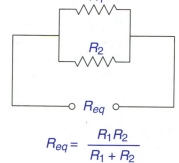

$$R_{eq} = \frac{R_1 R_2}{R_1 + R_2}$$

$$G_1 = \frac{1}{R_1} \qquad G_2 = \frac{1}{R_2}$$

$$G_{eq} = G_1 + G_2$$

FIGURE 3–12 Parallel resistances for problem 47.

50. The BASIC expression for the total power dissipated in the two series resistances in Figure 3–15 is given by

$$V1\text{\textasciicircum}2/R1\ +\ V2\text{\textasciicircum}2/R2$$

Rewrite this expression in BASIC as one fraction.

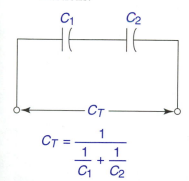

$$C_T = \frac{1}{\dfrac{1}{C_1} + \dfrac{1}{C_2}}$$

FIGURE 3–13
Capacitances in series for problem 48.

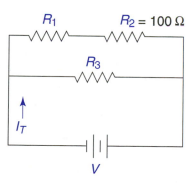

$$R_1 = R_2 \quad I_T = \frac{V}{R_1} + \frac{V}{R_1 + 100}$$

FIGURE 3–14 Series-parallel circuit for problem 49.

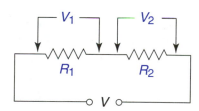

FIGURE 3–15 Series resistances for problem 50.

≡ CHAPTER HIGHLIGHTS

3.1 ALGEBRAIC TERMS

Study rules (3.1) through (3.4) for adding, subtracting, multiplying, and dividing signed numbers. Only *like terms* that have the same literal part can be combined. To combine like terms, add the coefficients only; do not change the literal part including the exponents.

3.2 RULES FOR EXPONENTS

The five *basic rules for exponents* are:

Addition Rule	$(x^m)(x^n) = x^{m+n}$	(3.6)

Subtraction Rule	$\dfrac{x^m}{x^n} = x^{m-n} \ (x \neq 0)$	(3.7)

Multiplication Rule	$(x^m)^n = x^{mn}$	(3.8)

Product Rule	$(xy)^n = x^n \, y^n$ and $\left(\dfrac{x}{y}\right)^n = \dfrac{x^n}{y^n}$	(3.9)

Division Rule	$\sqrt[n]{x^m} = x^{m/n}$	(3.10)

Any number to the *zero power* is one (except zero):

$$x^0 = 1 \ (x \neq 0) \qquad (3.11)$$

A *negative exponent* means invert and raise to the positive power:

$$x^{-n} = \left(\frac{1}{x}\right)^n = \frac{1}{x^n} \text{ and } \left(\frac{x}{y}\right)^{-n} = \left(\frac{y}{x}\right)^n (x, y \neq 0) \qquad (3.12)$$

To *multiply or divide* terms, multiply or divide the coefficients only, and apply the exponent rules to the literal parts.

3.3 POLYNOMIALS AND FACTORING

Multiply two binomials by using the *FOIL* method: First, Outside, Inside, and Last products.

Factor a polynomial by first removing the highest common factor.

Factor a trinomial into two binomials by applying the reverse of the FOIL method trying different combinations and checking by multiplying back the binomials.

A common type of factoring is the *difference of two squares*:

$$X^2 - Y^2 = (X + Y)(X - Y)$$

3.4 MULTIPLICATION AND DIVISION OF FRACTIONS

When *reducing fractions* make sure you only divide factors and not terms. See the Error Box at the end of this section.

The rules for *multiplying and dividing fractions* are

$$\frac{A}{B} \cdot \frac{C}{D} = \frac{AC}{BD}$$

$$\frac{A}{B} \div \frac{C}{D} = \frac{A}{B} \cdot \frac{D}{C} = \frac{AD}{BC}$$

Before multiplying or dividing, factor all numerators and denominators and divide out common factors.

3.5 ADDITION OF FRACTIONS

To *add fractions*, first factor every denominator. Then find the LCD, which is the smallest quantity that contains all the factors of each denominator. It is made up of the highest power of each prime factor. Change each denominator to the LCD by applying the fundamental rule of fractions.

Rule of Fractions	Multiplying or dividing the numerator and denominator by the same number (other than zero) does not change the value of a fraction.

For each fraction, divide its denominator into the LCD to determine the necessary factor to multiply into the fraction. Add the numerators multiplied by their factors over the LCD. Pay careful attention to any minus signs in front of a fraction. Simplify the resulting fraction reducing it if possible.

Study Examples 3.27 through 3.30. Adding fractions is best learned by doing many exercises.

≡ REVIEW QUESTIONS

In problems 1 through 26, perform the indicated operations.

1. $10 - (-15) + (-1)$

2. $(-0.5)(6) + (-10)(-0.1)$

3. $\dfrac{(-7)(-3)(-1)}{(4)(6) - 3}$

4. $\left(\dfrac{5}{8}\right)\left(-\dfrac{2}{5}\right) - 4\left(-\dfrac{3}{8}\right)$

5. $x^2 + 2x - 3x^2 + x$

6. $a^2b - ab^2 + 3a^2b + ab^2$

7. $5 - (8 - 3CV) + (2CV - 1)$

8. $(2V + IR) - [(2IR - V) + 2V]$

9. $(kP^2)(k^2P)(kP)$

10. $\dfrac{f^4 \cdot f^2}{f^3}$

11. $\dfrac{10^5 \cdot 10^{-2}}{10^0 \cdot 10^2}$

12. $(3)^{-3}\left(\dfrac{2}{9}\right)^{-2}$

13. $\dfrac{(6X_C^2)(-7X_C^3)}{2X_C}$

14. $\dfrac{(3V^2)^{-2}}{2V^{-3}}$

15. $\sqrt{1.21V^4}$

16. $\sqrt[3]{0.008H^{-3}}$

17. $\dfrac{\sqrt[3]{27 \times 10^{-9}}}{(2.5 \times 10^{-2})^3}$

18. $\left(\sqrt{2.25 \times 10^6}\right)(1.4 \times 10^3)^2$

19. $3\phi(\phi^2 - 2\phi + 1)$

20. $\dfrac{8x^3y + 12x^2y^2 - 4xy^3}{4xy}$

21. $(W - 4)(W + 2)$

22. $(0.2E_1 + 0.1)(0.2E_1 - 0.1)$

23. $3(2k - 3)^2$

24. $(2N)(N + 2)(2N - 1)$

25. $(I + 2)(I^2 - 4I + 4)$

26. $(R + 1)^2 (R - 1)$

In problems 27 through 34, factor each expression completely.

27. $8A^2B - 12AB^2$

28. $21z^3 + 15z^2 - 27z$

29. $V_R^2 + 4V_R - 21$

30. $16C^2 - L^2$

31. $x^2 + 0.3x + 0.02$

32. $4I^2R^2 - 12IR + 9$

33. $6u^2 - 17uv + 12v^2$

34. $12t^2 + 18t - 12$

In problems 35 and 36, reduce the fractions.

35. $\dfrac{3.3\pi r^2 - 1.1\pi r}{4.5\pi r - 1.5\pi}$

36. $\dfrac{(x - y)^2}{x^2 - y^2}$

In problems 37 through 44, perform the indicated operations.

37. $\dfrac{6\alpha V^2}{35} \div \dfrac{4\alpha^2 V}{21\alpha}$

38. $\dfrac{a}{b + d} \cdot \dfrac{ac + ad}{a^2}$

39. $\dfrac{3R_1 + R_2}{R_1 + R_2} \cdot \dfrac{R_1 + 2R_2}{3R_1 + R_2}$

40. $\dfrac{3P_T - 3P_1}{28P_1} \cdot \dfrac{7P_T + 7P_1}{6P_T{}^2 - 6P_1{}^2}$

41. $\dfrac{n^2 - d^2}{n^2 - 2nd + d^2} \div (n + d)$

42. $\dfrac{2x^2 + x}{2x^2 + 9x - 5} \cdot \dfrac{x^2 + 3x - 10}{2x + 1}$

43. $\dfrac{2}{3} \cdot \dfrac{3V - 3}{I + 1} \cdot \dfrac{5I + 5}{2V - 2}$

44. $\dfrac{\delta^4 - 1}{\delta^2 + \delta} \cdot \dfrac{\delta}{\delta^2 + 1}$

In problems 45 through 52, combine the fractions and simplify.

45. $\dfrac{5E_1}{6} + \dfrac{3E_1}{8}$

46. $\dfrac{a}{3xy} - \dfrac{b}{5x^2}$

47. $\dfrac{5}{4 \times 10^6} + \dfrac{1}{8 \times 10^5}$

48. $\dfrac{3.8}{2G} - \dfrac{1.9}{G^2 + G}$

49. $\dfrac{4}{Z - 1} - \dfrac{5}{Z^2 - 1}$

50. $\dfrac{m - 1}{5mv^2} + \dfrac{5}{3mv} + \dfrac{2}{15v^2}$

51. $\dfrac{2r^2}{r^2 + 4r + 4} - \dfrac{5r}{3r + 6}$

52. $1 + \dfrac{3}{W + 1} - \dfrac{W + 1}{W - 1}$

In problems 53 through 56, solve each applied problem.

53. The quickest rise in temperature ever recorded was in the morning at Spearfish, South Dakota on January 22, 1943. At 7:30 AM the temperature was -4°F, and at 7:32 AM it was 45°F. Express this change as a signed number.

54. The kinetic energy of one particle is given by

$$\frac{1}{2} m v_1{}^2$$

The kinetic energy of a second particle is given by

$$\frac{1}{8} m v_2{}^2$$

Factor the expression that represents the difference in energy between the two particles.

55. The area of a rectangular microprocessor circuit is given by $0.10h^2 - 0.30h + 0.20$ and the width by $0.10h - 0.10$. Find the expression for the length of the circuit.

56. The stress factor in a spring mechanism is given by

$$S_f = \frac{3.0}{K} + \frac{2.1K - 3.9}{1.3K + 2.1}$$

Simplify this expression by adding the fractions.

Applications to Electronics In problems 57 through 64 solve each applied problem.

57. The power output of one electric generator is given by $R(I_1{}^2 + 5I_2{}^2)$ and the power output of a second generator by $R(3I_1{}^2 - I_2{}^2)$. What formula gives the total power output of both generators?

58. Figure 3–16 shows an electric force field with two electric intensity vectors perpendicular to each other. The magnitude of the resultant electric intensity is given by

$$E_R = \sqrt{E_x^2 + E_y^2}$$

If $E_x = (5.1 \times 10^{-3})Q$ and $E_y = (6.8 \times 10^{-3})Q$, find the expression for E_R in terms of Q.

59. A circular coil of $n = 2.0 \times 10^2$ turns and radius $r = 0.10$ m, has a magnetic flux density of $B = 300 \times 10^{-6}$ T (tesla). The current I required to produce this flux density in a vacuum is given by

$$I = \frac{2rB}{4\pi n \times 10^{-7}}$$

Substitute the given values, and calculate the current in amperes to two significant digits.

60. Figure 3–17 shows a circuit containing a resistance and a capacitance in series. Under steady-state conditions, the impedance of the RC circuit is given by

$$Z = \sqrt{R^2 + \left(\frac{1}{2\pi fC}\right)^2}$$

Calculate Z in ohms when $R = 1.0$ k Ω, $C = 20$ nF, and the frequency $f = 12$ kHz. Give the answer to two significant digits.

61. Simplify the following expression, which comes from a problem in circuit design:

$$0.80(R_0 + x)(R_0 - x) - 1.2R_0(x - R_0)$$

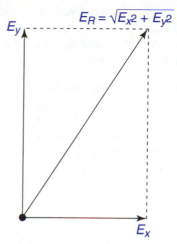

$$E_R = \sqrt{E_x^2 + E_y^2}$$

FIGURE 3–16 Electric intensity vectors for problem 58.

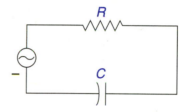

$$Z = \sqrt{R^2 + \left(\frac{1}{2\pi fC}\right)^2}$$

FIGURE 3–17 Steady-state resistance-capacitance circuit for problem 60.

62. In analyzing the electromagnetic force between two pairs of parallel wires that have a current flow, the following expression arises:

$$\pi[a^2 + b^2 + c^2] - 2\pi ab$$

By multiplying out and factoring part of this expression, show that it can be written

$$\pi[(a - b)^2 + c^2]$$

63. The power and current in a circuit as a function of time t are given by

$$P = \left(\frac{2t}{t - 1}\right)^2$$

$$I = \frac{t^2}{t^2 - 1}$$

Using the power law $V = P/I$ find the expression for the voltage as a function of t.

64. Figure 3–18 shows a series-parallel circuit containing a resistance R_1 connected in series with two parallel resistances R_2 and R_3. The total resistance of the circuit is given by

$$R_T = R_1 + \cfrac{1}{\cfrac{1}{R_2} + \cfrac{1}{R_3}}$$

Simplify this expression for R_T by combining all the terms into one fraction.

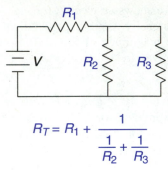

$$R_T = R_1 + \cfrac{1}{\cfrac{1}{R_2} + \cfrac{1}{R_3}}$$

FIGURE 3–18 Series-parallel circuit for problem 64.

4

Equations and Formulas

Courtesy of Arthur Kramer and New York City Technical College.

Technicians check and repair a computer and disk drive.

Science and technology depend on formulas and equations to solve all types of problems. This is especially important in electricity and electronics where formulas are used to express relationships and concepts. To be able to use and manipulate formulas, and know how to solve equations, is an important algebraic skill you need to master. This chapter shows you how to solve first-degree equations and shows you how to use formulas and solve them for any of their variables. Another name for a formula is a literal equation, which is an equation that contains several letters or variables. The last section studies the very important and critical skill of problem solving. It shows you a step-by-step procedure that will help you to analyze, set up, and solve a problem. There are many examples discussed and exercises given to help you develop this ability.

Chapter Objectives

In this chapter, you will learn:

- How to solve first-degree or linear equations.
- How to solve proportions and equations with fractions.
- How to solve formulas and literal equations for any of the variables.
- How to analyze and solve a verbal problem that leads to an algebraic equation.

≡ 4.1
FIRST-DEGREE EQUATIONS

A *first-degree equation*, or *linear equation*, is a statement of equality that contains only variables to the first power or degree. Three examples of linear equations in one variable are

$$3t + 7 = 16$$

$$27 = \frac{5}{9}(F - 32)$$

$$\frac{1}{R_T} = \frac{1}{7.5} + \frac{1}{5.0}$$

Note that the variables t, F and R_T all have an exponent of 1. To solve for a variable in an equation, you apply the following principle.

Equation Principle	Adding, subtracting, multiplying, or dividing each term on both sides of an equation by the same quantity does not change the equation. (Division by zero is not allowed.)

Study the next two examples, which show how to apply the equation principle.

EXAMPLE 4.1

Solve and check the linear equation $3t + 7 = 16$ for the unknown variable t.

Solution   To solve the equation, apply the equation principle twice to *isolate the unknown* on one side of the equation. First isolate the term $3t$ by subtracting 7 from both sides of the equation.

1. Subtract 7 from both sides:
$$3t + 7 - 7 = 16 - 7$$

2. Combine terms:
$$3t = 9$$

Then solve for t by dividing both sides by the coefficient of the unknown.

1. Divide both sides by 3:
$$\frac{\cancel{3}t}{\cancel{3}} = \frac{9}{3}$$

2. Simplify:
$$t = 3$$

To check the solution, substitute $t = 3$ into the original equation:
$$3\,(3) + 7 = 16 \quad ?$$
$$9 + 7 = 16 \quad \checkmark$$

▪

EXAMPLE 4.2

Solve and check the linear equation:
$$4P - 3 = 5 + 2P$$

Solution   Apply the equation principle to put all the terms containing the unknown on one side of the equation and all the constant terms on the other side.

First eliminate the 3 on the left and the $2P$ on the right.

1. Add 3 to both sides:
$$4P - 3 + 3 = 5 + 2P + 3$$

2. Simplify:
$$4P = 8 + 2P$$

3. Subtract $2P$ from both sides:
$$4P - 2P = 8 + 2P - 2P$$

4. Simplify:
$$2P = 8$$

This is the form to which you reduce each linear equation: A term containing the unknown on one side and a constant term on the other side. Then you divide both sides by the coefficient of the unknown.

1. Divide both sides by 2:
$$\frac{\cancel{2}P}{\cancel{2}} = \frac{8}{2}$$

2. Simplify:
$$P = 4$$

EXAMPLE 4.2 (Cont.)

To check the solution substitute back into the original equation:

$$4(4) - 3 = 5 + 2(4) \quad ?$$
$$16 - 3 = 5 + 8 \quad ?$$
$$13 = 13 \quad \checkmark$$

The process of solving an equation can be simplified by noting that adding or subtracting the same quantity on both sides of an equation is equivalent to *moving that quantity to the other side of the equation and changing its sign.* For example, look at the first step in Example 4.2, adding 3 to both sides. This is the same as moving the −3 on the left side to the right side and changing it to +3. The process is called *transposing*. Study the next example, which illustrates the method of transposing applied to Example 4.2.

EXAMPLE 4.3

Solve Example 4.2 using the method of transposing.

Solution

1. Transpose −3 to the right side:
$$4P - 3 = 5 + 2P \Rightarrow 4P = 5 + 2P + 3$$

2. Simplify:
$$4P = 8 + 2P$$

3. Transpose 2P to the left side:
$$4P = 8 + 2P \Rightarrow 4P - 2P = 8$$

4. Simplify:
$$2P = 8$$

5. Divide both sides by 2:
$$P = 4$$

Observe that you can transpose the −3 and the $2P$ in one step, speeding up the process of solution. You should use the method of transposing whenever possible.

Study the next example carefully. It illustrates the method of transposing and shows how to handle a simple fraction and parentheses in an equation.

EXAMPLE 4.4

The formula relating degrees Fahrenheit to degrees Celsius is

$$C = \frac{5}{9}(F - 32)$$

Find F when $C = 27°C$.

Solution Substitute $C = 27$ into the formula to give the linear equation

$$27 = \frac{5}{9}(F - 32)$$

EXAMPLE 4.4 (Cont.)

You should eliminate fractions as soon as possible in an equation to make the solution easier. Multiply both sides by 9 to eliminate the fraction:

$$(9)\,27 = (9)\frac{5}{9}(F - 32)$$

$$243 = 5\,(F - 32)$$

Note carefully that you only multiply one factor $\frac{5}{9}$ on the right by 9 and not the other factor $(F - 32)$.

If there is more than one fraction in an equation, you multiply by the lowest common denominator to eliminate all fractions. Now remove the parentheses, and solve for F.

1. Remove parentheses:

$$243 = 5F - 160$$

2. Transpose −160:

$$243 = 5F - 160 \Rightarrow 243 + 160 = 5F$$

3. Simplify:

$$403 = 5F \text{ or } 5F = 403$$

By convention the unknown term is written on the left. Divide by 5:

$$F = \frac{403}{5} = 80.6\,°$$

The solution check is:

$$27 = \frac{5}{9}(80.6 - 32) \ \ ?$$

$$27 = \frac{5}{9}(48.6) \ \ ?$$

$$27 = 27 \ \ \checkmark$$

You can also do this example by first solving for F and then substituting $C = 27$ as follows:

1. Multiply by 9:

$$(9)\,C = (9)\frac{5}{9}(F - 32) \Rightarrow 9C = 5\,(F - 32)$$

2. Remove parentheses:

$$9C = 5F - 160$$

3. Transpose 160 and write $5F$ on the left:

$$5F = 9C + 160$$

4. Divide by 5:

$$F = \frac{9}{5}C + 32$$

5. Substitute:

$$F = \frac{9}{5}(27) + 32 = 80.6\,°$$

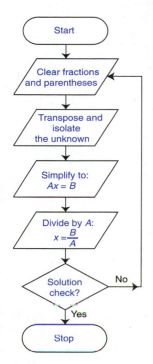

FIGURE 4–1 Flowchart for solving equations.

The step by step process or *algorithm* for solving a linear equation is as follows:

1. Eliminate fractions by multiplying by the lowest common denominator (LCD), and then remove the parentheses.
2. Combine similar terms, and transpose all terms containing the unknown to one side and all constant terms to the other side.
3. Simplify to put the equation in the form

$$Ax = B$$

where x = unknown and A and B are constants.
4. Divide both sides by the coefficient of x to obtain the solution:

$$x = \frac{B}{A}$$

5. Check the solution in the original equation.

See Figure 4–1, which illustrates the algorithm in a flowchart.

The following example applies all the steps in the algorithm.

EXAMPLE 4.5

Solve and check:

$$\frac{5V}{2} + \frac{2V}{5} = 3(V - 3)$$

Solution   Eliminate the fractions by multiplying each term by the LCD, which is 10:

$$(\cancel{10})\overset{5}{\underset{}{\frac{5V}{\cancel{2}}}} + (\cancel{10})\overset{2}{\underset{}{\frac{2V}{\cancel{5}}}} = (10)\,3\,(V - 3)$$

$$25V + 4V = 30(V - 3)$$

Remove parentheses:

$$25V + 4V = 30V - 90$$

Combine terms and transpose $30V$:

$$29V = 30V - 90 \Rightarrow 29V - 30V = -90$$

Combine terms:

$$-V = -90$$

You might be tempted to stop here, thinking you have the solution. The number −90, however, represents the *negative* value of the unknown. You must go one step further and divide, or multiply, both sides of the equation by −1. This changes the sign of both sides of the equation and results in the correct answer, which is the positive value of the unknown:

$$(-1)-V = (-1)-90 \Rightarrow V = 90$$

EXAMPLE 4.5 (Cont.)

The solution check is:

$$\frac{5\,(90)}{2} + \frac{2\,(90)}{5} = 3\,(90 - 3) \quad ?$$

$$225 + 36 = 3\,(87) \quad ?$$

$$261 = 261 \quad \checkmark$$

Many formulas in electricity and electronics require you to solve a linear equation when you apply the formula. The next two examples close the circuit and illustrate such applications of electrical formulas.

EXAMPLE 4.6

Close the Circuit

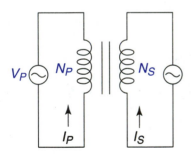

$$\frac{V_P}{V_S} = \frac{N_P}{N_S}$$

FIGURE 4–2 Iron-core transformer for Example 4.6

Figure 4–2 shows a transformer that steps down the voltage in a circuit, depending on the number of turns in the primary and secondary windings. The voltage ratio is equal to the turns ratio and is given by the proportion

$$\frac{V_P}{V_S} = \frac{N_P}{N_S}$$

A proportion is an equation where one ratio (fraction) equals another ratio. In this proportion, V_P = primary voltage, V_S = secondary voltage, N_P = number of turns in primary, and N_S = number of turns in secondary. Find V_P when $V_S = 110$ V, $N_P = 1000$, and $N_S = 700$.

Solution Substitute the values in the proportion to give the linear equation:

$$\frac{V_P}{110} = \frac{1000}{700}$$

Reduce the fraction on the right:

$$\frac{V_P}{110} = \frac{10}{7}$$

The LCD of 7 and 110 is 770. Multiply both sides by 770:

$$(\cancel{770}) \frac{V_P}{\cancel{110}}^{7} = (\cancel{770}) \frac{10}{\cancel{7}}^{110}$$

Then

$$7V_P = 1100$$

$$V_P = \frac{1100}{7} = 157 \text{ V}$$

A proportion can be solved more directly by *cross multiplication*. Cross multiplication means multiplying the left numerator by the right denominator and vice versa:

$$\frac{V_P}{110} \times \frac{10}{7}$$

$$7V_P = 1100$$

$$V_P = 157 \text{ V}$$

Cross multiplication is equivalent to multiplying both sides of the equation by the product of the denominators.

EXAMPLE 4.7

Close the Circuit

The total resistance R_T of the parallel circuit in Figure 4–3 containing R_1 and R_2 in parallel is given by

$$\frac{1}{R_T} = \frac{1}{R_1} + \frac{1}{R_2}$$

If $R_1 = 1.5 \text{ k}\Omega$ and $R_2 = 2.0 \text{ k}\Omega$, find the value of R_T and check the solution.

Solution   Substitute the values of R_1 and R_2 in the formula:

$$\frac{1}{R_T} = \frac{1}{1.5} + \frac{1}{2.0}$$

Eliminate the fractions by multiplying each fraction by the lowest common denominator, which is $6R_T$:

$$(6R_T)\frac{1}{R_T} = (6R_T)\overset{4}{\frac{1}{1.5}} + (6R_T)\overset{3}{\frac{1}{2.0}}$$

$$6 = 4R_T + 3R_T$$

$$7R_T = 6$$

$$R_T = \frac{6}{7}\text{ k}\Omega = 0.857\text{ k}\Omega \approx 860\,\Omega$$

The solution is rounded to two significant digits because the given values are accurate to two significant digits. To check the solution you can use the calculator:

$$\frac{1}{1.5} + \frac{1}{2.0} = \frac{1}{0.857} \quad ?$$

$$1.5 \boxed{1/x} \boxed{+} 2.0 \boxed{1/x} \boxed{=} \boxed{1/x} \rightarrow 0.857 \checkmark$$

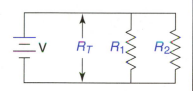

$$\frac{1}{R_T} = \frac{1}{R_1} + \frac{1}{R_2}$$

FIGURE 4–3 Parallel resistances for Example 4.7

EXERCISE 4.1

In problems 1 through 42, solve and check each linear equation.

1. $x + 2 = 7$

2. $a - 3 = 9$

3. $3b + 2 = 5$

4. $4y - 8 = 12$

5. $4 + 5I = I - 12$

6. $6V - 11 = V - 16$

7. $3 - 3t = 6$

8. $16 = 8 - 4w$

9. $\dfrac{u}{8} = \dfrac{1}{2}$

10. $\dfrac{I_P}{2} = \dfrac{3}{5}$

11. $5\alpha + 9 = 6\alpha + 3$

12. $3 - 7\beta = 3\beta - 2$

13. $R = \dfrac{1}{2}(4R + 2)$

14. $2K - \dfrac{1}{3} = 1 + \dfrac{2}{3}$

15. $0.2X_L + 3.6 = 1.4$

16. $1.2X_C + 1.3X_C = 5.5$

17. $0.7 = 2V_0 - 2.3$

18. $0.06 + 0.05T_1 = 0.03T_1$

19. $2(Y - 4) = Y - 5$

20. $6 + 2P = 2(2P + 3)$

21. $5 - (f + 2) = 5f$

22. $4 - (x - 4) = 2x$

23. $0.26 = 0.2(v - 1)$

24. $2(r - 1) = 6.5 + 3r$

25. $\dfrac{10}{V_S} = \dfrac{200}{50}$

26. $\dfrac{N_P}{500} = \dfrac{24}{120}$

27. $3Q - \dfrac{5}{9} = \dfrac{2}{9}Q$

28. $\dfrac{9}{5}C + 32 = 4C - 1$

29. $\dfrac{k}{3} + \dfrac{1}{3} = 2$

30. $h - \dfrac{h}{2} = \dfrac{3}{4}$

31. $\dfrac{3a}{4} + \dfrac{a}{6} = 11$

32. $13 - \dfrac{2b}{3} = \dfrac{5b}{12}$

33. $\dfrac{V_1 - 1}{2} = \dfrac{3V_1}{8}$

34. $\dfrac{2R_T + 1}{3} = \dfrac{3R_T + 2}{9}$

35. $\dfrac{3}{5} - \dfrac{4 - 15y}{10} = 2y$

36. $\dfrac{2x}{3} - \dfrac{3x - 1}{8} = 1$

37. $\dfrac{3}{2D} - \dfrac{9}{D^2} = \dfrac{6}{5D}$

38. $\dfrac{5}{4C} - \dfrac{7}{10C} = \dfrac{11}{2C^2}$

39. $\dfrac{5}{G_x + 3} - \dfrac{1}{G_x} = \dfrac{3}{2G_x}$

40. $\dfrac{5}{Z} + \dfrac{3}{Z - 4} = \dfrac{20}{Z}$

41. $\dfrac{2\omega - 1}{\omega + 3} + \dfrac{1}{\omega} = 2$

42. $\dfrac{3\lambda}{\lambda - 2} - 3 = \dfrac{5}{\lambda}$

In problems 43 through 48, solve each applied problem.

43. Mr. Leeds wants to give $1000 to three charities in the ratio 5:3:2. To solve this problem, he needs to solve the equation:

$$5C + 3C + 2C = 1000$$

where $5C$, $4C$, and $3C$ are the amounts to be given to each charity and C = common factor. How much should he give to each charity?

44. Diana wants to buy a calculator on sale for $18.60. The salesman tells her that the regular price was reduced 25% for the sale. To find the regular price, she has to solve the equation:

$$p - 25\%p = 18.60$$

What is the regular price p of the calculator?

45. The formula relating degrees Fahrenheit to degrees Celsius is

$$F = 1.8C + 32$$

Find C when $F = -13°F$.

46. The velocity of an object under free fall is given by

$$v = v_o + gt$$

where v_o = initial velocity, g = gravitational acceleration, and t = time. If $v_o = 38$ ft/s, find t when $v = 150$ ft/s. Use $g = 32$ ft/s^2.

47. For a hydraulic lift, the ratio of the applied force F_1 to the resulting force F_2 is equal to the ratio of the squares of the diameters of the pistons:

$$\frac{F_1}{F_2} = \frac{D_1^2}{D_2^2}$$

What applied force is necessary to lift a 1440-kg automobile if $D_1 = 10$ cm and $D_2 = 80$ cm?

48. The effect of a moving load on the stress in a beam results in the influence diagram in Figure 4–4. The two right triangles are similar, which means that corresponding sides are in proportion:

$$\frac{AB}{DE} = \frac{BC}{CE} = \frac{AC}{CD}$$

If $AB = 1.1$ cm, $BC = 1.8$ cm, $CD = 2.7$ cm, and $DE = 1.4$ cm, find CE and AC.

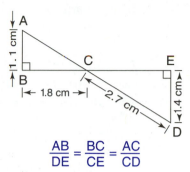

$$\frac{AB}{DE} = \frac{BC}{CE} = \frac{AC}{CD}$$

FIGURE 4–4 Influence diagram for stress in a beam for problem 48.

Applications to Electronics   In problems 49 through 60 solve each applied problem.

49. Figure 4–5 shows a series-parallel circuit containing a resistance R_1 in series with two parallel resistances R_2 and R_3. The total voltage V_T in the circuit is given by

$$V_T = I_1 R_1 + I_3 R_3$$

If $V_T = 4.0$ V, $I_1 = 160$ mA, $R_1 = 20$ Ω and $R_3 = 10$ Ω, find I_3 (in milliamps).

50. In the series parallel circuit in Figure 4–5, the voltage drop across R_2 equals the voltage drop across R_3:

$$I_2 R_2 = I_3 R_3$$

Find R_3 when $R_2 = 1.3$ kΩ, $I_2 = 350$ μA, and $I_3 = 0.50$ mA. (Note: The units must be consistent in the equation.)

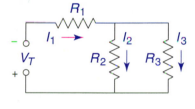

$$V_T = I_1 R_1 + I_3 R_3$$

$$I_2 R_2 = I_3 R_3$$

FIGURE 4–5 Series-parallel circuit for problems 49 and 50.

51. The change in resistance of a conductor is proportional to the temperature change expressed by:

$$\frac{R_2 - R_1}{R_4 - R_3} = \frac{T_2 - T_1}{T_4 - T_3}$$

If the resistance of an aluminum wire increases from $R_1 = 20$ mΩ to $R_2 = 40$ mΩ when the temperature increases from $T_1 = 0°C$ to $T_2 = 80°C$, how much will the resistance increase when the temperature increases from $T_3 = 0°C$ to $T_4 = 100°C$?

52. In the transformer in Figure 4–2, the voltage ratio is inversely proportional to the current ratio:

$$\frac{V_P}{V_S} = \frac{I_S}{I_P}$$

Find I_S when $V_S = 15$ V, $V_P = 110$ V, and $I_P = 90$ mA

53. In Example 4.7, find R_T when $R_1 = 30$ Ω and $R_2 = 20$ Ω.

54. In Example 4.7, find R_1 when $R_T = 1.25$ kΩ and $R_2 = 7.50$ kΩ.

55. Figure 4–6 shows two capacitances in series. The total capacitance C_T is given by

$$C_T = \frac{C_1 C_2}{C_1 + C_2}$$

Find C when $C_T = 20\ \mu F$ and $C_2 = 30\ \mu F$.

56. Find C_2 in problem 55 when $C_T = 100\ pF$ and $C_1 = 300\ pF$.

57. Figure 4–7 shows three inductances in parallel. Assuming no mutual inductance, the total inductance is given by

$$\frac{1}{L_T} = \frac{1}{L_1} + \frac{1}{L_2} + \frac{1}{L_3}$$

Find L_T when $L_1 = 5\ mH$, $L_2 = 5\ mH$, and $L_3 = 10\ mH$.

58. In Figure 4–7, find L_3 when $L_T = 500\ mH$, $L_1 = 1.0\ H$, and $L_2 = 2.0\ H$. (Note: The units must be consistent in the equation.)

59. Figure 4–8 shows a series-parallel circuit containing a resistance R_1 in series with two parallel resistances R_2 and R_3. The total resistance of the circuit is given by

$$R_T = R_1 + \frac{R_2 R_3}{R_2 + R_3}$$

Find R_1 when $R_T = 150\ \Omega$, $R_2 = 100\ \Omega$, and $R_3 = 150\ \Omega$.

60. In problem 59, find R_2 and R_3 when $R_T = 18\ \Omega$, $R_1 = 12\ \Omega$, and $R_3 = 2R_2$.

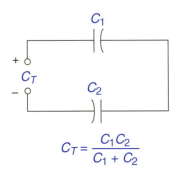

$$C_T = \frac{C_1 C_2}{C_1 + C_2}$$

FIGURE 4–6
Capacitances in series
for problems 55 and 56.

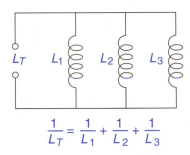

$$\frac{1}{L_T} = \frac{1}{L_1} + \frac{1}{L_2} + \frac{1}{L_3}$$

FIGURE 4–7 Inductances
in parallel for problems 57
and 58.

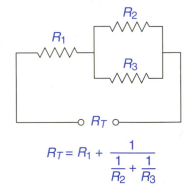

$$R_T = R_1 + \frac{1}{\dfrac{1}{R_2} + \dfrac{1}{R_3}}$$

FIGURE 4–8 Series-parallel
circuit for problems 59
and 60.

≡ 4.2

FORMULAS AND LITERAL EQUATIONS

Many important electrical formulas have been introduced in Chapters 1, 2, and 3, both in the examples and in the exercises. In this section, these formulas and others are studied as literal equations. A *literal equation* is an equation containing two or more variables (letters). By applying the equation principle to a formula or literal equation, you can solve it for any of the variables that you like, or derive a more useful formula. This can lead to an easier or more direct solution to a problem. It also allows you to change a formula to any of its various forms so that you do not have to memorize each form.

For example, suppose you are given Ohm's law solved for I:

$$I = \frac{V}{R}$$

With two algebraic steps you can solve it for V and R as follows:

1. First multiply both sides by R:

$$(R)\,I = \frac{V}{\cancel{R}}\,(\cancel{R})$$

2. This gives you Ohm's law solved for V:

$$IR = V$$

3. Then divide both sides by I:

$$\frac{\cancel{I}R}{\cancel{I}} = \frac{V}{I}$$

4. You now have Ohm's law solved for R:

$$R = \frac{V}{I}$$

Note that you can derive any form of Ohm's law from any other form by applying the equation principle.

This section contains many examples that close the circuit since these concepts are very useful in electricity and electronics. The first example applies the ideas to the formula for power.

EXAMPLE 4.8

Close the Circuit

Given the power formula $P = VI$ and Ohm's law solved for I, $I = \frac{V}{R}$, derive the following two formulas:

$$P = \frac{V^2}{R}$$

$$R = \frac{V^2}{P}$$

Solution Apply the method of substitution to Ohm's law and the power law. Start with Ohm's law:

$$I = \frac{V}{R}$$

Substitute this expression for I in $P = VI$ as follows:

$$P = VI = V\left(\frac{V}{R}\right)$$

You then obtain the first formula:

$$P = V\left(\frac{V}{R}\right) = \frac{V^2}{R}$$

Now take this formula and solve it for R. Multiply both sides by R:

$$P\,(R) = \frac{V^2}{\cancel{R}}\,(\cancel{R})$$

EXAMPLE 4.8 (Cont.)

Divide both sides by P:

$$\frac{\cancel{P}R}{\cancel{(P)}} = \frac{V^2}{P}$$

You now have the second formula:

$$R = \frac{V^2}{P}$$

◾

The following four examples illustrate the solutions of some important formulas and explain what they represent.

EXAMPLE 4.9

Given the formula for the velocity of a body with acceleration a,

$$v = v_0 + at$$

solve this formula for a.

Solution Transpose v_0, changing the sign:

$$v - v_0 = at$$

Divide both sides by t:

$$\frac{v - v_0}{t} = \frac{a\cancel{t}}{\cancel{t}}$$

The formula for a is then:

$$a = \frac{v - v_0}{t}$$

◾

The next example closes the circuit and involves a formula for the voltage in a series circuit.

EXAMPLE 4.10

Close the Circuit

Given the formula for the voltage of a series circuit with two resistances,

$$V_T = IR_1 + IR_2$$

solve this formula for I.

Solution Since you are solving for I, you must first separate I as a common factor:

$$V_T = I(R_1 + R_2)$$

You can then divide both sides of the equation by the quantity in parentheses, $(R_1 + R_2)$, which is the coefficient of I:

$$\frac{V_T}{(R_1 + R_2)} = \frac{I\,\cancel{(R_1 + R_2)}}{\cancel{(R_1 + R_2)}}$$

The formula for I is then:

$$I = \frac{V_T}{R_1 + R_2}$$

◾

EXAMPLE 4.11

Given the formula for the surface area of a cylinder,

$$A = \pi r(2h + r)$$

solve this formula for h.

Solution First multiply out the parentheses:

$$A = 2\pi rh + \pi r^2$$

Transpose πr^2:

$$A - \pi r^2 = 2\pi rh$$

Divide by $2\pi r$:

$$\frac{A - \pi r^2}{2\pi r} = \frac{2\pi rh}{2\pi r}$$

The formula for h is then:

$$h = \frac{A - \pi r^2}{2\pi r}$$

The next example closes the circuit and involves a formula for the resistance in a parallel circuit.

EXAMPLE 4.12

Close the Circuit

Given the formula for the total resistance of two resistances in parallel,

$$R_T = \frac{R_1 R_2}{R_1 + R_2}$$

solve this formula for R_1.

Solution First clear the fraction by multiplying both sides by the LCD $R_1 + R_2$:

$$R_T(R_1 + R_2) = \frac{R_1 R_2}{(R_1 + R_2)}(R_1 + R_2)$$

Multiply out the parentheses on the left:

$$R_T R_1 + R_T R_2 = R_1 R_2$$

Move the terms containing R_1 to one side:

$$R_T R_2 = R_1 R_2 - R_T R_1$$

Factor out R_1:

$$R_T R_2 = R_1(R_2 - R_T)$$

Divide by $(R_2 - R_T)$:

$$\frac{R_T R_2}{R_2 - R_T} = \frac{R_1(R_2 - R_T)}{R_2 - R_T}$$

The formula for R_1 is then:

$$R_1 = \frac{R_T R_2}{R_2 - R_T}$$

The next example that closes the circuit further shows the usefulness of combining formulas to produce a new electrical formula.

EXAMPLE 4.13

Close the Circuit

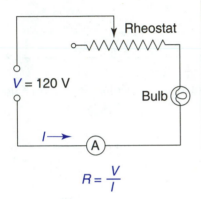

Rheostat

$V = 120\ V$

Bulb

$I \rightarrow$ (A)

$R = \dfrac{V}{I}$

FIGURE 4–9 Rheostat circuit for Example 4.13

Consider the circuit in Figure 4–9, which contains a variable resistor, or rheostat, in series with a light bulb and an ammeter to measure the current. As the rheostat is varied, the total resistance in the circuit will vary. This will change the current in the circuit, causing the light to vary in brightness. Suppose the voltage is constant at 120 V and the rheostat is varied so the total circuit resistance changes from R_1 to R_2. When the circuit resistance is R_1, the ammeter measures the current $I_1 = 1.5$ A. When the resistance is R_2, the ammeter measures the current $I_2 = 2.4$ A.

1. Using Ohm's law $R = \dfrac{V}{I}$, find the change in the resistance of the circuit. This is represented algebraically as $\Delta R = R_2 - R_1$, where Δ = delta for "difference."

2. Find a formula for ΔR in terms of the change in the circuit current: $\Delta I = I_2 - I_1$.

Solution

1. Applying Ohm's law $R = \dfrac{V}{I}$, the resistance R_1 of the circuit when $I_1 = 1.5$ A is

$$R_1 = \frac{V}{I_1} = \frac{120}{1.5} = 80\ \Omega$$

The resistance R_2 of the circuit when $I_2 = 2.4$ A is

$$R_2 = \frac{V}{I_2} = \frac{120}{2.4} = 50\ \Omega$$

The change in the total resistance is then
$$\Delta R = R_2 - R_1 = 50 - 80 = -30\ \Omega$$

or a decrease of 30 Ω.

2. To find a formula for ΔR in terms of ΔI, consider the formulas for R_1 and R_2 from Ohm's law:

$$R_1 = \frac{V}{I_1}$$

$$R_2 = \frac{V}{I_2}$$

Combining these formulas, the change in the total resistance in terms of V and I is then

$$\Delta R = R_2 - R_1 = \frac{V}{I_2} - \frac{V}{I_1}$$

Combine these fractions and factor out V to simplify the formula for ΔR:

$$\Delta R = \frac{VI_1 - VI_2}{I_1 I_2} = \frac{V(I_1 - I_2)}{I_1 I_2}$$

Now since $\Delta I = I_2 - I_1$ you have
$$-\Delta I = -(I_2 - I_1) = -I_2 + I_1 = I_1 - I_2$$

EXAMPLE 4.13 (Cont.)

Substituting $-\Delta I$ for $I_1 - I_2$ in the formula for ΔR gives you

$$\Delta R = \frac{V(-\Delta I)}{I_1 I_2} = \frac{-V \Delta I}{I_1 I_2}$$

You now have a useful formula for the change in resistance in terms of the change in current. The negative sign in this formula shows that for an increase in the current the resistance will decrease and vice versa.

The formula can be used to directly obtain the answer to part 1. Substituting $V = 120$ V, $I_1 = 1.5$ A, and $I_2 = 2.4$ A,

$$\Delta R = \frac{-120\,(2.4 - 1.5)}{(1.5)\,(2.4)} = \frac{-120\,(0.9)}{3.6} = -30\,\Omega$$

Deriving the formula for ΔR to find the change in resistance may at first appear to be more involved than the solution in part 1. However, if the change in resistance needs to be calculated several times for various current changes, then the derived formula provides a direct way to get the results. It also emphasizes the relationship between the changes in the resistance and the change in the current, which is sometimes of more importance than the actual values. ▪

EXERCISE 4.2

In problems 1 through 16, solve each formula for the indicated letter.

1. $I^2 = \dfrac{P}{R}$; R (I = current in circuit)

2. $a = \dfrac{F}{M}$; m (a = acceleration of a body)

3. $X_C = \dfrac{1}{2\pi f C}$; f (X_C = capacitive reactance)

4. $C = \dfrac{\varepsilon A}{d}$; ε (C = capacitance of two parallel plates)

5. $V = I\,(R + r_g)$; r_g (V = generator voltage)

6. $P = 2\,(l + w)$; w   (P = perimeter of rectangle)

7. $I = \dfrac{V - V_D}{R}$; V_D (I = current in LED circuit)

8. $V = B - \dfrac{V_0}{A}$; V_0 (V = transistor voltage)

9. $V_T = IR_1 + IR_2 + IR_3$; I (V_T = total voltage of series circuit)

10. $P = VI_1 + VI_2$; V (P = power in parallel circuit)

11. $I_C = \dfrac{\beta I_E}{\beta + 1}$; β (I_C = current in transistor)

12. $I_1 = \dfrac{I_T G_1}{G_1 + G_2}; G_1$ (I_1 = current in parallel circuit)

13. $\dfrac{1}{L_T} = \dfrac{1}{L_1} + \dfrac{1}{L_2}; L_1$ (L_T = Total inductance of L_1 and L_2 in parallel)

14. $R_T = \dfrac{R_1 R_2}{R_1 + R_2}; R_2$ (R_T = Total resistance of R_1 and R_2 in parallel)

15. $Z = \dfrac{1}{\dfrac{1}{\omega L} - \omega C}; L$ (Z = impedance of LC circuit)

16. $R_{AB} = \dfrac{R_A R_B}{R_A + R_B + R_C}; R_A$ (Star to delta resistance conversion)

In problems 17 through 22, do each problem by solving a literal equation.

17. The boiling point of water in degrees Fahrenheit at a height h feet above sea level is given approximately by the formula

$$T = 212 - \frac{h}{550}$$

(a) Solve this formula for h.
(b) What is the increase in height if T changes from 211 to 209°F?

18. The velocity of a falling body is given by

$$v = v_0 + gt$$

where v_0 = initial velocity, t = time, and g = gravitational acceleration.
(a) Solve this formula for t.
(b) If $v_0 = 7.5$ m/s, how long will it take for this velocity to triple? Use $g = 9.8$ m/s^2.

19. Given the temperature conversion formula

$$F = \frac{9}{5}C + 32$$

(a) Solve this formula for C.
(b) What is the increase in degrees Celsius if the temperature doubles from 45°F to 90°F?

20. In problem 19, if F changes from F_1 to F_2 when C changes from C_1 to C_2,
(a) Find the formula for the change in degrees Fahrenheit $\Delta F = F_2 - F_1$ in terms of $\Delta C = C_2 - C_1$. See Example 4.13.
(b) If the temperature falls 50° on the Celsius scale, how many degrees does it fall on the Fahrenheit scale?

21. The total time required to travel a round trip whose one way distance is D is given by

$$T = \frac{2D}{R_{avg}}$$

where R_{avg} = average rate for the entire trip. If R_1 = rate going and R_2 = rate returning, the total time is also given by

$$T = \frac{D}{R_1} + \frac{D}{R_2}$$

(a) Equate these two expressions for T, and find the formula for R_{avg} by solving the literal equation.
(b) If $R_1 = 40$ mi/h and $R_2 = 60$ mi/h, what is the average rate for the entire trip?

22. A certain steel rod of length l_0 at 0°C will increase in length to l_t at t°C according to the formula

$$l_t = l_0(1 + 10.7 \times 10^{-6}t)$$

(a) Solve this formula for t.

(b) At what temperature will the steel rod increase in length by 1%? (Hint: Let $l_t = 1.01l_0$.)

Applications to Electronics In applied problems 23 through 36, do each problem by solving a literal equation.

23. Using the power formula $P = VI$ and Ohm's law $V = IR$, derive the formula

$$P = I^2R$$

24. Using the power formula $P = I^2R$ and Ohm's law $I = \frac{V}{R}$, derive the formula

$$P = \frac{V^2}{R}$$

25. The battery voltage in the series circuit in Figure 4–10 is given by $V = V_0 + IR_1$ where V_0 is the voltage drop across R_0.

(a) Solve this formula for R_1.

(b) Find R_1 when $V = 12$ V, $V_0 = 4.8$ V, and $I = 1.2$ A.

26. (a) In problem 25, solve the voltage formula for I.

(b) If $V = 7.5$ V, $V_0 = 4.5$ V, and $R_1 = 150$ Ω, find I (in milliamps).

27. The total resistance of the parallel circuit in Figure 4–11 is related to R_1 and R_x by the formula

$$\frac{1}{R_T} = \frac{1}{R_1} + \frac{1}{R_X}$$

Solve this formula for the unknown resistance R_X.

28. The total resistance R_T of the parallel circuit in Figure 4–11 is given by

$$R_T = \frac{R_1 R_X}{R_1 + R_X}$$

Solve this formula for the unknown resistance R_X. The result should be the same as problem 27.

29. The resistance of a material is a linear function of its temperature t in degrees Celsius expressed by

$$R_t = R_0 + R_0\alpha t$$

where R_0 = resistance at 0°C and α = temperature coefficient of resistance.

(a) Solve this formula for α.

(b) Find α when $R_t = 25$ Ω, $R_0 = 20$ Ω, and $t = 80$°C.

30. (a) In problem 29, solve the resistance formula for R_0.

(b) Find R_0 if $R_t = 340$ Ω when $t = 20$°C and $\alpha = 0.0018$ per degree Celsius.

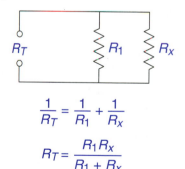

$$V = V_0 + IR_1$$

FIGURE 4–10 Series circuit for problems 25 and 26.

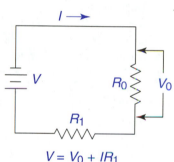

$$\frac{1}{R_T} = \frac{1}{R_1} + \frac{1}{R_X}$$

$$R_T = \frac{R_1 R_X}{R_1 + R_X}$$

FIGURE 4–11 Parallel circuit for problems 27 and 28.

31. The resistance of a wire is directly proportional to its length l and inversely proportional to its cross-sectional area A expressed by the formula

$$R = \frac{\rho l}{A}$$

The constant ρ (rho) is called the resistivity and has the units Ω-m (ohm-meter).

(a) Solve this formula for ρ.

(b) Find ρ when $R = 3.5$ mΩ, $l = 10$ m, and $A = 60 \times 10^{-6}$ m^2. Note: R must be in ohms.

32. Figure 4–12 shows a resistance R_1 in parallel with two series resistances R_2 and R_3. The total resistance R_T of the series-parallel circuit can be expressed by the formula

$$\frac{1}{R_T} = \frac{1}{R_1} + \frac{1}{R_2 + R_3}$$

(a) Given $R_2 = R_3$, solve this formula for R_T in terms of R_1 and R_2.

(b) Find R_T when $R_1 = 20$ kΩ and $R_2 = 10$ kΩ.

33. Figure 4–13 contains a rheostat in series with a light bulb whose resistance $R = 80$ Ω.

(a) Using Ohm's law $V = IR$, find the change in the voltage drop across the bulb when the current changes from $I_1 = 1.3$ A to $I_2 = 1.1$ A. Assume the resistance of the bulb is constant.

(b) Find the formula for the change in the voltage drop across the light bulb $\Delta V = V_2 - V_1$ in terms of the current change $\Delta I = I_2 - I_1$.

Check your answer to (a) using the formula for ΔI. (See Example 4.13)

34. In Figure 4–13, given V_S = source voltage:

(a) Using the power formula $P = VI$, find the formula for the power change $\Delta P = P_2 - P_1$ in the circuit in terms of the current change $\Delta I = I_2 - I_1$.

(b) Find ΔP when the current changes from $I_1 = 1.7$ A to $I_2 = 1.5$ A and $V_s = 110$ V. (See Example 4.13.)

35. **(a)** In problem 33, find the formula for ΔV if the current doubles from I to $2I$.

(b) Find the formula for ΔV if the current decreases by 10%.

36. **(a)** In problem 34, find the formula for ΔP if the current increases by 50% from I to $1.5I$.

(b) Find the formula for ΔP if the current decreases by one-third its value.

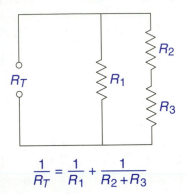

$$\frac{1}{R_T} = \frac{1}{R_1} + \frac{1}{R_2 + R_3}$$

FIGURE 4–12 Series-parallel circuit for problem 32.

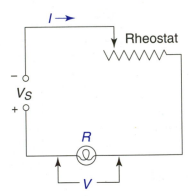

FIGURE 4–13 Rheostat with bulb for problems 33 through 36.

≡ 4.3
PROBLEM SOLVING

It is not an easy task to translate verbal problems into mathematics in order to find their solution. However, people must communicate ideas in English, Spanish, French, or some verbal language. Therefore, if you are to apply the skills you are learning, you must be able to translate verbal problems into mathematics. Verbal problems can be presented in many different ways and usually require that an equation be solved to obtain their solution. The necessary equation is not always directly given, which means that you have to formulate it yourself. That requires a certain skill. The best way to develop this skill is by *doing many problems*. Although this takes time, you should not be discouraged. Stay with it, and you will develop the skill to solve these problems. Furthermore, you will sharpen your thought processes and find working with mathematics more enjoyable. Study the first example carefully. It illustrates the basic procedures of problem solving. You may also find this problem useful in your other classes.

EXAMPLE 4.14

Suppose the grade on your first test in this course is 76 and the grade on your second test is 78. What grade do you need on the third test if you want to raise your average for the three tests to 80?

Solution   Observe the order of steps used in solving the problem. First, the unknown can usually be set equal to what you are asked to find. However, try to pick a representative letter. For example,

$$\text{Let } g = \text{Grade needed on the third test}$$

Then ask yourself, "What should the equation say?" Here is the "verbal equation":

$$\frac{(\text{First grade}) + (\text{Second grade}) + (\text{Third grade})}{3} = \text{Average of three tests}$$

The average is the sum of the three tests divided by 3. This is what you should be thinking about in reference to the problem. You now need an algebraic expression for the average of the three tests in terms of g:

$$\frac{76 + 78 + g}{3} = \text{Average}$$

The algebraic equation is then

$$\frac{76 + 78 + g}{3} = 80$$

The solution, multiplying both sides by 3, is

$$(3)\frac{76 + 78 + g}{3} = 80\,(3)$$

$$76 + 78 + g = 240$$

$$g = 240 - 76 - 78$$

$$g = 86$$

Check the answer in the original equation:

$$\frac{76 + 78 + 86}{3} = \frac{240}{3} = 80 \quad ✓$$

Even more importantly, *determine whether the answer makes sense*. This last step is essential. Very often you can quickly discover an incorrect answer to a verbal problem by seeing that it does not fit the facts. For example, if you obtained an answer of 70, you would know right away that something is wrong. If you want to raise your test average to 80, you need a grade higher than 80 to balance your lower grades.

▪

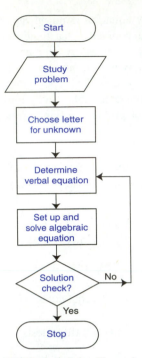

FIGURE 4–14 Flowchart for problem solving.

Example 4.14 illustrates a step-by-step method or algorithm for problem solving:

1. Study the problem several times, and determine what is given and what you are trying to find.
2. Choose a meaningful letter for the unknown, and write down clearly what it represents. For example, t for time, p for price, d for distance, I for current, P for power.
3. Determine what the equation should say in words, that is, the verbal equation, and write down the other quantities in terms of the unknown that are needed for the equation.
4. Formulate and solve the algebraic equation.
5. Check that the answer not only satisfies the equation but also logically satisfies the conditions of the problem.

See Figure 4–14, which illustrates this algorithm as a flowchart.

The next example is another useful problem. It illustrates the steps in the problem-solving algorithm.

EXAMPLE 4.15

Last week you bought a graphing calculator for \$84.50. The cost included 8% sales tax. Today, while traveling in another state, you discover that you can buy the same calculator for \$84.50 but pay only 3% sales tax. How much could you have saved if you bought the calculator at the 3% tax rate?

Solution   You need to find the *price of the calculator before tax,* and then the saving can be determined. Choose a meaningful letter for the unknown:

$$\text{Let } p = \text{Price of the calculator before tax}$$

The verbal equation is

$$\text{Price} + \text{Tax} = \text{Total cost}$$

The tax is 8% times the price. Therefore the Tax $= 0.08p$. The algebraic equation is then

$$p + 0.08p = 84.50$$

The solution is as follows: p is changed to $1.00p$ to clarify the arithmetic:

$$1.00p + 0.08p = 84.50$$
$$1.08p = 84.50$$
$$p = \frac{84.50}{1.08} = 78.2407\ldots = \$78.24 \text{ (nearest cent)}$$

The answer seems logically correct and checks the equation:

$$78.24 + 0.08(78.24) = 78.24 + 6.26 = 84.50 \ \checkmark$$

EXAMPLE 4.15 (Cont.)

The cost of the calculator at the 3% rate would then be

$$78.24 + 0.03(78.24) = \$80.59$$

Therefore, you could have saved $84.50 – $80.59 = $3.91. This saving is clearly not worth traveling to another state. However, if you were buying a more expensive item, or were making the trip for another reason, it might be worth your while.

■

The next example illustrates an application of electrical ideas where you need to set up and solve an algebraic equation.

EXAMPLE 4.16

A copper wire 1 m long is to be cut in such a way that the electrical resistance of one piece is two-thirds the electrical resistance of the other piece. Given that electrical resistance is directly proportional to length, how long should each piece be in centimeters?

Solution   Since electrical resistance is directly proportional to length, the length of the shorter piece should be two-thirds the length of the longer piece. Since the length of the short piece is given *in terms* of the long piece, choose the unknown to be the long piece:

Let l = Length of the long piece

Then,

$$\frac{2}{3}l = \text{Length of the short piece}$$

The verbal equation is

Long piece + Short piece = 1 m = 100 cm

Using centimeters, the algebraic equation is then:

$$l + \frac{2}{3}l = 100$$

The solution is, clearing the fraction first,

$$(3)\,l + (3)\frac{2}{3}l = (3)\,100$$
$$3l + 2l = 300$$
$$5l = 300$$
$$l = 60\text{ cm}$$

The length of the short piece is then

$$\frac{2}{3}l = \frac{2}{3}(60) = 40\text{ cm}$$

The solution check is

$$60 + \frac{2}{3}(60) = 60 + 40 = 100 \ \checkmark$$

As mentioned previously, skill in problem solving comes through constant practice. As in baseball, it is not difficult to understand the rules, but only through constant practice can you really play ball well!

ERROR BOX

A common error to watch out for is incorrectly changing a verbal phrase into algebra by applying the words directly. For example, the phrase "The voltage of one battery V_1 is 2 V more than the voltage of a second battery V_2," may lead to the equation: $V_1 + 2 = V_2$. This is incorrect. The first battery has the *higher* voltage and the correct equation is: $V_1 = V_2 + 2$. Another example is the phrase "One current I_2 is 50% more than I_1," may lead to the equation $I_1 + 50\% = I_2$. This is incorrect; 50% more means $I_1 + 50\%(I_1) = I_2$ or $I_1 + 0.50 I_1 = I_2$. See if you can apply these ideas and do each of the practice problems correctly.

Practice Problems: For each phrase, write an algebraic equation:

1. The first resistance R_1 is 50 Ω less than R_2.
2. The second voltage drop V_2 is 10 V more than V_1.
3. The current in the third resistor I_3 is twice I_2.
4. The power P_1 dissipated in R_1 is 25% less than P_2.
5. Capacitance C_1 is 10% more than twice capacitance C_2.
6. If resistor R_2 is twice R_1 and R_3 is one-third R_2, how is R_3 related to R_1?

Answers:

1. $R_1 + 50 = R_2$ 2. $V_2 = 10 - V_1$ 3. $I_3 = 2(I_2)$

4. $P_1 = P_2 - 0.25(P_2)$ 5. $C_1 = 2C_2 + 0.10(2C_2)$ 6. $R_3 = \dfrac{2R_1}{3}$

The next example closes the circuit and shows an application to a series circuit where you need to set up and solve an equation.

EXAMPLE 4.17

Close the Circuit

Figure 4–15 shows three resistors connected in series where the total voltage $V_T = 70$ V. The resistance of R_2 is 50% of R_1 and the resistance of R_3 is 25% of R_1. Ohm's law says that in a series circuit the voltage drop is directly proportional to the resistance. Kirchhoff's voltage law says that the total voltage equals the sum of the voltage drops. Applying these laws, what is the voltage drop across each resistance?

Solution   First apply Ohm's law. If the resistance of R_2 is 50% of R_1, then the voltage drop across R_2 will be 50% of the voltage drop across R_1. Similarly, the voltage drop across R_3 will be 25% of the voltage drop across R_1. Since the voltage drops across R_2 and R_3 are given in terms of R_1, start with

$$\text{Let } V_1 = \text{Voltage drop across } R_1$$

Then,

$$V_2 = 50\% \, V_1 = 0.50 V_1$$
$$V_3 = 25\% \, V_1 = 0.25 \, V_1$$

EXAMPLE 4.17 (Cont.)

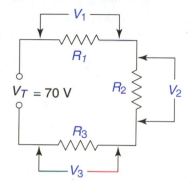

$$V_1 + V_2 + V_3 = V_T$$

FIGURE 4–15   Resistances for series for Example 4.17

Applying Kirchhoff's law, the sum of the voltage drops in a series circuit equals the total voltage, the verbal equation is

Voltage across R_1 + Voltage across R_2 + Voltage across R_3 = V_T

The algebraic equation is then

$$V_1 + V_2 + V_3 = V_T$$

Substituting $0.50V_1$ for V_2, $0.25V_1$ for V_3 and 70 V for V_T the equation becomes

$$V_1 + 0.50V_1 + 0.25V_1 = 70$$

The solution is

$$1.75 V_1 = 70$$
$$V_1 = 40 \text{ V}$$
$$V_2 = 0.50 V_1 = 20 \text{ V}$$
$$V_3 = 0.25 V_1 = 10 \text{ V}$$

The last example shows how to solve a "motion" problem.

EXAMPLE 4.18

A sailboat sails 5 nmi/h (nautical miles per hour, or knots) without any current (1 nmi = 1.15 statute or land miles). If the boat sails upstream in 3 h and takes 2 h to return downstream the same distance, what is the speed of the current in nautical miles per hour?

Solution   The relationship between the distance traveled D, the average rate of speed R, and the time T, is given by the distance formula:

$$D = RT$$

Let c = speed of the current in knots. What can be used to formulate the equation? Since the *distance upstream is equal to the distance downstream*, the verbal equation is

Distance upstream = Distance downstream

Then, since the speed of the boat is *decreased* by the current going *upstream,*

$$5 - c = \text{Rate upstream}$$

Similarly, since the speed of the boat is *increased* by the current going *downstream,*

$$5 + c = \text{Rate downstream}$$

Applying the distance formula $D = RT$, the distances are

Distance upstream = $(5 - c)(3)$
Distance downstream = $(5 + c)(2)$

The algebraic equation and solution are then

$$(5 - c)(3) = (5 + c)(2)$$
$$15 - 3c = 10 + 2c$$
$$5 = 5c$$
$$c = 1 \text{ knot}$$

The connection between baseball and mathematics can be carried further. One cannot develop much skill in playing baseball by watching a lot of games. Likewise, studying a lot of text examples will increase problem-solving skills only a little. You must participate and *solve as many problems yourself as possible*. Mathematics is not a spectator sport!

EXERCISE 4.3

In problems 1 through 16,

   (a) Set up and solve an algebraic equation to find the answer to the problem.
   (b) Check that the answer makes logical sense and that it satisfies the equation.

1. Lucia and Romero together earn $66,000 a year. If Lucia earns $11,000 more than Romero, how much does each earn?

2. The cost of renting a computer and a laser printer is $600 per month. If the cost for the computer is three times that for the printer, what is the monthly cost for each?

3. Your boss is raising your salary to $38,500 and indicates this is a 10% raise. Based on that figure, what is your present salary?

4. Maria paid $32.29 for a calculator that includes 8% sales tax. What is the price of the calculator before tax?

5. An electronics store gives one-third off list prices. If Rodd paid $28 for a voltmeter before tax, what is the list price?

6. An insurance company gives you $4300 to replace your stolen car and informs you that the "book" value is only 40% of the original value. What is the original value according to the insurance company?

7. To be eligible for a government job, you must average at least 80% on three tests. If your first two scores are 73% and 78%, what do you need on the third test to be eligible for a government job?

8. The average of your class tests is 77. The instructor indicates that your final average equals two-thirds of your class average plus one-third of your final exam grade. What do you need on the final exam to raise your average to 80?

9. A laptop computer costs $3000, which includes the price of a software package. This cost is a 15% price increase over the cost of the computer without the software package. However, next week there is a sale, and there will only be a 5% price increase for the computer with the software package. How much will you save if you buy the computer next week? (See Example 4.15.)

10. Mary Ellen paid $864 for a stereo, which includes 8% sales tax. How much could she have saved if she bought the stereo in another state at a sales tax rate of 5%? (See Example 4.15.)

11. A new high-speed laser printer operates 3.5 times faster than an older model. Together, they print 2475 lines per minute. How many lines per minute does the new model print?

12. In a phone order business, one computer system completes an order in two-thirds the time of another system. If the total time for both systems to complete an order is exactly 4 min, how long does it take each system to complete its order?

13. A sailboat sails 6 nmi/h (knots) with no current. If the sailboat takes 4 h to sail upsteam and 2 h to sail downstream, how fast is the current in nmi/h? (See Example 4.18.)

14. A train leaves Los Angeles averaging 30 mi/h. A car leaves 1 h later, averaging 50 mi/h. After how many hours will the car catch up to the train? (Hint: Distances are equal.)

15. A private jet flying between New York and Chicago, a distance of 700 miles, encounters favorable winds *both* ways. The trip west takes 2 h with a tail wind of 20 mi/h. The trip east takes 1 h 45 min flying with the jet stream of 80 mi/h.

    (a) Was the plane flying faster, relative to the air, going west or east? Find the airspeed each way.

    (b) How much greater was the faster airspeed?

16. A boat explosion is heard 8 s sooner through the water than through the air. If the speed of sound is 1.5 km/s through the water and 0.30 km/s through the air,

    (a) How long did the sound take to travel through the air?

    (b) How far away was the explosion?

Applications to Electronics In applied problems 17 through 28, (a) set up an algebraic equation to solve the problem, (b) check that the answer makes logical sense and that it satisfies the equation.

17. Figure 4–16 shows a circuit containing two inductances L_1 and L_2 in series. The total inductance $L_T = L_1 + L_2$. If $L_T = 11$ mH and L_2 is 10% greater than L_1, find each inductance.

18. In problem 17, find L_1 and L_2 if $L_T = 360$ μH and L_1 is 20% less than L_2.

19. Figure 4–17 shows a circuit containing three parallel resistors. The total current I_T in the circuit equals the sum of the currents through the resistors: $I_T = I_1 + I_2 + I_3$. The current in the second resistor is 1.5 times the current in the first resistor. The current in the third resistor is 3 times the current in the first resistor. If the total current is 275 mA, how much current flows through each resistor?

20. In problem 19, given that I_2 equals one-third of I_T and I_3 equals one-half of I_2, if $I_1 = 36$ mA, find I_T, I_2, and I_3.

21. Figure 4–18 shows two of the same type batteries connected in series aiding so that the total voltage across both batteries is equal to the sum of their individual voltages. However, it is found that the second battery does not contain as much charge as the first and its voltage is 5% less than that of the first battery. If the total voltage is 11.7 V, what is the voltage of each battery?

22. The total output of three "identical" steam turbine generators in a power plant is 20,000 MW. An electrical technician finds that two of the generators have the same power output but that the third generator puts out 20% more than one of the other two. What is the power output of each generator?

23. An aluminum wire 3.5 m long is to be cut so that the resistance of one piece is 75% of the resistance of the other piece. How many centimeters should each piece be? (See Example 4.16.)

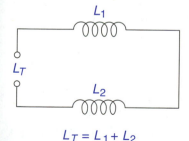

$$L_T = L_1 + L_2$$

FIGURE 4–16 Inductances in series for problems 17 and 18.

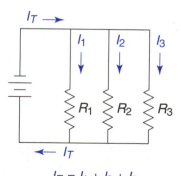

$$I_T = I_1 + I_2 + I_3$$

FIGURE 4–17 Three parallel resistors for problems 19 and 20.

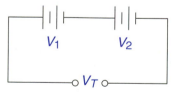

$$V_T = V_1 + V_2$$

FIGURE 4–18 Series-aiding voltages for problem 21.

24. A copper wire 7 ft long is to be cut into three pieces. The second piece is to have twice the resistance of the first piece, and the third piece is to have twice the resistance of the second piece. How many inches should each piece be? (See Example 4.16.)

25. In Example 4.17, find the voltage drop across each resistance if R_2 is twice the resistance of R_1 and R_3 is one-third the resistance of R_1.

26. In Example 4.17, find the voltage drop across each resistor to the nearest volt if R_1 is half of R_2 and R_3 is 50% greater than R_2.

27. Figure 4–19 shows a variable resistor in series with a bulb. The voltage drop across the bulb is 1.5 times the voltage drop across the resistor. If the applied voltage is 12 V, what are the voltage drops across the bulb and the resistor? Apply Kirchhoff's voltage law, which says that applied voltage equals the sum of the voltage drops.

28. In the circuit in Figure 4–19, the power dissipated in the variable resistor is 40% less than the power dissipated in the bulb. If the total power dissipated is 9.6 W, how much power is dissipated in the bulb?

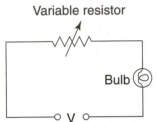

FIGURE 4–19 Variable resistor with bulb for problems 27 and 28.

CHAPTER HIGHLIGHTS

4.1 FIRST-DEGREE EQUATIONS

To solve an equation, isolate the unknown on one side of the equation by applying:

Equation Principle	Adding, subtracting, multiplying, or dividing each term on both sides of an equation by the same quantity does not change the equation.

Study the algorithm after Example 4.4 for solving a linear equation and how it is applied in Examples 4.3 through 4.7.

4.2 FORMULAS AND LITERAL EQUATIONS

Formulas and literal equations can be solved for one of the variables by applying the equation principle in Section 4.1. To solve for a variable appearing in more than one term, move the terms containing the variable to one side of the equation and then separate the variable as a common factor before dividing by the other factor to isolate it.

Study the beginning of Section 4.2 and Example 4.8 to see how to derive different forms of Ohm's law and the power law. Also study Examples 4.10 and 4.12.

4.3 PROBLEM SOLVING

Problem solving is an important and essential skill in electricity and electronics. It can best be learned through *constant practice*. Study the algorithm after Example 4.14 and the flowchart in Figure 4–14 for the steps to follow when solving a verbal problem. Then study how it is applied in Examples 4.15, 4.16, and 4.17. Study the Error Box after Example 4.16 and solve as many problems as possible in Exercise 4.3.

≡ REVIEW QUESTIONS

In problems 1 through 10, solve each equation.

1. $2x - 2 = 10$

2. $4(w + 2) = 5 - w$

3. $\frac{5}{2}R_X = 2 - \frac{3}{2}R_X$

4. $3.5 - 2.5m = 0.50 - m$

5. $\frac{4.5}{I_s} = \frac{300}{40}$

6. $\frac{C_T}{0.4} + 1.1\,C_T = 18$

7. $\frac{x}{8} + \frac{3x}{4} = \frac{7}{2}$

8. $\frac{2h + 1}{4} - \frac{h - 1}{4} = 1$

9. $\frac{G_2 + 1}{G_2} = \frac{2}{5G_2} + \frac{1}{2G_2}$

10. $\frac{3}{2V_A} - \frac{2}{V_A + 1} = \frac{1}{6V_A}$

In problems 11 through 16, solve each problem.

11. The distance s an object falls in time t due to the force of gravity is given by

$$s = \frac{1}{2}g\,t^2 + v_0 t + s_0$$

where g = gravitational acceleration = 9.8 m/s², v_0 = initial velocity, and s_o = initial distance. If $s = 40$ m when $t = 1.5$ s, and $s_0 = 0$, find v_0 to two significant digits.

12. Figure 4–20 shows a rope passing over a pulley with a force F on one side and a load L on the other side. If the force F is not equal to the load L, then the rope moves with a constant acceleration given by

$$a = \frac{(F - L)\,g}{F + L}$$

Solve this literal equation for F.

13. An old model cellular phone is selling for $325, which is 35% less than the original price. What was the original price?

14. A high-volume discount store makes only 8% profit on sales. A shoplifter walks out with a $30 calculator. How much does the store have to sell to make up the loss?

15. The average of three voltmeter readings is 39 V. If two of the readings are 40 V and 35 V, what is the third reading?

16. Two trains 180 mi apart leave at the same time and travel toward each other. If the average rate of one train is 40 mi/h and that of the other is 50 mi/h, how long will it take for the trains to meet?

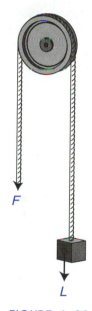

F

L

FIGURE 4–20
Rope and
pulley for
problem 12.

Applications to Electronics   In problems 17 through 20, solve each formula for the indicated letter.

17. $E = \frac{k\,Q}{r_2}$; Q (E = electric intensity)

18. $B = \frac{\mu I}{2\pi r}$; r (B = magnetic flux density)

19. $P = V_1 I + V_2 I$; I (P = power in a series circuit)

20. $V_{TH} = \left[\dfrac{R_L}{R_S + R_L}\right] V_S;\ R_L$ (V_{TH} = Thevenin voltage)

In problems 21 through 28, solve each applied problem.

21. The formula for the inductance L of a coil of n turns is given by

$$L = \frac{n\Phi}{I}$$

where Φ = magnetic flux. If $L = 0.05$ H, $n = 500$, and $\Phi = 3 \times 10^{-4}$ Wb (weber), find the value of the current I.

22. Figure 4–21 shows a series circuit containing two resistances and two opposing voltages. By Kirchhoff's voltage law, battery voltage V_2 is related to the other voltages in the circuit by the equation

$$V_2 = IR_1 + IR_2 + V_1$$

Find R_1 when $V_2 = 24$ V, $I = 0.80$ A, $R_2 = 12\ \Omega$, and $V_1 = 8.4$ V.

23. Figure 4–22 shows a circuit containing two capacitances in series connected to a third capacitance in parallel. The formula for the total capacitance of the circuit is

$$C_T = C_1 + \frac{C_2 C_3}{C_2 + C_3}$$

(a) Given $C_2 = C_3$, solve this formula for C_2 in terms of C_T and C_1.
(b) Find C_2 when $C_1 = 50$ μF and $C_T = 80$ μF.

24. For the series-parallel circuit in Figure 4–22, show that when $C_2 = C_1$ and $C_T = 1.5C$ then $C_3 = C_1$.

25. Figure 4–23 shows a variable resistor R_V in parallel with a fixed resistor R_0. The variable resistor is adjusted so that the current through the variable resistor remains constant at $I_V = 0.30$ A when the voltage V changes. The total current in the circuit is then given by

$$I_T = 0.30 + \frac{V}{R_0}$$

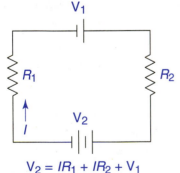

$$V_2 = IR_1 + IR_2 + V_1$$

FIGURE 4–21 Series-opposing voltages for problem 22.

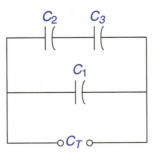

$$C_T = C_1 + \frac{C_2 C_3}{C_2 + C_3}$$

FIGURE 4–22 Capacitances in series and parallel for problems 23 and 24.

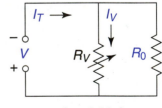

$I_V = 0.30$ A

FIGURE 4–23 Constant current in parallel circuit for problems 25 and 26.

(a) If I_T changes from I_1 to I_2 when V changes from V_1 to V_2, find the formula for $\Delta I_T = I_2 - I_1$ in terms of ΔV.

(b) If $R_0 = 100 \ \Omega$, find ΔI_T when V changes from 11 V to 14 V.

26. For the circuit in Figure 4–23, find the formula for ΔI_T if the voltage increases by one-third of its value from V_1 to $V_2 = V_1 + \frac{1}{3}V_1$.

27. A new electric heater consumes only one-fourth of the power of an old model. If the total power consumption of both heaters is 3000 W, what is the power consumption of each?

28. A copper wire 10 ft long is to be cut into three pieces. The resistance of the shortest piece is to be one-half the resistance of the longest piece. The resistance of the middle piece is to be 20% more than the resistance of the shortest piece. How long should each piece be to the nearest tenth of a foot?

5

DC Circuits

Courtesy of Hewlett Packard Company.

A technician checks a circuit board at a calculator and test equipment repair center in Mississauga, Ontario.

This chapter, like the examples that close the circuit, applies many of the important ideas from Chapters 1 through 4 to direct current or dc circuits. The chapter begins with Ohm's law, one of the most fundamental laws in electricity. The power formula is then presented with its several forms. Series circuits and parallel circuits are studied applying Ohm's law and the power formula. The last section analyzes series-parallel circuits, which bring together many of the important ideas of dc circuits.

Close the Circuit

Chapter Objectives

In this chapter, you will learn:

- Ohm's law and how to apply it.
- The meaning of direct and inverse proportion.
- The power formula and how to derive its several forms.
- Series circuits: how to find their total resistance and how to solve problems involving series circuits.
- Parallel circuits: how to find their total resistance and how to solve problems involving parallel circuits.
- The meaning of conductance and how to apply it in a parallel circuit.
- Series-parallel circuits: how to analyze them, find their total resistance, and solve problems involving series-parallel circuits.

≡ 5.1
OHM'S LAW

In 1828 George Simon Ohm, a German physicist, discovered one of the fundamental laws of electricity. It explains the relationship between the voltage, current, and resistance in an electrical circuit. Consider the basic circuit in Figure 5–1 consisting of a dc voltage source V, such as a battery, connected to a resistance R, such as a light bulb or a heating element. *Ohm's law* states

$$I = \frac{V}{R} \qquad (5.1)$$

where I = current in amps
V = voltage in volts
R = resistance in ohms

That is, the current equals the voltage divided by the resistance.

Note in Figure 5–1 that *electron current* going from (−) to (+) is used. Electron flow is used throughout the text instead of conventional current which goes from (+) to (−). Ohm's law can be expressed another way in terms of the voltage by multiplying both sides of (5.1) by R:

$$I(R) = \frac{V}{R}(R)$$
$$V = IR \qquad (5.1a)$$

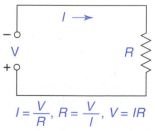

$$I = \frac{V}{R}, \ R = \frac{V}{I}, \ V = IR$$

FIGURE 5–1   Ohm's law and direct proportion for Example 5.1.

Ohm's law can also be expressed a third way in terms of the resistance by dividing both sides of (5.1a) by I:

$$R = \frac{V}{I}$$

(5.1b)

Study the following examples, which illustrate Ohm's law.

EXAMPLE 5.1

For the circuit in Figure 5–1, given that the resistance is constant at $R = 100 \ \Omega$ and the initial voltage $V = 6.0$ V, find the current I:

1. At the initial voltage.

2. When the voltage doubles to 12 V.

Solution

1. For the initial voltage of 6.0 V, apply Ohm's law (5.1), letting $V = 6.0$ V and $R = 100 \ \Omega$:

$$I = \frac{V}{R} = \frac{6.0 \text{ V}}{100 \ \Omega} = 0.060 \text{ A} = 60 \text{ mA}$$

2. When the voltage doubles to 12V, apply Ohm's law again to obtain

$$I = \frac{V}{R} = \frac{12 \text{ V}}{100 \ \Omega} = 0.12 \text{ A} = 120 \text{ mA}$$

Observe in Example 5.1 that when the voltage doubles from 6.0. V to 12 V the current also doubles from 60 mA to 120 mA. This is because when the *resistance is constant,* Ohm's law tells us that the *current is directly proportional to the voltage.* This means that the current and the voltage change by the same ratio or percentage. For example, if the voltage increases by 10%, then the current will increase by 10%. If the voltage decreases to half its value then the current will decrease to half its value.

Direct Proportion	When two quantities are directly proportional and one changes by a certain ratio, then the other changes by the same ratio.

Ohm's law also tells us that when the *current is constant,* the *voltage drop* across the resistance *is directly proportional to the resistance.* The voltage drop across a resistance is the voltage, or *potential difference,* that exists between the ends of the resistance. Suppose in a circuit that the resistance increases by 10% from 100 Ω to 110 Ω and the voltage changes so that the current remains constant at 0.10 A. By Ohm's law, the voltage drop across the resistance will increase from

$$V = IR = (0.10\text{A})(100\Omega) = 10 \text{ V}$$

to

$$V = (0.10\text{A})(110\Omega) = 11 \text{ V}$$

Applying (1.7) from Section 1.3, the percent increase in the voltage is

$$\text{Percent change} = \frac{(11-10)\,(100\%)}{10} = \frac{(1)\,(100\%)}{10} = 10\%$$

The voltage increases in the same percentage as the resistance, 10%. Therefore, the voltage is directly proportional to the resistance. Similarly, if the resistance decreases to half its value, from 100 Ω to 50 Ω, and the current does not change, the voltage will decrease to half its value, from 10 V to 5 V.

Study the next example, which illustrates further properties of Ohm's law.

EXAMPLE 5.2

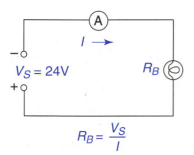

$V_S = 24V$

R_B

$R_B = \dfrac{V_S}{I}$

FIGURE 5–2 Ohm's law and inverse proportion for Examples 5.2 and 5.3.

Figure 5–2 shows a circuit containing a light bulb resistance R_B, an ammeter, and a constant dc voltage source $V_S = 24$ V. Find the current I through the light bulb:

1. When the resistance is 120 Ω.

2. When the resistance doubles to 240 Ω.

Solution

1. Apply Ohm's law (5.1), letting $V = V_S = 24$ V and $R = R_B = 120$ Ω:

$$I = \frac{V_S}{R_B} = \frac{24\text{ V}}{120\,\Omega} = 0.20\text{ A} = 200\text{ mA}$$

2. When the resistance doubles to 240 Ω, the current will become

$$I = \frac{V_S}{R_B} = \frac{24\text{ V}}{240\,\Omega} = 0.10\text{ A} = 100\text{ mA}$$

Observe in Example 5.2 that when the resistance doubles from 120 Ω to 240 Ω, the current decreases by one half, from 200 mA to 100 mA. This is because when the voltage is constant, Ohm's law tells us that the current is *inversely proportional* to the resistance. This means that the current and resistance change in the inverse or reciprocal ratio. The resistance increases in the ratio of $\frac{2}{1}$, and the current decreases by the reciprocal ratio $\frac{1}{2}$.

> **Inverse Proportion** When two quantities are inversely proportional and one changes by a certain ratio, then the other changes by the reciprocal ratio.

Study the following example, which is another illustration of inverse proportion using percentage change.

EXAMPLE 5.3

Suppose in Figure 5–2 that V_S is constant at 24 V while $I = 250$ mA and decreases by 20%. Show that the current changes in inverse proportion to the resistance.

Solution If the current decreases by 20%, this means that it changes to 80% of its value, or from 250 mA to 80% (250 mA) = 200 mA. Apply Ohm's law to find that the resistance changes from

EXAMPLE 5.3 (Cont.)

$$R = \frac{V_S}{I} = \frac{24\text{ V}}{0.25\text{ A}} = 96\,\Omega$$

to

$$R = \frac{24\text{ V}}{0.20\text{ A}} = 120\,\Omega$$

Note that for V in volts I must be in amps when you use Ohm's law.

To show that the current changes in inverse proportion to the resistance, compare the ratio that the current changes to the ratio that the resistance changes. The current changes in the ratio

$$\frac{200\text{ mA}}{250\text{ mA}} = \frac{4}{5}$$

while the resistance changes in the ratio

$$\frac{120\,\Omega}{96\,\Omega} = \frac{5}{4}$$

The current ratio is the reciprocal of the resistance ratio. Therefore, the current is inversely proportional to the resistance.

Ohm's law, therefore, contains three proportional relationships:

1. The current is directly proportional to the voltage when the resistance is constant.

2. The voltage is directly proportional to the resistance when the current is constant.

3. The current is inversely proportional to the resistance when the voltage is constant.

EXERCISE 5.1

In problems 1 through 10, apply Ohm's law to find the missing quantity. Give answers to two significant digits.

1. $V_1 = 100$ V, $R_1 = 300\ \Omega$

2. $V_2 = 9.0$ V, $R_2 = 18\ \Omega$

3. $V_2 = 36$ V, $I_2 = 48$ mA

4. $V_T = 1.1$ kV, $I_T = 500$ mA

5. $I_T = 200$ mA, $R_T = 75\ \Omega$

6. $I_2 = 250$ mA, $R_2 = 180\ \Omega$

7. $V_T = 13$ V, $R_T = 36\ \Omega$

8. $V_3 = 2.4$ V, $R_3 = 15\ \Omega$

9. $I_3 = 250$ mA, $R_3 = 68\ \Omega$

10. $I_T = 80$ mA, $R_T = 330\ \Omega$

In problems 11 through 22, solve each applied problem to two significant digits.

11. Figure 5–3 shows a dc circuit with a battery voltage source and a constant resistance $R = 56\ \Omega$. Find the current I when

(a) $V_B = 28$ V.

(b) V_B decreases to one-half its value.

12. Given the dc circuit in Figure 5–3 with $R = 200\ \Omega$, find the current I when

(a) $V_B = 6.0$ V.

(b) V_B increases by 50%.

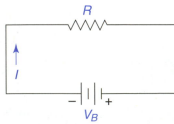

FIGURE 5–3 A dc circuit for problems 11 and 12.

13. Figure 5–4 shows a circuit with a constant dc voltage source V_S and a resistance R. If $V_S = 12$ V, find the current I when
    (a) $R = 30$ Ω.
    (b) R increases to two times its value.
14. Given the dc circuit in Figure 5–4 with $V_S = 12$ V, find the current I when
    (a) $R = 200$ Ω.
    (b) R decreases by 50%.
15. In a dc circuit with a constant resistance, $I = 25$ mA. Apply the concept of direct proportion, and find I when the voltage doubles.
16. In a dc circuit with a constant resistance, $I = 120$ mA. Apply the concept of direct proportion, and find I when the voltage decreases by 25%.
17. In a dc circuit with a constant voltage, $I = 50$ mA. Apply the concept of inverse proportion, and find I when R decreases by 50%.
18. In a dc circuit with a constant voltage, $I = 1.5$ A. Apply the concept of inverse proportion, and find I when R triples in value.
19. The resistance of a wire is directly proportional to its length. If $R = 30$ mΩ for a wire 10 cm long, find the resistance when the length is 15 cm.
20. In problem 19, find the resistance of the wire if the length increases by 40%.
21. If the voltage drop across a resistance is constant and the current decreases to half its value, by what ratio has the resistance changed?
22. If the voltage drop across a resistance is constant and the resistance increases by 50%, by what ratio will the current change?

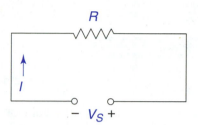

FIGURE 5–4   A dc circuit for problems 13 and 14.

≡ 5.2
ELECTRIC POWER

Electric power is a measure of the electrical energy being used every second. One watt of electric power is equal to the energy produced every second by a voltage of one volt and a current of one ampere. This gives rise to the following power formula:

$$P = VI \tag{5.2}$$

where P = power in watts
       V = voltage in volts
       I = current in amps

For example, if a voltage of 20 V produces 1.5 A in a circuit, then the power is

$$P = (20 \text{ V})(1.5 \text{ A}) = 30 \text{ W}$$

Formula (5.2) can also be expressed in terms of I or V:

$$I = \frac{P}{V}$$

$$V = \frac{P}{I} \tag{5.2a}$$

The following example shows how to apply the power formula.

EXAMPLE 5.4

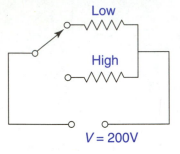

FIGURE 5–5 Current in an electric heater for Example 5.4.

Figure 5–5 shows an electric heater connected to a 200 V power line. The heater is rated at 500 W on the low setting and 1000 W on the high setting. How much current does it use on each setting?

Solution Apply formula (5.2a) for I:

$$\text{Low: } I = \frac{P}{V} = \frac{500 \text{ W}}{200 \text{ V}} = 2.5 \text{ A}$$

$$\text{High: } I = \frac{1000 \text{ W}}{200 \text{ V}} = 5.0 \text{ A}$$

Observe that the current doubles from 2.5 A to 5.0 A when the power doubles from 500 W to 1000 W. The current is directly proportional to the power when the voltage is constant; that is, power and current change in the same ratio when the voltage is constant.

When Ohm's law is combined with the power formula, two more formulas for power arise in terms of resistance:

$$P = I^2R$$

$$P = \frac{V^2}{R} \tag{5.3}$$

The first formula comes from Ohm's law $V = IR$ by substituting IR for V in the power formula:

$$P = VI = (IR)I = I^2R$$

The derivation of the second formula is left as problem 41 in Exercise 5.2.

Each of the formulas in (5.3) can be expressed two other ways in terms of R and I, or R and V:

$$R = \frac{P}{I^2} \text{ and } R = \frac{V^2}{P} \tag{5.3a}$$

$$I = \sqrt{\frac{P}{R}} \text{ and } V = \sqrt{PR} \tag{5.3b}$$

It is not necessary to memorize all of these formulas. If you apply some basic algebra, you can derive (5.3a) and (5.3b) from (5.3).

For example, if you divide both sides of $P = I^2R$ by I^2,

$$\frac{P}{I^2} = \frac{I^2R}{I^2}$$

you obtain the first of formulas (5.3a):

$$R = \frac{P}{I^2}$$

Then if you solve for I^2, you have

$$I^2 = \frac{P}{R}$$

Now you can eliminate the square by taking the square root of both sides of the equation. In the same way that you can multiply or divide both sides of an equation by the same quantity and not change the equation, you can also take the square root of both sides of an equation. When you take the square root, you usually take both the positive and negative square roots, as there can be two solutions. In this case, however, if you assume the current is positive, it is only necessary to take the positive square root:

$$\sqrt{I^2} = \sqrt{\frac{P}{R}}$$

The square root of a squared quantity is just the quantity itself. That is, the square root cancels the square and you have the first of formulas (5.3b):

$$I = \sqrt{\frac{P}{R}}$$

The derivation of the second formulas of (5.3a) and (5.3b) is left as problem 42 in Exercise 5.2. See also the Error Box in this section.

Study the next two examples, which show how to apply these formulas.

EXAMPLE 5.5

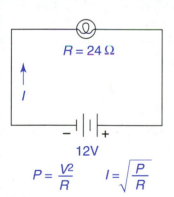

$R = 24\,\Omega$

I

12V

$P = \dfrac{V^2}{R}$ $I = \sqrt{\dfrac{P}{R}}$

FIGURE 5–6 Power and current in a circuit for Example 5.5.

Figure 5–6 shows a dc circuit containing a 12-V battery and a bulb whose resistance is 24 Ω. What is the power and the current in the circuit?

Solution   To find the power in the circuit, apply formula (5.3) with $V = 12$ V and $R = 24$ Ω:

$$P = \frac{V^2}{R} = \frac{(12)^2}{24} = \frac{144}{24} = 6.0\,\text{W}$$

You can find the current using formula (5.3b) for I or by using Ohm's law. Both formulas yield the same answer:

$$I = \sqrt{\frac{P}{R}} = \sqrt{\frac{6.0}{24}} = \sqrt{0.25} = 0.50\,\text{A} = 500\,\text{mA}$$

$$I = \frac{V}{R} = \frac{12}{24} = 0.50\,\text{A} = 500\,\text{mA}$$

Power, Energy, and Work

Power was explained earlier as the amount of energy consumed (or work done) per second. More specifically,

$$\text{Power} = \frac{\text{Energy}}{\text{Time}}$$

All energy—mechanical, electrical, or heat—is measured in joules, and 1 W = 1 J/s. Similarly,

$$\text{Energy} = (\text{Power})(\text{Time})$$

and

$$1J = (1 \ W)(1 \ s) = 1 \ W \cdot s \ (\text{watt-second})$$

A familiar unit of electrical energy, the *kilowatt-hour* (kWh), is used to measure electrical consumption in the home. One kilowatt-hour is equal to 1000 W of power consumed for one hour:

$$1 \ kWh = (1000 \ W)(1 \ h)$$

or an equivalent product, such as

$$1 \ kWh = (200 \ W)(5 \ h) = (100 \ W)(10 \ h)$$

The relationship between joules and kilowatt-hours is

$$1 \ J = 2.778 \times 10^{-7} \ kWh$$
$$1 \ kWh = 3.600 \times 10^{6} \ J$$

ERROR BOX

You do not need to memorize all nine ways of writing the power formula. You can solve any problem if you know Ohm's law, the basic power formula $P = VI$, and either one of the following formulas in (5.3b):

$$I = \sqrt{\frac{P}{R}} \ \text{ or } \ V = \sqrt{PR}$$

You may have to do one or two additional calculations, but there is less to remember and less chance of making an error. Consider Example 5.5. Given $V = 12$ V and $R = 24 \ \Omega$, first find I using Ohm's law:

$$I = \frac{V}{R} = \frac{12}{24} = 0.50 \ A = 500 \ mA$$

Then find P:

$$P = VI = (12)(0.50) = 6.0 \ W$$

Consider Example 5.6. Suppose you only know the formula for $I = \sqrt{\frac{P}{R}}$, and are given $P = 1.5$ kW and $R = 60 \ \Omega$.

First find I:

$$I = \sqrt{\frac{P}{R}} = \sqrt{\frac{1500}{60}} = \sqrt{25} = 5.0 \ A$$

Then find V:

$$V = \frac{P}{I} = \frac{1500}{5.0} = 300 \ V$$

Practice Problems: Using Ohm's law, the power formula $P = VI$, and only one of the formulas in (5.3b), find the indicated quantity for each:
1. $V = 9V$, $R = 27 \ \Omega$; P 2. $I = 10$ mA, $R = 15 \ \Omega$; P 3. $P = 100$ W, $R = 25 \ \Omega$; I
4. $P = 200$ W, $R = 50 \ \Omega$; V

Answers: 1. $P = 3$ W   2. $P = 1.5$ mW   3. $I = 2.0$ A   4. $V = 100$ V

EXAMPLE 5.6

Figure 5–7 shows a resistance of 60 Ω in a dc circuit. If the resistance consumes 1.5 kW of power, find

1. The voltage in the circuit.

2. The energy consumption for 8.0 h in kilowatt-hours and joules.

Solution

1. Careful attention must be paid to the units involved. The formulas require that the units be consistent. Therefore, to find the voltage V using formula (5.3b), set $P = 1.5$ kW $= 1.5 \times 10^3$ W or 1500 W, and $R = 60$ Ω. Then,

$$V = \sqrt{PR} = \sqrt{(1500)(60)} = \sqrt{90000} = 300 \text{ V}$$

2. The energy consumption for 8.0 h is

$$\text{Energy} = (1.5 \text{ kW})(8.0 \text{ h}) = 12 \text{ kWh}$$

The energy consumption in joules to two significant digits is

$$12 \text{ kWh} (3.600 \times 10^6 \text{ J/kWh}) = 43 \times 10^6 \text{ J} = 43 \text{ MJ}$$

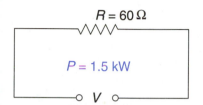

$R = 60\,\Omega$

$P = 1.5$ kW

V

Energy = (Power) x (Time)

FIGURE 5–7 Energy in a circuit for Example 5.6.

EXERCISE 5.2

In problems 1 through 26, use the power formulas to find the indicated quantity to two significant digits.

1. $V_1 = 14$ V, $I_1 = 500$ mA; P_1
2. $V_2 = 120$ V, $I_2 = 1.5$ A; P_2
3. $P_2 = 12$ W, $V_2 = 8.0$ V; I_2
4. $P_1 = 6.5$ W, $V_1 = 10$ V; I_1
5. $P_T = 20$ W, $I_T = 1.5$ A; V_T
6. $P_T = 15$ W, $I_T = 200$ mA; V_T
7. $V_D = 12$ V, $R_D = 10$ Ω; P_D
8. $V_D = 20$ V, $R_D = 100$ Ω; P_D
9. $I_1 = 200$ mA, $R_1 = 160$ Ω; P_1
10. $I_2 = 400$ mA, $R_2 = 68$ Ω; P_2
11. $V_3 = 32$ V, $P_3 = 100$ W; R_3
12. $V_3 = 9.0$ V, $P_3 = 10$ W; R_3
13. $I_2 = 800$ mA, $P_2 = 14$ W; R_2

14. $I_1 = 1.6$ A, $P_1 = 1000$ W; R_1
15. $P_1 = 1.2$ kW, $R_1 = 300$ Ω; I_1
16. $P_2 = 6.0$ W, $R_2 = 20$ Ω; I_2
17. $P_2 = 25$ W, $R_2 = 56$ Ω; V_2
18. $P_1 = 10$ W, $R_1 = 33$ Ω; V_1
19. $V_T = 14$ V, $I_T = 120$ μA; P_T
20. $P_T = 1.5$ W, $I_T = 75$ mA; V_T
21. $I_X = 45$ mA, $R_X = 1.5$ kΩ; P_X
22. $V_y = 720$ mV, $R_y = 27$ Ω; P_y
23. $V_a = 1.5$ V, $P_a = 110$ mW; R_a
24. $I_a = 25$ mA, $P_a = 15$ mW; R_a
25. $P_b = 55$ mW, $R_b = 150$ Ω; I_b
26. $P_b = 250$ mW, $R_b = 3.3$ kΩ; V_b

In problems 27 through 42, solve each problem, paying careful attention to the units involved. Give answers to two significant digits.

27. An electric stove connected to a 240 V power line is rated at 400 W on the lowest setting and 1200 W on the highest setting. How much current does the stove use on each setting?

28. A radar unit on a boat draws 4.2 A when operating and 1.2 A when standing by. If the voltage is 12 V, what is the power consumption when operating and when standing by?

29. An automobile lightbulb uses 420 mA in a constant voltage dc circuit. If the wattage of the bulb is doubled, how much current will it use?

30. A 60 W bulb is replaced by a smaller bulb in a constant voltage dc circuit. If the current decreases by one-third, what is the wattage of the smaller bulb?

31. The voltage in a high-power circuit is measured at 3.5 kV and the current at 4.3 A.
    (a) What is the power in the circuit in watts and kilowatts?
    (b) What is the resistance of the circuit?

32. In a video circuit, a resistor draws 100 mA and dissipates 20 W.
    (a) What is the voltage across the resistor?
    (b) What is the value of the resistance?

33. In an amplifier circuit, a 7.5-Ω resistor draws 150 mA of current.
    (a) How much power does the resistor dissipate?
    (b) What is the voltage drop across the resistor?

34. The voltage drop over a long-distance, high-voltage power line is 200 V. If the power dissipated is 4.4 kW,
    (a) What is the total resistance of the power line?
    (b) Find the current in the line.

35. Figure 5–8 shows a circuit containing two resistors connected in parallel. If $R_1 = 27\ \Omega$, $R_2 = 36\ \Omega$, and the voltage drop across the circuit $V = 9.2$V, find the power dissipated by each resistor, P_1 and P_2. Note: The voltage drop across each resistor equals the voltage drop across the circuit.

36. In the parallel circuit in Figure 5–8, $I_1 = 85$ mA and $I_2 = 58$ mA. If $R_1 = 150\ \Omega$ and $R_2 = 220\ \Omega$, find the total power dissipated in the circuit.

37. A circuit whose total resistance is 130 Ω dissipates 50 W of power.
    (a) Find the voltage across the circuit.
    (b) Find the energy consumption in kilowatt-hours and joules for 24 h.

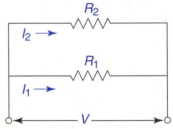

FIGURE 5–8 Power dissipation in a parallel circuit for problems 35 and 36.

38. A 430-Ω resistor dissipates 1.1 kW of power.
    (a) Find the current through the resistor.
    (b) Find the energy consumption in kilowatt-hours and joules for 12 h.

39. Figure 5–9 shows a series circuit containing two resistors R_1 and R_2. The current in the circuit $I = 750$ mA. If $R_1 = 3.3$ kΩ and $R_2 = 1.5$ kΩ, find the total power dissipated through both resistors. Note: The current through each resistor equals the current in the circuit.

40. In Figure 5–9, find the total power dissipated when $I = 900\ \mu$A, $R_1 = 1.2$ kΩ, and $R_2 = 910\ \Omega$.

41. Using Ohm's law and the power formula, derive formula (5.3):

$$P = \frac{V^2}{R}$$

FIGURE 5–9 Power dissipation in a series circuit for problems 39 and 40.

42. Using formula (5.3), $P = \frac{V^2}{R}$, derive formulas (5.3a) and (5.3b):

$$R = \frac{V^2}{P}$$

$$V = \sqrt{PR}$$

5.3
SERIES CIRCUITS

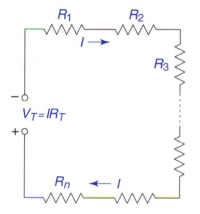

$$R_T = R_1 + R_2 + R_3 + \ldots + R_n$$

FIGURE 5–10 Series circuit.

A *series circuit* is a circuit in which all the components are connected on the same line so the *same current* flows on one path through each component. Figure 5–10 shows a circuit containing n resistances R_1, R_2, R_3, ... , R_n connected in series. If R_T is the total resistance of the circuit, then, applying Ohm's law, the total applied voltage $V_T = IR_T$. The applied voltage IR_T is equal to the sum of the voltage drops across each resistance. The voltage drop or potential difference across a resistance is the voltage that exists between the two ends of the resistance. This can be measured experimentally by attaching a voltmeter to each end of the resistance. It can also be calculated using Ohm's law. Therefore, equating the applied voltage to the sum of the voltage drops gives the following formula:

$$V_T = IR_T = IR_1 + IR_2 + IR_3 + \ldots + IR_n$$

Dividing each term by I leads to the basic relationship for resistances connected in series:

$$\frac{\cancel{I}R_T}{\cancel{I}} = \frac{\cancel{I}R_1}{\cancel{I}} + \frac{\cancel{I}R_2}{\cancel{I}} + \frac{\cancel{I}R_3}{\cancel{I}} + \ldots + \frac{\cancel{I}R_n}{\cancel{I}}$$

$$R_T = R_1 + R_2 + R_3 + \ldots + R_n \qquad (5.4)$$

> **Rule** The total resistance of two or more resistances connected in series is equal to the sum of the resistances.

Study the following example, which applies Ohm's law to resistances in series.

EXAMPLE 5.7

Given the series circuit in Figure 5–11 where $R_1 = 10\ \Omega$, $R_2 = 30\ \Omega$, and the applied voltage $V_T = 3.0$ V,

1. Find the current in the circuit.

2. Find the voltage drop across each resistance.

Solution

1. To find the current in the circuit, first find the total resistance R_T in the circuit. Apply (5.4):

$$R_T = R_1 + R_2 = 10\ \Omega + 30\ \Omega = 40\ \Omega$$

EXAMPLE 5.7 (Cont.)

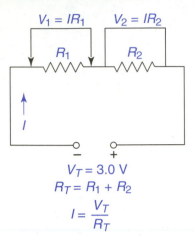

FIGURE 5–11 Series circuit and voltage drop for Example 5.7.

Then apply Ohm's law for $V_T = 3.0$ V and $R_T = 40$ Ω:

$$I = \frac{V_T}{R_T} = \frac{3.0\text{ V}}{40\text{ Ω}} = 0.075\text{ A} = 75\text{ mA}$$

2. Apply Ohm's law again to find the voltage drop V_1 across R_1:

$$V_1 = IR_1 = (0.075\ \text{A})(10\ \text{Ω}) = 0.75\ \text{V}$$

The voltage drop V_2 across R_2 is calculated in the same way:

$$V_2 = IR_2 = (0.075\ \text{A})(30\ \text{Ω}) = 2.25\ \text{V}$$

Observe that the sum of the voltage drops

$$V_1 + V_2 = 0.75\ \text{V} + 2.25\ \text{V} = 3.0\ \text{V}$$

equals the total voltage V_T. Note that $R_2 = 30$ Ω is *three times* $R_1 = 10$ Ω and the voltage drop across R_2 || 2.25 V is *three times* the voltage drop across R_1 || 0.75 V. This is due to Ohm's law, which says that the *voltage drop is directly proportional to the resistance* when the current is constant.

■

The next example illustrates how voltage drop is used to reduce the voltage across a load in a series circuit.

EXAMPLE 5.8

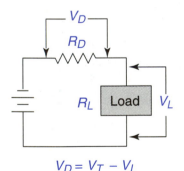

FIGURE 5–12 Voltage dropping resistor for Example 5.8.

Figure 5–12 shows a series circuit containing a load, whose resistance is R_L, in series with a voltage dropping resistor R_D. The load operates at a voltage of 10 V and uses 250 mW of power. If the total battery voltage $V_T = 14$ V, what should be the resistance of R_D?

Solution Since the sum of the voltage drops equals the total voltage and the voltage across the load $V_L = 10$ V, then the voltage drop across the resistor

$$V_D = V_T - V_L = 14\ \text{V} - 10\ \text{V} = 4.0\ \text{V}.$$

If the load uses 250 mW at 10 V, apply the power formula to find the current in the load:

$$I = \frac{P}{V_L} = \frac{0.25\text{ W}}{10\text{ V}} = 0.025\text{ A} = 25\text{ mA}.$$

The current through R_D must also be 25 mA since the current is the same throughout a series circuit. Apply Ohm's law to find R_D:

$$R_D = \frac{V_D}{I} = \frac{4.0\text{ V}}{0.025\text{ A}} = 160\ \text{Ω}$$

■

Aiding and Opposing Voltages

When two series voltages V_1 and V_2 are connected in *series aiding,* the resultant or total voltage is equal to their sum:

$$V_T = V_1 + V_2.$$

This occurs when the positive terminal of one voltage is connected to the negative terminal of the other voltage and each voltage causes the current to flow in the same direction. See Figure 5–13a.

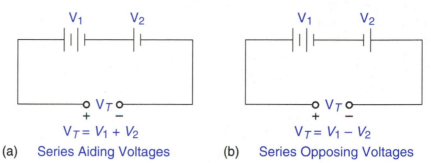

$$V_T = V_1 + V_2$$

(a)   Series Aiding Voltages

$$V_T = V_1 - V_2$$

(b)   Series Opposing Voltages

FIGURE 5–13   Aiding and opposing voltages.

When two series voltages are connected in *series opposing,* the resultant or total voltage is equal to their difference

$$V_T = V_1 - V_2.$$

This occurs when the negative terminal of one voltage is connected to the negative terminal of the other voltage (or positive to positive) and each voltage tends to cause the current to flow in the opposite direction. See Figure 5–13b. The larger voltage determines the resultant current flow.

EXAMPLE 5.9

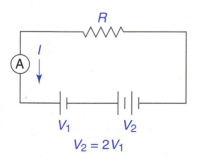

FIGURE 5– 14   Series-aiding voltages for Example 5.9.

Figure 5–14 shows two series-aiding voltages connected to a resistance $R = 75\ \Omega$. An ammeter in the circuit reads 100 mA. If $V_2 = 2V_1$, find V_1, V_2, and the power dissipated in the resistance.

Solution   The total voltage in the circuit $V_T = V_1 + V_2$. Since $V_2 = 2V_1$, then $V_T = V_1 + 2V_1 = 3V_1$. Apply Ohm's law to find V_1:

$$V_T = 3V_1 = IR = (0.10\ \text{A})(75\ \Omega) = 7.5\ \text{V}$$

$$V_1 = \frac{7.5\ \text{V}}{3} = 2.5\ \text{V}$$

Then,

$$V_2 = 2V_1 = 2(2.5\ \text{V}) = 5.0\ \text{V}$$

The power dissipated in the resistance is

$$P = V_T I = (7.5\text{V})(0.10\text{A}) = 0.75\ \text{W} = 750\ \text{mW}$$

EXERCISE 5.3

In problems 1 through 14, solve each problem to two significant digits.

1. Figure 5–15 shows a series circuit with two resistances R_1 and R_2 connected in series to a source whose total voltage is V_T. If $R_1 = 100\ \Omega$, $R_2 = 200\ \Omega$, and $V_T = 24$ V,
   (a) Find the current I.
   (b) Find the voltage drop across each resistance.

2. Given the series circuit in Figure 5–15 with $R_1 = 10\ \Omega$, $R_2 = 30\ \Omega$, and $V_T = 12$ V,
   (a) Find the current I.
   (b) Find the voltage drop across each resistance.

3. In the series circuit in Figure 5–15, given that $R_2 = 3R_1$ and $R_T = 300\ \Omega$, find R_1 and R_2.

4. In the series circuit in Figure 5–15, given that R_2 is 50% greater than R_1 and $V_T = 20$ V, find V_1 and V_2.

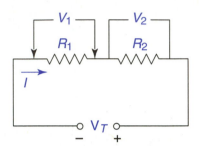

FIGURE 5–15   Series circuit for problems 1 through 4.

5. Figure 5–16 shows a series circuit containing three resistances in series connected to a battery whose voltage $V_T = 8.0$ V. If $R_1 = 20\ \Omega$, $R_2 = 24\ \Omega$, and $R_3 = 36\ \Omega$,
   (a) Find the current in the circuit.
   (b) Find the power P_1 dissipated in R_1.

6. Given the series circuit in Figure 5–16 with $R_1 = 120\ \Omega$, $R_2 = 130\ \Omega$, $R_3 = 150\ \Omega$, and the current $I = 50$ mA,
   (a) Find the battery voltage.
   (b) Find the total power dissipated in the circuit.

7. In the series circuit in Figure 5–16, given that $R_1 = R_2$ and $R_3 = 2R_1$, if the voltage drop $V_1 = 3.0$ V,
   (a) Find the voltage drop across R_2 and R_3.
   (b) Find V_T.

8. In the series circuit in Figure 5–16, given that $R_1 = R_3$ and $R_2 = 2R_3$,
   (a) If $V_T = 36$ V, find V_1, V_2, and V_3.
   (b) If the total power dissipated $P_T = 24$ W, what is the total resistance R_T of the circuit?

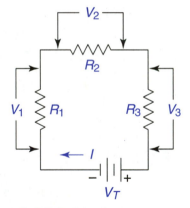

FIGURE 5–16   Series circuit for problems 5 through 8.

9. In Example 5.7, if the load operates at 8.0 V and uses 200 mW of power, find R_D.

10. A voltage dropping resistor is in series with a load whose resistance $R_L = 330\ \Omega$. The total voltage drop across the load and the resistor is 100 V. If the load uses 14 W of power, what should be the value of the dropping resistor?

11. In Example 5.8, if the ammeter reads 600 mA, find V_1, V_2, and the power dissipated in the resistance.

12. Two voltages V_1 and V_2 are connected in series opposing to a resistance $R = 16\ \Omega$. An ammeter in the circuit reads 750 mA. If $V_1 = 22$ V, what are the two possible values for V_2?

13. The total resistance of a circuit $R_T = 2.7$ kΩ. Assuming the voltage is constant, how much resistance must be added in series to reduce the current to one-half its value?

14. Given the circuit in problem 13, how much resistance must be added in series to reduce the current by 25% of its value?

☰ 5.4
PARALLEL CIRCUITS

Figure 5–17 shows a circuit containing n resistances $R_1, R_2, R_3, \ldots, R_n$ in parallel. Each resistance is connected separately to the voltage source V. If R_T is the total resistance of the circuit and I_T is the total current in the circuit, then by Ohm's law:

$$I_T = \frac{V}{R_T}$$

The total current I_T must also equal the sum of the currents through all the resistances:

$$I_T = \frac{V}{R_T} = I_1 + I_2 + I_3 + \ldots + I_n \tag{5.5}$$

The voltage across each resistance is equal to the applied voltage V because the ends of each resistance are connected directly to the voltage source. The current through each resistance is given by Ohm's law:

$$I_1 = \frac{V}{R_1}, \; I_2 = \frac{V}{R_2}, \; I_3 = \frac{V}{R_3}, \; \ldots, \; I_n = \frac{V}{R_n}$$

Equation (5.5) can then be written as follows:

$$\frac{V}{R_T} = \frac{V}{R_1} + \frac{V}{R_2} + \frac{V}{R_3} + \ldots + \frac{V}{R_n}$$

Dividing each term by V leads to the basic relationship for resistances in parallel:

$$\frac{\cancel{V}}{\cancel{V}R_T} = \frac{\cancel{V}}{\cancel{V}R_1} + \frac{\cancel{V}}{\cancel{V}R_2} + \frac{\cancel{V}}{\cancel{V}R_3} + \ldots + \frac{\cancel{V}}{\cancel{V}R_n}$$

$$\frac{1}{R_T} = \frac{1}{R_1} + \frac{1}{R_2} + \frac{1}{R_3} + \ldots + \frac{1}{R_n} \tag{5.6}$$

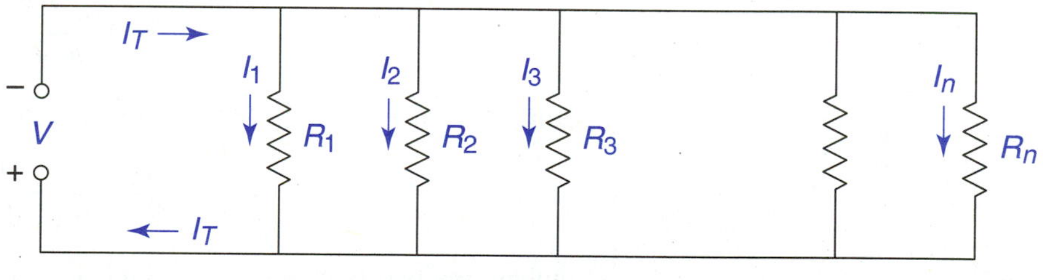

$$I_T = \frac{V}{R_T} = I_1 + I_2 + I_3 + \ldots + I_n$$

$$\frac{1}{R_T} = \frac{1}{R_1} + \frac{1}{R_2} + \frac{1}{R_3} + \ldots + \frac{1}{I_n}$$

FIGURE 5–17 Parallel circuit.

> **Rule** The reciprocal of the total resistance of two or more resistances in parallel equals the sum of the reciprocals of the resistances.

Note that R_T *will always be less* than the smallest of the parallel resistances. Parallel circuits provide additional current paths and therefore result in less total resistance.

A special case of formula (5.6) is when all the parallel resistances are equal to the same value R. If there are n resistances in parallel, each equal to R, then

$$\frac{1}{R_T} = \frac{1}{R} + \frac{1}{R} + \frac{1}{R} + \ldots + \frac{1}{R} \quad (\text{n fractions})$$

Combining the n fractions,

$$\frac{1}{R_T} = \frac{1 + 1 + 1 + \ldots + 1}{R} = \frac{n}{R}$$

When two fractions are equal, their reciprocals are equal. Therefore,

$$R_T = \frac{R}{n} \tag{5.7}$$

Formula (5.7) tells you that if there are two equal resistances in parallel, the equivalent resistance is one-half of one of the resistances. If there are three equal resistances in parallel the equivalent resistance is one-third of one of the resistances, and so on. For example, two 100-Ω resistances in parallel are equivalent to one resistance of 50 Ω. Three 30-Ω resistances in parallel are equivalent to one resistance of 10 Ω.

Study the next three examples, which show a convenient formula for two resistances in parallel.

EXAMPLE 5.10

Show that the total resistance of two resistances in parallel can be expressed as

$$R_T = \frac{R_1 R_2}{R_1 + R_2} \tag{5.8}$$

Solution Apply formula (5.6) for two parallel resistances:

$$\frac{1}{R_T} = \frac{1}{R_1} + \frac{1}{R_2}$$

Combine the two fractions on the right side over the LCD $R_1 R_2$:

$$\frac{1}{R_T} = \frac{1\,(R_2)}{R_1\,(R_2)} + \frac{1\,(R_1)}{R_2\,(R_1)}$$

$$\frac{1}{R_T} = \frac{R_2 + R_1}{R_1 R_2}$$

EXAMPLE 5.10 (Cont.)

If two fractions are equal, then their reciprocals must be equal. Therefore, invert the fractions on both sides of the equation to produce the following result:

$$R_T = \frac{R_1 R_2}{R_1 + R_2}$$

Observe that the terms in the denominator are arranged in order. Formula (5.8) is a convenient formula to calculate the equivalent resistance of two parallel resistances. One way to remember it is "the product divided by the sum."

Example 5.9 can also be done by multiplying all the terms in the equation by the LCD = $R_T R_1 R_2$. This is left as problem 23 in Exercise 5.4.

EXAMPLE 5.11

Figure 5–18 shows a circuit containing two resistances R_1 and R_2 in parallel connected to a voltage source V_T. If $R_1 = 150\ \Omega$, $R_2 = 200\ \Omega$, and $V_T = 20$ V, find R_T and I_T.

Solution The total resistance R_T can be found two ways: using formula (5.8) or formula (5.6). The first method, using formula (5.8) and substituting the given values, yields to two significant digits.

$$R_T = \frac{R_1 R_2}{R_1 + R_2} = \frac{(150)(200)}{150 + 200} = \frac{30,000}{350} = 85.7\ \Omega \approx 86\ \Omega$$

The second method, using formula (5.6) for two resistances, gives you

$$\frac{1}{R_T} = \frac{1}{150} + \frac{1}{200}$$

This calculation for R_T can be done on the calculator in one series of steps using the reciprocal key $\boxed{1/x}$:

$$150\ \boxed{1/x}\ \boxed{+}\ 200\ \boxed{1/x}\ \boxed{=}\ \boxed{1/x}\ \rightarrow\ 85.7\ \Omega\ \approx\ 86\ \Omega$$

Now using the value for R_T, find the total current using Ohm's law:

$$I_T = \frac{V_T}{R_T} = \frac{20\ \text{V}}{86\ \Omega} = 230\ \text{mA}$$

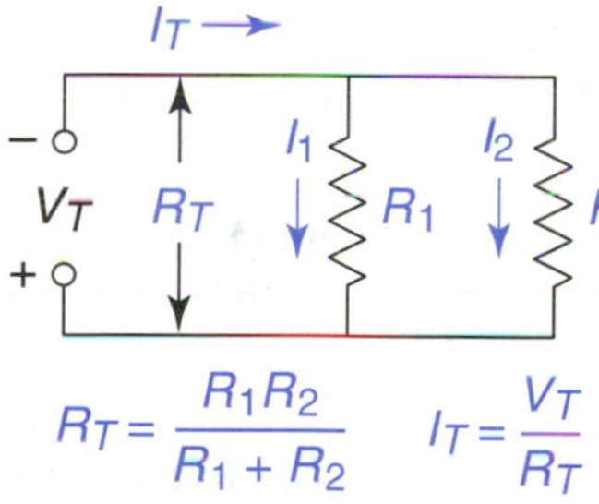

FIGURE 5–18 Two parallel resistances for Example 5.11 and 5.12.

$$R_T = \frac{R_1 R_2}{R_1 + R_2} \qquad I_T = \frac{V_T}{R_T}$$

In Figure 5–18, if you know the value of one resistance R_2 and the total resistance R_T, you can find the value of the other resistance R_1 using formula (5.8) or (5.6) solved for R_1. Study the next example, which shows the procedure.

EXAMPLE 5.12

In Figure 5–18, if $R_2 = 75\ \Omega$, find R_1 so that $R_T = 50\ \Omega$.

Solution Using formula (5.8), it is first necessary to solve the formula for R_1. The solution is left as problem 24 in Exercise 5.4. The result is given here:

$$R_1 = \frac{R_2 R_T}{R_2 - R_T} \tag{5.9}$$

EXAMPLE 5.12 (Cont.)

Substituting the values for $R_2 = 75\ \Omega$ and $R_T = 50\ \Omega$ in formula (5.9),

$$R_1 = \frac{(75)(50)}{75 - 50} = \frac{3750}{25} = 150\ \Omega$$

Using formula (5.6) and solving for $\frac{1}{R_1}$, you have

$$\frac{1}{R_1} = \frac{1}{R_T} - \frac{1}{R_2}$$

Therefore, R_1 can be found on the calculator as follows:

$$50\ \boxed{1/x}\ \boxed{-}\ 75\ \boxed{1/x}\ \boxed{=}\ \boxed{1/x}\ \rightarrow\ 150\ \Omega$$

Conductance

A more direct way to express Equation (5.6) is in terms of *conductance*. The *conductance G* of a resistance *R* is defined as its reciprocal:

$$G = \frac{1}{R} \tag{5.10}$$

Conductance is measured in siemens (S). Applying (5.10) to the parallel circuit in Figure 5–17,

$$G_T = \frac{1}{R_T}, \; G_1 = \frac{1}{R_1}, \; G_2 = \frac{1}{R_2}, \; G_3 = \frac{1}{R_3}, \; ..., \; G_n = \frac{1}{G_n}$$

Substituting into (5.6), the formula for the total conductance is

$$G_T = G_1 + G_2 + G_3 + ... + G_n \tag{5.11}$$

EXAMPLE 5.13

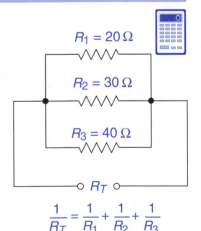

$R_1 = 20\ \Omega$

$R_2 = 30\ \Omega$

$R_3 = 40\ \Omega$

$\circ\ R_T\ \circ$

$$\frac{1}{R_T} = \frac{1}{R_1} + \frac{1}{R_2} + \frac{1}{R_3}$$

$$G_T = G_1 + G_2 + G_3$$

FIGURE 5–19 Total conductance of a parallel bank for Example 5.13.

Figure 5–19 shows a *parallel bank* of three resistances. Two or more resistances in parallel is called a parallel bank. If $R_1 = 20\ \Omega$, $R_2 = 30\ \Omega$, and $R_3 = 40\ \Omega$, find the total conductance G_T and the total resistance R_T to two significant digits.

Solution The first method, using formula (5.11) for the total conductance, yields

$$G_T = \frac{1}{R_T} = \frac{1}{20} + \frac{1}{30} + \frac{1}{40} = \frac{6 + 4 + 3}{120}$$

$$G_T = \frac{13}{120} = 0.11\,\text{S} = 110\,\text{mS}$$

Then inverting,

$$R_T = \frac{1}{G_T} = \frac{120}{13} = 9.2\ \Omega$$

With the second method, using formula (5.6), you can compute G_T and then R_T directly on the calculator using the reciprocal key $\boxed{1/x}$:

$$20\ \boxed{1/x}\ \boxed{+}\ 30\ \boxed{1/x}\ \boxed{+}\ 40\ \boxed{1/x}\ \boxed{=}\ \rightarrow\ 0.11\text{S}\ \boxed{1/x}\ \rightarrow\ 9.2\ \Omega$$

EXAMPLE 5.13 (Cont.)

Therefore, $G_T = 110$ mS and $R_T = 9.2$ Ω to two significant digits. Observe that R_T is less than the smallest resistance of 20 Ω.

You can also find the total resistance of three resistances in parallel by applying formula (5.8) twice. First find the equivalent resistance of R_1 and R_2, which is R_{12}. Then find the equivalent resistance of R_{12} and R_3.

■

EXERCISE 5.4

For all problems round answers to two significant digits.

In problems 1 through 6, the two resistances are connected in parallel. Find R_T for each parallel circuit.

1. $R_1 = 10$ Ω, $R_2 = 30$ Ω
2. $R_1 = 510$ Ω, $R_2 = 430$ Ω
3. $R_1 = 56$ Ω, $R_2 = 56$ Ω

4. $R_1 = 18$ Ω, $R_2 = 36$ Ω
5. $R_1 = 750$ Ω, $R_2 = 1.2$ kΩ
6. $R_1 = 1.0$ MΩ, $R_2 = 820$ kΩ

In problems 7 through 12, the three resistances are connected in parallel. Find G_T and R_T for each parallel circuit.

7. $R_1 = 15$ Ω, $R_2 = 30$ Ω, $R_3 = 30$ Ω
8. $R_1 = 330$ Ω, $R_2 = 330$ Ω, $R_3 = 330$ Ω
9. $R_1 = 220$ Ω, $R_2 = 300$ Ω, $R_3 = 360$ Ω

10. $R_1 = 47$ Ω, $R_2 = 62$ Ω, $R_3 = 24$ Ω
11. $R_1 = 1.3$ kΩ, $R_2 = 820$ Ω, $R_3 = 1.8$ kΩ
12. $R_1 = 1.6$ kΩ, $R_2 = 910$ Ω, $R_3 = 750$ Ω

In problems 13 through 16, R_1 and R_2 are connected in parallel. Find the value of R_1 that will produce the given value of R_T.

13. $R_2 = 150$ Ω, $R_T = 75$ Ω
14. $R_2 = 30$ Ω, $R_T = 10$ Ω

15. $R_2 = 1.5$ kΩ, $R_T = 670$ Ω
16. $R_2 = 330$ Ω, $R_T = 200$ Ω

In problems 17 and 18, R_1, R_2, and R_3 are connected in parallel. Find the value of R_1 that will produce the given value of R_T.

17. $R_2 = 30$ Ω, $R_3 = 20$ Ω, $R_T = 7.5$ Ω

18. $R_2 = 200$ Ω, $R_3 = 200$ Ω, $R_T = 50$ Ω

In problems 19 through 22 solve each applied problem.

19. In Figure 5–18, given $R_1 = 24$ Ω, $R_2 = 30$ Ω, and $V = 12$ V, find I_1, I_2, I_T, and R_T.
20. In Figure 5–18, given $R_1 = 220$ Ω, $R_2 = 300$ Ω, and $V = 36$ V, find I_1, I_2, I_T, and R_T.
21. Figure 5–20 shows three resistances in parallel connected to a battery whose voltage is V_B. If $V_B = 50$ V, $R_1 = 150$ Ω, $R_2 = 200$ Ω, and $R_3 = 180$ Ω, find I_1, I_2, I_3, I_T, and R_T.
22. Given the parallel circuit in Figure 5–20 with $V = 15$ V, $R_1 = 20$ Ω, $R_2 = 24$ Ω, and $R_3 = 30$ Ω, find I_1, I_2, I_3, I_T, and R_T.

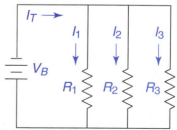

FIGURE 5–20   Parallel circuit for problems 21 and 22.

In problems 23 through 26, derive the indicated formula.

23. Starting with formula (5.6) for two parallel resistances,

$$\frac{1}{R_T} = \frac{1}{R_1} + \frac{1}{R_2}$$

derive formula (5.8),

$$R_T = \frac{R_1 R_2}{R_1 + R_2}$$

by multiplying all the terms by the LCD $= R_T R_1 R_2$.

24. Starting with formula (5.8),

$$R_T = \frac{R_1 R_2}{R_1 + R_2}$$

derive formula (5.9),

$$R_1 = \frac{R_2 R_T}{R_2 - R_T}$$

25. Given a circuit with two resistances in parallel, if $R_2 = 2R_1$, using the formula (5.8) show that the equivalent resistance is

$$R_{eq} = \frac{2R_1}{3}$$

26. Given a circuit with three resistances in parallel, R_1, R_2, and R_3, if $R_3 = R_1$, using formula (5.6), show that the equivalent resistance is

$$R_{eq} = \frac{R_1 R_2}{R_1 + 2R_2}$$

Courtesy of Hewlett Packard Company.

A technician at Ford Electrical and Electronics division tracks radio test information.

≡ **5.5** SERIES-PARALLEL CIRCUITS

Series circuits and parallel circuits are combined in various ways to create many different types of series-parallel circuits. Some of the basic combinations are considered here. Two or more resistances in series is called a *series branch*. Two or more resistances in parallel is called a *parallel bank*.

To analyze a series-parallel circuit, it is necessary to consider the series branches and the parallel banks separately. You reduce each series branch to an equivalent resistance, and you reduce each parallel bank to an equivalent resistance. The equivalent resistances are then combined to obtain the total resistance of the circuit. Study the following examples that illustrate the techniques needed to solve series-parallel circuits.

EXAMPLE 5.14

Figure 5–21(a) shows a series-parallel circuit containing a resistance R_1 in series with a parallel bank of two resistances R_2 and R_3.

1. Find the formula for the total resistance of the circuit R_T.

2. Find R_T when $R_1 = 15\ \Omega$, $R_2 = 20\ \Omega$, and $R_3 = 10\ \Omega$ to two significant digits.

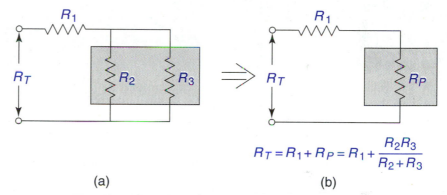

$$R_T = R_1 + R_P = R_1 + \frac{R_2 R_3}{R_2 + R_3}$$

(a)   (b)

FIGURE 5–21   Series-parallel circuit for Example 5.14.

Solution

1. First find the equivalent resistance R_P of the parallel bank of R_2 and R_3. Apply formula (5.8) to R_2 and R_3:

$$R_P = \frac{R_2 R_3}{R_2 + R_3}$$

This reduces the circuit to two resistances in series: R_1 and R_P, as shown in Figure 5–21(b).

The formula for the total resistance is then

$$R_T = R_1 + R_P = R_1 + \frac{R_2 R_3}{R_2 + R_3}$$

2. Substitute in the formula to find the value of R_T:

$$R_T = 15 + \frac{(20)(10)}{20 + 10} = 15 + \frac{200}{30} = 21.7\ \Omega \approx 22\ \Omega$$

EXAMPLE 5.15

Figure 5–22(a) shows a series-parallel circuit that contains a series branch of R_1 and R_2 connected to the parallel bank of R_3 and R_4.

1. Find the formula for the total resistance R_T.
2. Calculate R_T given $R_1 = 200\ \Omega$, $R_2 = 240\ \Omega$, $R_3 = 150\ \Omega$, and $R_4 = 300\ \Omega$.

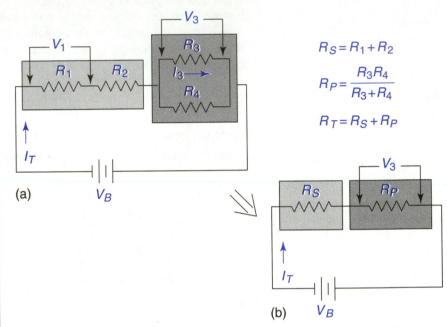

$$R_S = R_1 + R_2$$

$$R_P = \frac{R_3 R_4}{R_3 + R_4}$$

$$R_T = R_S + R_P$$

FIGURE 5–22 Series-parallel circuit for Examples 5.15 and 5.16.

Solution

1. Find the equivalent resistance of the series branch and the equivalent resistance of the parallel bank separately. The equivalent resistance of the series branch is

$$R_S = R_1 + R_2$$

Apply formula (5.8) to R_3 and R_4 to find the equivalent resistance of the parallel bank:

$$R_P = \frac{R_3 R_4}{R_3 + R_4}$$

The circuit can now be reduced to a series circuit containing R_S and R_P as shown in Figure 5–22(b). The total resistance R_T is then

$$R_T = R_S + R_P = R_1 + R_2 + \frac{R_3 R_4}{R_3 + R_4}$$

2. To find R_T, substitute the values $R_1 = 200\ \Omega$, $R_2 = 240\ \Omega$, $R_3 = 150\ \Omega$, and $R_4 = 300\ \Omega$ into the formula for R_T:

$$R_T = 200 + 240 + \frac{(150)(300)}{150 + 300} = 440\ \Omega + 100\ \Omega = 540\ \Omega$$

The next example applies the results of Example 5.14 to find voltages and currents in the series-parallel circuit.

EXAMPLE 5.16

In Example 5.16, if the battery voltage $V_B = 24$ V, find the total current I_T, the voltage drops V_1 and V_3, and the current I_3 to two significant digits.

Solution   From Example 5.16, the total resistance $R_T = 540$ Ω. Apply Ohm's law to find the total current:

$$I_T = \frac{V_T}{R_T} = \frac{24 \text{ V}}{540 \text{ Ω}} = 0.0444 \text{ A} \approx 44 \text{ mA}$$

To find the voltage drop V_1 across R_1, note that the total current goes through R_1. Therefore, by Ohm's law,

$$V_1 = I_T R_1 = (0.0444 \text{ 4A})(200 \text{ Ω}) = 8.88 \text{ V} \approx 8.9 \text{ V}$$

To find the voltage drop V_3 across R_3, note that V_3 is the same as the voltage drop across the equivalent parallel resistance R_P in the series circuit of Figure 5–22(b). First calculate R_P:

$$R_P = \frac{R_3 R_4}{R_3 + R_4} = \frac{(150)(300)}{150 + 300} = 100 \text{ Ω}$$

The current through R_P is the total current I_T. Apply Ohm's law to find the voltage drop across R_P:

$$V_3 = I_T R_P = (0.0444 \text{ A})(100 \text{ Ω}) = 4.4 \text{ V}$$

To find I_3, apply Ohm's law again to V_3 and R_3:

$$I_3 = \frac{V_3}{R_3} = \frac{4.4 \text{ V}}{150 \text{ Ω}} = 29 \text{ mA}$$

EXAMPLE 5.17

Figure 5–23(a) shows a series-parallel circuit that contains a resistance R_1 in series with a parallel bank. The parallel bank consists of a series branch of two resistances R_2 and R_3 in parallel with R_4.

1. Find the formula for the total resistance of the circuit.

2. Calculate R_T given $R_1 = 330$ Ω, $R_2 = 510$ Ω, $R_3 = 620$ Ω, and $R_4 = 1.0$ kΩ to two significant digits.

EXAMPLE 5.17 (Cont.)

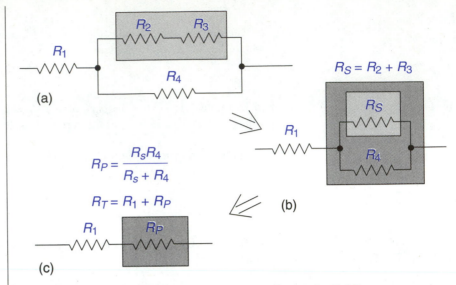

(a)

$R_S = R_2 + R_3$

$R_P = \dfrac{R_S R_4}{R_S + R_4}$

$R_T = R_1 + R_P$

(b)

(c)

FIGURE 5–23   Series-parallel circuit for Example 5.17.

Solution

1. You must first reduce the series branch before you can reduce the parallel bank. The equivalent series resistance of R_2 and R_3 is

$$R_S = R_2 + R_3$$

This reduces the circuit to Figure 5–23 (b). Then apply formula (5.8) to the parallel bank of R_S and R_4:

$$R_P = \frac{R_S R_4}{R_S + R_4}$$

Substitute $R_2 + R_3$ for R_S to obtain:

$$R_P = \frac{(R_2 + R_3)R_4}{(R_2 + R_3) + R_4}$$

This reduces the circuit to the series circuit in Figure 5–23(c). The total resistance of the circuit is then

$$R_T = R_1 + R_P = R_1 + \frac{(R_2 + R_3) R_4}{(R_2 + R_3) + R_4}$$

2. To calculate R_T given $R_1 = 330\ \Omega$, $R_2 = 510\ \Omega$, $R_3 = 620\ \Omega$, and $R_4 = 1.0\ k\Omega$, substitute the given values:

$$R_T = 330 + \frac{(510 + 620)(1000)}{(510 + 620) + 1000} = 330 + \frac{(1130)(1000)}{2130} = 861\ \Omega \approx 860\ \Omega$$

Example 5.17 illustrated a basic procedure for a series-parallel circuit: You should always reduce any series branch that is part of a parallel bank before reducing the parallel circuit.

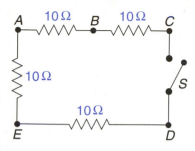

FIGURE 5–24 Series and parallel circuits.

ERROR BOX

You must always clearly identify when resistances are in series or when they are in parallel.

For two resistances to be in series, the *same current* must travel through both resistances. As you follow the current through the two resistances, the end of one resistance must be connected to the beginning of the other resistance with no branch points in between.

For two resistances to be in parallel, the *same voltage drop* must be across each resistance. The ends of one resistance must be connected to the ends of the other resistance with no branch points in between. The same *total* current must enter and leave both resistances.

See if you can correctly identify series and parallel resistances in the practice problems.

Practice Problems: In the circuit in Figure 5–24, find the equivalent resistance between the two points indicated. The equivalent resistance would be the total resistance of the circuit if a voltage were applied across the two points.
With switch S open, find: 1. R_{AB} 2. R_{CD}
With switch S closed, find: 3. R_{AB} 4. R_{AC}

Answers: Ʊ ΟΙ 'Ɖ Ʊ ϛ'L 'Ɛ Ʊ ΟƉ 'Ʒ Ʊ ΟΙ 'Ι

EXERCISE 5.5

In problems 1 through 18, solve each circuit problem to two significant digits.

1. In the series-parallel circuit of Figure 5–21(a), find R_T when $R_1 = 20 \ \Omega$, $R_2 = 30 \ \Omega$, and $R_3 = 15 \ \Omega$.

2. In the series-parallel circuit of Figure 5–22(a), find R_T when $R_1 = 36 \ \Omega$, $R_2 = 27 \ \Omega$, $R_3 = 30 \ \Omega$, and $R_4 = 20 \ \Omega$.

3. In the series-parallel circuit of Figure 5–21(a), if $R_1 = 12 \ \Omega$, and $R_2 = 15 \ \Omega$, what should be the value of R_3 if R_T is to be 22 Ω? Substitute in the formula for R_T, and solve for R_3.

4. In the series-parallel circuit of Figure 5–22(a), if $R_1 = 20 \ \Omega$, $R_2 = 30 \ \Omega$, and $R_3 = 40 \ \Omega$, what should be the value of R_4 if R_T is to be 58 Ω? Substitute in the formula for R_T, and solve for R_4.

5. In the circuit of Figure 5–25, find the formula for R_T. Calculate R_T given $R_1 = 120 \ \Omega$, $R_2 = 240 \ \Omega$, and $R_3 = 100 \ \Omega$.

6. In the circuit of Figure 5–26, find the formula for R_T. Calculate R_T given $R_1 = 30 \ \Omega$, $R_2 = 10 \ \Omega$, $R_3 = 20 \ \Omega$, and $R_4 = 75 \ \Omega$.

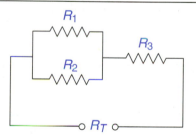

FIGURE 5–25 Series-parallel circuit for problems 5 and 11.

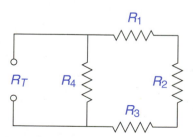

FIGURE 5–26 Series-parallel circuits for problems 6 and 12.

7. In Example 5.16, if the battery voltage $V_B = 14$ V, find I_T, V_2, V_4, and I_4.

8. In the circuit of Figure 5–22(a), $R_1 = 18\ \Omega$, $R_2 = 22\ \Omega$, $R_3 = 75\ \Omega$, and $R_4 = 150\ \Omega$. If $V_B = 9.0$ V, find R_T, I_T, V_2, V_3, and I_3.

9. In the circuit of Figure 5–23(a), find R_T if $R_1 = 120\ \Omega$, $R_2 = 180\ \Omega$, $R_3 = 120\ \Omega$, and $R_4 = 200\ \Omega$.

10. In the circuit of Figure 5–23(a), if $R_1 = 10\ \Omega$, $R_2 = 5\ \Omega$, and $R_3 = 5\ \Omega$, what should be the value of R_4 if R_T is to be 15 Ω?

11. In the circuit of Figure 5–25, if $R_2 = 12\ \Omega$ and $R_3 = 7.5\ \Omega$, what should be the value of R_1 if R_T is to be 15 Ω?

12. In the circuit of Figure 5–26, find R_1 if $R_2 = 1.5$ kΩ, $R_3 = 1.5$ kΩ, $R_4 = 1.2$ kΩ, and R_T is to be 1.0 kΩ.

13. In the circuit of Figure 5–27, find R_T if $R_1 = 20\ \Omega$, $R_2 = 10\ \Omega$, $R_3 = 20\ \Omega$, $R_4 = 30\ \Omega$, and $R_5 = 15\ \Omega$.

14. In the circuit of Figure 5–28, find R_T if $R_1 = 120\ \Omega$, $R_2 = 100\ \Omega$, $R_3 = 200\ \Omega$, and $R_4 = 100\ \Omega$.

15. In the circuit of Figure 5–27, if $R_1 = 100\ \Omega$, $R_2 = 50\ \Omega$, $R_3 = 50\ \Omega$, $R_4 = 150\ \Omega$, $R_5 = 100\ \Omega$, and $V_T = 12$ V, find the total current I_T in the circuit and the voltage drops V_1 and V_2.

16. In the circuit of Figure 5–28, if $R_1 = 43\ \Omega$, $R_2 = 120\ \Omega$, $R_3 = 120\ \Omega$, $R_4 = 120\ \Omega$, and $V_T = 24$ V, find the total current I_T, and the voltage drops V_1 and V_2.

17. In the circuit of Figure 5–29, find the total power P_T in the circuit if $R_1 = 30\ \Omega$, $R_2 = 15\ \Omega$, $R_3 = 15\ \Omega$, $R_4 = 10\ \Omega$ and $V = 6.0$ V. (Note: $P_T = V^2/R_T$)

18. In the circuit of Figure 5–29, find the total voltage V_T if $I_T = 1.2$ A, $R_1 = 10\ \Omega$, $R_2 = 30\ \Omega$, $R_3 = 20\ \Omega$, and $R_4 = 20\ \Omega$.

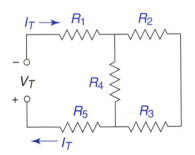

FIGURE 5–27 Series-parallel circuit for problems 13 and 15.

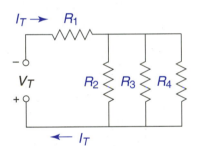

FIGURE 5–28 Series-parallel circuit for problems 14 and 16.

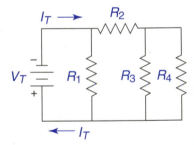

FIGURE 5–29 Series-parallel circuit for problems 17 and 18.

≡ CHAPTER HIGHLIGHTS

5.1 OHM'S LAW

Ohm's law can be written three ways:

$$I = \frac{V}{R}$$

$$V = IR$$

$$R = \frac{V}{I} \qquad (5.1)$$

where I = current in amps
V = voltage in volts
R = resistance in ohms

Direct Proportion	If two quantities are directly proportional and one changes by a certain ratio, then the other changes by the same ratio.

Ohm's law tells us that for the voltage across a resistance,

1. The current is directly proportional to the voltage when the resistance is constant.
2. The voltage is directly proportional to the resistance when the current is constant.

Inverse Proportion	If two quantities are inversely proportional and one changes by a certain ratio, then the other changes by the reciprocal ratio.

Ohm's law also tells us:

3. The current is inversely proportional to the resistance when the voltage is constant.

5.2 ELECTRIC POWER

The basic formula for electric power can be expressed three ways:

$$P = VI$$
$$I = \frac{P}{V}$$
$$V = \frac{P}{I} \tag{5.2}$$

The formulas for power in terms of resistance are

$$P = I^2R$$
$$P = \frac{V^2}{R} \tag{5.3}$$

Formulas (5.3) can be written two other ways:

$$R = \frac{P}{I^2} \text{ and } R = \frac{V^2}{P} \tag{5.3a}$$

$$I = \sqrt{\frac{P}{R}} \text{ and } V = \sqrt{PR} \tag{5.3b}$$

The units for these formulas must be consistent: watts for P, volts for V, amps for I, and ohms for R.

It is not necessary to memorize all of the formulas (5.1) to (5.3). You can derive any of the formulas from $P = VI$ and $V = IR$. See the Error Box in Section 5.2.

The kilowatt-hour is a measure of energy and is equal to 1000 W of power consumed per hour: 1 kWh = (1000 W)(1 h).

5.3 SERIES CIRCUITS

A series circuit is a circuit in which all the components are connected on the same line so that the same current flows on one path through each component.

The total resistance of two or more resistances in series is equal to the sum of the resistances:

$$R_T = R_1 + R_2 + R_3 + \dots + R_n \tag{5.4}$$

The sum of the voltage drops or IR drops equals the applied voltage:

$$IR_T = IR_1 + IR_2 + IR_3 + \dots + IR_n$$

Voltage drop is directly proportional to resistance.

When two voltages are connected positive to negative, they are series aiding, and the resultant voltage is equal to their sum. When two voltages are connected negative to negative or positive to positive, they are series opposing, and the resultant voltage is equal to their difference.

5.4 PARALLEL CIRCUITS

A parallel circuit is a circuit in which each component is connected on a separate line to the voltage source. The same voltage is across each component, but a different current flows through each component. The total current equals the sum of the currents through all the resistances:

$$I_T = \frac{V}{R_T} = I_1 + I_2 + I_3 + \dots + I_n \tag{5.5}$$

The reciprocal of the total resistance of two or more resistances in parallel equals the sum of the reciprocals of the resistances:

$$\frac{1}{R_T} = \frac{1}{R_1} + \frac{1}{R_2} + \frac{1}{R_3} + \dots + \frac{1}{R_n} \tag{5.6}$$

If n parallel resistances are all equal to the same value R, then

$$R_T = \frac{R}{n} \tag{5.7}$$

A very useful formula for the total resistance R_T of two resistances R_1 and R_2 in parallel is given by the "product divided by the sum":

$$R_T = \frac{R_1 R_2}{R_1 + R_2} \tag{5.8}$$

Another formula that lends itself to the calculator is a special case of (5.6) for two parallel resistances:

$$\frac{1}{R_T} = \frac{1}{R_1} + \frac{1}{R_2}$$

Add the reciprocals of R_1 and R_2 on the calculator, and then take the reciprocal of the result.

For two parallel resistances, if you know R_2 and R_T and want to find R_1, one formula is

$$R_1 = \frac{R_2 R_T}{R_2 - R_T} \qquad (5.9)$$

Another formula that lends itself to the calculator is

$$\frac{1}{R_1} = \frac{1}{R_T} - \frac{1}{R_2}$$

The *conductance* G measured in siemens (S) is the reciprocal of the resistance:

$$G = \frac{1}{R} \qquad (5.10)$$

In a parallel circuit the total conductance is

$$G_T = G_1 + G_2 + G_3 + \cdots + G_n \qquad (5.11)$$

5.5 SERIES-PARALLEL CIRCUITS

To solve a series-parallel circuit, reduce each series branch and each parallel bank separately to an equivalent resistance. When a series branch is part of a parallel bank, reduce the series branch first. Then combine the equivalent resistances to obtain the total resistance of the circuit. Study examples 5.14 to 5.17, and the Error Box at the end of Section 5.5.

☰ REVIEW QUESTIONS

In problems 1 through 24 solve each problem to two significant digits.

1. Given the dc circuit in Figure 5-30 with constant resistance $R = 36\ \Omega$, find the current I when the battery voltage (a) $V_B = 18$ V (b) V_B increases by 50%. What is the percent increase in the current when V_B increases by 50%? Why?

2. Given the dc circuit in Figure 5–30 with a constant voltage $V_B = 12$ V, find the current I when (a) $R = 20\ \Omega$ (b) R doubles. By what ratio does the current change when R doubles? Why?

3. Given the dc circuit in Figure 5–31 with $I = 1.2$ A and a voltage source V_S, find V_S when (a) $R = 7.5\ \Omega$ (b) R doubles and the current does not change.

4. Given the circuit in Figure 5–31 with $I = 200$ mA, find the new current
   (a) If the voltage increases by 50% and the resistance remains constant.
   (b) If the resistance doubles and the voltage remains constant.

5. The change in the *resistance R* of a tungsten bulb filament is directly proportional to the *temperature change*. If $R = 10\ \Omega$ at 20°C and 12 Ω at 30°C, what will be the resistance at 60°C?

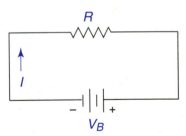

FIGURE 5–30 A dc circuit for problems 1 and 2.

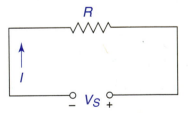

FIGURE 5–31 A dc circuit for problems 3 and 4.

6. The capacitance C of a parallel plate capacitor is inversely proportional to the distance d between the plates. If $C = 10$ μF when $d = 1.0$ mm, what will be the capacitance when $d = 0.25$ mm?

In problems 7 through 10, use the power formulas to find the indicated quantity.

7. $V_1 = 110$ V, $I_1 = 0.8$ A; P_1

8. $P_2 = 2.2$ W, $V_2 = 3.5$ V; I_2

9. $I_T = 34$ mA, $R_T = 1.3$ kΩ; P_T

10. $P_3 = 10$ W, $R_3 = 33$ Ω; V_3

11. A 1.2-kΩ resistor dissipates 50 W of power.
  (a) Find the current through the resistor.
  (b) Find the energy consumption in kilowatt-hours for 72 h.

12. A high-voltage video circuit draws 560 μA and uses 100 mW of power.
  (a) What is the voltage across the circuit?
  (b) If the current doubles and the voltage remains constant, how much power does the circuit use?

13. Given the series circuit in Figure 5–32 with $R_1 = 33$ Ω, $R_2 = 18$ Ω and $V_T = 24$ V,
  (a) Find the current I.
  (b) Find V_1 and V_2.

14. Given the series circuit in Figure 5–32 with $V_T = 24$ V,
  (a) If $R_1 = 2R_2$, find V_1 and V_2.
  (b) If $R_1 = 120$ Ω and $R_2 = 240$ Ω, find the total power dissipated in the circuit.

15. Figure 5–33 shows a voltage dropping variable resistor in series with a bulb and a 9-V battery.
  (a) If the bulb uses 5.0 W of power and draws 800 mA, what should be the value of R_D?
  (b) What should be the value of R_D to decrease the voltage across the bulb by one-half of the voltage in part (a)?

16. Three resistances R_1, R_2, and R_3 are connected in series to a battery. The total resistance of the circuit is 1.1 kΩ. If $V_2 = V_1$ and $V_3 = \frac{1}{2}V_2$, find each resistance.

17. Two resistances $R_1 = 180$ Ω and $R_2 = 220$ Ω are connected in a parallel bank.
  (a) What is the resistance of the bank?
  (b) How much resistance should be added in parallel to reduce the resistance of the bank by one-half?

18. Three resistances R_1, R_2, and R_3 are connected in parallel across a voltage of 40 V. If $R_1 = 33$ Ω, $R_2 = 68$ Ω, and $R_3 = 100$ Ω, find I_1, I_2, I_3, I_T, R_T, and G_T for the circuit.

19. For the series-parallel circuit in Figure 5–34,
  (a) Find the formula for R_T.
  (b) Calculate R_T given $R_1 = 120$ Ω, $R_2 = 130$ Ω, and $R_3 = 150$ Ω.

20. For the series-parallel circuit in Figure 5–34, find R_3 if $R_1 = 3.0$ kΩ, $R_2 = 2.0$ kΩ, and R_T is to be 3.0 kΩ.

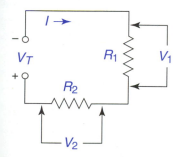

FIGURE 5–32 Series-circuit for problems 13 and 14.

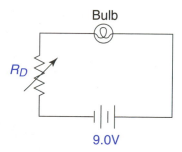

FIGURE 5–33 Voltage-dropping resistor for problem 15.

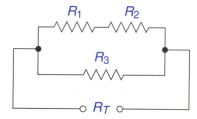

FIGURE 5–34 Series-parallel circuit for problems 19 and 20.

21. For the series-parallel circuit in Figure 5–35, given $R_1 = 100\ \Omega$, $R_2 = 200\ \Omega$, $R_3 = 110\ \Omega$, $R_4 = 90\ \Omega$, and $I_T = 500$ mA, find R_T, V_1 and V_3.

22. For the series-parallel circuit in Figure 5–35, given $R_1 = 15\ \Omega$, $R_2 = 50\ \Omega$, $R_3 = 50\ \Omega$, $R_4 = 75\ \Omega$, and $V_T = 6.5$ V, find I_1, V_3, and the total power dissipated.

23. For the series-parallel circuit in Figure 5–36, given $R_1 = 240\ \Omega$, $R_2 = 1.1$ kΩ, $R_3 = 1.3$ kΩ, $R_4 = 1.6$ kΩ, and $I_T = 35$ mA, find R_T and V_T.

24. For the series-parallel circuit in Figure 5–36, given $R_1 = 100\ \Omega$, $R_2 = 100\ \Omega$, $R_3 = 200\ \Omega$, $R_4 = 200\ \Omega$, and $V_T = 50$ V, find R_T and I_T.

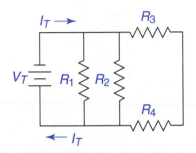

FIGURE 5–35 Series-parallel circuit for problems 21 and 22.

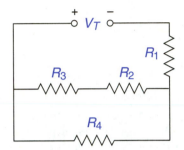

FIGURE 5–36 Series-parallel circuit for problems 23 and 24.

Courtesy of Siemens Corporation

A technician examines the design of integrated circuits.

This chapter continues with the methods of equation solving presented in Chapter 4 and that are important in electricity and electronics. It begins with the study of first-degree equations in two variables and the graphs of these equations, which are straight lines. This leads to the important concept of a linear function. The algebraic and graphical solution of systems of two linear equations and their applications are then studied. This is followed by the important concepts of determinants. The algorithms for solving systems of two and three linear equations using determinants is shown. These algorithms are convenient because they can be readily programmed on a calculator or computer. Systems of equations occur often in electricity and electronics and one of their most important applications is in solving circuits and networks.

Chapter Objectives

In this chapter you will learn:

- How to plot the graph of a linear equation in two variables.
- The concept of a linear function and how to apply it to formulas that are linear functions.
- How to solve two equations with two unknowns graphically and algebraically.
- How to expand a 2×2 and a 3×3 determinant.
- How to solve a linear system of two equations using determinants.
- How to solve a linear system of three equations using determinants.

6.1
GRAPHS AND LINEAR FUNCTIONS

Rectangular Coordinate System

A rectangular coordinate system uses a horizontal or x axis and a vertical or y axis. See Figure 6–1. Although x and y are the convention, any variables can be used. Each point is located using two coordinates. The first coordinate, x, is the distance in the horizontal direction. The second coordinate, y, is the distance in the vertical direction. To the right and up is positive; to the left and down is negative. The following example shows the procedure for plotting points on a rectangular coordinate system and is important in understanding many types of graphs. If you are already familiar with rectangular coordinates, this example will provide a good review.

EXAMPLE 6.1

Draw a rectangular coordinate system and plot the following points: A (6,5), B (–3,2), C (0,4), D (4,0), E (–5,–6), F (–5,0), G (2,–3), and H (0,–5).

Solution   First draw the x and y axes. This divides the graph into four quadrants I, II, III, and IV, as shown in Figure 6–1. The scales are usually shown on each axis but can be omitted if all the boxes are equal to one unit. Locate each point as follows: Measure the distance of the first coordinate along the horizontal or x axis. Move to the right for (+) and to the left for (–). Then measure the distance of the second coordinate in the vertical or y direction. Move up for (+) and down for (–). Study Figure 6–1 to see how each point is plotted.

Observe the following: The origin where the axes intersect has the coordinates (0,0). Every point on the x axis has a y coordinate of zero. Every point on the y axis has an x coordinate of zero. In quadrant I, x and y are both positive. In quadrant II, x is negative and y is positive. In quadrant III, x and y are both negative, and in quadrant IV, x is positive and y is negative.

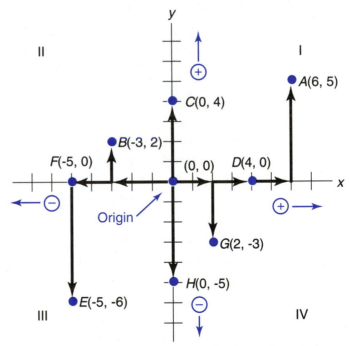

FIGURE 6–1   Rectangular coordinate system for Example 6.1.

The next example shows how to plot the graph of a first degree or linear equation in two variables, which is a straight line.

EXAMPLE 6.2

Draw the graph of the following linear equation:
$$2x - y = -1$$

Solution   It is necessary to find values of x and y (points) that satisfy the equation. To help calculate values of y for certain values of x, first solve for y:

$$2x - y = -1$$
$$2x = -1 + y$$
$$2x + 1 = y$$
$$y = 2x + 1$$

In this form, y is expressed as a *linear or first degree function* of x. This concept is explained further after this example. To graph this function, you can now assign any value to x (the independent variable) and calculate the corresponding value of y (the dependent variable). Two points are the minimum necessary to graph a linear function or straight line. However, if one point is incorrect, the entire line is wrong. Therefore, you should plot at least three points or more. You should choose convenient positive, zero, and negative values of x, such as those shown in Figure 6–2: $x = -2, -1, 0, 1,$ and 2. Substitute each value of x in the function and calculate the corresponding value of y. Figure 6–2 shows the table of values of x and the calculated values of y, the five plotted points, and the graph of the line $y = 2x + 1$.

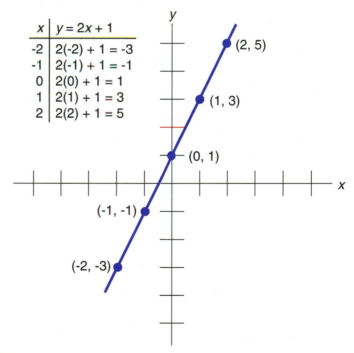

FIGURE 6–2   Linear function for Example 6.2.

Linear Function

The concept of a *function* is an important idea. There are many examples of different types of functions in the chapters that follow. Observe in the preceeding example that each value of *x* determines exactly one value of *y*. When that relationship exists between two variables, the second variable (*y*) is said to be a function of the first variable (*x*).

> **Function**   *y* is a function of *x* if, given a set of values for *x*, there corresponds one and only one value of *y* for each value of *x*.   (6.1)

In Example 6.2, since the graph of $y = 2x + 1$ is a straight line, *y* is called a *linear function* of *x*. Later chapters introduce other types of functions that do not graph as straight lines, such as quadratic functions, trigonometric functions, exponential functions, and logarithmic functions.

A linear function is a *first-degree* equation in *x* and *y*. That is, there are no exponents greater than one or negative exponents in a linear equation. A first-degree equation, when solved for *y*, expresses *y* as a linear function of *x*:

$$y = mx + b \qquad (m,b \text{ constants}) \tag{6.2}$$

Equation (6.2) always graphs as a straight line. The constant *m* is called the slope and is a measure of the steepness of the line. The constant *b* is the *y* intercept and is the value of *y* where the line crosses the *y* axis. When $b = 0$, Equation (6.2) becomes $y = mx$ and states that *y* is directly proportional to *x*. The variable *m* is the constant of proportion. Direct proportion was introduced in Section 5.1.

The next example shows some applications of linear functions and direct proportion to electricity and other technical areas.

EXAMPLE 6.3

Illustrate some applications of linear functions and direct proportion.

Solution   The first five cases are linear functions that are also relations of direct proportion.

1. $V = IR$. Ohm's law in this form states that when the resistance *R* of a circuit is constant, the voltage *V* is a linear function of the current *I* and *V* is directly proportional to *I*.

2. $C = 2\pi r$. The circumference *C* of a circle is a linear function of the radius *r* and *C* is directly proportional to *r*.

3. $P = VI$. When the voltage *V* of a circuit is constant, the power *P* is a linear function of the current *I* and *P* is directly proportional to *I*. See Example 6.6.

4. $F = ma$. Newton's second law of motion states that when the mass *m* of a body is constant, the force *F* exerted on the body is a linear function of the acceleration *a*. The force *F* is directly proportional to the acceleration.

EXAMPLE 6.3 (Cont.)

5. $R = \frac{\rho L}{A}$. The resistance R of a wire is a linear function of the length L; ρ is a constant called the resistivity, which depends on the type of wire; and A is the cross-sectional area of the wire. The variable R is directly proportional to the length L.

   The next three cases are linear functions where the constant b is not zero. In this case, it is the *change* in the variables that is proportional rather than the variables themselves.

6. $F = 1.8C + 32$. This is the familiar temperature conversion formula for changing Celsius to Fahrenheit. It states that F is a linear function of C. The *change* in the Celsius temperature is proportional to the *change* in the Fahrenheit temperature. See Example 6.5.

7. $R_t = R_0 + R_0\alpha t$. Assuming α is constant, the resistance R_t of a material is a linear function of the temperature t in degrees Celsius. Here, R_0 is the resistance of the material at 0°C and α is the temperature coefficient of resistance at 0°C. This formula can also be expressed as $R_t = R_0(1 + \alpha t)$, where R_0 is the common factor. See Example 6.7.

8. $v = v_0 + at$. The velocity v of a body with constant acceleration a is a linear function of the time t, where v_0 is the velocity at $t = 0$.

The graph of a linear function can also be drawn by leaving the equation in *standard form*:

$$Ax + By = C \tag{6.3}$$

The next example shows how to plot the graph of a linear function in standard form using the *x* intercept, which is the point where the line crosses the *x* axis, and the *y* intercept, which is the point where the line crosses the *y* axis.

EXAMPLE 6.4

Draw the graph of $2x - y = -1$ by plotting the *x* and *y* intercepts.

Solution   This is the same equation as in Example 6.2. Working with the equation in the given standard form you can assign values to either *x* or *y*. The values $x = 0$ and $y = 0$ are easy to work with and give the points where the line intercepts each axis as follows:

y intercept	*x* intercept
Let $x = 0$	Let $y = 0$
$0 - y = -1$	$2x - 0 = -1$
$y = 1$	$x = -\dfrac{1}{2}$

The point $(0,1)$ is where the line crosses the *y* axis and is the *y* intercept. The point $(-\frac{1}{2}, 0)$ is where the line crosses the *x* axis and is the *x* intercept. Since the *x* intercept falls between two units on the graph, it is necessary to estimate its position halfway between two units or change the scale. See Figure 6–3. Now choose one more value for *x* as a check point and solve for *y*. Using $x = 1$ as the check point,

$$2(1) - y = -1$$
$$-y = -3$$
$$y = 3$$

EXAMPLE 6.4 (Cont.)

The three points $(0,1)$, $(-\frac{1}{2}, 0)$, and $(1,3)$, and the line $2x - y = -1$ are shown in Figure 6–3.

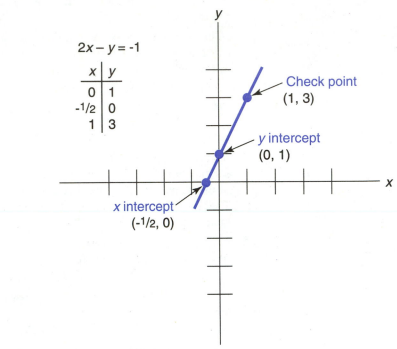

FIGURE 6–3 Intercept method for Example 6.4.

The next example shows an application of the temperature conversion formula from Example 6.3, Case 6.

EXAMPLE 6.5

Given the linear function for temperature conversion,

$$F = 1.8C + 32$$

construct a table of values from –30°C to 40°C using values every 10°, and draw the graph of the function. This range of values is what people experience in temperate climates around the world during the course of a year.

Solution Choose values of C every 10°, and calculate the corresponding values of F to produce the following table:

C	(°C)	– 30	– 20	–10	0	10	20	30	40
F	(°F)	– 22	– 4	14	32	50	68	86	104

For example, when $C = -10°$,

$$F = 1.8(-10) + 32 = -18 + 32 = 14°$$

EXAMPLE 6.5 (Cont.)

The table enables you to compare certain temperatures in degrees Celsius with those in degrees Fahrenheit. However, when you draw the graph, it will provide you with a picture of how the temperature scales compare for all values of C and F. In Figure 6–4, the horizontal axis is used for Celsius values because in the linear function $F = 1.8C + 32$, C takes the place of the independent variable x. Similarly, the vertical axis is used for Fahrenheit values.

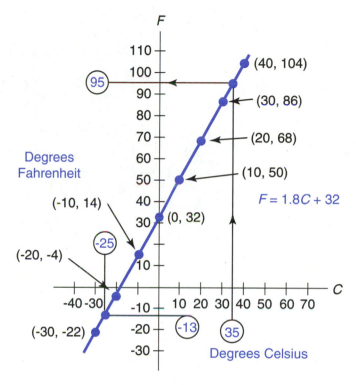

FIGURE 6–4 Fahrenheit versus Celsius for Example 6.5.

Each box on the graph is set equal to 10 units to accommodate the range of values. The values from the table are then plotted on the graph. From the graph you can determine any value of F for a given value of C, or vice versa. For example, when $C = 35°$, the arrows on the graph show that $F = 95°$. Similarly, when $F = -13°$, the arrows show that $C = -25°$.

Sometimes it is useful to quickly determine the value of F for a certain value of C. An approximate formula that can be used is $F \approx 2C + 30$. This gives results within a few degrees of error for the preceding range of values.

The next two examples close the circuit and show electrical applications of linear functions.

EXAMPLE 6.6

Close the Circuit

Example 6.3, case 3 states that the power in a circuit is directly proportional to the current when the voltage is constant. Figure 6–5(a) shows a series-parallel circuit containing a variable resistor R_V in series with a parallel bank R_1 and R_2. If the battery voltage is constant at 24 V and R_V varies from 10 Ω to 30 Ω, graph the total power P_T versus the total current I_T showing at least five points.

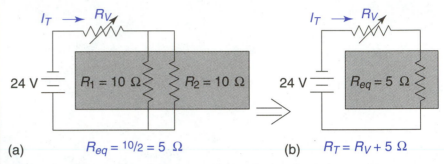

(a) $R_{eq} = {}^{10}/2 = 5\ \Omega$ (b) $R_T = R_V + 5\ \Omega$

FIGURE 6–5   Series-parallel circuit for Example 6.6.

Solution   You first need to calculate the total resistance R_T for five values of R_V. Then you can calculate the total current I_T and the total power P_T. For the parallel bank $R_1 = R_2 = 10\ \Omega$. Therefore, the equivalent resistance of the parallel bank is

$$R_{eq} = \frac{10}{2} = 5\ \Omega$$

The circuit can then be reduced to the series circuit in Figure 6–5(b) and the total resistance is

$$R_T = R_V + 5\ \Omega$$

If R_V varies from 10 Ω to 30 Ω, a difference of 20 Ω, choose values of R_V every 5 Ω which will provide the five points: 10 Ω, 15 Ω, 20 Ω, 25 Ω and 30Ω. For each value of R_V, you calculate I_T using Ohm's law and then P_T using the power formula. For example, when $R_V = 25\ \Omega$,

$$R_T = 25 + 5 = 30\ \Omega$$

$$I_T = \frac{V_T}{R_T} = \frac{24\ \text{V}}{25\ \Omega} = 0.96\text{A}$$

$$P_T = V_T I_T = (24\ \text{V})(0.96\,\text{A}) = 23.04\text{W} \approx 23\ \text{W}$$

The table of values is then

R_V (Ω)	10	15	20	25	30
R_T (Ω)	15	20	25	30	35
I_T (A)	1.6	1.2	0.96	0.80	0.69
P_T (W)	38	29	23	19	16

EXAMPLE 6.6 (Cont.)

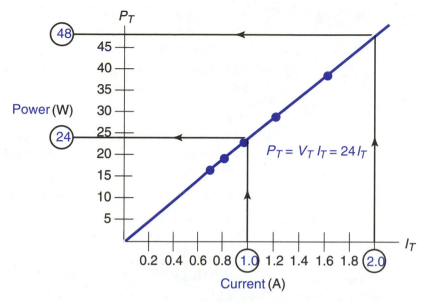

FIGURE 6–6   Power versus current in a circuit for Example 6.6.

To graph P_T versus I_T, you use the vertical axis for P_T and the horizontal axis for I_T. You must use separate scales for I_T and P_T that will clearly show all the values in the table and the relationship between P_T and I_T. To decide on the scale to use for I_T, take the difference between the largest and the smallest value of I_T to be shown on the graph. Assuming the graph will start at $I_T = 0$, this difference is 1.6 – 0 = 1.6 A. Divide it by 10, which is 0.16. Round this off to one digit 0.2. You can then use 0.1 A, 0.2 A, or 0.3 A for each box on the horizontal axis, and this will give you approximately 10 boxes to spread the points on the graph. Choose the one that makes it easiest to estimate the points; 0.1 A and 0.2 A are good choices. Here, 0.2 A is used for the graph in Figure 6–6. The difference between the largest and the smallest value of P_T to be shown on the graph is 38 – 0 = 38. Divided by 10 this is 3.8 ≈ 4. You can choose 3 W, 4 W or 5 W for each box. Each box is set equal to 5 W to make it easier to estimate the points. Observe that the graph shows P_T is directly proportional to I_T. When $I_T = 1.0$ A, the arrows on the graph show $P_T = 24$ W. When the current doubles to 2.0 A, the arrows show that P_T doubles to 48 W. ▪

Study the next example, which closes the circuit and is an important application of linear functions to the electrical resistance of a material.

EXAMPLE 6.7

Close the Circuit

As shown in Example 6.3, case 7, the resistance R_t of a material is a linear function of the temperature t given by:

$$R_t = R_0 + R_0 \alpha t$$

where R_t = resistance at temperature t in °C
R_0 = resistance at 0°C
α = temperature coefficient of resistance at 0°C

EXAMPLE 6.7 (Cont.)

Given an aluminum wire where $\alpha = 4.00 \times 10^{-3}$ per °C and $R_0 = 6.00 \ \Omega$, assuming α is constant, graph R_t versus t from $t = -10°C$ to $t = 25°C$. Note that the units for α are 1/°C or °C^{-1}.

Solution Substitute the given values in the linear function to obtain the formula for the aluminum wire:

$$R_t = 6.00 + 6.00(0.00400)t = 6.00 + 0.0240t$$

Choose values of t every 5° to provide enough points for the graph. Substitute each value of t in the formula to obtain the corresponding value of R_t. For example, when $t = 15°$,

$$R_t = 6.00 + 0.0240(15) = 6.00 + 0.36 = 6.36 \ \Omega$$

The table of values is then

t (°C)	−10	−5	0	5	10	15	20	25
R_t (Ω)	5.76	5.88	6.00	6.12	6.24	6.36	6.48	6.60

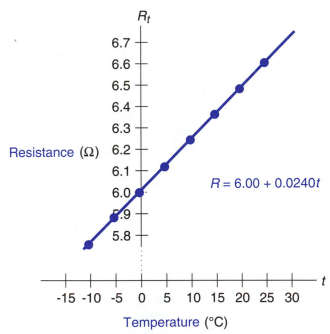

FIGURE 6−7 Resistance versus temperature for an aluminum wire for Example 6.7.

For the graph of this function, you must choose an appropiate scale for each axis. The difference between the largest and smallest values of R_t to be graphed is 6.60 − 5.76 = 0.84. Divided by 10 this is 0.084 ≈ 0.1. Each box is set equal to 0.1 Ω on the R_t (vertical) axis. See the graph in Figure 6–7. Observe how only the

EXAMPLE 6.7 (Cont.)

portion of the vertical axis that is needed is shown. The vertical axis is broken from below the graph to zero. Each box is set equal to 5 units on the t (horizontal) axis. The graph shows that R_t is a linear function for the given range of values. Over a larger range of temperature values, α may not be constant. This would mean that R_t is not a linear function of t over the larger range.

▪

EXERCISE 6.1

In problems 1 through 10, plot each point on a rectangular coordinate system.

1. (3,1)
2. (1,4)
3. (0,5)
4. (−5,0)
5. (−3,−2)

6. (−3,1)
7. (−5,−5)
8. (0,0)
9. (3.5,−0.5)
10. (1.5,−2.5)

In problems 11 through 18, draw the graph of the linear function. Solve for y first if necessary.

11. $y = x$
12. $y = 3x$
13. $y = 4x - 3$
14. $y = 0.5x + 1$

15. $2x + y - 4 = 0$
16. $x + 2y + 2 = 0$
17. $x - 4y = 4$
18. $5x - 5y = 8$

In problems 19 through 24, draw each graph using the intercept method. See Example 6.4. Plot the variable in parentheses on the horizontal axis.

19. $x - y = 3$
20. $x + y = -5$
21. $4C + 2L = 8$ (C)

22. $3A - 4B = 12$ (A)
23. $3R_1 - 4R_2 = 6$ (R_1)
24. $3V_1 + 6V_2 = 2$ (V_1)

In problems 25 through 30, draw each graph, adjusting scales if necessary. Plot the variable in parentheses on the horizontal axis. Plot at least five points for each graph.

25. $P = 3.5$ (I)
26. $V = 12.6$ (I)
27. $V_T = 3.2R + 2.3$ (R)

28. $L = 5.5n - 1.5$ (n)
29. $1.3X_C + X_L = 6.5$ (X_C)
30. $2\alpha + 10\beta = 15$ (α)

In problems 31 through 36, draw the graph of each linear function, plotting at least five points for each line.

31. Hooke's law says that the force exerted on a spring is a linear function of the distance x the spring stretches, given by $F = kx$. The force, F, is directly proportional to x since this function is like Equation (6.2), where $b = 0$. If $k = 1.5$ lb/ft for a given spring, graph F versus x from $x = 0$ ft to $x = 6$ ft.

32. Given Newton's second law of motion, $F = ma$, from Example 6.3, case 4, where $m = 1.1$ kg, graph F versus a from $a = 0$ m/s² to $a = 12$ m/s². Use the horizontal axis for a and note that F is measured in newtons (N).

33. Given the linear function $v = v_0 + at$ from Example 6.3, case 8, where $v_0 = 100$ cm/s and $a = 50$ cm/s², graph v versus t from $t = -3$ s to $t = 5$ s. Use the horizontal axis for t and different scales for v and t.

34. The velocity of sound in air in meters per second is a linear function of the temperature t in degrees Celsius given by

$$v = 0.607t + 332$$

Graph v versus t from $t = -20°$ to $t = 40°$. Use the horizontal axis for t and different scales for v and t. Show the v axis from $v = 310$ m/s. See Example 6.7.

35. The cost C for the use of a powerful computer system is a linear function of the time t in minutes given by

$$C = 150t + 1200$$

   (a) Graph this function from $t = 5$ min to $t = 60$ min. Use the horizontal axis for t.
   (b) What is the minimum cost for the use of the computer?

36. (a) Extend the graph in Fig. 6–4 for values of $C < -30$, and find the point where $F = C$.
   (b) Calculate this point algebraically by letting $F = C$ in the formula and solving for C.

Applications to Electronics In problems 37 through 44, draw the graph for each application. Use appropriate scales and the horizontal axis for the second variable.

37. Figure 6–8 shows a circuit containing a resistance $R_0 = 150$ Ω. Assuming that the resistance is constant, graph the voltage drop V_0 versus the current I for $I = 0$ mA to $I = 30$ mA. Use the horizontal axis for I.

38. In problem 37, show that the power P in the circuit is directly proportional to the square of the current I by graphing P in milliwatts versus I^2 in (mA)^2 for $I = 0$ mA to $I = 30$ mA. Use the horizontal axis for I^2.

39. A brass wire has a temperature coefficient of resistance $\alpha = 2.00 \times 10^{-3}$ per °C at 0°C. If the resistance at 0°C, $R_0 = 15.0$ Ω, graph resistance versus temperature from $t = -10°$C to $t = 20°$C. Use the horizontal axis for t. See Example 6.7.

40. A copper wire has a resistivity $\rho = 1.55 \times 10^{-8}$ Ω–m and a diameter $d = 2.00$ mm. Graph resistance versus length from $L = 1.0$ m to $L = 6.0$ m. Use the horizontal axis for L and different scales for R and L. See Example 6.3, case 5, and note that $A = \frac{\pi d^2}{4}$.

41. Figure 6–9 shows a dc circuit containing a series voltage dropping resistor R_S and a load R_L. The voltage across R_L is a linear function of the current I given by

$$V_L = V_B - V_S$$

where V_B = battery voltage and V_S = voltage drop across R_S. If $V_B = 9.0$ V and $R_S = 100$ Ω, graph V_L versus I from $I = 10$ mA to $I = 40$ mA. Use the horizontal axis for I.

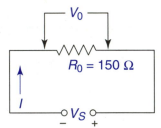

FIGURE 6–8 Voltage drop versus current in a resistance for problems 37 and 38.

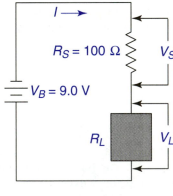

$$V_L = V_B - V_S$$

FIGURE 6–9 Series voltage dropping resistor for problem 41.

42. Figure 6–10 shows a circuit containing two series opposing voltages V_X and V_0. The positive terminals of V_X and V_0 are connected so that the voltages oppose each other. If V_0 is a constant voltage and V_X is a variable voltage, the total voltage V_T is a linear function of V_X, given by

$$V_T = V_0 - V_X$$

If $V_0 = 12$ V, graph V_T versus V_X from $V_X = 0.0$ V to $V_X = 3.0$ V. Use the horizontal axis for V_X.

43. Figure 6–11 shows a series-parallel circuit with a rheostat R in series with a parallel bank of two bulbs R_1 and R_2. If $R_1 = 600$ Ω, $R_2 = 400$ Ω, and the applied voltage $V_A = 120$ V, graph the total power P_T versus I_T when R varies from 50 Ω to 250 Ω. Use the horizontal axis for I_T.

44. In an amplifier circuit, the dc collector voltage V_C is related to the collector current I_C by the linear function

$$V_C = V_{CC} - I_C R_L$$

where V_{CC} is the fixed voltage supply for the collector current and R_L is the load resistor. For $R_L = 2.5$ kΩ and $V_{CC} = 16$ V, graph V_C versus I_C from $I_C = 1$ mA to $I_C = 5$ mA. Use the horizontal axis for I_C.

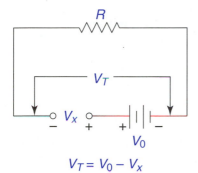

$$V_T = V_0 - V_X$$

FIGURE 6–10 Series-opposing voltages for problem 42.

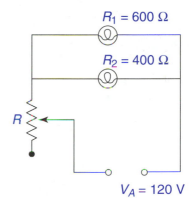

$V_A = 120$ V

$$P_T = 120\, I_T$$

FIGURE 6–11 Rheostat in series with parallel bulb bank for problem 43.

≡ 6.2
SIMULTANEOUS EQUATIONS

Problems involving electrical networks and other technical applications lead to linear equations containing two or more unknowns. These equations are called linear systems and must be solved simultaneously, that is, at the same time. There are several methods of solution that can be used. Two algebraic methods and a graphical method are shown in this section.

Study the first example carefully. It illustrates all three methods of solution for two simultaneous equations.

EXAMPLE 6.8

Solve the linear system of two equations algebraically and graphically:

$$2x - y = -5$$
$$3x + 2y = 3$$

EXAMPLE 6.8 (Cont.)

Solution Two algebraic solutions and the graphical solution are shown here.

Algebraic Solution—Addition Method. To solve by the addition method, you multiply one or both equations so that the coefficients of x, or y, are the same, but opposite in sign. Then you add the equations and eliminate x or y. To eliminate x in this example, you find the least common multiple (LCM) of the coefficients 2 and 3, which is 6. Then, if you multiply the first equation by 3, and the second equation by –2, one coefficient will become 6 and the other –6. However, you can eliminate y easier by just multiplying the first equation by 2 and adding equations. Therefore, y is eliminated as follows:

$$2(2x - y = -5) \rightarrow 4x - 2y = -10$$
$$3x + 2y = 3$$
$$\overline{}$$
$$7x = -7$$
$$x = -1$$

To find y, substitute $x = -1$ into the first (or second) equation:

$$2x - y = -5$$
$$2(-1) - y = -5$$
$$-y = -3$$
$$y = 3$$

The solution is, therefore, $(-1,3)$.

Algebraic Solution—Substitution Method. To solve by substitution, first solve either of the equations for one of the variables. Then substitute this expression into the other equation. The first equation, $2x - y = -5$, can be readily solved for y and substituted into the second equation:

$$2x - y = -5$$

Transpose $2x$:

$$-y = -2x - 5$$

Multiply by –1:

$$y = \boxed{2x + 5}$$

Now substitute $2x + 5$ for y in the other equation:

$$3x + 2y = 3$$
$$3x + 2(2x + 5) = 3$$

Simplify:

$$3x + 4x + 10 = 3$$

Solve for x:

$$7x = -7$$
$$x = -1$$

Then find y by substituting –1 for x in the equation solved for y:

$$y = 2x + 5 = 2(-1) + 5 = 3$$

The substitution method is helpful when variables cannot be easily eliminated by adding equations and may be the only method of solution.

EXAMPLE 6.8 (Cont.)

Graphical Solution. Each equation is a linear function whose graph is a straight line. The intersection of the two lines represents the value of x and y that satisfy both equations simultaneously. It is, therefore, the solution of the system. For each equation, construct a table of values for at least three points, and carefully draw each line on the same graph. Figure 6–12 shows the table of values and the graph for each line. Each table contains the intercepts and a check point for each line. The point of intersection $(-1,3)$ is the graphical solution that agrees with the algebraic solution.

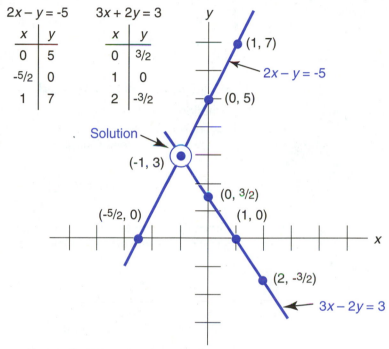

FIGURE 6–12 Graphical solution of simultaneous equations for Example 6.8

Study the next example which closes the circuit and shows an application to voltages connected in series.

EXAMPLE 6.9

Close the Circuit

Two unknown voltages V_1 and V_2 are connected in series to a resistance $R = 12\ \Omega$. When the voltages are series-aiding with the positive terminal of one connected to the negative terminal of the other, the ammeter in the circuit reads 1.5 A. See Figure 6–13(a). When the voltages are series-opposing with the negative terminal of one connected to the negative terminal of the other, the ammeter reads 500 mA. See Figure 6–13(b). What are the two voltages?

EXAMPLE 6.9 (Cont.)

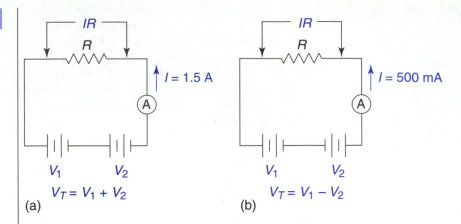

FIGURE 6–13 Series-aiding and series-opposing voltages for Example 6.9.

Solution Apply the problem-solving techniques from Section 4.3. Let V_1 = the larger voltage and V_2 = the smaller voltage. When two voltages V_1 and V_2 are series-aiding, the total voltage V_T equals their sum:

$$V_T = V_1 + V_2$$

When two voltages are series-opposing, the total voltage equals their difference:

$$V_T = V_1 - V_2$$

Since the IR voltage drop across the resistance equals the total voltage for each circuit, you can write two simultaneous equations:

$$V_1 + V_2 = IR = (1.5)(12) = 18$$

$$V_1 - V_2 = (0.5)(12) = 6.0$$

You can solve this linear system of two equations algebraically or graphically. Both solutions are shown next.

Algebraic Solution. The two equations represent a linear system that can be solved algebraically using the method of addition, or substitution, to eliminate one of the unknowns. The addition method is shown here. Adding the two equations as given eliminates V_2:

$$V_1 + V_2 = 18$$

$$V_1 - V_2 = 6.0$$

$$\overline{2V_1 + 0 = 24}$$

$$V_1 = 12 \text{ V}$$

You now solve for V_2, substituting $V_1 = 12$ into either of the original equations. Using the first equation,

$$V_1 + V_2 = 18$$

$$12 + V_2 = 18$$

$$V_2 = 6.0 \text{ V}$$

EXAMPLE 6.9 (Cont.)

Graphical Solution. Each equation is a linear function whose graph is a straight line. The intersection of the two lines represents the values of V_1 and V_2 that satisfy both equations simultaneously and is therefore the solution of the system. For each equation, construct a table of values for at least three points, and carefully draw each line on the same graph. Figure 6–14 shows the table of values and the graph for each line. Each table contains the intercepts and a check point for each line. The point of intersection (12,6) is the graphical solution that agrees with the algebraic solution.

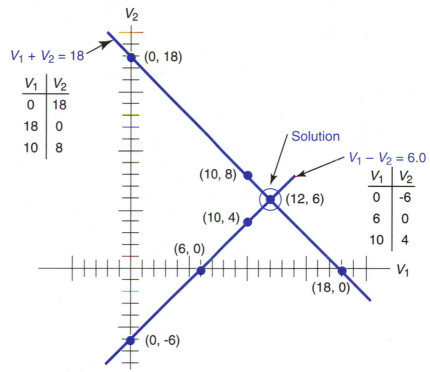

FIGURE 6–14 Graphical solution of simultaneous equations for Example 6.9.

Graphical methods of solving problems are generally not as precise as algebraic methods. However, graphs help to further the understanding and solution of a problem and serve as a check on the algebra. In many technical and engineering problems, graphical methods provide a more direct and quicker solution and adapt more readily to programming on a computer. Sometimes they are the only way to solve a problem.

EXAMPLE 6.10

Solve the linear system algebraically and graphically to the nearest tenth:
$$5R + 2T = 4$$
$$3R + 5T = 6$$

Solution *Algebraic Solution.* It does not make too much difference whether you eliminate R or T first. Suppose you choose R. Multiply the first equation by –3 and the second equation by 5. (You could also use 3 and –5.) This pair of numbers produces the LCM of 3 and 5, which is 15, as the coefficient of R:

$$(-3)(5R + 2T) = 4) \rightarrow -15R - 6T = -12$$
$$(5)(3R + 5T = 6) \rightarrow \underline{15R + 25T = 30}$$
$$19T = 18$$
$$T = \frac{18}{19} = 0.94 \approx 0.9$$

Substitute into the first equation:
$$5R + 2(0.94) = 4$$
$$5R = 2.12$$
$$R = 0.42 \approx 0.4$$

The solution is, therefore, (0.4, 0.9) to the nearest tenth. Note that both answers are calculated to the nearest hundredth first and then rounded off.

Graphical Solution. Plot R on the x axis and T on the y axis. Find the intercepts and one check point for each equation. When plotting points, you estimate decimal values. See the graph in Figure 6–15.

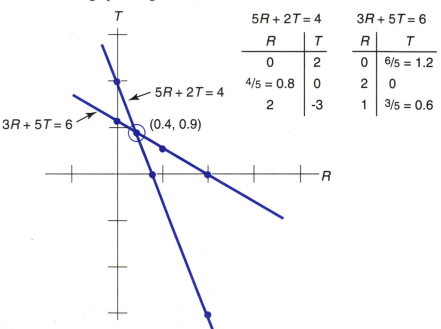

$5R + 2T = 4$		$3R + 5T = 6$	
R	T	R	T
0	2	0	6/5 = 1.2
4/5 = 0.8	0	2	0
2	-3	1	3/5 = 0.6

FIGURE 6–15 Graphical solution of simultaneous equations for Example 6.10.

EXAMPLE 6.10 (Cont.)

Note that for the equation $5R + 2T = 4$, $R = 1$ is *not* chosen as a check point. It is too close to the intercept $R = 0.8$ to be a good check. The coordinates of the point of intersection are estimated on the graph. You can always make the graphical solution as accurate as you want by using more boxes for each unit. ∎

The next two examples show applications of linear systems to a geometric and an electrical problem. The solutions of these verbal problems apply the problem-solving techniques discussed in Section 4.3. The solutions shown are algebraic. See the flowchart in Figure 6–16 that illustrates the procedure.

EXAMPLE 6.11

Adam bought 320 ft of fencing to enclose the front and two sides of his house within a rectangular fence. See Figure 6–17. He decides that the front section will be three times as long as each side. What should be the dimensions of the rectangle in feet and meters?

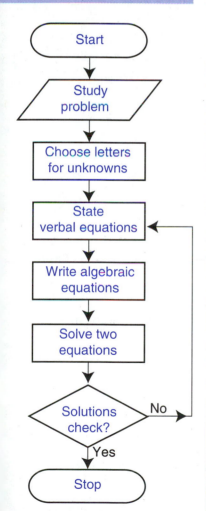

FIGURE 6–16 Algorithm for solving verbal problems with two unknowns.

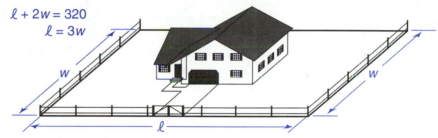

$\ell + 2w = 320$
$\ell = 3w$

FIGURE 6–17 House fence for Example 6.11.

Solution   This problem has two unknowns. Let l = length and w = width. From the first sentence you have the verbal equation:

$$(\text{Length}) + (\text{Width}) + (\text{Width}) = \text{Total fencing}$$

The first algebraic equation is then:

$$l + w + w = 320$$

The second sentence says

$$\text{Length} = 3 \times \text{Width}$$

which gives the second algebraic equation:

$$l = 3w$$

These equations can be solved by the addition method or the substitution method. Since the second equation is already solved for l, the substitution method lends itself readily to the solution. Substitute $3w$ for l in the first equation:

$$l = 3w$$
$$l + w + w = 320$$
$$3w + w + w = 320$$

Then,

$$5w = 320$$
$$w = 64 \text{ ft}$$
$$l = 3(64) = 192 \text{ ft}$$

EXAMPLE 6.11 (Cont.)

Check by substituting the values for w and l into the first equation:

$$192 + 2(64) = 192 + 128 = 320 \checkmark$$

Using the conversion 1 m = 3.28 ft from section 2.4, the answers in meters are

$$w = 64\,\text{ft}\left(\frac{1\,\text{m}}{3.281\,\text{ft}}\right) = 19.5\,\text{m}$$

$$l = 192\,\text{ft}\left(\frac{1\,\text{m}}{3.281\,\text{ft}}\right) = 58.5\,\text{m}$$

To solve this example by the addition method, it is first necessary to put both equations in the standard form:

$$l + 2w = 320$$
$$l - 3w = 0$$

Complete the solution by the addition method and check that you get the same answers.

■

Study the next example which closes the circuit and shows an application of Ohm's law to a series circuit.

EXAMPLE 6.12

Close the Circuit

Patricia has two types of unknown resistors, an ammeter and a 12-V battery supply. To find the value of the resistors, she first connects two resistors of the first type and one resistor of the second type in series with the 12-V battery. See Figure 6–18(a). The ammeter in the circuit reads 250 mA. Then she connects two resistors of the second type and one resistor of the first type in series, and the ammeter reads 200 mA. See Figure 6–18(b). Using this data, what are the values of the two resistors?

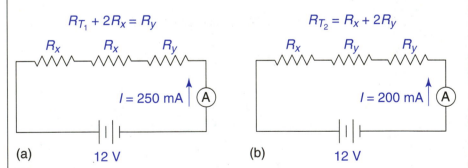

$$R_{T_1} + 2R_x = R_y \qquad\qquad R_{T_2} = R_x + 2R_y$$

(a) $I = 250\,\text{mA}$ 12 V

(b) $I = 200\,\text{mA}$ 12 V

FIGURE 6–18 Resistors in series for Example 6.12.

Solution Let R_x = first resistance and R_y = second resistance. Because the resistances are connected in series, the verbal equation that applies to both circuits is

Sum of the resistances = Total resistance

EXAMPLE 6.12 (Cont.)

The algebraic equations are then

$$2R_X + R_Y = R_{T_1}$$
$$R_X + 2R_Y = R_{T_2}$$

where R_{T_1} is the first total resistance and R_{T_2} is the second total resistance. Before you can solve the equations, it is necessary to find these total resistances using Ohm's law. In the first circuit,

$$R_{T_1} = \frac{V}{I} = \frac{12\,V}{0.250\,A} = 48\,\Omega$$

Note that 0.250 A must be used for 250 mA in Ohm's law.

In the second circuit,

$$R_{T_2} = \frac{12\,V}{0.200\,A} = 60\,\Omega$$

The equations for each circuit are then

$$2R_X + R_Y = 48$$
$$R_X + 2R_Y = 60$$

The addition method is shown for the solution of the two equations. To solve the system by the addition method, multiply the first equation by -2 and add the equations:

$$-2(2R_X + R_Y = 48) \rightarrow -4R_X - 2R_Y = -96$$
$$R_X + 2R_Y = 60$$
$$\overline{-3R_X = -36}$$
$$R_X = 12\,\Omega$$
$$\text{and } R_Y = 48 - 2(12) = 24\,\Omega$$

EXERCISE 6.2

In problems 1 through 12, solve each linear system algebraically and graphically. Show that the answers agree. Find decimal answers to the nearest tenth.

1. $2x - y = 4$
$x + y = 5$

2. $3x + y = 6$
$x - 2y = 2$

3. $I_1 + 2I_2 = 4$
$2I_1 + I_2 = 5$

4. $3V_1 - V_2 = 7$
$V_1 - 2V_2 = 9$

5. $3R + 2X = -1$
$6R - 4X = -6$

6. $5C - 2L = 20$
$C + 4L = 15$

7. $2I_a - 3I_b - 3 = 0$
$3I_a - 2I_b + 3 = 0$

8. $I_1 + 2I_2 - 3 = 0$
$3I_1 + 4I_2 - 5 = 0$

9. $3P_1 = 4 - 8P_2$
$P_1 = 4P_2$

10. $R_1 = 1 - R_2$
$R_2 = 4R_1 + 5$

11. $3r - t = 2.1$
$r + 2t = 1.4$

12. $\dfrac{b}{2} + n = 3$
$b - n = 0$

In problems 13 through 20, solve each linear system by any method (or as directed by your instructor).

13. $3A - B = 1$
$A + 2B = 12$

14. $X_1 - 2X_2 = 7$
$2X_1 + 3X_2 = 0$

15. $x = y + 4$
$y = 2x - 7$

16. $y = x + 6$
$x = 2 - y$

17. $2G_1 - 3G_2 = 4$
$3G_1 - 2G_2 = -2$

18. $6\alpha - 5\beta = -6$
$4\alpha + 4\beta = 7$

19. $0.5V_x + V_y = 4$
$V_x - 0.5V_y = -5$

20. $1.5\theta_1 - 2.0\theta_2 = -1.0$
$3.0\theta_1 + 2.5\theta_2 = 3.2$

In problems 21 through 28, solve each problem algebraically by setting up and solving two simultaneous equations.

21. The total number of fire alarms in Los Angeles in one week is 103. If there are 11 more false alarms than real alarms, how many of each kind are there?

22. Jose and Luz pull in the same direction on an object with a total force of 18 N. Reny joins them and pushes on the object in the same direction with a force equal to Luz. This results in a total force of 26 N. What is the force exerted by each person?

23. Jorge has 108 m of fencing to enclose his house on all four sides in the shape of a rectangle. If the width is to be half the length, what should be the dimensions of the rectangle in meters and feet? See Example 6.11.

24. Betty bought 105 ft of fencing to enclose a vegetable garden in her backyard. The garden is to be a rectangle, twice as long as it is wide, with a fence across the middle parallel to the width. What should the dimensions of the garden be in feet and meters? See Example 6.11.

25. Reginald lives 16 mi from his job. One morning he decides to get some exercise by walking part of the way and taking the bus the rest of the way. He estimates he can average 4 mi/h walking and 20 mi/h on the bus. If he allows one hour to get to work, how many minutes should he walk and how many should he ride? [Note: Distance = (Rate)(Time)].

26. A tug pushes a barge 10 mi upriver in 2.5 h and downriver the same distance in only 1 h. What is the speed of the tug relative to the water, and what is the speed of the current? [Note: Speed upriver = Tug speed + Current. Speed downriver = Tug speed − Current.]

27. Mr. Lee and his wife together had an unpaid balance of $500 on their credit cards for one month. The bank that issued Mr. Lee's credit card charges $1\frac{1}{4}\%$ a month interest on the unpaid balance; his wife's bank charges $1\frac{1}{2}\%$ a month. If the total interest charge was $6.75, how much was each unpaid balance? [Note: $1\frac{1}{4}\% = 0.0125$, $1\frac{1}{2}\% = 0.015$]

28. Carol invests $1000 in two CDs (certificates of deposit). The total interest after one year is $50.50. If the interest rate is $5\frac{1}{2}\%$ on one CD and $4\frac{1}{2}\%$ on the other, how much was invested in each CD?

Applications to Electronics In problems 29 through 40, solve each applied problem algebraically by setting up and solving two simultaneous equations.

29. The following equations come from a problem involving the intensity of an electrical field:

$$0.75E_x + 0.25E_y = 5.0$$
$$0.25E_x + 0.75E_y = 4.0$$

Solve this linear system for the electric intensities E_x and E_y in newtons/coulomb (N/C).

30. Solve Example 6.9 using the substitution method.

31. Applying Kirchhoff's voltage law to the series-parallel circuit in Figure 6–19, the following equations result for the currents I_1 and I_2:

$$3.5I_1 + 1.5I_2 = 0.50$$
$$1.5I_1 + 4.5I_2 = 0.50$$

Solve this linear system for the currents I_1 and I_2 in amps to two significant digits.

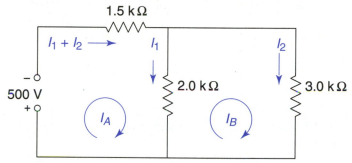

FIGURE 6–19 Kirchhoff's laws applied to a series-parallel circuit for problems 31 and 32.

32. Applying mesh equations and Kirchhoff's voltage law to the circuit in Figure 6–19, the following equations result:

$$3.5I_A - 2I_B = 0.50$$
$$5I_B - 2I_A = 0$$

Solve this system for I_A and I_B. Then find I_1 and I_2 using the two relationships $I_2 = I_B$ and $I_1 = I_A - I_B$. The answers should be the same as problem 31 to two significant digits.

33. In Figure 6–20(a), the total conductance of two resistors R_1 and R_2 in parallel equals 90 mS. In Figure 6–20(b), when another resistor R_3 whose resistance is equal to twice that of R_2 is added in parallel, the total conductance equals 110 mS. Find the conductances and the resistances of R_1 and R_2. Note: Conductance $G = \dfrac{1}{R}$ and $\dfrac{1}{2R} = \left(\dfrac{1}{2}\right)\left(\dfrac{1}{R}\right) = \dfrac{1}{2}G$

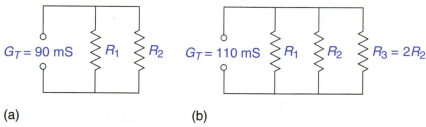

(a) (b)

FIGURE 6–20 Total conductance of resistors in parallel for problem 33.

34. When two inductances are connected in series so they are series-aiding, as shown in Figure 6–21(a), the total inductance is given by

$$L_T = L_1 + L_2 + L_M$$

where L_M is the mutual inductance. When the same two inductances are connected so they are series-opposing as shown in Figure 6–21(b), the total inductance is given by

$$L_T = L_1 + L_2 - L_M$$

If the total inductance is 1.20 mH when they are series-aiding and 500 μH when they are series-opposing, find L_2 and L_M if $L_1 = 400$ μH.

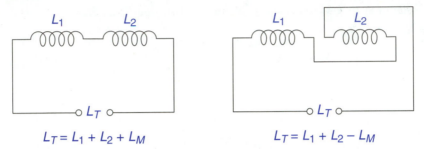

$$L_T = L_1 + L_2 + L_M \qquad\qquad L_T = L_1 + L_2 - L_M$$

FIGURE 6–21 Series-aiding and series-opposing inductances for problem 34.

35. Two unknown voltages are connected in series to a resistance R. When the voltages are series-aiding, as shown in Figure 6–22(a), the current is 1.7 A. When the voltages are series-opposing, as shown in Figure 6–22(b), the current is 500 mA in the same direction. If R is measured at 20 Ω, what are the two voltages? See Example 6.9.

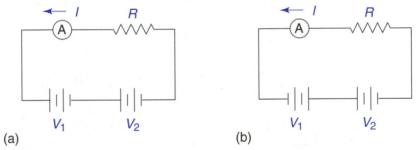

(a)   (b)

FIGURE 6–22 Series-aiding and series-opposing voltages for problems 35 and 36.

36. In Figure 6–22 when the voltages are series-aiding, the ammeter in the circuit reads 750 mA. When the voltages are series-opposing, the ammeter reads 150 mA. If $V_1 = 9.0$ V, how much is V_2 and the resistance R? See Example 6–9.

37. A powerful computer takes 60 ns (nanoseconds) to perform 5 operations of one type and 9 operations of a second type. The same computer takes 81 ns to perform 7 operations of the first type and 12 operations of the second type. How many nanoseconds are required for each operation?

38. When an unknown resistor R_X is connected in series with a resistance of 10 Ω to a battery, an ammeter in the circuit reads 250 mA as shown in Figure 6–23(a). When R_X is connected in series with a resistance of 20 Ω to the same battery, the ammeter reads 200 mA as shown in Figure 6–23(b). What is the value of the resistor and the voltage of the battery?

39. When two unknown resistors R_X and R_Y are connected in series to a 12-V battery, an ammeter in the circuit reads 100 mA. See Figure 6–24(a). When a third resistor, equal to R_X, is added in series to the circuit, the ammeter reads 60 mA. See Figure 6–24(b). What are the values of the two resistors? See Example 6.12.

40. In Problem 39 if the voltage of the battery is 20 V and the current readings are first 80 mA and then 50 mA, what are the values of the two resistors?

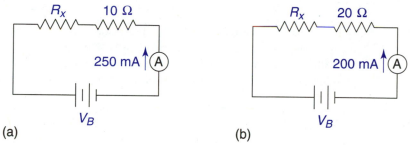

(a)   (b)

FIGURE 6–23   Unknown resistor in series for problem 38.

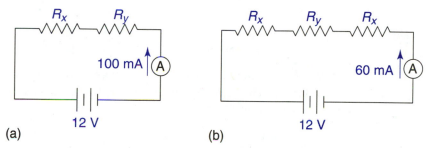

(a)   (b)

FIGURE 6–24   Unknown resistors in series for problem 39.

☰ 6.3
DETERMINANTS

Linear systems become more difficult to solve when the coefficients are decimals that are not simple numbers or when there are three or more equations in three or more variables. Matrices and determinants provide direct methods for solving all linear systems, and these methods adapt to programming on calculators and computers. Matrices were first introduced in 1858 and proved to be an important tool in many applications in electricity, electronics, and other technical areas.

Matrices

A *matrix* is a rectangular array of numbers, called elements, arranged in rows and columns. A matrix is enclosed in parentheses or brackets, and its size is given by (Number of rows) × (Number of columns). Some examples of matrices and their sizes are as follows:

$$\begin{pmatrix} 3 & -4 & 5 & -1 \\ 7 & 5 & 8 & 3 \end{pmatrix} 2 \times 4 \qquad \begin{pmatrix} 9 \\ 2 \\ 6 \end{pmatrix} 3 \times 1 \qquad \begin{pmatrix} 4.3 & 2.1 \\ -6.8 & 3.3 \\ 5.4 & -7.7 \end{pmatrix} 3 \times 2$$

The following matrices have the same number of rows as columns and are called *square matrices:*

$$\begin{pmatrix} -4 & -2 & 5 \\ 3 & -1 & 8 \\ -9 & 10 & -2 \end{pmatrix}_{3 \times 3} \qquad \begin{pmatrix} 2 & 4 \\ -2 & 3 \end{pmatrix}_{2 \times 2}$$

Every square matrix has a number associated with it called its *determinant*. Determinants of square matrices provide a key to solving linear systems. The determinant of a square matrix is represented by parallel lines adjacent to the elements as illustrated in (6.4). The determinant of a 2×2 matrix is calculated as follows:

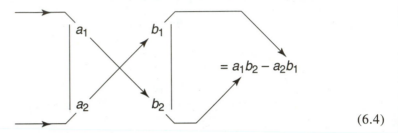

$$= a_1 b_2 - a_2 b_1 \tag{6.4}$$

That is, you multiply the elements in the upper-left diagonal and subtract the product of the elements in the lower-left diagonal. The determinant of the 2×2 matrix shown above is, therefore,

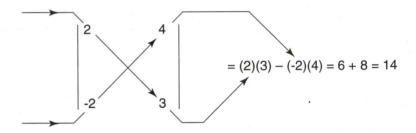

$$= (2)(3) - (-2)(4) = 6 + 8 = 14$$

The next example shows how to evaluate more 2×2 determinants.

EXAMPLE 6.13

Evaluate each 2×2 determinant:

$$\begin{vmatrix} 8 & -5 \\ -3 & 0 \end{vmatrix} = (8)(0) - (-3)(-5) = -15$$

$$\begin{vmatrix} 0.50 & -0.60 \\ 0.25 & -0.44 \end{vmatrix} = (0.50)(-0.44) - (0.25)(-0.60) = -0.070$$

Solution of Two Linear Equations

The following example shows the procedure for solving two linear equations using determinants. Study this example. The general formulas are presented after the example.

EXAMPLE 6.14

Solve the linear system of Example 6.8 using determinants:

$$2x - y = -5$$

$$3x + 2y = 3$$

Solution In Example 6.8 the solution of $(-1,3)$ is first found by the addition method. The addition method of solution is equivalent to the direct method shown here using determinants. The equations must first be in the standard form $Ax + By = C$. The solution is then as follows:

Constants ⟶

$$x = \frac{\begin{vmatrix} -5 & -1 \\ 3 & 2 \end{vmatrix}}{\begin{vmatrix} 2 & -1 \\ 3 & 2 \end{vmatrix}} = \frac{(-5)(2) - (3)(-1)}{(2)(2) - (3)(-1)} = \frac{-10 + 3}{4 + 3} = \frac{-7}{7} = -1$$

Coefficient matrix

Constants

$$y = \frac{\begin{vmatrix} 2 & -5 \\ 3 & 3 \end{vmatrix}}{\begin{vmatrix} 2 & -1 \\ 3 & 2 \end{vmatrix}} = \frac{(2)(3) - (3)(-5)}{(2)(2) - (3)(-1)} = \frac{6 + 15}{4 + 3} = \frac{21}{7} = 3$$

Coefficient matrix

Study these expressions closely. The *denominator for x and y is the same*. It is called the *coefficient matrix*. It is made up of the four coefficients of x and y arranged exactly as they appear in the equations in standard form. The numerator for x is then obtained from the coefficient matrix by replacing the column of x coefficients by the constants as they appear on the right side of the equations in standard form. The numerator for y is obtained from the coefficient matrix by replacing the column of y coefficients by the constants.

These procedures are expressed algebraically by *Cramer's Rule* (no relation to the author) as follows. Given a linear system of two equations in standard form,

$$a_1x + b_1y = k_1 \quad (k_1, k_2 \text{ constants})$$
$$a_2x + b_2y = k_2$$

the solutions for x and y are

$$x = \frac{\begin{vmatrix} k_1 & b_1 \\ k_2 & b_2 \end{vmatrix}}{\begin{vmatrix} a_1 & b_1 \\ a_2 & b_2 \end{vmatrix}} \quad y = \frac{\begin{vmatrix} a_1 & k_1 \\ a_2 & k_2 \end{vmatrix}}{\begin{vmatrix} a_1 & b_1 \\ a_2 & b_2 \end{vmatrix}} \tag{6.5}$$

Observe how the constants k_1 and k_2 replace the coefficients in each numerator. Remember that the equations *must be in standard form* when Cramer's rule is applied.

The next example closes the circuit and shows an application of determinants to an electrical problem.

EXAMPLE 6.15

Close the Circuit

The resistance of 1000 ft of no. 14 gauge copper wire at 20°C is measured to be 2.6 Ω. At 75°C it is measured to be 3.1 Ω. Find the resistance at 0°C and the temperature coefficient of resistance at 0°C for the wire.

Solution From Example 6.3, case 7, the resistance of a material is a linear function of the temperature given by

$$R_t = R_0 + R_0\alpha t$$

where R_t = resistance at t °C
R_0 = resistance at 0°C
α = temperature coefficient of resistance at 0°C

This equation can also be written as

$$R_t = R_0 + at$$

where $a = R_0\alpha$. In this form, the formula more closely resembles the linear function $y = mx + b$, where a takes the place of m and R_0 takes the place of b. Substituting the two sets of values of R_t and t, a linear system of two equations results where the unknowns are R_0 and a:

$$2.6 = R_0 + a(20)$$
$$3.1 = R_0 + a(75)$$

Write the equations in standard form for a and R_0:

$$20a + R_0 = 2.6$$
$$75a + R_0 = 3.1$$

Then apply Cramer's rule (6.5):

$$a = \frac{\begin{vmatrix} 2.6 & 1 \\ 3.1 & 1 \end{vmatrix}}{\begin{vmatrix} 20 & 1 \\ 75 & 1 \end{vmatrix}} = \frac{(2.6)(1) - (3.1)(1)}{(20)(1) - (75)(1)} = \frac{-0.50}{-55} = 0.0091\,\Omega/°C$$

$$R_0 = \frac{\begin{vmatrix} 20 & 2.6 \\ 75 & 3.1 \end{vmatrix}}{\begin{vmatrix} 20 & 1 \\ 75 & 1 \end{vmatrix}} = \frac{(20)(3.1) - (75)(2.6)}{(20)(1) - (75)(1)} = \frac{-133}{-55} = 2.4\,\Omega$$

Clearly, it is not necessary to calculate the denominator twice since it is the same. Then, since $a = R_0\alpha$,

$$\alpha = \frac{a}{R_0} = \frac{0.0091}{2.4} = 0.0038 \text{ per degree Celsius}$$

Solution of Three Linear Equations

A system of three linear equations can be solved using determinants of 3×3 matrices similar to the procedure for solving two linear equations. One method for evaluating the determinant of a 3×3 matrix uses six diagonals as follows. Write the 3×3 determinant repeating the first two columns on the right:

$$\begin{vmatrix} a_1 & b_1 & c_1 \\ a_2 & b_2 & c_2 \\ a_3 & b_3 & c_3 \end{vmatrix} \begin{matrix} a_1 & b_1 \\ a_2 & b_2 \\ a_3 & b_3 \end{matrix}$$

Draw three diagonals from the *upper* left. These are the *positive* products. Then draw three diagonals from the *lower* left. These are the *negative* products:

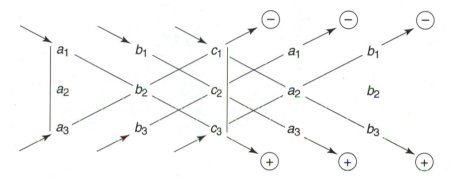

The value of the determinant is equal to the sum of the three positive products and the three negative products:

$$a_1b_2c_3 + b_1c_2a_3 + c_1a_2b_3 - a_3b_2c_1 - b_3c_2a_1 - c_3a_2b_1 \qquad (6.6)$$

Study the next example, which shows how to apply formula (6.6).

EXAMPLE 6.16

Evaluate the following 3×3 determinant:

$$\begin{vmatrix} 2 & -1 & 3 \\ 3 & 2 & -2 \\ 1 & -3 & 4 \end{vmatrix}$$

Solution

Repeat the first two columns and expand the determinant applying formula (6.6):

$$
\begin{aligned}
\begin{vmatrix} 2 & -1 & 3 \\ 3 & 2 & -2 \\ 1 & -3 & 4 \end{vmatrix} \begin{matrix} 2 & -1 \\ 3 & -3 \\ 1 & -3 \end{matrix} = \ & (2)(2)(4) + (-1)(-2)(1) + (3)(3)(-3) \\
& - (1)(2)(3) - (-3)(-2)(2) - (4)(3)(-1) \\
= \ & 16 + 2 - 27 - 6 - 12 + 12 = 30 - 45 = -15
\end{aligned}
$$

Cramer's rule also applies to the solution of a linear system of three equations. The expressions for x, y, and z are similar to those for x and y in (6.5) for two equations. Given a linear system of three equations in standard form,

$$a_1x + b_1y + c_1z = k_1$$
$$a_2x + b_2y + c_2z = k_2 \quad (k_1, k_2, k_3 \text{ constants})$$
$$a_3x + b_3y + c_3z = k_3$$

the solutions for x, y, and z are

$$x = \frac{\begin{vmatrix} k_1 & b_1 & c_1 \\ k_2 & b_2 & c_2 \\ k_3 & b_3 & c_3 \end{vmatrix}}{\begin{vmatrix} a_1 & b_1 & c_1 \\ a_2 & b_2 & c_2 \\ a_3 & b_3 & c_3 \end{vmatrix}} \quad y = \frac{\begin{vmatrix} a_1 & k_1 & c_1 \\ a_2 & k_2 & c_2 \\ a_2 & k_2 & c_2 \end{vmatrix}}{\begin{vmatrix} a_1 & b_1 & c_1 \\ a_2 & b_2 & c_2 \\ a_3 & b_3 & c_3 \end{vmatrix}} \quad z = \frac{\begin{vmatrix} a_1 & b_1 & k_1 \\ a_2 & b_2 & k_2 \\ a_2 & b_2 & k_2 \end{vmatrix}}{\begin{vmatrix} a_1 & b_1 & c_1 \\ a_2 & b_2 & c_2 \\ a_3 & b_3 & c_3 \end{vmatrix}} \tag{6.7}$$

Look at the formulas closely. The denominator for x, y, and z is again the coefficient matrix. The numerator for each variable is the coefficient matrix with the coefficients of the variable being calculated replaced by the constants.

Study the next example, which shows how to apply Cramer's rule (6.7).

EXAMPLE 6.17

Solve the linear system for x, y, and z:

$$2x + 3y + z = 5$$
$$3x - 2y + 4z = -3$$
$$5x + y + 2z = -1$$

Solution Apply Cramer's rule (6.7) for x, and evaluate the numerator and denominator using formula (6.6):

$$x = \frac{5(-2)(2) + 3(4)(-1) + 1(-3)(1) - (-1)(-2)(1) - 1(4)(5) - 2(-3)(3)}{2(-2)(2) + 3(4)(5) + 1(3)(1) - 5(-2)(1) - 1(4)(2) - 2(3)(3)}$$

$$x = \frac{-20 - 12 - 3 - 2 - 20 + 18}{-8 + 60 + 3 + 10 - 8 - 18} = \frac{-39}{39} = -1$$

Once you have computed the value of the denominator, you do not have to do it again for y and z:

$$y = \frac{\begin{vmatrix} 2 & 5 & 1 \\ 3 & -3 & 4 \\ 5 & -1 & 2 \end{vmatrix}}{39} = \frac{78}{39} = 2 \qquad z = \frac{\begin{vmatrix} 2 & 3 & 5 \\ 3 & -2 & -3 \\ 5 & 1 & -1 \end{vmatrix}}{39} = \frac{39}{39} = 1$$

You should evaluate these determinants and check that you get the values shown. You can then check the results by substituting the values for x, y, and z in each of the original equations. The set of values must satisfy all three equations.

◼

The next example closes the circuit and shows an application of determinants to a series-parallel circuit.

EXAMPLE 6.18

Close the Circuit

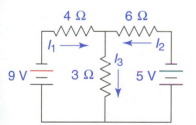

FIGURE 6–25 Kirchhoff's laws for Example 6.18.

Figure 6–25 shows a series-parallel circuit containing two voltage sources. The following three equations result from applying Kirchhoff's laws to the circuit:

$$I_1 + I_2 - I_3 = 0$$
$$4I_1 + 3I_3 - 9 = 0$$
$$6I_2 + 3I_3 - 5 = 0$$

Solve this linear system for the three currents I_1, I_2, and I_3.

Solution First put the three equations into standard form:

$$I_1 + I_2 - I_3 = 0$$
$$4I_1 + \quad 3I_3 = 9$$
$$6I_2 + 3I_3 = 5$$

Apply Cramer's rule (6.7) for I_1, and evaluate the determinants. The solution for I_1 to two significant digits is:

$$I_1 = \frac{\begin{vmatrix} 0 & 1 & -1 \\ 9 & 0 & 3 \\ 5 & 6 & 3 \end{vmatrix}}{\begin{vmatrix} 1 & 1 & -1 \\ 4 & 0 & 3 \\ 0 & 6 & 3 \end{vmatrix}} = \frac{-66}{-54} = 1.2 \, \text{A}$$

Note that a missing term is represented by a zero element in the determinant. Then compute I_2 and I_3:

$$I_2 = \frac{\begin{vmatrix} 1 & 0 & -1 \\ 4 & 9 & 3 \\ 0 & 5 & 3 \end{vmatrix}}{-54} = \frac{-8}{-54} = 0.15 \, \text{A} = 150 \, \text{mA}$$

EXAMPLE 6.18 (Cont.)

$$I_3 = \frac{\begin{vmatrix} 1 & 1 & 0 \\ 4 & 0 & 9 \\ 0 & 6 & 5 \end{vmatrix}}{-54} = \frac{-74}{-54} = 1.4 \text{ A}$$

The solution must satisfy all three equations. As a check for the first equation you have

$$I_1 + I_2 - I_3 = 0$$
$$1.2 + 0.15 - 1.4 = -0.02 \approx 0$$

Since the answers are rounded off, the check may not always satisfy the equation exactly.

From these examples you can see how the solution of linear systems using determinants can be readily programmed on a calculator or computer. The value of a 3×3 determinant can also be computed using the elements in any row or column and 2×2 determinants.

EXERCISE 6.3

In problems 1 through 6, evaluate each determinant.

1. $\begin{vmatrix} 2 & 3 \\ 1 & 4 \end{vmatrix}$

2. $\begin{vmatrix} -1 & 5 \\ 3 & 1 \end{vmatrix}$

3. $\begin{vmatrix} 0 & 7 \\ -3 & 5 \end{vmatrix}$

4. $\begin{vmatrix} \frac{1}{2} & \frac{1}{2} \\ -\frac{1}{2} & -\frac{1}{2} \end{vmatrix}$

5. $\begin{vmatrix} 3.2 & 2.5 \\ 1.2 & -1.5 \end{vmatrix}$

6. $\begin{vmatrix} 0.45 & -0.20 \\ 0.75 & 0.40 \end{vmatrix}$

In problems 7 through 12, evaluate each determinant. See Example 6.16.

7. $\begin{vmatrix} 2 & 1 & -2 \\ 3 & 2 & -1 \\ 1 & 1 & 3 \end{vmatrix}$

8. $\begin{vmatrix} 1 & -2 & 4 \\ 3 & 2 & -1 \\ 4 & -1 & -2 \end{vmatrix}$

9. $\begin{vmatrix} 4 & 0 & -3 \\ -2 & 1 & 0 \\ 5 & -3 & 1 \end{vmatrix}$

10. $\begin{vmatrix} -2 & 1 & 0 \\ 0 & 8 & -3 \\ 4 & -6 & 5 \end{vmatrix}$

11. $\begin{vmatrix} 1.2 & 2.0 & 0.80 \\ 0.0 & -1.0 & 3.0 \\ 2.0 & 0.0 & 0.50 \end{vmatrix}$

12. $\begin{vmatrix} 2.6 & 1.5 & 4.0 \\ 0.0 & 3.5 & 5.0 \\ 1.2 & 4.4 & 0.0 \end{vmatrix}$

In problems 13 through 38, solve each linear system using determinants.

13. $3X - Y = 8$
    $2X + 3Y = 9$

14. $a + 3b = 6$
    $4a + b = 2$

15. $2R + 3V = 0$
    $4R - 6V = -4$

16. $8C - 2L = 3$
    $C + 4L = 11$

17. $3I_1 - 2I_2 - 2 = 0$
    $2I_1 - 3I_2 - 8 = 0$

18. $V_1 + 2V_2 - 1 = 0$
    $4V_1 + 3V_2 + 6 = 0$

19. $2R_A = 8R_T - 2$
    $R_T = 4R_A - 2$

20. $X_1 = 1 - X_2$
    $X_2 = 10 - 6X_1$

21. $2\alpha - \beta = 1.5$
    $3\alpha + 2\beta = 6.1$

22. $\dfrac{\lambda}{2} + n = 4$
    $2\lambda - n = 0$

23. $I_A - 0.5I_B = 0$
    $0.5I_A + I_B + 5.5 = 0$

24. $2.5\phi_1 - 2.0\phi_2 + 1.1 = 0$
    $3.0\phi_1 + 2.5\phi_2 - 2.6 = 0$

25. $\dfrac{X_C}{5} - \dfrac{X_L}{3} = 1.2$

    $\dfrac{X_C}{7} + \dfrac{2X_L}{5} = 1.1$

26. $1.0i + 0.1v = 0.7$
    $2.0i + 0.7v = -3.1$

27. $x + y + z = 6$
    $x - y + 2z = 5$
    $2x - y - z = -3$

28. $3a + b - c = 2$
    $a - 2b - 2c = 6$
    $4a + b + 2c = 7$

29. $2I_1 + I_2 - 3I_3 = -2$
    $I_1 + 3I_2 - 2I_3 = -5$
    $3I_1 + 2I_2 - I_3 = 7$

30. $2I_1 + 3I_2 + I_3 = 4$
    $I_1 + 5I_2 - 2I_3 = -1$
    $3I_1 - 2I_2 + 4I_3 = 3$

31. $8I_A + I_B + I_C = 1$
    $7I_A - 2I_B + 9I_C = -3$
    $4I_A - 6I_B + 8I_C = -5$

32. $2I_A + 3I_B + 6I_C = 3$
    $3I_A + 4I_B - 5I_C = 2$
    $4I_A - 2I_B - 2I_C = 1$

33. $2p + q + 3r = -2$
    $5p + 2q - 5 = 0$
    $2q - 3r + 7 = 0$

34. $2.3R = 1.3V + 1.7$
    $2.5V = 2.5I + 5.0$
    $0.6I = 1.2R - 2.4$

35. $2.0\,V_1 - 3.0\,V_2 = -1.0$
    $3.0V_1 + 5.0V_3 = 4.5$
    $1.0V_1 + 6.0V_2 - 10V_3 = -1.5$

36. $0.2R_1 + 0.1R_2 + 0.2R_3 = 0.8$
    $0.4R_1 + 0.2R_2 - 0.4R_3 = 0.4$
    $0.4R_1 - 0.3R_2 + 0.6R_3 = 1.4$

37. $3I_1 + 0.5I_2 + I_3 - 1.5 = 0$
    $0.5I_1 + 3I_2 + 2.5I_3 - 2 = 0$
    $4I_1 + 1.5I_2 - 2I_3 - 4.5 = 0$

38. $3I_1 - 2I_2 + 3I_3 = 0$
    $2I_1 - I_2 = 0$
    $I_1 + I_2 + I_3 - 1.1 = 0$

In problems 39 through 44, solve each applied problem by setting up and solving a linear system using determinants.

39. A shipment of 650 computer parts contains two types of parts. The first type costs $1.50 each, and the second part costs $1.70 each. If the total cost for the shipment is $1045.00, how many parts of each type are in the shipment?

40. An electronics firm produces two types of integrated circuits using robots. A problem involving the most effective programming of the robots for production leads to the following linear system:

$$1.8C_1 + 2.4C_2 = 5400$$
$$2.8C_1 - 1.2C_2 = 1000$$

where C_1 = the number of one type produced and C_2 = the number of the other type produced. Solve this system for C_1 and C_2.

41. A small plane flying against the wind travels 225 mi in $2\frac{1}{2}$ h. On the return trip, with the wind, the plane travels the same distance in only $1\frac{1}{2}$ h. Assuming the speeds relative to the air are the same each way, what is the air speed of the plane, and what is the wind speed?

42. A car traveling over a mountain pass 10 mi long climbs for 21 min and descends for 6 min. On the return trip, the car climbs for 9 min and descends for 14 min. If the speeds climbing and descending are the same each way, find the two speeds in miles per hour.

43. Roman wants to determine the volume of three oddly shaped bottles. All he has is a gallon container filled with water. After much experimenting, he finds the following. The gallon container fills the first and the second container. The first and the third container exactly fill the second container. Three times the volume of the first container fills the second. What is the volume of each container?

44. Three pumps can empty a swimming pool that holds 144,000 L in 8 h. When only the first and second pumps are operating, it takes 12 h to empty the pool. It is found that the first pump is clogged and is pumping at only half its normal rate. With the defective first pump, and the second and third pumps operating, it takes 9 h to empty the pool. What is the normal pumping rate in liters per hour for each pump?

Applications to Electronics   In problems 45 through 58, solve each applied problem by setting up and solving a linear system using determinants.

45. Applying Kirchhoff's voltage law to the circuit in Figure 6–26 results in the following equations:

$$70I_1 - 50I_2 = 8$$
$$30I_1 + 80I_2 = 14$$

Solve this linear system for I_1 and I_2.

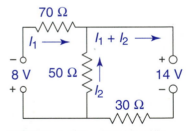

FIGURE 6–26   Kirchhoff's voltage law applied to series-parallel circuit for problem 45.

46. Two electric forces in an electric field are found to satisfy the following equations:

$$6.4F_1 - 4.6F_2 = 36 \times 10^{-6}$$
$$3.2F_1 + 2.0F_2 = 74 \times 10^{-6}$$

Solve this linear system for F_1 and F_2 in newtons (N).

47. Analysis of the voltages in an amplifier circuit produces the following equations:

$$0.2V_A + 0.1V_B = 30$$
$$0.2V_B - 0.1V_C = 25$$
$$0.1V_A + 0.2V_C = 35$$

Solve this linear system for V_A, V_B, and V_C.

48. A computer analysis produces the following equations:

$$2\alpha_1 - \alpha_2 + 4\alpha_3 = 1.9$$
$$\alpha_1 + 5\alpha_2 - \alpha_3 = -1.6$$
$$\alpha_1 - 2\alpha_2 - \alpha_3 = 0.50$$

Solve this linear system for α_1, α_2, and α_3.

49. In Figure 6–27, if $V_1 = 12$ V and $V_2 = 16$ V, application of Kirchhoff's laws to the circuit produces the following equations:

$$I_1 - I_2 + I_3 = 0$$
$$10I_1 + 10I_2 - 16 = 0$$
$$10I_2 + 20I_3 - 12 = 0$$

Solve this system for I_1, I_2, and I_3.

50. In Figure 6–27, if $V_1 = 230$ mV and $V_2 = 160$ mV, application of Kirchhoff's laws produces the following equations:

$$I_1 - I_2 + I_3 = 0$$
$$10I_1 - 20I_3 - 0.07 = 0$$
$$10I_2 + 20I_3 - 0.23 = 0$$

Solve this system for I_1, I_2, and I_3 to the nearest milliamp.

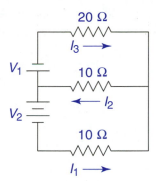

FIGURE 6–27
Kirchoff's laws for problems 49 and 50.

51. The resistance of 1000 ft of no. 21 aluminum wire at 50°C is measured to be 23.3 Ω. At 100°C, it is measured to be 27.1 Ω. Find the resistance at 0°C and the temperature coefficient of resistance at 0°C. (See Example 6.15.)

52. The resistance of a carbon rod at 20°C is measured to be 200 Ω. At 80°C it is found to *decrease* to 196 Ω. Find the resistance at 0°C and the temperature coefficient of resistance at 0°C. (See Example 6.15.)

53. Two unknown voltages V_1 and V_2 are connected in series to a resistance of 50 Ω. When the voltages are series-aiding, the current in the circuit is 1.12 A. When the voltages are series-opposing, the current is 160 mA. Find the two voltages.

54. The total current in two circuits is 4.5 A. If the difference of the currents in the two circuits equals half of the larger current, find the current in each circuit.

55. The equivalent resistance of two resistances R_1 and R_2 in parallel is 10 Ω. See Figure 6–28a. When R_1 is reduced by half to $R_1/2$, the equivalent resistance decreases to 7.5 Ω. See Figure 6–28b. Find the two resistances. Hint: Treat $1/R_1$ and $1/R_2$ as the unknowns, and note that $\frac{1}{R_1/2} = 2\left(\frac{1}{R_1}\right)$.

56. Figure 6–29 shows three resistances R_1, R_2, and R_3 in series connected to a 9V battery. The current in the circuit is measured to be 20 mA. If R_2 is 50 Ω more than R_3 and R_1 is 50 Ω more than R_2, what are the three resistances?

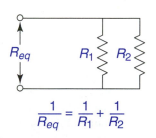

$$\frac{1}{R_{eq}} = \frac{1}{R_1} + \frac{1}{R_2}$$

(a)

$$\frac{1}{R_{eq}} = \frac{1}{R_{1/2}} + \frac{1}{R_2}$$

(b)

FIGURE 6–28   Parallel resistances for problem 55.

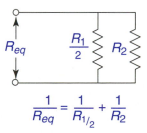

FIGURE 6–29   Series resistances for problem 56.

57. In Figure 6–30, $R_3 = 4\ \Omega$ and $R_4 = 5\ \Omega$. When Kirchhoff's laws are applied to the circuit, the following three equations result:

$$5I_1 + 5I_2 - 9I_3 = 0$$
$$3I_2 + 4I_3 - 10 = 0$$
$$3I_1 + 4I_3 - 12 = 0$$

Solve this linear system for the three currents I_1, I_2, and I_3 to two significant digits.

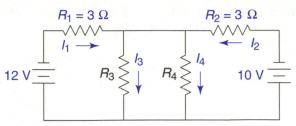

FIGURE 6–30 Kirchoff's laws for problems 57 and 58.

58. In Figure 6–30, if $R_3 = R_4 = 4\ \Omega$, Kirchhoff's laws produce the following four equations:

$$I_1 + I_2 - I_3 - I_4 = 0$$
$$3I_1 + 4I_3 - 12 = 0$$
$$3I_2 + 4I_4 - 10 = 0$$
$$4I_3 - 4I_4 = 0$$

Solve this linear system by first reducing it to three equations. Solve the last equation for I_4, and substitute in the other equations.

≡ CHAPTER HIGHLIGHTS

6.1 GRAPHS AND LINEAR FUNCTIONS

In a rectangular coordinate system, the x axis is the horizontal axis and the y axis is the vertical axis. The first or x coordinate of a point is the horizontal distance; the second or y coordinate is the vertical distance. See Figure 6–1.

Function	y is a function of x if, given a set of values for x, there corresponds one and only one value of y for each value of x. (6.1)

In particular, y is a linear function of x if it is of the form:

$$y = mx + b \qquad (6.2)$$

This first-degree equation always graphs as a straight line. When $b = 0$, $y = mx$ states that y is directly proportional to x. See Example 6.3 for four important linear functions in electricity.

To graph a linear function, you need at least three points. When the linear function is in the standard form

$$Ax + By = C \qquad (6.3)$$

two important points you can quickly calculate are the x intercept, where $y = 0$, and the y intercept, where $x = 0$.

6.2 SIMULTANEOUS EQUATIONS

Two equations in two unknowns can be solved by algebraic and graphical methods. To solve graphically, draw the line for each equation on the same graph, and determine the point of intersection.

To solve algebraically, the addition method or the substitution method can be used. The addition method is done by multiplying the equations so the coefficients for one unknown cancel when the equations are added. The substitution method is done by solving one equation for one of the unknowns and substituting this expression into the other equation. Study Example 6.8 for the three methods.

6.3 DETERMINANTS

A matrix is a rectangular array of numbers arranged in rows and columns. Every square matrix has a number associated with it called its determinant. The determinant of a 2×2 matrix is

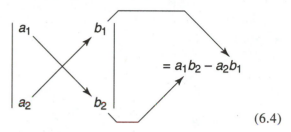

$$= a_1 b_2 - a_2 b_1$$

$$(6.4)$$

Given a linear system of two equations in standard form,

$$a_1 x + b_1 y = k_1$$
$$a_2 x + b_2 y = k_2$$

the determinant solutions for x and y are

$$x = \frac{\begin{vmatrix} k_1 & b_1 \\ k_2 & b_2 \end{vmatrix}}{\begin{vmatrix} a_1 & b_1 \\ a_2 & b_2 \end{vmatrix}} \quad y = \frac{\begin{vmatrix} a_1 & k_1 \\ a_2 & k_2 \end{vmatrix}}{\begin{vmatrix} a_1 & b_1 \\ a_2 & b_2 \end{vmatrix}}$$

$$(6.5)$$

The determinant of a 3×3 matrix is calculated by repeating the first two columns and computing six triple products, three positive and three negative:

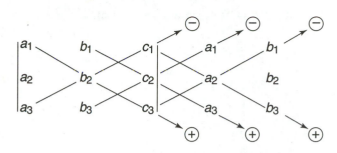

$$= a_1 b_2 c_3 + b_1 c_2 a_3 + c_1 a_2 b_3$$
$$- a_3 b_2 c_1 - b_3 c_2 a_1 - c_3 a_2 b_1$$

$$(6.6)$$

Given a linear system of three equations in standard form,

$$a_1 x + b_1 y + c_1 z = k_1$$
$$a_2 x + b_2 y + c_2 z = k_2$$
$$a_2 x + b_3 y + c_3 z = k_3$$

the determinant solutions for x, y, and z are

$$x = \frac{\begin{vmatrix} k_1 & b_1 & c_1 \\ k_2 & b_2 & c_2 \\ k_3 & b_3 & c_3 \end{vmatrix}}{D} \quad y = \frac{\begin{vmatrix} a_1 & k_1 & c_1 \\ a_2 & k_2 & c_2 \\ a_3 & k_3 & c_3 \end{vmatrix}}{D} \quad z = \frac{\begin{vmatrix} a_1 & b_1 & k_1 \\ a_2 & b_2 & k_2 \\ a_3 & b_3 & k_3 \end{vmatrix}}{D}$$

$$(6.7)$$

where D is the determinant of the coefficient matrix (6.6).

≡ REVIEW QUESTIONS

In problems 1 through 6, draw the graph of each linear function. Plot the variable in parentheses on the horizontal axis.

1. $y = 2x - 3$

2. $y = -3x + 1$

3. $V_1 + V_2 = 4 \ (V_1)$

4. $R_1 - 3R_2 + 6 = 0 \ (R_1)$

5. $P_T = 9.4 I_T \ (I_T)$

6. $V = 120 - 100I \ (I)$

In problems 7 through 10, evaluate each determinant.

7. $\begin{vmatrix} 12 & 15 \\ 18 & -22 \end{vmatrix}$

8. $\begin{vmatrix} -0.056 & -0.044 \\ 0.012 & 0.088 \end{vmatrix}$

9. $\begin{vmatrix} 1 & 3 & -4 \\ 2 & -3 & 1 \\ 0 & 5 & -6 \end{vmatrix}$

10. $\begin{vmatrix} 1.2 & -3.5 & 2.0 \\ 7.0 & 0.0 & 6.1 \\ -3.3 & 4.1 & -4.2 \end{vmatrix}$

In problems 11 through 20, solve each linear system.

11. $3x - y = -5$
$x + 4y = 7$

12. $A - 2B = 0$
$3A + 6B = 3$

13. $20I_1 - 10I_2 = -1$
$4I_1 + 6I_2 = 7$

14. $V_R = V_L + 2$
$V_L = 1 - 3V_R$

15. $1.6X_L - 1.4X_C + 0.9 = 0$
$3.2X_L + 0.4X_C - 9.4 = 0$

16. $6.6I_a = 10I_b - 4.2$
$12I_b = 8.2I_a + 4.9$

17. $x_1 - x_2 + 2x_3 = 6$
$2x_1 + x_2 - x_3 = 4$
$x_1 + 3x_2 + 3x_3 = 3$

18. $2\alpha_1 + 2\alpha_2 - 3\alpha_3 = 5$
$4\alpha_1 - 3\alpha_2 + 2\alpha_3 = -4$
$6\alpha_1 + 5\alpha_2 - 4\alpha_3 = 13$

19. $I_1 + I_2 - I_3 = 0$
$2I_1 + 3I_2 - 9 = 0$
$3I_2 + 4I_3 - 20 = 0$

20. $4.5R = 3.0V + 1.5$
$4.8V = 6.0I - 6.0$
$1.2I = 1.1R + 1.4$

In problems 21 through 26, solve each applied problem.

21. Charles's law says that the volume of a gas is a linear function of its absolute temperature in Kelvin when the pressure is constant:

$$V = kT$$

Here, V is also directly proportional to T. Graph V versus T for hydrogen where $k = 0.034$ m^3/K, from $T = 0$ K to $T = 300$ K.

22. The "normal" weight of a person is approximately a linear function of their height given by the formula:

$$W = 0.97H - 100$$

where W is in kilograms and H is in centimeters. Graph W versus H from $H = 150$ cm (4 ft 11 in) to $H = 200$ cm (6 ft 7 in).

23. Jack, a resourceful car salesman, offers some specials on a new model. The standard model with a compact disc player is priced at $11,700. The standard model with a compact disc player and an air conditioner is priced at $12,800. If the air conditioner is five and a half times the value of the disc player, how much is the standard car alone worth?

24. The length of a rectangular field is 150% of its width. If the fencing around the perimeter is 155 m, what are the dimensions of the field?

25. A test plane, starting at an initial velocity of 50 mi/h, takes 30 s at a constant acceleration to reach its takeoff velocity. In another test, the plane starting at 75 mi/h, takes 25 s at the same constant acceleration to reach its takeoff velocity. What is the acceleration (in miles per hour per second) and the takeoff velocity (in miles per hour) of the plane? Use the linear function $v = v_0 + at$ from Example 6.3, case 8.

26. A sailboat leaves a harbor at 0600 (6 A.M.) and motors under power with no wind. At 1000 (10 A.M.), the wind picks up and the skipper, Michael, shuts off the boat's engine and sails until 1200 (noon). At this time, Michael determines that the boat has traveled only 26 nmi (nautical miles) since departing the harbor, so he decides to use both power and sail to move faster. The sailboat arrives in port under power and sail at 1800 (6 P.M.) after traveling a total of 74 nmi. If the speed under power and sail equals the speed under power plus the speed under sail, how fast did the boat travel under power alone and under sail alone?

Applications to Electronics In problems 27 through 34 solve each applied problem.

27. Figure 6–31 shows a parallel circuit with a variable resistor R_V in parallel with a fixed resistor $R_0 = 40$ Ω. If the applied voltage is 60 V, graph the total power P_T versus I_T when R_V varies from 20 Ω to 60 Ω. Use the horizontal axis for I_T.

28. The thermoelectric power of copper with respect to lead is a linear function of the temperature t given by $Q = 2.76 + 1.22t$, where Q = power in $\mu V/°C$. Graph Q versus t from $t = -10°C$ to $t = 10°C$. Use the horizontal axis for t.

29. In order to accurately find the voltage V_B of a battery and its small internal resistance r_i, Eduardo connects a small resistance of 3.0 Ω to the battery and accurately measures the current with an ammeter to be 1.88 A. He then repeats the experiment with a resistance of 5.1 Ω and measures the current to be 1.15 A. What is the voltage of the battery and its internal resistance?

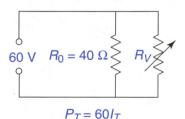

$$P_T = 60I_T$$

FIGURE 6–31 Power versus current in a parallel circuit for problem 27.

30. The total conductance of two resistors R_1 and R_2 in parallel is 28 mS. When a third resistor, whose resistance is equal to R_1 is added in parallel, the total conductance is 38 mS. Find the conductance and the resistance of each resistor.

31. A computer analysis gives rise to the following linear system of three equations:
$$2.2\lambda_2 + 3.4\lambda_3 = 1.56$$
$$2.6\lambda_1 + 5.2\lambda_3 = 2.60$$
$$3.8\lambda_1 - 5.4\lambda_2 = 0.12$$
Solve this system for λ_1, λ_2, and λ_3.

32. The total resistance of three unknown resistances connected in series is found to be 51 Ω. When two resistances of the first type and one resistance of the second type are connected in series, the total resistance is found to be 60 Ω. When two resistances of the second type and one resistance of the third type are connected in series, the total resistance is found to be 39 Ω. Find the value of each resistance.

33. For the circuit in Figure 6–32, if $R_0 = 10$ Ω, application of Kirchhoff's laws produce the two equations:
$$40I_1 + 10I_2 = 9$$
$$10I_1 + 40I_2 = 6$$
Solve this linear system for I_1 and I_2.

34. For the circuit in Figure 6–32, if $R_0 = 20$ Ω, application of Kirchhoff's laws produce the three equations:
$$I_1 + I_2 = I_3$$
$$30I_1 + 20I_3 = 9$$
$$30I_2 + 20I_3 = 6$$
Solve this linear system for I_1, I_2, and I_3.

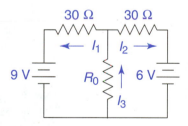

FIGURE 6–32 Applications of Kirchoff's laws for problems 33 and 34.

7

Exponents and Radicals

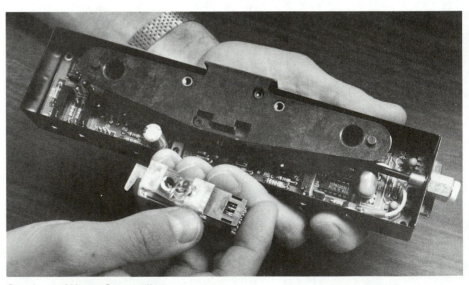

Courtesy of Xerox Corporation.

A copier optical scanning unit has been microfabricated with snap-together connections for the electronics and optics units.

Positive powers and roots were introduced in Chapter 2, and negative exponents were introduced in Chapter 3 along with the rules for exponents. In this chapter, we continue to build on this foundation and introduce fractional exponents, which are actually roots or radicals. All radicals can be expressed in terms of exponents. The rules for radicals are presented, and the study of exponents and radicals leads to quadratic equations. Quadratic equations are equations that contain the square of the unknown, such as x^2 or I^2. These equations occur often in electronic applications, many of which are shown throughout the chapter. Some quadratic equations can be solved by factoring, but most require the quadratic formula, which uses exponents and radicals.

Chapter Objectives

In this chapter, you will learn:

- How to evaluate and work with fractional exponents.
- The rules for working with radicals.
- How to simplify and combine radicals.
- How to solve a quadratic equation by factoring and taking roots.
- How to solve a quadratic equation by the quadratic formula.

≡ 7.1
FRACTIONAL EXPONENTS

From chapter 3, Section 3.2, five basic rules for exponents are, for x and $y > 0$:

- Addition rule

$$(x^m)(x^n) = x^{m+n} \qquad (3.6)$$

- Subtraction rule

$$\frac{x^m}{x^n} = x^{m-n} \qquad (3.7)$$

- Multiplication rule

$$(x^m)^n = x^{mn} \qquad (3.8)$$

- Product rule

$$(xy)^n = x^n y^n \text{ and } \left(\frac{x}{y}\right)^n = \frac{x^n}{y^n} \qquad (3.9)$$

- Division rule

$$\sqrt[n]{x^m} = x^{m/n} \qquad (3.10)$$

In the division rule, when n divides into m evenly, such as

$$\sqrt[3]{x^6} = x^{6/3} = x^2$$

the result is an exponent that is a whole number or integer. However, when n does not divide into m evenly, the result is a fractional exponent:

$$\sqrt[3]{x^2} = x^{2/3}$$

The division rule is used in this case to define a fractional exponent. For example, if $x = 8$, then

$$8^{2/3} = \sqrt[3]{8^2} = \sqrt[3]{64} = 4$$

The calculation can also be done by taking the root first:

$$8^{2/3} = \left(\sqrt[3]{8}\right)^2 = (2)^2 = 4$$

Taking the root first gives the same result and is simpler.

A fractional exponent is, therefore, defined as

$$x^{m/n} = \left(\sqrt[n]{x}\right)^m \text{ or } \sqrt[n]{x^m} \quad (x > 0) \tag{7.1}$$

The *denominator* of the fraction is the *root* and the *numerator* is the *power.* For example,

$$27^{4/3} = \left(\sqrt[3]{27}\right)^4 = (3)^4 = 81$$

A special case of (7.1) is

$$x^{1/n} = \sqrt[n]{x} \quad (x > 0)$$

It illustrates that any radical can be expressed as a fractional exponent, and vice versa. For example,

$$\sqrt{x} = x^{1/2}$$

$$\sqrt[3]{I} = I^{1/3}$$

$$\sqrt[4]{10} = 10^{1/4}$$

See Example 2.24 to see how to use the fractional exponent $\frac{1}{3}$ on the calculator to find a cube root. All the rules for positive and negative exponents also apply to fractional exponents. For example,

$$16^{-3/4} = \frac{1}{16^{3/4}} = \frac{1}{\left(\sqrt[4]{16}\right)^3} = \frac{1}{2^3} = \frac{1}{8}$$

Note that the negative fractional exponent means the same as a negative integral exponent: invert and change to a positive exponent.

Another example is

$$(V_T^{0.5})(V_T^{2.0}) = V_T^{(0.5 + 2.0)} = V_T^{2.5} \text{ or } V_T^{5/2} = \left(\sqrt{V}\right)^5$$

Observe that the addition rule for exponents applies here. A decimal exponent such as 0.5 is the same as the fractional exponent $\frac{1}{2}$. The next two examples further illustrate the exponent rules applied to fractional exponents.

EXAMPLE 7.1

Evaluate:

$$(4 \times 10^{-4})^{1.5}$$

$$(4 \times 10^{-4})^{1.5} = 4^{1.5} \times 10^{(-4)(1.5)} = 8 \times 10^{-6}$$

Solution Observe that the product rule and the multiplication rule are both used here. Also note that $4^{1.5} = 4^{3/2} = (\sqrt{4})^3 = 8$.

One way this example can be done on the calculator is

$$4 \boxed{EXP} \ 4 \boxed{+/-} \boxed{x^y} \ 1.5 \boxed{=} \ \rightarrow 8 \ -06$$

■

EXAMPLE 7.2

Simplify:

$$(0.25I^2R)^{1/2}$$

Solution Use the product rule and apply the exponent 1/2 to each factor in the parentheses:

$$(0.25\,I^2R)^{1/2} = (0.25)^{1/2}\,I^{(2)(1/2)}R^{(1/2)} = \sqrt{0.25}\,I^{(1)}\sqrt{R} = 0.5\,I\sqrt{R}$$

■

Study the next example, which closes the circuit and shows an application of exponents to an *RC* (resistance-capacitance) circuit.

EXAMPLE 7.3

Close the Circuit

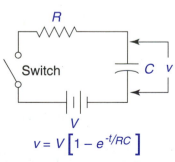

$$R$$

Switch C v

$$V$$

$$v = V\left[1 - e^{-t/RC}\right]$$

FIGURE 7–1 Resistance-capacitance circuit for Example 7.3.

Figure 7–1 shows an *RC* circuit containing a resistance *R* and a capacitance *C* in series with a switch *S*. The voltage across the capacitance *t* seconds after the switch is closed is given by

$$v = V\left(1 - e^{-t/RC}\right)$$

where *V* = battery voltage and *e* is the constant 2.718. If *V* = 25 V, *R* = 200 Ω and *C* = 100 μF, find *v* to two significant digits when *t* = 10 ms and *t* = 20 ms.

Solution To find *v* when *t* = 10 ms, first compute the exponent of *e* by substituting the values of *R*, *C*, and *t* = 0.010 s:

$$\frac{-t}{RC} = \frac{-0.010}{(200)(100 \times 10^{-6})} = \frac{-0.010}{0.020} = -0.50$$

Then compute *v* by substituting the values of *e* and *V* into the given formula:

$$v = 25\left[1 - 2.718^{-0.50}\right] = 25\left[1 - 0.607\right] = 9.8 \text{ V}$$

This calculation can be done on the calculator as follows:

$$1 \boxed{-} \ 2.718 \boxed{x^y} \ 0.50 \boxed{+/-} \boxed{=} \boxed{\times} \ 25 \boxed{=} \ \rightarrow 9.8$$

You can also use the $\boxed{e^x}$ key on the calculator instead of 2.718 and the $\boxed{x^y}$ key, as follows:

$$1 \boxed{-} \ 0.50 \boxed{+/-} \boxed{e^x} \boxed{=} \boxed{\times} \ 25 \boxed{=} \ \rightarrow 9.8$$

EXAMPLE 7.3 (Cont.)

To find v when $t = 20$ ms, proceed the same as for $t = 10$ ms. The exponent is

$$\frac{-t}{RC} = \frac{-0.020}{(200)(100 \times 10^{-6})} = -1$$

and v is

$$v = 25\left[1 - 2.718^{-1}\right] = 16 \text{ V}$$

▪

EXERCISE 7.1

In problems 1 through 30, evaluate or simplify each expression.

1. $25^{1/2}$
2. $16^{1/4}$
3. $(-27)^{2/3}$
4. $-81^{3/2}$
5. $100^{-1/2}$
6. $(1/8)^{-1/3}$
7. $9^{1.5}$
8. $64^{-0.5}$
9. $0.027^{1/3}$
10. $0.16^{1/2}$
11. $(1.44 \times 10^2)^{0.5}$
12. $(125 \times 10^{-3})^{1/3}$
13. $0.001^{-2/3}$
14. $0.0016^{3/4}$
15. $\left(\dfrac{25}{16}\right)^{3/2}$
16. $\left(\dfrac{-27}{8}\right)^{-1/3}$
17. $4^{1/3}4^{1/6}$

18. $8^{1/2}8^{-1/6}$
19. $I_T^{-0.3} I_T^{0.7}$
20. $P_Z^{0.3}P_Z^{0.2}$
21. $(-27\omega^6)^{1/3}$
22. $\left(144X_C^4\right)^{0.5}$
23. $(0.04L^4X^2)^{-1/2}$
24. $(0.001I^3R^3)^{2/3}$
25. $\left(\dfrac{V_x^2}{10^{-6}}\right)^{1/2}$
26. $\left(\dfrac{R_1^{-2}}{16}\right)^{-1/4}$
27. $(2s^{1/2})^2(3s^{2/3})^3$
28. $(-5k^{1/3})^{-3}(10k^{3/2})^2$
29. $\left(\dfrac{9\alpha^{-2}}{4\beta^4}\right)^{-1/2}$
30. $\left(\dfrac{64t_0^3}{t_0^6}\right)^{1/3}$

In problems 31 and 32 solve each applied problem to two significant digits.

31. The maximum deflection δ (delta) of a beam with a concentrated off-center load is given by

$$\delta = (3.1 \times 10^{-6})(b/l)(l^2 - b^2)^{3/2}$$

Find δ when $l = 25$ m and $b = 20$ m.

32. At a certain reservoir, the time t in minutes that it takes for the runoff of rainfall to reach a maximum is given by

$$t = 2.53\left(\frac{d}{Sr^2}\right)^{(1/3)}$$

where d = distance in feet to the most remote area supplying the reservoir, S = slope, and r = rain intensity in inches per hour. Find t when $d = 1000$ ft, $S = 0.20$, and $r = 3.2$ in/h.

Applications to Electronics   In problems 33 through 38 solve each applied problem to two significant digits.

33. The bandwidth of a three-stage dc amplifier is given by

$$B = B_1(2^{1/3} - 1)^{1/2}$$

where B_1 = bandwidth of one stage. Find B when $B_1 = 6.0$ kHz.

34. A circular coil of $n = 2.0 \times 10^2$ turns and radius $r = 0.10$ m has a magnetic flux density of $B = 1.0 \times 10^{-3.5}$ T (tesla). The current in amperes required to produce this flux density in a vacuum is given by

$$I = \frac{2rB}{(4\pi \times 10^{-7})(n)}$$

Using the given data, calculate the current.

35. The behavior of iron in a magnetic field gives rise to the equation

$$W = nB^{1.6}$$

Calculate W when $n = 1.3$ and $B = 0.072$.

36. In calculating the heat dissipation from a circuit component, the following expression arises:

$$(QR)^{0.6}\left(\frac{Q}{R}\right)^{0.4}$$

Simplify this expression.

37. In the *RC* circuit in Example 7.3, the current t seconds after the switch is closed is given by

$$I = \frac{V}{R}e^{-t/RC}$$

where the constant $e = 2.718$. Find I when $t = 300$ ms, $V = 50$ V, $R = 2.0$ MΩ, and $C = 200$ nF.

38. In the *RC* circuit in Example 7.3, the charge of the capacitor t seconds after the switch is closed is given by

$$Q = CV(1 - e^{-t/RC})$$

where the constant $e = 2.718$. Find Q in coulombs (C) when $t = 100$ ms, $V = 10$ V, $R = 1.0$ MΩ, and $C = 5.0$ μF.

≡ 7.2
RADICALS

In Section 7-1, it can be seen that fractional exponents and radicals are two ways of expressing the same operation. At times, radical notation is more convenient, especially when you work with the most common radicals—square roots and cube roots. Three rules for radicals are shown here. They follow from the rules for exponents. The first rule is for any nth root,

$$\left(\sqrt[n]{x}\right)^n = \sqrt[n]{x^n} = x \quad (x > 0) \tag{7.2}$$

For example,

$$\left(\sqrt[3]{10\,Q}\right)^3 = 10\,Q$$

Observe that the exponent and the radical cancel each other.

The second rule is

$$\sqrt[n]{xy} = \left(\sqrt[n]{x}\right)\left(\sqrt[n]{y}\right) \quad (x, y > 0) \tag{7.3}$$

For example,

$$5\sqrt{18} = 5\sqrt{(9)(2)} = 5\sqrt{9}\sqrt{2} = 5(3)\sqrt{2} = 15\sqrt{2}$$

The perfect square root, $\sqrt{9}$, is factored out to simplify the radical. The perfect square roots that can be factored out are $\sqrt{4} = 2, \sqrt{9} = 3, \sqrt{16} = 4, \sqrt{25} = 5$, and so on. It would not help to factor $\sqrt{18}$ as $\sqrt{6}\sqrt{3}$; these are not perfect square roots.

Two more examples of (7.3) are

$$\sqrt{P^2 R} = \sqrt{P^2}\sqrt{R} = P\sqrt{R}$$

$$\sqrt{9 \times 10^3} = \sqrt{9}\sqrt{10^2}\sqrt{10} = (3)(10)\sqrt{10} = 30\sqrt{10}$$

The third rule is

$$\sqrt[n]{\frac{x}{y}} = \frac{\sqrt[n]{x}}{\sqrt[n]{y}} \qquad (x,y > 0) \tag{7.4}$$

For example,

$$\frac{\sqrt{6V}}{\sqrt{24}} = \sqrt{\frac{6V}{24}} = \sqrt{\frac{V}{4}} = \frac{\sqrt{V}}{2}$$

Rules (7.4) and (7.3) work both ways. Products or quotients of radicals can be combined or separated.

Consider the following fraction with a radical denominator:

$$\frac{6}{\sqrt{3}}$$

The radical denominator can be eliminated and the fraction simplified by multiplying top and bottom by $\sqrt{3}$ and then dividing top and bottom by 3:

$$\frac{6}{\sqrt{3}}\frac{(\sqrt{3})}{(\sqrt{3})} = \frac{\overset{2}{\cancel{6}}\sqrt{3}}{\cancel{3}} = 2\sqrt{3}$$

Eliminating radicals (irrational numbers) from the denominator is called *rationalizing the denominator*. Rationalizing the denominator can simplify expressions and calculations as the next two examples show.

EXAMPLE 7.4

Rationalize the denominator:

$$\sqrt{\frac{3}{5}}$$

Solution You can work inside the radical or outside. In each method, you multiply top and bottom by the quantity that makes the radical in the denominator a perfect root.

Inside Method.

$$\sqrt{\frac{3}{5}} = \sqrt{\frac{3(5)}{5(5)}} = \sqrt{\frac{15}{25}} = \frac{\sqrt{15}}{\sqrt{25}} = \frac{\sqrt{15}}{5} \text{ or } \frac{1}{5}\sqrt{15}$$

Outside Method.

$$\sqrt{\frac{3}{5}} = \frac{\sqrt{3}}{\sqrt{5}} = \frac{\sqrt{3}(\sqrt{5})}{\sqrt{5}(\sqrt{5})} = \frac{\sqrt{15}}{5} \text{ or } \frac{1}{5}\sqrt{15}$$

There is usually more than one way to simplify a fraction with a radical denominator as the next example shows.

EXAMPLE 7.5

Rationalize the denominator:

$$\sqrt{\frac{25}{18}}$$

Solution   You can multiply top and bottom by 18 and obtain the perfect square 324 in the denominator; however, it is easier to multiply by 2 and obtain the perfect square 36:

$$\sqrt{\frac{25(2)}{18(2)}} = \sqrt{\frac{50}{36}} = \frac{5\sqrt{2}}{6}$$

You can also first simplify the numerator and denominator by separating perfect roots and then rationalize the denominator:

$$\sqrt{\frac{25}{18}} = \frac{\sqrt{25}}{\sqrt{18}} = \frac{5}{3\sqrt{2}} = \frac{5(\sqrt{2})}{3\sqrt{2}(\sqrt{2})} = \frac{5\sqrt{2}}{3(2)} = \frac{5\sqrt{2}}{6}$$

When radicals are to be added or subtracted you must proceed carefully.

Rule   Radicals can be combined only if they are similar.

For example, $\sqrt{2} + \sqrt{3}$ cannot be added or simplified in any way. It can be computed only as a decimal on the calculator. However, $4\sqrt{5} + 2\sqrt{5}$ can be combined by adding the coefficients:

$$4\sqrt{5} + 2\sqrt{5} = 6\sqrt{5} \qquad \text{[Think of: } 4x + 2x = 6x]$$

Similar radicals are combined like similar algebraic terms, by combining the coefficients. It is helpful to think of a radical as an unknown x or y. Two more examples are

$$\sqrt{2} + \sqrt{3} + \sqrt{2} = 2\sqrt{2} + \sqrt{3} \qquad \text{[Think of: } x + y + x = 2x + y]$$

$$5\sqrt{7} + 2\sqrt{5} - 5\sqrt{5} = 5\sqrt{7} - 3\sqrt{5} \qquad \text{[Think of: } 5x + 2y - 5y = 5x - 3y]$$

Observe that only the similar radicals can be combined. Radicals that are not similar remain as separate terms like $5x - 3y$. Sometimes radicals that are not similar can be changed to similar radicals, and combined, as the following examples show.

EXAMPLE 7.6

Simplify and combine similar radicals:

$$2\sqrt{12} + 4\sqrt{3} - \sqrt{27}$$

Solution   First simplify each radical as much as possible by separating perfect square roots:

$$2\sqrt{12} + 4\sqrt{3} - \sqrt{27} = 2\sqrt{4}\sqrt{3} + 4\sqrt{3} - \sqrt{9}\sqrt{3}$$

Then combine the coefficients:

$$2(2)\sqrt{3} + 4\sqrt{3} - 3\sqrt{3} = 4\sqrt{3} + 4\sqrt{3} - 3\sqrt{3} = 5\sqrt{3}$$

EXAMPLE 7.7

Simplify and combine:

$$\sqrt{80 \times 10^6} - \sqrt{20 \times 10^6} + \sqrt{5 \times 10^6}$$

Solution   Separate perfect roots and even powers of 10:

$$\sqrt{80 \times 10^6} - \sqrt{20 \times 10^6} + \sqrt{5 \times 10^6}$$
$$= (\sqrt{16 \times 10^6})\sqrt{5} - (\sqrt{4 \times 10^6})\sqrt{5} + (\sqrt{10^6})\sqrt{5}$$
$$= (4 \times 10^3)\sqrt{5} - (2 \times 10^3)\sqrt{5} + (10^3)\sqrt{5}$$

Then combine the coefficients and simplify:

$$(4 \times 10^3)\sqrt{5} - (2 \times 10^3)\sqrt{5} + (10^3)\sqrt{5} = (3 \times 10^3)\sqrt{5} \text{ or } 6.71 \times 10^3$$

EXAMPLE 7.8

Simplify and combine:

$$\sqrt{50x^3} + 3\sqrt{72x^3} - 5\sqrt{8x^2}$$

Solution   Separate perfect roots and even powers of x:

$$\sqrt{50x^3} + 3\sqrt{72x^3} - 5\sqrt{8x^2} = (\sqrt{25x^2})\sqrt{2x} + (3\sqrt{36x^2})\sqrt{2x} - (5\sqrt{4x^2})\sqrt{2}$$
$$= 5x\sqrt{2x} + 3(6x)\sqrt{2x} - 5(2x)\sqrt{2}$$

Then combine similar radicals:

$$5x\sqrt{2x} + 3(6x)\sqrt{2x} - 5(2x)\sqrt{2} = 5x\sqrt{2x} + 18x\sqrt{2x} - 10x\sqrt{2}$$
$$= 23x\sqrt{2x} - 10x\sqrt{2}$$

Note that $\sqrt{2x}$ and $\sqrt{2}$ are not similar and cannot be combined.

EXAMPLE 7.9

Simplify and combine:

$$\sqrt{\frac{2}{3}} + \frac{\sqrt{3}}{\sqrt{8}}$$

Solution   Rationalize the denominators:

$$\sqrt{\frac{2}{3}} + \frac{\sqrt{3}}{\sqrt{8}} = \sqrt{\frac{2(3)}{3(3)}} + \frac{\sqrt{3}(\sqrt{2})}{\sqrt{8}(\sqrt{2})}$$

$$= \frac{\sqrt{6}}{\sqrt{9}} + \frac{\sqrt{6}}{\sqrt{16}} = \frac{\sqrt{6}}{3} + \frac{\sqrt{6}}{4}$$

You could multiply top and bottom of the second fraction by $\sqrt{8}$ to rationalize the denominator; however, it is easier to multiply by the smaller quantity $\sqrt{2}$. Now combine over the lowest common denominator:

$$\frac{\sqrt{6}}{3} + \frac{\sqrt{6}}{4} = \frac{\sqrt{6}(4)}{3(4)} + \frac{\sqrt{6}(3)}{4(3)} = \frac{7\sqrt{6}}{12}$$

Square roots and higher roots of numbers that are not perfect roots, such as $\sqrt{5}$, $\sqrt{13}$, and $\sqrt[3]{10}$, are irrational numbers, which are infinite decimals with no pattern of repeating digits:

$$\sqrt{5} = 2.236067977 \ldots$$

$$\sqrt{13} = 3.605551275 \ldots$$

$$\sqrt[3]{10} = 2.154434690 \ldots$$

Computers and calculators approximate irrational numbers using 8 to 10 digits. Radicals that are irrational numbers can be calculated to as many digits as desired by applying the method of successive approximations or the iterative method. This is a very useful procedure employed in computer programs. One such iteration formula that can be used for square roots is Newton's approximation formula:

$$X_{n+1} = \frac{1}{2}\left(X_n + \frac{N}{X_n}\right)$$

(7.5)

where N = radicand (number under the radical), X_n = n^{th} approximation and X_{n+1} = $(n+1)^{th}$ approximation. The following example illustrates the use of Newton's formula.

EXAMPLE 7.10

Approximate $\sqrt{5}$ to three decimal places, using Newton's approximation formula 7.5).

Solution   The first approximation X_1 can be any number; however, it is convenient to choose $X_1 = N = 5$, which is the radicand. The second approximation to three decimal places is then

$$X_2 = \frac{1}{2}\left(5 + \frac{5}{5}\right) = 3.000$$

EXAMPLE 7.10 (Cont.)

The successive approximations are

$$X_3 = \frac{1}{2}\left(3.000 + \frac{5}{3.000}\right) = 2.333$$

$$X_4 = \frac{1}{2}\left(2.333 + \frac{5}{2.333}\right) = 2.238$$

$$X_5 = \frac{1}{2}\left(2.238 + \frac{5}{2.238}\right) = 2.236$$

$$X_6 = \frac{1}{2}\left(2.236 + \frac{5}{2.236}\right) = 2.236$$

Since X_6 is equal to X_5 to three decimal places, the calculation ends. Newton's formula produces accurate results after only a few approximations and can be easily programmed on a computer to produce any degree of accuracy. There are also Newton approximation formulas for third-order and higher order radicals. See problems 67 and 68 in Exercise 7.2.

Most programming languages have a *library function* that calculates the square root. For example, in BASIC it is SQR(X). In Fortran it is SQRT(X).

EXERCISE 7.2

In problems 1 through 28, simplify by applying the rules for radicals.

1. $\sqrt{(6x)^2}$

2. $(\sqrt{3y})^2$

3. $\sqrt{0.04I^2}$

4. $\sqrt{\dfrac{R^4}{49}}$

5. $\sqrt{4C_1{}^4 C_2{}^2}$

6. $\sqrt{9\alpha^2\beta^6}$

7. $\sqrt{12}$

8. $\sqrt{50}$

9. $2\sqrt{27}$

10. $5\sqrt{200}$

11. $\dfrac{2\sqrt{18}}{3}$

12. $\dfrac{5\sqrt{32}}{4}$

13. $\sqrt{r^5\theta^3}$

14. $\rho\sqrt{\rho\varphi^3}$

15. $\sqrt{32X_L{}^2 X_T{}^5}$

16. $\sqrt{64P^3Q^3}$

17. $\sqrt[3]{24}$

18. $\sqrt[3]{0.002}$

19. $\sqrt{5}\sqrt{45}$

20. $\sqrt{6}\sqrt{18}$

21. $\dfrac{\sqrt{2}\sqrt{27}}{\sqrt{6}}$

22. $\dfrac{\sqrt{50}}{\sqrt{5}\sqrt{40}}$

23. $\sqrt{0.2I^2R}\sqrt{0.05I^2R}$

24. $\sqrt{0.5E}\sqrt{2E^3}$
25. $\sqrt{12\times10^6}$
26. $\sqrt{0.16\times10^{-6}}$

27. $\sqrt{1.44\times10^{-6}}$
28. $\sqrt{1.69\times10^4}$

In problems 29 through 38, rationalize each denominator and simplify.

29. $\sqrt{\dfrac{1}{3}}$

30. $\dfrac{1}{\sqrt{2}}$

31. $\dfrac{\sqrt{5}}{\sqrt{6}}$

32. $\sqrt{\dfrac{2}{5}}$

33. $\dfrac{2}{\sqrt{3}}$

34. $\dfrac{\sqrt{7}}{\sqrt{12}}$

35. $\sqrt{\dfrac{9}{8}}$

36. $\sqrt{\dfrac{8}{27}}$

37. $\dfrac{\sqrt{0.02V_x}}{\sqrt{0.5V_y}}$

38. $\sqrt{\dfrac{1}{R_T}+\dfrac{1}{4}}$

In problems 39 through 60, simplify and combine similar radicals.

39. $3\sqrt{5}+2\sqrt{7}-\sqrt{5}$
40. $5\sqrt{3}-7\sqrt{2}+6\sqrt{2}$
41. $\sqrt{45}+2\sqrt{20}$
42. $4\sqrt{24}-5\sqrt{6}$
43. $3\sqrt{12}-\sqrt{48}+2\sqrt{27}$
44. $8\sqrt{8}+\sqrt{50}-3\sqrt{2}$
45. $\sqrt{125}+2\sqrt{5}-\sqrt{500}$
46. $3\sqrt{63}-8\sqrt{28}+7\sqrt{7}$

47. $\dfrac{1}{2}\sqrt{32}+2\sqrt{\dfrac{1}{2}}$

48. $6\sqrt{\dfrac{1}{3}}-2\sqrt{12}$

49. $\sqrt{75\times10^{12}}+\sqrt{3\times10^{12}}$
50. $\sqrt{24\times10^{-4}}-\sqrt{6\times10^{-4}}$

51. $\sqrt{32\times10^{-6}}+\sqrt{50\times10^{-6}}-\sqrt{8\times10^{-6}}$
52. $\sqrt{2\times10^6}-\sqrt{8\times10^6}+\sqrt{18\times10^6}$
53. $\sqrt[3]{16}+2\sqrt[3]{2}$
54. $\sqrt[3]{54}-\sqrt[3]{2}$
55. $\sqrt{18Y^3}+2\sqrt{2Y^3}$
56. $\sqrt{12PV^2}-\sqrt{48PV^2}$

57. $\sqrt{\dfrac{1}{2}}+\sqrt{\dfrac{1}{8}}$

58. $\dfrac{2}{\sqrt{8}}-\dfrac{1}{2\sqrt{2}}$

59. $\dfrac{3\sqrt{3}}{2G}+\dfrac{2\sqrt{12}}{3G}$

60. $\dfrac{3\sqrt{200}}{10R_T}-\dfrac{2\sqrt{50}}{5R_T}$

In problems 61 through 64, find the indicated root to three decimal places using Newton's formula (7.5) and the method of successive approximations. See Example 7.10.

61. $\sqrt{8}$
62. $\sqrt{13}$

63. $\sqrt{51}$
64. $\sqrt{37}$

In problems 65 through 68, solve each applied problem.

65. The velocity V in miles per hour of a boat is given by

$$V = 1.1\sqrt{L}\ \sqrt[3]{\frac{1000H}{D}}$$

where L = waterline length, H = horsepower, and D = displacement or weight. Calculate V to two significant digits when L = 32 ft, H = 64 hp, and D = 27,000 lb.

66. The velocity V of a small water wave is given by

$$V = \sqrt{\frac{\pi}{4\lambda d}} + \sqrt{\frac{4\pi}{\lambda d}}$$

Simplify and combine the radicals, and express V with a rational denominator.

67. Newton's approximation formula for cube roots is

$$X_{n+1} = \frac{1}{3}\left(2X_n + \frac{N}{X_n^2}\right)$$

Use this formula to approximate $\sqrt[3]{2}$ to three decimal places. See Example 7.10.

68. Using the formula in problem 67, approximate $\sqrt[3]{0.5}$ to three decimal places.

Applications to Electronics   In problems 69 through 74, solve each applied problem.

69. The resultant electric intensity of the two electric charges shown in Figure 7–2 is given by the Pythagorean formula:

$$E_R = \sqrt{\left(E_x\right)^2 + \left(E_y\right)^2}$$

Find E_R when $E_x = -2.29 \times 10^2$ N/C (newtons/coulomb) and $E_y = -1.55 \times 10^2$ N/C. Round answer to three significant digits.

70. In problem 69, if E_x is given by $\frac{3e+4}{2}$ and E_y by $\frac{4e-3}{2}$, express E_R in simplest radical form.

71. In the *RC* (resistance-capacitance) series circuit shown in Figure 7–3, the impedance Z is given by

$$Z = \sqrt{R^2 + \left(\frac{1}{2\pi fC}\right)^2}$$

where f = frequency. Simplify this expression for Z by combining the terms in the radical and rationalizing the denominator.

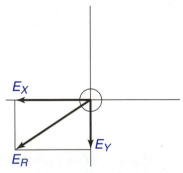

$$E_R = \sqrt{(E_X)^2 + (E_Y)^2}$$

FIGURE 7–2   Resultant electric intensity for problems 69 and 70.

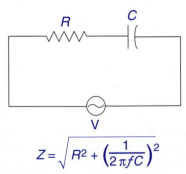

$$Z = \sqrt{R^2 + \left(\frac{1}{2\pi fC}\right)^2}$$

FIGURE 7–3   Resistance-capacitance series circuit for problems 71 and 72.

72. In problem 71, calculate Z in ohms to two significant digits when $R = 100\ \Omega$, $C = 20\ \mu F$, and $f = 60\ Hz$. (Note: C must be in farads in the formula.)

73. The impedance Z of the RL (resistance-inductance) parallel circuit in Figure 7–4 is given by

$$Z = \frac{1}{\sqrt{\left(\dfrac{1}{X_L}\right)^2 + \left(\dfrac{1}{R}\right)^2}}$$

where X_L = inductive reactance. Simplify this expression for Z by first combining the fractions in the radical and then rationalizing the denominator.

74. In problem 73, calculate Z in ohms to two significant digits when $X_L = 500\ \Omega$ and $R = 750\ \Omega$.

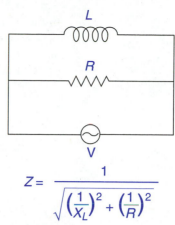

$$Z = \frac{1}{\sqrt{\left(\dfrac{1}{X_L}\right)^2 + \left(\dfrac{1}{R}\right)^2}}$$

FIGURE 7–4 Resistance-inductance parallel circuit for problems 73 and 74.

7.3
QUADRATIC EQUATION

Consider the following problem. A printed rectangular circuit board is to have an area of 48 cm². The length is to be 2 cm longer than the width. What should be the dimensions of the board? Apply the steps for problem solving. First write the verbal equation, which is

$$(\text{Length})(\text{Width}) = \text{Area}$$

Since the length is given in terms of the width, let w = width. Then $w + 2$ = length, and the algebraic equation is

$$w(w + 2) = 48$$

Multiplying out the parentheses and moving all the terms to the left side, you obtain the equation

$$w^2 + 2w - 48 = 0$$

This is an example of a *quadratic equation*. Quadratic equations first occurred many years ago in problems involving the areas of squares and other figures. The prefix "quad" and the expression "x squared" for x^2, come from their early use in problems involving squares. A *quadratic equation* is a second-degree equation and contains variables to the second power but no higher power. The general quadratic equation in x is

$$ax^2 + bx + c = 0 \tag{7.6}$$

where a and b are coefficients and c is the constant term. You can readily solve a quadratic equation in the general form if the left side can be factored. Consider the preceding equation for the circuit board $w^2 + 2w - 48 = 0$, which can be factored as follows:

$$(w + 8)(w - 6) = 0$$

In factored form, the equation states that the product of two numbers, $(w + 8)$ and $(w - 6)$, equals zero. A product of two numbers can only be zero if *one (or both) of the numbers equals zero*. Therefore, if you can set each factor equal to zero, you can find two possible solutions for w by solving each linear equation:

$$w + 8 = 0 \quad \text{and} \quad w - 6 = 0$$
$$w = -8 \qquad\qquad w = 6$$

Both answers satisfy the original equation $w(w + 2) = 48$:

$$w = -8 \rightarrow (-8)(-8 + 2) = (-8)(-6) = 48$$
$$w = 6 \rightarrow (6)(6 + 2) = (6)(8) = 48$$

However, since the width must be positive, the only acceptable answer is $w = 6$ cm. The length is then $w + 2 = 8$ cm.

EXAMPLE 7.11

Solve the quadratic equation by factoring:
$$2x^2 + 5x - 3 = 4x + 12$$

Solution To solve a quadratic equation by factoring, you must first put the equation in the general quadratic form (7.6). Move all the terms to the left side:
$$2x^2 + 5x - 3 - 4x - 12 = 0$$
$$2x^2 + x - 15 = 0$$

Factor the left side, and set each factor equal to zero:
$$(2x - 5)(x + 3) = 0$$
$$2x - 5 = 0 \quad x + 3 = 0$$

Solve the equations to obtain the two solutions:
$$x = \frac{5}{2} \text{ or } 2.5 \quad x = -3$$

Observe that it is possible to factor the left side of the *original* equation $2x^2 + 5x - 3 = (2x - 1)(x + 3)$. However, you cannot set each factor equal to zero and solve the equation unless you have a zero on the right side of the equation.

When the constant term $c = 0$ in the general quadratic equation (7.6), it becomes $ax^2 + bx = 0$. The equation can then be solved by separating the common factor of the unknown, as the next example shows.

EXAMPLE 7.12

Solve the quadratic equation by factoring:
$$23V^2 = 240V$$

Solution Factor out V:
$$23V^2 - 240V = 0$$
$$V(23V - 240) = 0$$

EXAMPLE 7.12 (Cont.)

Set each factor = 0:

$$V = 0 \qquad 23V - 240 = 0$$

$$V = \frac{240}{23} = 10.4$$

Observe that one of the solutions is zero. You might be tempted to divide each term by V to simplify the original equation. If you did this, the equation would become $23V = 240$, and there would only be one solution $V = 10.4$. What has happened is that you have divided by the solution $V = 0$, and this is not permitted. You must factor out the unknown and obtain both solutions, one of which is zero.

When $b = 0$ in the general quadratic equation (7.6), the first-degree term is missing, and it becomes $ax^2 + c = 0$. The solution can be found directly by first solving for the square of the unknown:

$$ax^2 + c = 0$$

$$ax^2 = -c$$

$$x^2 = -\frac{c}{a}$$

and then taking the square root of each side of the equation:

$$\sqrt{x^2} = x = \pm\sqrt{\frac{-c}{a}}$$

or

$$x = \sqrt{\frac{-c}{a}} \text{ and } x = -\sqrt{\frac{-c}{a}}$$

When you take the square root of both sides of an equation the positive and the negative square roots must be considered. In a practical problem, the negative answer sometimes does not apply.

Study the next example, which closes the circuit and illustrates this type of solution in an application to electronics.

EXAMPLE 7.13

Close the Circuit

The circuit in Figure 7–5(a) contains a resistance R in series with a capacitance C (series RC circuit). The relationship between R, the capacitive reactance X_C, and the total impedance Z_T is given by the pythagorean relation:

$$Z_T{}^2 = R^2 + X_C{}^2$$

This is shown by the phasor diagram in Figure 7–5(b). If $Z_T = 6.0$ kΩ and the capacitive reactance is twice the resistance, find R and X_C.

EXAMPLE 7.13 (Cont.)

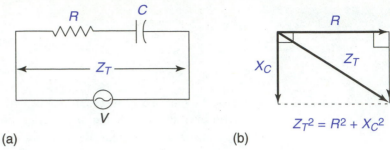

(a)

(b)

FIGURE 7–5   Series *RC* circuit for Example 7.13.

Solution   Let $X_C = 2R$ and substitute this value and $Z_T = 6.0 \times 10^6$ Ω into the equation:

$$(6 \times 10^6)^2 = R^2 + (2R)^2$$

Then:

$$36 \times 10^{12} = R^2 + 4R^2$$

There is no first-degree term, so you can solve directly for R^2:

$$5R^2 = 36 \times 10^{12}$$
$$R^2 = 7.2 \times 10^{12}$$

Now take the square root of both sides of the equation. The answer to two significant digits is then:

$$\sqrt{R^2} = \pm\sqrt{7.2 \times 10^{12}}$$
$$R = 2.7 \times 10^6 \text{ and } R = -2.7 \times 10^6$$

Since *R* must be positive the only acceptable answer is $R = 2.7$ kΩ. Then $X_C = 2(2.7 \text{ kΩ}) = 5.4$ kΩ.   ▪

The next example closes the circuit and shows how to solve a literal quadratic equation or formula containing the square of a variable by factoring.

EXAMPLE 7.14

Close the Circuit

Consider the series circuit in Figure 7–6, consisting of a voltage source *V* and two resistors R_1 and R_2 in series. The total power in the circuit equals the sum of the powers dissipated in each resistor. Applying Ohm's law and the power formula, this is expressed by the equation:

$$VI = I^2R_1 + I^2R_2$$

Solve this equation for *I*.

Solution   This equation is quadratic in *I*. First, put all the terms containing *I* on one side of the equation in order of descending power:

$$I^2R_1 + I^2R_2 - VI = 0$$

There is no constant term, so you can factor out *I*:

$$I(IR_1 + IR_2 - V) = 0$$

EXAMPLE 7.14 (Cont.)

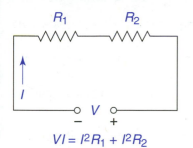

FIGURE 7–6 Power in a series circuit for Example 7.14.

Set each factor equal to zero:
$$I = 0$$
$$IR_1 + IR_2 - V = 0$$

Solve for I in the right factor by factoring again:
$$IR_1 + IR_2 = V$$
$$I(R_1 + R_2) = V$$
$$I = \frac{V}{R_1 + R_2}$$

There are two solutions; however, $I = 0$ usually does not apply.

EXERCISE 7.3

In problems 1 through 22, solve each quadratic equation by factoring or taking square roots.

1. $x^2 - 4 = 0$
2. $2y^2 = 32$
3. $q^2 - 3q - 10 = 0$
4. $t^2 + 6t + 8 = 0$
5. $3I_1{}^2 - 8I_1 + 4 = 0$
6. $5V_T{}^2 + 2V_T = 7$
7. $56P^2 = 8P$
8. $10C^2 + 18C = 0$
9. $7.28R_T{}^2 + 3.15R_T = 0$
10. $0.38X_L{}^2 = 0.59X_L$
11. $0.72 - 0.50\omega^2 = 0$

12. $9.3\lambda^2 - 8.6 = 0$
13. $29 - 53Z^2 = 0$
14. $50R^2 - 32 = 0$
15. $10a^2 + 5a = 5$
16. $6\alpha^2 - 20\alpha + 16 = 0$
17. $3E^2 + E = E^2 - 4E + 3$
18. $5e^2 - 10 = e^2 - 3e$
19. $p(p + 3) = 40$
20. $(x - 2)(x + 3) = 6$
21. $(3n - 1)^2 - n^2 = 0$
22. $2k^2 + 2 = (k + 1)^2$

In problems 23 through 28, solve each literal quadratic equation for the indicated variable. See Example 7.14.

23. $I_T{}^2 R_T = P_1 + P_2$; I
24. $V_1 I - V_2 I = I^2 R_T$; I
25. $4R_1{}^2 - 5R_1 R_2 + R_2{}^2 = 0$; R_1

26. $2I_1{}^2 - I_1 I_2 - I_2{}^2 = 0$; I_1
27. $t^2 + v_0{}^2 = 2v_0 t$; t
28. $2ax^2 + bx^2 = c$; x

In problems 29 through 34, solve each problem to two significant digits by setting up and solving a quadratic equation.

29. The distance s above the ground of a missile is given by
$$s = v_0 t - \frac{1}{2}gt^2$$
where v_0 = initial vertical velocity, g = gravitational acceleration, and t = time. How many seconds does it take for a missile fired at $v_0 = 196$ m/s to return to the ground? Use $g = 9.8$ m/s², and let $s = 0$.

30. In problem 29, if $v_0 = 550$ ft/s and $g = 32$ ft/s², how many seconds does it take for the missile to return to the ground?

31. One leg of a right triangular structure is 3 m longer than the other leg. If the area of the structure is 35 m², how long is each side? (Note: Area = Half the product of the legs.)

32. One leg of a right triangular structure is 7 ft more than the other leg. If the hypotenuse is 17 ft long, how long is each leg? (Hint: Use the Pythagorean theorem).

33. A rectangular IC (integrated circuit) is designed to have an area of 10 mm². If the length is to be twice the width, what should be the dimensions of the circuit?

34. A rectangular solar panel has an area of 1.0 m². If the length is 150 cm more than the width, find the length and width.

Applications to Electronics   In problems 35 through 42, solve each applied problem by setting up and solving a quadratic equation. Round answers to two significant digits.

35. Figure 7–7 shows an *RLC* circuit containing a resistance, an inductance, and a capacitance in series. Resonance occurs when the inductive reactance equals the capacitive reactance given by the relationship:

$$2\pi fL = \frac{1}{2\pi fC}$$

Solve this formula for the frequency f.

36. Figure 7–8 shows two resistances R_1 and R_2 in series. The total power dissipation is given by

$$P_T = I^2R_1 + I^2R_2$$

(a) Solve this equation for I. (See Example 7.14.)
(b) If $P_T = 110$ W, $R_1 = 2.0$ kΩ, and $R_2 = 1.5$ kΩ, calculate the current.

37. Figure 7–9(a) shows an *RL* circuit containing a resistance and an inductance in series. The relationship between R, the inductive reactance X_L, and the total impedance Z_T is shown in Figure 7–9(b) and is given by

$$Z_T^2 = R^2 + X_L^2$$

If $Z_T = 5.0$ kΩ and the resistance equals half the inductive reactance, find R and X_L (See Example 7.13).

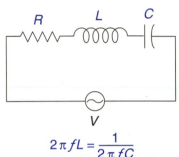

$$2\pi fL = \frac{1}{2\pi fC}$$

FIGURE 7–7   Resonance in an *RLC* circuit for problem 35.

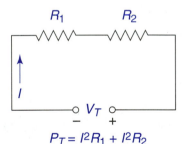

$$P_T = I^2R_1 + I^2R_2$$

FIGURE 7–8   Total power in a series circuit for problem 36.

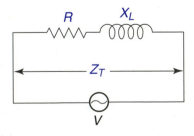

(a)

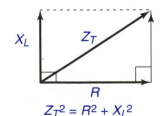

$$Z_T^2 = R^2 + X_L^2$$

(b)

FIGURE 7–9   Series RL circuit for problems 37 and 38.

38. In problem 37, if Z_T = 5.0 kΩ and R is 1.0 kΩ more than X_L, find the resistance and the reactance.

39. Figure 7–10 shows a generator with internal resistance r_i connected to a variable load R_V. The power P delivered to the load is given by

$$P = V_G I - I^2 r_i$$

where I = current in the circuit and V_G = voltage output of the generator. If V_G = 12 V, r_i = 2.0 Ω and P = 16 W, find the two possible values of the current.

40. In the circuit of Figure 7–10, the power delivered to the load will be a maximum when $R_V = r_i$. If the maximum power P = 18 W, find the current that produces this maximum power.

41. A bulb is connected to a battery in series with a voltage dropping resistor R_D (Figure 7–11). If the voltage of the battery V_B = 12 V, R_D = 6.0 Ω, and the power dissipation in the bulb is 6.0 W, what is the current I in the circuit? Use Ohm's law and the power formula to set up a quadratic equation.

42. In Figure 7–11, if V_B = 24 V, R_D = 12 Ω, and the power dissipation in the bulb is 9.0 W, what are the two possible values of the current I?

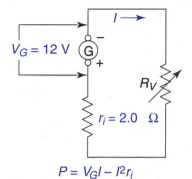

$$P = V_G I - I^2 r_i$$

FIGURE 7–10 Generator power for problems 39 and 40.

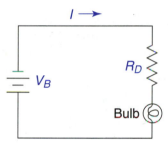

FIGURE 7–11 Voltage-dropping resistor for problems 41 and 42.

≡ 7.4
QUADRATIC FORMULA

When a quadratic equation cannot be solved by factoring or by taking square roots, it is necessary to apply the quadratic formula, which can be used to solve any quadratic equation. Given the general quadratic equation:

$$ax^2 + bx + c = 0$$

The two solutions for x are, by the *quadratic formula*

$$x = \frac{-b \pm \sqrt{b^2 - 4ac}}{2a} \tag{7.7}$$

Observe that the numerator in the formula contains both $-b$ *and* the radical. This basic and important formula comes from the general quadratic equation by changing it to a perfect square (completing the square) and then solving for x by taking square roots. The proof is as follows:

Given:

$$ax^2 + bx + c = 0$$

Isolate x:

$$ax^2 + bx = -c$$

Multiply by $4a$:

$$4a^2x^2 + 4abx = -4ac$$

Add b^2 to both sides to make the left side a perfect square:

$$4a^2x^2 + 4abx + b^2 = b^2 - 4ac$$

Factor the left side:

$$(2ax + b)^2 = b^2 - 4ac$$

Take the square root of both sides, which cancels the square on the left side:

$$2ax + b = \pm\sqrt{b^2 - 4ac}$$

Transpose b and divide by $2a$ to obtain the two solutions for x:

$$2ax = -b \pm \sqrt{b^2 - 4ac}$$

$$x = \frac{-b \pm \sqrt{b^2 - 4ac}}{2a}$$

You need to memorize the quadratic formula. It can be used to solve any quadratic equation whether it can be factored or not. Study the following three examples, which show how to use the quadratic formula.

EXAMPLE 7.15

Solve the quadratic equation $5x^2 + 8x = 6x + 3$ by the quadratic formula.

Solution To use the quadratic formula, you must first put the equation in the general form by moving all the terms to one side in order:

$$5x^2 + 8x = 6x + 3$$

$$5x^2 + 8x - 6x - 3 = 0$$

$$5x^2 + 2x - 3 = 0$$

Then write down the values of a, b, and c:

$$a = 5, \; b = 2, \; c = -3$$

Substitute the values in the quadratic formula, and simplify the radical:

$$x = \frac{-b \pm \sqrt{b^2 - 4ac}}{2a}$$

$$x = \frac{-2 \pm \sqrt{(2)^2 - 4(5)(-3)}}{2(5)}$$

$$x = \frac{-2 \pm \sqrt{4 + 60}}{10} = \frac{-2 \pm \sqrt{64}}{10} = \frac{-2 \pm 8}{10}$$

Note that under the radical, you multiply the minus sign in front of $-4(5)(-3)$ before combining this product with $(2)^2$.

EXAMPLE 7.15 (Cont.)

Separate the solution into two answers, one using the plus sign and one using the minus sign:

$$x = \frac{-2+8}{10} = \frac{6}{10} = \frac{3}{5} = 0.6 \text{ and } x = \frac{-2-8}{10} = \frac{-10}{10} = -1$$

Observe that the solutions are integers or fractions, which are rational roots. A quadratic equation has rational roots when the *discriminant*, $b^2 - 4ac$, is a perfect square. Under these conditions, the equation can also be solved by factoring to check the answers:

$$5x^2 + 2x - 3 = 0$$

$$(5x - 3)(x + 1) = 0$$

$$x = \frac{3}{5}$$

$$x = -1 \checkmark$$

The next example shows an equation that cannot be solved by factoring and that has irrational roots.

EXAMPLE 7.16

Solve the quadratic equation:

$$3I^2 - 2I = 2$$

Express the roots in simplest radical form and as decimals to the nearest hundredth.

Solution   Put the equation in the general form:

$$3I^2 - 2I - 2 = 0$$

Note that the equation cannot be factored. Use the quadratic formula with I in place of x and $a = 3$, $b = -2$, $c = -2$:

$$I = \frac{-b \pm \sqrt{b^2 - 4ac}}{2a}$$

$$I = \frac{-(-2) \pm \sqrt{(-2)^2 - 4(3)(-2)}}{2(3)}$$

$$I = \frac{2 \pm \sqrt{4 + 24}}{6} = \frac{2 \pm \sqrt{28}}{6}$$

Observe that when b is negative, the first term in the numerator is positive. You can simplify the radical by separating a perfect root:

$$I = \frac{2 \pm \sqrt{4}\sqrt{7}}{6} = \frac{2 \pm 2\sqrt{7}}{6}$$

EXAMPLE 7.16 (Cont.)

This fraction can be reduced by dividing out the common factor of 2. However, the fraction line is like parentheses, so the factor should be separated first before dividing to make sure each term is divided by 2:

$$I = \frac{2(1 \pm \sqrt{7})}{6} = \frac{(1 \pm \sqrt{7})}{3}$$

Then the two irrational roots in simplest radical form are

$$I = \frac{1 + \sqrt{7}}{3} \text{ and } I = \frac{1 - \sqrt{7}}{3}$$

When the discriminant $b^2 - 4ac$ is not a perfect square, as in this example, the roots are irrational and the equation cannot be solved by factoring. Find the roots in decimal form to the nearest hundredth with the calculator:

$$I = \frac{1 + \sqrt{7}}{3} = \frac{1 + 2.65}{3} = 1.22$$

$$I = \frac{1 - \sqrt{7}}{3} = \frac{1 - 2.65}{3} = -0.55$$

The decimal roots can also be calculated without simplifying the radical. To check the decimal roots, you can use the calculator and substitute in the original equation:

$$3(1.22)^2 - 2(1.22) - 2 = 4.47 - 2.44 - 2 = 0.03 \approx 0$$

$$3(-0.55)^2 - 2(-0.55) - 2 = 0.91 - 1.10 - 2 = 0.01 \approx 0$$

The results are close enough to zero to verify the answers. Since the roots are approximate to the nearest hundredth, you cannot expect to get exactly zero.

■

The next example shows a quadratic equation that has *complex* roots.

EXAMPLE 7.17

Solve the quadratic equation

$$Z^2 + 2 = 2Z$$

Solution   Put the equation in the general form:

$$Z^2 - 2Z + 2 = 0$$

The equation cannot be factored. Use the formula with Z in place of x and $a = 1$, $b = -2$, and $c = 2$:

$$Z = \frac{-(-2) \pm \sqrt{(-2)^2 - 4(1)(2)}}{2(1)} = \frac{2 \pm \sqrt{-4}}{2}$$

The number under the radical (radicand) is negative and $\sqrt{-4}$ is called an *imaginary number*. The roots consist of a real number combined with an imaginary number. This is called a *complex* number. Complex numbers are studied in Chapter 10 and are important in ac circuits. In this chapter, it is noted that such roots exist and are left in the above radical form without simplifying.

■

Complex roots occur when the discriminant $b^2 - 4ac$ is negative. The different types of roots a quadratic equation may have are summarized in Table 7.1. They are determined by the discriminant, which is denoted by a capital greek delta:

$$\Delta = b^2 - 4ac$$

TABLE 7.1   Roots of a Quadratic Equation

$\Delta = b^2 - 4ac$	Type of Roots
Positive: perfect square	Real, rational, and unequal
Positive: not perfect square	Real, irrational, and unequal
Zero	Real, rational, and equal
Negative	Complex (real + imaginary)

The quadratic equation adapts easily to a programmed solution on a calculator or computer using the quadratic formula.

ERROR BOX

A common error to watch out for when using the quadratic formula is simplifying the fraction incorrectly. The first step is to multiply out $4ac$ and b^2 separately. Then combine these terms and simplify the radical. Now, if there is a common factor in the numerator and the denominator, make sure you divide *both terms* in the numerator by the common factor. For example, for the quadratic equation $x^2 - 4x + 2 = 0$, the quadratic formula yields

$$x = \frac{4 \pm \sqrt{8}}{2}$$

There is a tendency to divide the 2 in the denominator into only the 4 and not the $\sqrt{8}$. This is wrong because the fraction line is like parentheses, and the radical must be divided by 2 also. You must first simplify the radical and *then* divide both terms if possible:

$$\frac{4 \pm 2\sqrt{2}}{2} = \frac{\overset{2}{4} \pm \overset{1}{2}\sqrt{2}}{\underset{1}{2}} = 2 \pm \sqrt{2}$$

See if you can correctly solve and simplify the irrational roots of the following quadratic equations.

Practice Problems   Find the roots in *simplest radical form*:

1. $R^2 - 4R + 1 = 0$ 2. $2I^2 + 6I - 1 = 0$ 3. $3X^2 + 3X - 1 = 0$ 4. $5V^2 - 5V - 1 = 0$

Answers: 1. $2 \pm \sqrt{3}$ 2. $\dfrac{-3 \pm \sqrt{11}}{2}$ 3. $\dfrac{-3 \pm \sqrt{21}}{6}$ 4. $\dfrac{5 \pm 3\sqrt{5}}{6}$

Study the next two examples, which close the circuit and illustrate applications of the quadratic formula to problems in electronics.

EXAMPLE 7.18

Close the Circuit

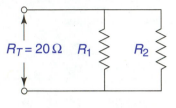

$R_2 = R_1 + 20 \; \Omega$

FIGURE 7–12
Parallel resistances for
Example 7.18.

The total resistance R_T of two parallel resistances R_1 and R_2 is to be 20 Ω, as shown in Figure 7–12. If R_2 is to be 20 Ω more than R_1, what should be the values of R_1 and R_2 to two significant digits?

Solution   The total resistance of two resistances in parallel as given by formula (5.8) is

$$R_T = \frac{R_1 R_2}{R_1 + R_2}$$

Substitute $R_T = 20$ and $R_2 = R_1 + 20$ in the formula:

$$20 = \frac{R_1(R_1 + 20)}{R_1 + (R_1 + 20)} = \frac{R_1{}^2 + 20R_1}{2R_1 + 20}$$

To solve this equation, you must first multiply both sides by $2R_1 + 20$ to clear the fraction:

$$(2R_1 + 20)20 = (2R_1 + 20)\frac{R_1{}^2 + 20R_1}{2R_1 + 20}$$

$$40R_1 + 400 = R_1{}^2 + 20R_1$$

Put the quadratic equation in the general form:

$$R_1{}^2 - 20R_1 - 400 = 0$$

This equation cannot be factored. Apply the quadratic formula with R_1 in place of x and $a = 1$, $b = -20$, and $c = -400$:

$$R_1 = \frac{20 \pm \sqrt{(-20)^2 - 4\,(1)(-400)}}{2(1)}$$

$$R_1 = \frac{20 \pm \sqrt{400 + 1600}}{2} = \frac{20 \pm \sqrt{2000}}{2}$$

The radical can be simplified by separating $\sqrt{400}$:

$$R_1 = \frac{20 \pm \sqrt{400}\sqrt{5}}{2} = \frac{20 \pm 20\sqrt{5}}{2}$$

You can reduce the fraction by factoring out and dividing by the common factor of 2:

$$R_1 = \frac{2(10 \pm 10\sqrt{5})}{2} = 10 \pm 10\sqrt{5}$$

There are two answers; however, $10 - 10\sqrt{5}$ is negative. Therefore, the only acceptable answer is

$$R_1 = 10 + 10\sqrt{5} = 32 \; \Omega$$

Then

$$R_2 = 32 + 20 = 52 \; \Omega$$

EXAMPLE 7.19

Close the Circuit

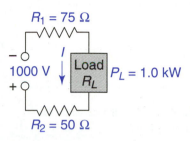

$R_1 = 75\ \Omega$

1000 V   Load R_L $P_L = 1.0$ kW

$R_2 = 50\ \Omega$

$R_T = R_1 + R_2 + R_L$

FIGURE 7–13 Series circuit with a load R_L for Example 7.19.

Figure 7–13 shows a load connected in series to two resistances and a 1000-V source. If $R_1 = 75\ \Omega$, $R_2 = 50\ \Omega$, and the power used by the load is $P_L = 1.0$ kW, what is the current I in the circuit and the resistance of the load R_L? Round answers to two significant digits.

Solution In a series circuit, the total resistance is equal to the sum of the resistances:

$$R_T = R_1 + R_2 + R_L$$

Expressing the total resistance by Ohm's law as $\frac{V}{I}$ and substituting the given values produces the equation:

$$\frac{1000}{I} = 75 + 50 + R_L$$

Clearing the fraction and putting all the terms on one side, this equation can be written:

$$125I - 1000 + IR_L = 0$$

To find an expression for the voltage drop IR_L in terms of I, you can use the power formula $P_L = I^2 R_L$. Since the power dissipated in the load is 1.0 kW = 1000 W

$$I^2 R_L = 1000$$

Divide both sides by I:

$$IR_L = \frac{1000}{I}$$

Substitute this last expression for IR_L into the preceding equation:

$$125I - 1000 + \frac{1000}{I} = 0$$

Multiply each term by I to clear the fraction, and you have a quadratic equation in the general form:

$$125I^2 - 1000I + 1000 = 0$$

Now simplify the equation by dividing each term by the common factor of 125:

$$I^2 - 8I + 8 = 0$$

Then apply the quadratic formula:

$$I = \frac{8 \pm \sqrt{64 - 32}}{2} = \frac{8 \pm \sqrt{32}}{2}$$

Simplify the radical, and reduce the fraction:

$$I = \frac{8 \pm 4\sqrt{2}}{2} = \frac{2(4 \pm 2\sqrt{2})}{2} = 4 \pm 2\sqrt{2}$$

There are two possible solutions for I:

$$I = 4 + 2(1.414) = 6.83\ \text{A} \approx 6.8\ \text{A}$$

$$I = 4 - 2(1.414) = 1.17\ \text{A} \approx 1.2\ \text{A}$$

EXAMPLE 7.19 (Cont.)

Each solution for I corresponds to one possible solution for R_L. Apply the power formula $R_L = P_L/I^2$ to find R_L:

$$I = 6.83 \text{ A}: \quad R_L = \frac{1000}{(6.83)^2} = 21 \, \Omega$$

$$I = 1.17 \text{ A}: \quad R_L = \frac{1000}{(1.17)^2} = 730 \, \Omega$$

Depending on the whether the load resistance needs to be high or low, only one answer may apply in a particular circuit.

■

EXERCISE 7.4

In problems 1 through 22, solve each equation by using the quadratic formula. Check rational roots by factoring. Express irrational roots in radical or decimal form as directed by the instructor.

1. $P^2 + 4P + 3 = 0$
2. $C^2 + 2C = 8$
3. $3y^2 - y = 10$
4. $2w^2 - 7w + 3 = 0$
5. $2\lambda^2 + 4 = 9\lambda$
6. $3\beta^2 - \beta = 2$
7. $V^2 - 2V = 1$
8. $2R^2 - 1 = 2R$
9. $I - 4I_2 = I_2{}^2$
10. $6G_1 - G_1{}^2 = 7$
11. $2e^2 - 6 = e^2 + 2e$
12. $4i^2 + 4i = 2 - 2i$
13. $2S_R{}^2 + 0.5 = 3S_R$
14. $V_T{}^2 - 3V_T = 2.5$
15. $0.75f_c{}^2 + 0.5f_c = 0.25$
16. $0.25t_0{}^2 + 1.0t_0 = -0.25$
17. $1.2R_L{}^2 - 3.4 = 6.8R_L$
18. $0.66\mu = 0.33 - 0.84\mu^2$
19. $Z^2 + 2Z + 2 = 0$
20. $3X^2 - X + 2 = 0$
21. $6X_L{}^2 = 3X_L{}^2 - 3$
22. $Z_T{}^2 + 21 = 9Z_T$

In problems 23 through 28, solve each applied problem by setting up and solving a quadratic equation. Round answers to two significant digits except as stated.

23. A rectangular garden is to have an area of 100 ft². If the width is to be 10 ft shorter than the length, what should be the dimensions of the garden to the nearest foot?

24. A rectangular window is to have an area of half a square meter. If the length is to be 100 cm more than the width, what should be the dimensions of the window?

25. The smallest leg of a right triangular computer component is 2.0 mm shorter than the other leg. If the area of the component is 8.0 mm², what is the length of each of the three sides? [Note: Area of a right triangle $= \frac{1}{2}$ (product of legs).]

26. A solar cell collector on the roof of a house is to have the shape shown in Figure 7–14. If the area of the collector must be 30 ft² to collect sufficient energy to meet the needs of the house, what should the dimension w be? Observe that w cannot be greater than 10.

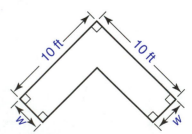

FIGURE 7–14 Solar collector cell for problem 26.

27. A car leaves Los Angeles traveling north at an average speed of 60 mi/h. At the same time, a second car leaves Los Angeles traveling east at an average speed of 50 mi/h. After how many hours will the cars be 200 mi apart? (Use the Pythagorean theorem $a^2 + b^2 = c^2$).

28. At noon a ship sails due south from New York at an average speed of 6.0 mi/h. One hour later, another ship sails due east at the same average speed. *At what time* will the ships be 30 mi apart? (Use the Pythagorean theorem: $a^2 + b^2 = c^2$).

Applications to Electronics In problems 29 through 36, solve each applied problem by setting up and solving a quadratic equation. Round answers to two significant digits except as stated.

29. The power output P of a generator is given by

$$P = VI - RI^2$$

where V = generator voltage, R = internal resistance, and I = current. Find I when $P = 40$ W, $V = 13$ V, and $R = 1.0\ \Omega$.

30. In a dc circuit, when a resistance R, an inductance L, and a capacitance C are connected in series, the following equation must be solved to determine the instantaneous value of the current:

$$m^2 + \left(\frac{R}{L}\right)m + \frac{1}{LC} = 0$$

where R is in ohms, L is in henrys, and C is in farads. Find the value of m when $R = 200\ \Omega$, $L = 1.0$ H, and $C = 500\ \mu$F. Round answers to three significant digits.

31. The total resistance R_T of two resistances R_1 and R_2 in parallel is $7.0\ \Omega$. If R_2 is $2.0\ \Omega$ more than R_1, find the values of R_1 and R_2. (See Example 7.18.)

32. Figure 7–15 shows two capacitances in series. The total capacitance is given by

$$C_T = \frac{C_1 C_2}{C_1 + C_2}$$

If $C_T = 7.5$ pF and $C_2 = C_1 + 20$ pF, find C_1 and C_2. (See Example 7.18.)

33. Figure 7–16 shows a generator in series with a resistance of $50\ \Omega$ and a power load of 100 W. If the terminal voltage of the generator $V_T = 200$ V, what are the two possible combinations of current I and the load resistance R_L that will satisfy the given conditions? (See Example 7.19.)

34. In Example 7.19, find I and R_L if the voltage source is 100 V, $P_L = 10$ W, $R_1 = 50\ \Omega$ and $R_2 = 50\ \Omega$.

35. Figure 7–17 shows a battery with an internal resistance $r_i = 30\ \Omega$ delivering 15 W to a variable load. Assuming the source emf of the battery $V_S = 50$ V, find the two possible values for the current in the circuit and the terminal voltage V_T of the battery. (See Example 7.19.)

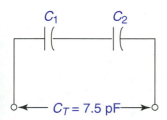

FIGURE 7–15
Capacitances in series
for problem 32.

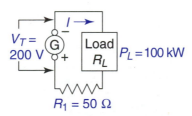

FIGURE 7–16   Generator
circuit for problem 33.

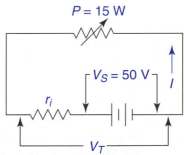

FIGURE 7–17   Battery with
internal resistance for
problem 35.

36. Figure 7–18 shows a fixed resistance R_0 in parallel with a variable resistance R_V, that varies from 1.0 Ω to 10 Ω. What should be the value of R_0 so that the *maximum* change in the total resistance R_T of the circuit is 3.0 Ω? (Hint: Let 3.0 Ω equal the difference in the maximum and the minimum values of R_T.)

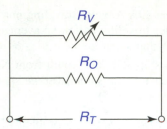

FIGURE 7–18 Parallel circuit with variable resistance for problem 36.

☰ CHAPTER HIGHLIGHTS

7.1 FRACTIONAL EXPONENTS

A fractional exponent is defined as

$$x^{m/n} = \left(\sqrt[n]{x}\right)^m \text{ or } \sqrt[n]{x^m} \quad (x>0) \quad (7.1)$$

A special case of (7.1) is

$$x^{1/n} = \sqrt[n]{x} \quad (x>0)$$

Study Examples 7.1, 7.2, and 7.3, which illustrate definition (7.1).

7.2 RADICALS

Three basic rules for radicals are for x and $y > 0$:

$$\left(\sqrt[n]{x}\right)^n = \sqrt[n]{x^n} = x \quad (7.2)$$

$$\sqrt[n]{xy} = \left(\sqrt[n]{x}\right)\left(\sqrt[n]{y}\right) \quad (7.3)$$

$$\sqrt[n]{\frac{x}{y}} = \frac{\sqrt[n]{x}}{\sqrt[n]{y}} \quad (7.4)$$

You can rationalize the denominator of a fraction by multiplying the top and bottom by a quantity that makes the radicand in the denominator a perfect square, as shown in Examples 7.4 and 7.5. You *cannot* combine radicals unless they have the same radicands. Study Examples 7.6 and 7.8.

Square roots that are irrational can be approximated on the computer using the method of successive approximations and Newton's approximation formula:

$$X_{n+1} = \frac{1}{2}\left(X_n + \frac{N}{X_n}\right) \quad (7.5)$$

where N = radicand (number under the radical), $X_n = n^{th}$ approximation and $X_n+1 = (n+1)^{th}$ approximation.

7.3 QUADRATIC EQUATION

The *general quadratic equation* in x is

$$ax^2 + bx + c = 0 \quad (7.6)$$

To solve a quadratic equation by factoring, first put it in this general form. Then factor the left side (if possible), set each factor equal to zero, and solve each linear equation for the one of the roots.

When $c = 0$, the general quadratic equation can always be solved by factoring out the common factor of the unknown:

$$ax^2 + bx = 0$$
$$x(ax + b) = 0$$
$$x = 0$$
$$x = -\frac{b}{a}$$

One of the roots is always zero.

A quadratic equation where $b = 0$ is readily solved by isolating the square of the unknown and then taking the square root of both sides of the equation:

$$ax^2 + c = 0$$
$$ax^2 = -c$$
$$x^2 = -\frac{c}{a}$$
$$x = \pm\sqrt{\frac{-c}{a}}$$

7.4 QUADRATIC FORMULA

Given the general quadratic equation (7.6), the *quadratic formula* is

$$x = \frac{-b \pm \sqrt{b^2 - 4ac}}{2a}$$

(7.7)

The discriminant $\Delta = b^2 - 4ac$ determines the types of roots as follows

$\Delta = b^2 - 4ac$	Type of Roots
Positive: perfect square	Real, rational, and unequal
Positive: not perfect square	Real, irrational, and unequal
Zero	Real, rational, and equal
Negative	Complex (real + imaginary)

Examples 7.15, 7.16, and 7.17 illustrate this table and the use of the quadratic formula. Study Examples 7.18 and 7.19 that close the circuit between math and electronics.

▤ REVIEW QUESTIONS

In problems 1 through 8, evaluate or simplify each expression.

1. $27^{2/3}$

2. $-16^{3/4}$

3. $0.001^{-1/3}$

4. $100^{-3/2}$

5. $(8a^3b^6)^{1/3}$

6. $\left(\dfrac{V_1^2 V_2^2}{16}\right)^{1/2}$

7. $(R^{1/2})(R^{1/2} + R^{-1/2})$

8. $(1.44 \times 10^4 I^2)^{3/2}$

In problems 9 through 24, perform the indicated operations or rationalize denominators and simplify.

9. $\sqrt{18}$

10. $\dfrac{5\sqrt{12}}{2}$

11. $2X\sqrt{75X^3}$

12. $\sqrt[3]{2C}\ \sqrt[3]{32C^2}$

13. $10\sqrt{1.21 \times 10^{-4}}$

14. $\dfrac{\sqrt{96}}{\sqrt{3}}$

15. $\dfrac{1}{\sqrt{8}}$

16. $\sqrt{\dfrac{12}{7}}$

17. $\sqrt{24} - \sqrt{3} + 3\sqrt{6}$

18. $2\sqrt{50} - \sqrt{8} - \sqrt{18}$

19. $\sqrt{72 \times 10^6} - \sqrt{50 \times 10^6}$

20. $\sqrt{\dfrac{3}{2}} - \sqrt{\dfrac{3}{8}}$

21. $\dfrac{2f}{\sqrt{3}} + \dfrac{f}{\sqrt{12}}$

22. $\sqrt{0.16\lambda} + \sqrt{\dfrac{\lambda}{0.25}}$

23. $3t\sqrt{tR_a^3} - 2R_a\sqrt{t^3 R_a}$

24. $\sqrt[3]{2} + 2\sqrt[3]{16}$

In problems 25 and 26, approximate each radical to three decimal places using Newton's formula (7.5).

25. $\sqrt{7}$ 26. $\sqrt{23}$

In problems 27 through 38, solve each quadratic equation.

27. $m^2 - 4m - 5 = 0$

28. $3t^2 + t = 2$

29. $7P_T = 8P_T^2$

30. $2R_X^2 - 32 = 0$

31. $(\delta + 1)^2 + 2\delta^2 = 2$

32. $16E^2 + 9 = 24E$

33. $I_C^2 + 3I_C + 1 = 0$

34. $4G_m^2 - 4G_m = 1$

35. $2s_R^2 + 2s_R + 1 = 10$

36. $v^2 - 0.3v = 0.5$

37. $5X_L + 2 = 3X_L^2 - 4$

38. $2Z_T^2 + 3Z_T + 3 = 0$

In problems 39 and 40, solve the literal equation for the indicated variable.

39. $IV_1 + IV_2 = I^2 R_T;$ I

40. $m^2 + \dfrac{R}{L}m + \dfrac{1}{L} = 0;$ m

In problems 41 through 44, solve each applied problem to two significant digits.

41. A power cable supported only at the ends will lie in a curve called a *catenary* which is given by the function

$$y = \frac{e^x + e^{-x}}{2}$$

where the base of natural logarithms $e \approx 2.718$.
   (a) Simplify the expression for y when $x = 0.5$.
   (b) Find the value of y when $x = 0.5$.
   (c) Find the value of y when $x = -0.5$.

42. The area of a rectangular video monitor is to be 600 cm². If the length is to be 5 cm more than the width, what should be the dimensions of the monitor to the nearest centimeter?

43. From Einstein's theory of relativity, the relationship between the rest mass m_0 and the mass m_v of a body at a high velocity v close to the speed of light is given by

$$\left(\frac{m_0}{m_v}\right)^2 = 1 - \frac{v^2}{c^2}$$

where the speed of light $c = 3.0 \times 10^8$ m/s. Find v when the mass m_v is twice the rest mass m_0. That is, $m_v = 2m_0$.

44. A piece of metal 25 cm wide is to be bent into an open rectangular trough of cross-sectional area 75 cm², as shown in Figure 7–19. What are the *two* possible values for the height h of the trough?

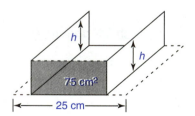

FIGURE 7–19   Trough for problem 44.

Applications to Electronics   In problems 45 through 50, solve each applied problem to two significant digits.

45. In an ac circuit, the root mean square or effective value of the voltage is related to the maximum voltage by

$$V_{\text{rms}}^2 = \frac{V_{\text{max}}^2}{2}$$

   (a) Solve and simplify the expression for V_{rms}.
   (b) Find V_{rms} when $V_{\text{max}} = 170$ V.

46. The height of a ship's radar target in *meters* is given by

$$h = \frac{R_M{}^2}{9} - 4(R_M - 9)$$

where R_M is the maximum detectable range of the target in *kilometers*. A target 40 m high will first appear on the radar screen at what maximum range? Find R_M for the given value of h.

47. The total resistance of two resistances in parallel is 6.0 Ω. If one resistance is 5.0 Ω more than the other resistance, what are the values of the two resistances?

48. The parallel circuit in Figure 7–20(a) contains a resistance and an inductance. The total current is related to the currents through the resistance and the inductance by the phasor diagram shown in Figure 7–20(b). This yields the Pythagorean relation:

$$I_T{}^2 = I_R{}^2 + I_L{}^2$$

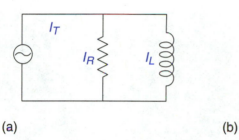
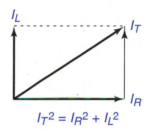

(a) (b)

FIGURE 7–20 Parallel circuit for problem 48.

If $I_T = 4.5$ A and I_R is 50% greater than I_L, find the currents through the resistance and the inductance.

49. Figure 7–21 shows a 15-V battery connected to 5-W bulb and a fixed resistance $R_0 = 5$ Ω. Find the current I and the bulb resistance R_B.

50. Figure 7–22 shows a series circuit containing a generator and three resistances. Given the circuit conditions shown with the generator operating and producing 14 V, find the current in the circuit and resistances R_2 and R_3.

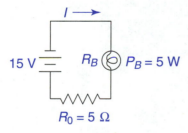

FIGURE 7–21 Bulb resistance and current in a circuit for problem 49.

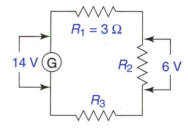

FIGURE 7–22 Generator series circuit for problem 50.

8

Network Analysis

Courtesy of Siemens Corporation.

A technician hand assembles a telecommunication switch system.

As a circuit becomes more complex, such as one with two voltage sources, it is necessary to have additional tools to analyze the circuit. Kirchhoff's laws, voltage and current divider formulas, and three important network theorems are discussed in this chapter. These ideas, along with Ohm's law and the power formulas, enable you to find currents and voltage drops in many complex circuits. Applying Kirchhoff's current and voltage laws leads to a linear system of equations, which are solved by the methods shown in Chapter 6. Each of the three network theorems—Superposition, Thevenin's theorem, and Norton's theorem—analyze a circuit in a similar way. They replace the original circuit with one or more simpler equivalent circuits that can be solved using the current and voltage divider formulas, Ohm's law, or the power formulas. The individual results are then combined to find the currents and voltage drops of the original circuit.

Chapter Objectives

In this chapter, you will learn:

- Kirchhoff's current law and how to apply it.
- Kirchhoff's voltage law and how to apply it.
- How to calculate proportional voltage drops in a series circuit.
- How to calculate proportional currents in a parallel circuit.
- How to apply the Superposition theorem to find currents and voltage drops in a circuit.
- The meaning of a Thevenin equivalent circuit.
- How to obtain the Thevenin voltage and Thevenin resistance for a load in a circuit.
- The meaning of a Norton equivalent circuit.
- How to obtain the Norton current and Norton resistance for a load in a circuit.

≡ 8.1
KIRCHHOFF'S LAWS

In addition to Ohm's law and the power formulas, two other basic electrical laws that apply to any circuit were formulated in 1847 by the German physicist Gustav R. Kirchhoff. In studying these laws, we will use electron current, which flows from (−) to (+); this convention is used throughout the text. Conventional current, which flows from (+) to (−) can also be used to apply Kirchhoff's laws.

Kirchoff's first law follows.

Kirchhoff's Current Law	The algebraic sum of all currents entering and leaving any point in a circuit equals zero. (8.1)

Kirchhoff's current law demonstrates that charge cannot accumulate at any point in a conductor. To apply the law, you assign a positive value to currents *entering* a branch point and a negative value to currents *leaving* a branch point.

For example, consider the circuit in Figure 8–1, which consists of two voltage sources and three resistances. The three currents shown go from (–) to (+) and illustrate electron current. At branch point **X**, I_1 and I_2 are entering and are therefore assigned (+); I_3 is leaving and is therefore assigned (–). Kirchhoff's current law then gives the equation:

$$I_1 + I_2 - I_3 = 0 \text{ or } I_1 + I_2 = I_3$$

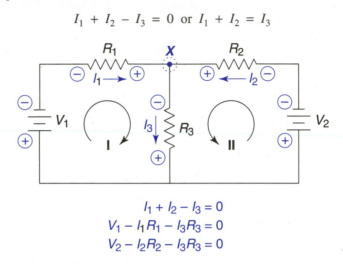

$$I_1 + I_2 - I_3 = 0$$
$$V_1 - I_1 R_1 - I_3 R_3 = 0$$
$$V_2 - I_2 R_2 - I_3 R_3 = 0$$

FIGURE 8–1 Kirchhoff's laws and electron current for Example 8.1.

Kirchhoff's second law follows.

Kirchhoff's Voltage Law   The algebraic sum of the voltages around any closed path equals zero.   (8.2)

Kirchhoff's voltage law demonstrates that electrical potential is conserved throughout a closed path or loop. That is, whatever potential you start at in a closed path, you must return to the same potential after traveling around the path. As a result the total voltage increase must equal the sum of the voltage drops. Apply the law by assigning a positive value to a voltage whose *positive* terminal is encountered first and a negative sign to a voltage whose *negative* terminal is encountered first. Consider the circuit in Figure 8–1. Polarities are shown by each resistance. Using electron current flowing from (–) to (+), the end of the resistor that the current *enters* is assigned a negative sign, and the other end is assigned a positive sign. If the end of a resistor is connected directly to a voltage source, that end is assigned the same polarity as the voltage source. Now, starting at the positive terminal of V_1 and traveling *clockwise* around loop **I** containing R_1 and R_3, Kirchhoff's voltage law gives the equation:

$$V_1 - I_1R_1 - I_3R_3 = 0$$

The voltages I_1R_1 and I_3R_3 are expressed using Ohm's law $V = IR$. Moving the terms I_1R_1 and I_3R_3 to the other side, the equation can also be written

$$V_1 = I_1R_1 + I_3R_3$$

This shows that the total voltage increase V_1 equals the sum of the voltage drops $I_1R_1 + I_3R_3$.

You can also travel around the same loop *counterclockwise* and apply Kirchhoff's voltage law. This reverses the sign of each voltage but results in the same equation. Study the next two examples, which show how to apply Kirchhoff's laws.

EXAMPLE 8.1

Given the circuit in Figure 8–1 with $V_1 = 14$ V, $V_2 = 12$ V, $R_1 = 6$ Ω, $R_2 = 5$ Ω, and $R_3 = 4$ Ω, find the currents I_1, I_2, and I_3.

Solution   Since there are three unknowns I_1, I_2, and I_3, it is necessary to write three equations containing the currents to find their values. First apply Kirchhoff's current law at point **X** to give the current equation that was shown above:

$$I_1 + I_2 - I_3 = 0$$

Then apply Kirchhoff's voltage law to loop **I:** $[V_1, R_1, R_3]$. Moving *clockwise* gives the second equation:

$$V_1 - I_1R_1 - I_3R_3 = 0$$

To obtain the third equation, apply Kirchhoff's voltage law to loop **II:** $[V_2, R_2, R_3]$. Starting at the positive terminal of V_2 and moving *counterclockwise* around the loop gives the equation:

$$V_2 - I_2R_2 - I_3R_3 = 0$$

The three equations represent the following linear system:

$$I_1 + I_2 - I_3 = 0$$
$$V_1 - I_1R_1 - I_3R_3 = 0$$
$$V_2 - I_2R_2 - I_3R_3 = 0$$

When the given values are substituted into this system, you have three equations in three unknowns:

$$I_1 + I_2 - I_3 = 0$$
$$14 - 6I_1 - 4I_3 = 0$$
$$12 - 5I_2 - 4I_3 = 0$$

There is more than one way to solve this linear system. Two useful methods will be shown. The first is by determinants as shown in Section 6.3. The second is an application of the substitution method shown in Section 6.2.

Solution by Determinants.   First put the equations into standard form with the variables on one side and the constants on the other side:

$$I_1 + I_2 - I_3 = 0$$
$$6I_1 + 4I_3 = 14$$
$$5I_2 + 4I_3 = 12$$

EXAMPLE 8.1 (Cont.)

Then solve for I_1, I_2, and I_3 using Cramer's rule (6.7) and determinants. Insert zeros for missing terms. The solutions to two significant digits are then:

$$I_1 = \frac{\begin{vmatrix} 0 & 1 & -1 \\ 14 & 0 & 4 \\ 12 & 5 & 4 \end{vmatrix}}{\begin{vmatrix} 1 & 1 & -1 \\ 6 & 0 & 4 \\ 0 & 5 & 4 \end{vmatrix}} = \frac{0 + 48 - 70 - 0 - 0 - 56}{0 + 0 - 30 - 0 - 20 - 24}$$

$$= \frac{-78}{-74} = 1.05 \, \text{A} \approx 1.1 \, \text{A}$$

$$I_2 = \frac{\begin{vmatrix} 1 & 0 & -1 \\ 6 & 14 & 4 \\ 0 & 12 & 4 \end{vmatrix}}{-74} = \frac{56 + 0 - 72 - 0 - 48 - 0}{-74}$$

$$= \frac{-64}{-74} = 0.86 \, \text{A} \approx 860 \, \text{mA}$$

$$I_3 = \frac{\begin{vmatrix} 1 & 1 & 0 \\ 6 & 0 & 14 \\ 0 & 5 & 12 \end{vmatrix}}{-74} = \frac{0 + 0 + 0 - 0 - 70 - 72}{-74}$$

$$= \frac{-142}{-74} = 1.92 \, \text{A} \approx 1.9 \, \text{A}$$

Solution by Substitution Method. First reduce the linear system to two equations in two unknowns by substitution. The first equation is easiest to solve for one of the unknowns. Solve the first equation for I_3:

$$I_1 + I_2 - I_3 = 0 \Rightarrow I_3 = I_1 + I_2$$

Then substitute $(I_1 + I_2)$ for I_3 in the other two equations:

$$14 - 6I_1 - 4(I_1 + I_2) = 0$$
$$12 - 5I_2 - 4(I_1 + I_2) = 0$$

Multiply out the parentheses and combine similar terms:

$$14 - 6I_1 - 4I_1 - 4I_2 = 0 \Rightarrow 14 - 10I_1 - 4I_2 = 0$$
$$12 - 5I_2 - 4I_1 - 4I_2 = 0 \Rightarrow 12 - 4I_1 - 9I_2 = 0$$

This gives you two equations in two unknowns. Put them in standard form:

$$10I_1 + 4I_2 = 14$$
$$4I_1 + 9I_2 = 12$$

EXAMPLE 8.1 (Cont.)

You can now solve for I_1 and I_2 by determinants or by the addition method. Using determinants, the solution is

$$I_1 = \frac{\begin{vmatrix} 14 & 4 \\ 12 & 9 \end{vmatrix}}{\begin{vmatrix} 10 & 4 \\ 4 & 9 \end{vmatrix}} = \frac{78}{74} = 1.05\,\text{A} \approx 1.1\,\text{A}$$

$$I_2 = \frac{\begin{vmatrix} 10 & 14 \\ 4 & 12 \end{vmatrix}}{74} = \frac{64}{74} = 0.86\,\text{A} \approx 860\,\text{mA}$$

Then I_3 is found by substituting back into $I_3 = I_1 + I_2$:

$$I_3 = 1.05 + 0.86 = 1.91\ \text{A} \approx 1.9\ \text{A}$$

The solutions for the three currents are shown to be all positive. This indicates that the directions assumed in Figure 8–1 are correct. If the solution for a current is negative, this indicates that the direction assumed for the current is actually the reverse. This can happen in a circuit like Figure 8–1 that contains two or more voltage sources, if the voltages differ by a significant amount. See the Error Box at the end of this section and problems 15 and 16 in Exercise 8.1.

■

The next example shows how to solve a circuit with four currents.

EXAMPLE 8.2

Given the circuit in Figure 8–2, with $V_1 = 9$ V, $V_2 = 6$ V, $R_1 = 3\ \Omega$, $R_2 = 3\ \Omega$, $R_3 = 4\ \Omega$, and $R_4 = 2\ \Omega$, find the currents I_1, I_2, I_3 and I_4.

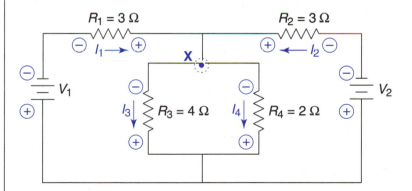

$$I_1 + I_2 - I_3 - I_4 = 0$$
$$V_1 - I_3 R_3 - I_1 R_1 = 0$$
$$V_2 - I_4 R_4 - I_2 R_2 = 0$$
$$I_3 R_3 - I_4 R_4 = 0$$

FIGURE 8–2 Kirchhoff's laws and electron current for Example 8.2.

EXAMPLE 8.2 (Cont.)

Solution Apply Kirchhoff's current law at point **X** in the circuit to give the current equation:

$$I_1 + I_2 - I_3 - I_4 = 0$$

Then apply Kirchhoff's voltage law to the three loops: $[V_1, R_1, R_3]$, $[V_2, R_2, R_4]$ and $[R_3, R_4]$. This gives the three equations:

$$V_1 - I_1 R_1 - I_3 R_3 = 0$$
$$V_2 - I_2 R_2 - I_4 R_4 = 0$$
$$I_3 R_3 - I_4 R_4 = 0$$

Study these equations carefully to make sure you understand how they are obtained.

Note that you will get the same equations whether you move clockwise or counterclockwise around a loop. In the third equation, observe that the loop does not need to include a voltage source. Substitute the given values to yield four equations in standard form:

$$I_1 + I_2 - I_3 - I_4 = 0$$

$$3I_1 \qquad + 4I_3 \qquad\quad = 9$$

$$3I_2 \qquad + 2I_4 = 6$$

$$4I_3 - 2I_4 = 0$$

This linear system of four equations can be reduced to three equations by substitution. You need to solve one of the equations for a variable and then substitute this expression into the other equations. The fourth equation is easiest to work with. Solve it for I_4:

$$4I_3 - 2I_4 = 0$$

$$2I_4 = 4I_3$$

$$I_4 = 2I_3$$

Then substitute $2I_3$ for I_4 in the first and third equations:

$$I_1 + I_2 - I_3 - 2I_3 = 0 \Rightarrow I_1 + I_2 - 3I_3 = 0$$
$$3I_2 + 2(2I_3) = 6 \Rightarrow 3I_2 + 4I_3 = 6$$

You now have three equations in three unknowns. In standard form they are

$$I_1 + I_2 - 3I_3 = 0$$

$$3I_1 \qquad + 4I_3 = 9$$

$$3I_2 + 4I_3 = 6$$

This system can be solved using determinants as shown in Example 8.1. It can also be solved by further reduction to two equations using substitution again.

EXAMPLE 8.2 (Cont.)

The solution is left as problem 20 in Exercise 8.2. The solutions to two significant digits are:

$$I_1 = 1.83\,\text{A} \approx 1.8\,\text{A}$$

$$I_2 = 0.82\,\text{A} = 820\,\text{mA}$$

$$I_3 = 0.88\,\text{A} = 880\,\text{mA}$$

$$I_4 = 1.76\,\text{A} \approx 1.8\,\text{A}$$

The answers are all positive, so the directions of the currents are correctly shown in Figure 8–2.

▪

ERROR BOX

A common error can occur when you assign polarities to resistances in a circuit for Kirchhoff's laws. *As long as the polarities agree with the direction of the current, the equations will be correct.* If a current actually flows opposite to your assumed direction, then the answer for that current will be negative. See if you can do the practice problem and discover the negative current.

Practice Problem: Given the circuit in Figure 8–3 with the currents as shown, find I_1, I_2, and I_3.

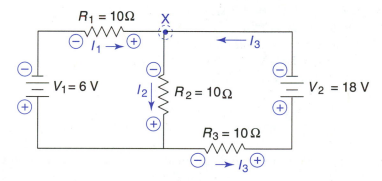

FIGURE 8–3 Polarities and Kirchhoff's laws.

Answers: $I_1 = -200\,\text{mA}, \; I_2 = 800\,\text{mA}, \; I_3 = 1.0\,\text{A}.$

EXERCISE 8.1

1. State Kirchhoff's current law.
2. State Kirchhoff's voltage law.

In problems 3 through 6, refer to the circuit in Figure 8-4, and write the equation that expresses each of the following.

3. Kirchhoff's current law at point **X**.
4. Kirchhoff's voltage law for the loop $[V_1, R_1, R_2]$.
5. Kirchhoff's voltage law for the loop $[V_1, R_1, R_3]$.
6. Kirchhoff's voltage law for the loop $[R_2, R_3]$.

In problems 7 through 10, refer to Figure 8-5, and write the equation that expresses each of the following.

7. Kirchhoff's current law at point **X**.
8. Kirchhoff's voltage law for the loop $[V_1, V_2, R_1, R_2]$.
9. Kirchhoff's voltage law for the loop $[V_1, R_1, R_3]$.
10. Kirchhoff's voltage law for the loop $[V_2, R_2, R_3]$.

In problems 11 through 22 solve each problem applying Kirchoff's laws. Round answers to two significant digits.

11. Given the circuit in Figure 8-4 with $V_1 = 9$ V, $R_1 = 3\ \Omega$, $R_2 = 5\ \Omega$, and $R_3 = 6\ \Omega$, find the currents I_1 I_2, and I_3.

12. Given the circuit in Figure 8-4 with $V_1 = 24$ V, $R_1 = 4\ \Omega$, $R_2 = 2\ \Omega$, and $R_3 = 8\ \Omega$, find the currents I_1, I_2, and I_3.

13. Given the circuit in Figure 8-5 with $V_1 = 24$ V, $V_2 = 12$ V, $R_1 = 10\ \Omega$, $R_2 = 3\ \Omega$, and $R_3 = 5\ \Omega$, find the currents I_1, I_2, and I_3.

14. Given the circuit in Figure 8-5 with $V_1 = 100$ V, $V_2 = 80$ V, $R_1 = 2.0$ kΩ, $R_2 = 3.0$ kΩ, and $R_3 = 2.0$ kΩ, find the currents I_1, I_2 and I_3.

15. Given the circuit in Figure 8-6 with $V_1 = 6$ V, $V_2 = 12$ V, $R_1 = 4\ \Omega$, $R_2 = 6\ \Omega$, and $R_3 = 3\ \Omega$, find the currents I_1, I_2, and I_3. Note the polarity of the voltage sources.

16. Given the circuit in Figure 8-6 with $V_1 = 32$ V, $V_2 = 6$ V, $R_1 = 20\ \Omega$, $R_2 = 10\ \Omega$, and $R_3 = 10\ \Omega$, find the currents I_1, I_2, and I_3.

17. An unknown voltage V_x is connected in series to an unknown resistance R_x and a resistor of 10 Ω. An ammeter in the circuit reads 200 mA. See Figure 8-7a. When the 10-Ω resistor is replaced by a 30-Ω resistor, the ammeter reads 100 mA. See Figure 8-7b. Use Kirchhoff's voltage law to find V_x and R_x.

18. A battery is connected to a resistance of 300 Ω. When a 20-V battery and a resistance of 1.0 kΩ are added in series to the circuit so that the voltages are series-aiding, the current remains the same. See Figure 8-8. Use Kirchhoff's voltage law to find V_x and I_x.

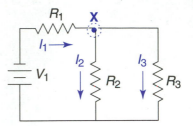

FIGURE 8-4 Kirchhoff's laws for problems 3 through 6, 11, and 12.

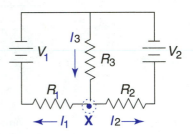

FIGURE 8-5 Kirchhoff's laws for problems 7 through 10, 13, and 14.

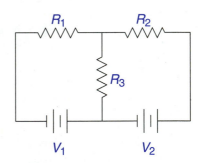

FIGURE 8-6 Kirchhoff's laws for problems 15 and 16.

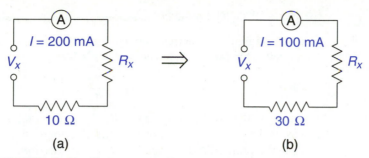

FIGURE 8–7 Unknown voltage and resistance for problem 17.

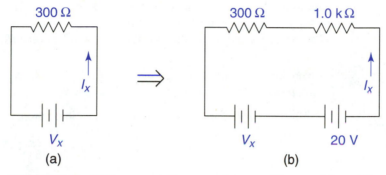

FIGURE 8–8 Unknown voltage and current for problem 18.

19. Given the circuit in Figure 8–2 from Example 8.2 with $V_1 = 3$ V, $V_2 = 6$ V, $R_1 = 2$ Ω, $R_2 = 3$ Ω, $R_3 = 6$ Ω, and $R_4 = 2$ Ω, find the currents I_1, I_2, I_3, and I_4.

20. In Example 8.2 solve the equations for I_1, I_2, I_3, and I_4, and verify the answers.

21. Given the circuit in Figure 8–9 with $V_1 = 12$ V, $V_2 = 18$ V, $R_1 = 10$ Ω, $R_2 = 20$ Ω, $R_3 = 20$ Ω, and $R_4 = 10$ Ω, find the current through each resistor.

22. Given the circuit in Figure 8–9 with $V_1 = 6.0$ V, $V_2 = 12$ V, $R_1 = 2.0$ kΩ, $R_2 = 1.0$ kΩ, $R_3 = 2.0$ kΩ, and $R_4 = 1.0$ kΩ, find the current through each resistor.

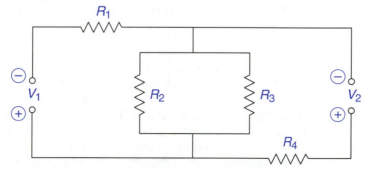

FIGURE 8–9 Kirchhoff's laws for problems 21 and 22.

≡ 8.2
VOLTAGE AND CURRENT DIVIDERS

Series Voltage Dividers

In a series circuit, the current through any resistance R_X equals the total current in the circuit. This can be expressed by Ohm's law:

$$I_T = \frac{V_X}{R_X} = \frac{V_T}{R_T}$$

where V_X is the voltage drop across R_X, V_T is the total voltage across all the resistances, and R_T is the total resistance. Multiplying both sides of the equation by R_X leads to the proportional voltage or voltage divider formula:

$$V_X = \frac{R_X}{R_T}(V_T) \tag{8.3}$$

This formula is another way of expressing that, in a series circuit, the voltage drop across a resistance is directly proportional to the resistance. For example, if R_1 and R_2 are in series and R_2 is twice R_1, then IR_2 will be twice IR_1.

EXAMPLE 8.3

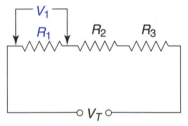

$$V_1 = \frac{R_1}{R_T}(V_T)$$

FIGURE 8–10 Voltage division in a series circuit for Example 8.3.

Given the series circuit in Figure 8–10 with $R_1 = 15\ \Omega$, $R_2 = 30\ \Omega$, $R_3 = 20\ \Omega$, and an applied voltage $V_T = 10\ V$, find V_1, V_2, and V_3.

Solution Apply formula (8.3) to find the voltage drops:

$$V_1 = \frac{R_1}{R_T}(V_T) = \frac{15}{15+30+20}(10\ V) = \frac{15}{65}(10\ V) = 2.3\ V$$

$$V_2 = \frac{R_2}{R_T}(V_T) = \frac{30}{65}(10\ V) = 4.6\ V$$

$$V_3 = \frac{R_3}{R_T}(V_T) = \frac{20}{65}(10\ V) = 3.1\ V$$

Note the proportional relationship between the resistances and the voltage drops. Here, R_2 is double R_1 and V_2 is double V_1. Also, R_2 is 1.5 times R_3 and V_2 is 1.5 times V_3.

Parallel Current Dividers

Ohm's law tells us that when the voltage is constant, the current through a resistance is inversely proportional to the resistance. That is, if the resistance doubles, the current will be reduced by half. Consider two resistances R_1 and R_2 in parallel where the voltage across each resistance is the same as shown in Figure 8–11.

If R_2 is twice R_1, the current through R_2 will then be *half* the current through R_1. Looking at it another way, the current through R_1 will be *twice* the current through R_2. That is, the part of the total current that passes through R_1 depends directly on the other parallel resistance R_2. This is expressed by the parallel *current division formula*:

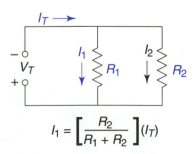

$$I_1 = \left[\frac{R_2}{R_1 + R_2}\right](I_T)$$

FIGURE 8–11 Current division in a parallel circuit for Example 8.4.

$$I_1 = \left[\frac{R_2}{R_1 + R_2}\right](I_T) \tag{8.4}$$

Formula (8.4) can also be written in terms of I_2:

$$I_2 = \left[\frac{R_1}{R_1 + R_2} \right] (I_T)$$

$$(8.4)$$

Do not confuse formula (8.4) with formula (8.3). Formula (8.3) applies to *voltage in any series circuit*. Formula (8.4) applies to *current only in a parallel circuit containing two resistances*. Also, in (8.3) the resistance on the top of the fraction corresponds to the voltage on the left, but in (8.4) the resistance on the top of the fraction does not correspond to the current on the left.

Formula (8.4) comes from the formula for two parallel resistances:

$$R_T = \frac{R_1 R_2}{R_1 + R_2}$$

Since $V_T = I_1 R_1 = I_T R_T$:

$$I_1 R_1 = (I_T) \left(\frac{R_1 R_2}{R_1 + R_2} \right)$$

Dividing both sides of this equation by R_1 results in formula (8.4). Study the next example, which shows how to apply (8.4).

EXAMPLE 8.4

In Figure 8–11, if $R_1 = 150 \ \Omega$, $R_2 = 300 \ \Omega$, and $I_T = 60$ mA, find I_1 and I_2.

Solution Apply formula (8.4):

$$I_1 = \frac{300}{150 + 300}(0.060 \text{ A}) = \frac{300}{450}(0.060 \text{ A}) = 0.040 \text{ A} = 40 \text{ mA}$$

$$I_2 = \frac{150}{450}(0.060 \text{ A}) = 0.020 \text{ A} = 20 \text{ mA}$$

Note that R_2 is twice R_1 and I_1 is twice I_2. Also, I_2 can be found by applying Kirchhoff's current law and subtracting I_1 from I_T.

For three or more resistances in parallel, it is easier to work with the conductances to calculate the currents. The conductance G is inversely proportional to the resistance:

$$G = \frac{1}{R}$$

Therefore, for any resistance, the current is directly proportional to the conductance. For a resistance R_1, in a parallel bank of resistances, this is expressed by the formula:

$$I_1 = \frac{G_1}{G_T}(I_T)$$

$$(8.5)$$

where G_1 is the conductance of R_1 and G_T is the sum of all the parallel conductances. Formula (8.5) is similar to formula (8.3) for voltage where R is replaced by G and V is replaced by I.

Study the next example which shows how to apply formula (8.5).

EXAMPLE 8.5

Figure 8–12 shows three resistances in parallel where $R_1 = 10\ \Omega$, $R_2 = 20\ \Omega$, $R_3 = 50\ \Omega$ and $I_T = 1.5$ A. Find the current through each resistance.

Solution First find the conductance of each resistance:

$$G_1 = \frac{1}{10\ \Omega} = 0.10\,\text{S} = 100\,\text{ms}$$

$$G_2 = \frac{1}{20\ \Omega} = 0.05\,\text{S} = 50\,\text{ms}$$

$$G_3 = \frac{1}{50\ \Omega} = 0.02\,\text{S} = 20\,\text{ms}$$

Then apply formula (8.5):

$$I_1 = \frac{0.10}{0.10 + 0.05 + 0.02}(1.5\,\text{A}) = \frac{0.10}{0.17}(1.5\,\text{A}) = 0.88\,\text{A} = 880\,\text{mA}$$

$$I_2 = \frac{0.05}{0.17}(1.5\,\text{A}) = 0.44\,\text{A} = 440\,\text{mA}$$

$$I_3 = \frac{0.02}{0.17}(1.5\,\text{A}) = 0.18\,\text{A} = 180\,\text{mA}$$

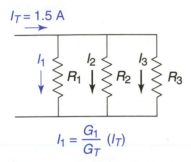

$I_T = 1.5$ A

$$I_1 = \frac{G_1}{G_T}\ (I_T)$$

FIGURE 8–12 Current division using conductance for Example 8.5.

The next example shows how you can use both formula (8.3) and (8.4) in a series-parallel circuit.

EXAMPLE 8.6

Given the series-parallel circuit in Figure 8–13(a) with an applied voltage $V_T = 12$ V, find I_3 (current through R_3) and V_4 (voltage drop across R_4).

Solution It is first necessary to find R_T and I_T. The series branch containing R_2 and R_4 can be added:

$$R_2 + R_4 = R_{2-4} = 10\ \Omega + 20\ \Omega = 30\ \Omega$$

and the circuit represented as in Figure 8–13(b). Now, $R_3 = 20\ \Omega$ is in parallel with R_{2-4}. Therefore, the equivalent resistance for R_3 and R_{2-4} is

$$R_{2-3-4} = \frac{(20)(30)}{20 + 30} = 12\ \Omega$$

The circuit can now be represented as in Figure 8–13(c). Here, R_1 is in series with R_{2-3-4}. Therefore, the total resistance of the circuit is

$$R_T = 8\ \Omega + 12\ \Omega = 20\ \Omega$$

Using Ohm's law the total current is

$$I_T = \frac{V_T}{R_T} = \frac{12\ \text{V}}{20\ \Omega} = 0.60\,\text{A} = 600\,\text{mA}$$

Now apply the current division formula (8.4) for the two parallel branches in Figure 8–13(b) to find I_3:

$$I_3 = \frac{R_{2-4}}{R_3 + R_{2-4}}\ (I_T) = \frac{30}{20 + 30}(0.60\,\text{A}) = 0.36\,\text{A} = 360\,\text{mA}$$

EXAMPLE 8.6 (Cont.)

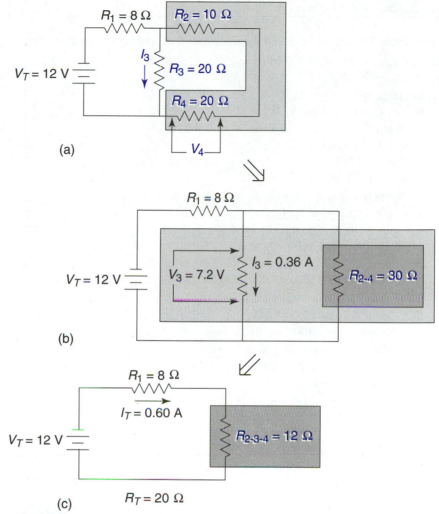

(a)

(b)

(c)

$R_T = 20 \ \Omega$

FIGURE 8–13 Current and voltage in a series-parallel circuit for Example 8.6.

One way to find V_4 using the voltage division formula (8.3) is to first find the voltage across $R_{2\text{-}4}$. This voltage is the same as V_3. Using Ohm's law,

$$V_3 = I_3 R_3 = (0.36 \text{ A})(20 \text{ A}) = 7.2 \text{ V}$$

Then apply formula (8.3) to R_2 and R_4 in Figure 8–13(a) using V_3 for V_T:

$$V_4 = \frac{R_4}{R_2 + R_4}(V_3) = \frac{20}{10 + 20} = (7.2 \text{ V}) = 4.8 \text{ V}$$

Here, V_4 can also be found by first finding the current through R_4:

$$I_4 = I_T - I_3 = 0.60 \text{ A} - 0.36 \text{ A} = 0.24 \text{ A} = 240 \text{ mA}$$

Then:

$$V_4 = I_4 R_4 = (0.24 \text{ A})(20 \ \Omega) = 4.8 \text{ V}$$

EXERCISE 8.2

In problems 1 through 16 solve each problem rounding answers to two significant digits.

1. In Figure 8–14, $V_T = 9.0$ V, $R_1 = 25$ Ω, $R_2 = 15$ Ω, and $R_3 = 30$ Ω. Find V_1, V_2, and V_3.

2. In Figure 8–14, $V_T = 32$ V, $R_1 = 300$ Ω, $R_2 = 750$ Ω, and $R_3 = 200$ Ω. Find V_1, V_2, and V_3.

3. In Figure 8–15, $I_T = 10$ mA, $R_1 = 2.0$ kΩ, and $R_2 = 1.5$ kΩ. Find I_1 and I_2.

4. In Figure 8–15, $I_T = 1.5$ A, $R_1 = 100$ Ω, and $R_2 = 68$ Ω. Find I_1 and I_2.

5. In Figure 8–16, $I_T = 200$ mA, $R_1 = 10$ Ω, $R_2 = 30$ Ω, and $R_3 = 15$ Ω. Find I_1, I_2, and I_3.

6. In Figure 8–16, $I_T = 60$ mA, $R_1 = 1.5$ kΩ, $R_2 = 1.0$ kΩ, and $R_3 = 2.0$ kΩ. Find I_1, I_2, and I_3.

7. For the series-parallel circuit in Figure 8–17, $R_1 = 300$ Ω, $R_2 = 400$ Ω, and $R_3 = 200$ Ω.

   (a) Given only that $V_T = 12$ V, find V_2 and V_3.

   (b) Given only that $I_T = 1.5$ A, find I_1 and I_2.

8. For the series-parallel circuit in Figure 8–17, $R_1 = 10$ kΩ, $R_2 = 10$ kΩ, and $R_3 = 30$ kΩ.

   (a) Given only that $V_T = 200$ V, find V_2 and V_3.

   (b) Given only that $I_T = 50$ mA, find I_1 and I_2.

9. For the series-parallel circuit in Figure 8–18, $V_T = 20$ V, $R_1 = 10$ Ω, $R_2 = 7.5$ Ω, $R_3 = 7.5$ Ω, and $R_4 = 10$ Ω. Find I_1 and V_2.

10. For the series-parallel circuit in Figure 8–18, $V_T = 60$ V, $R_1 = 20$ Ω, $R_2 = 10$ Ω, $R_3 = 30$ Ω, and $R_4 = 10$ Ω. Find I_2 and V_1.

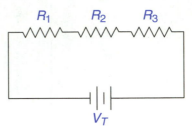

FIGURE 8–14 Series voltage dividers for problems 1 and 2.

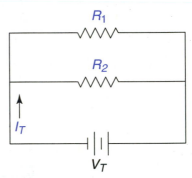

FIGURE 8–15 Current division for problems 3 and 4.

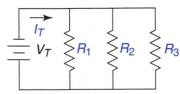

FIGURE 8–16 Current division for problems 5 and 6.

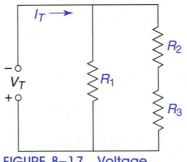

FIGURE 8–17 Voltage and current division for problems 7 and 8.

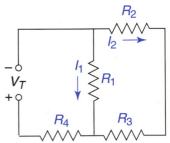

FIGURE 8–18 Voltage and current division for problems 9 and 10.

11. For the series-parallel circuit in Figure 8–19, $R_1 = 10$ kΩ, $R_2 = 20$ kΩ, $R_3 = 2.0$ kΩ, and $R_4 = 3.0$ kΩ.
    (a) Given only that $I_T = 2.0$ A, find I_1, I_2, and I_3.
    (b) Given only that $V_T = 200$ V, find V_2, V_3, and V_4.

12. For the series-parallel circuit in Figure 8–19, $R_1 = 12$ Ω, $R_2 = 15$ Ω, $R_3 = 2.0$ Ω, and $R_4 = 10$ Ω.
    (a) Given only that $I_T = 8.0$ mA, find I_1, I_2, and I_3.
    (b) Given only that $V_T = 50$ mV, find V_2, V_3, and V_4.

13. Two resistors R_1 and R_2 are in series. If $R_1 = 100$ Ω, what should be the value of R_2 so that V_2 equals 20% of the total voltage drop across both resistors?

14. Two resistors R_1 and R_2 are in parallel. If $R_1 = 100$ Ω, what should be the value of R_2 so that I_2 equals one-fourth of the total current through both resistors?

15. Given the series-parallel circuit in Figure 8–20 with $R_2 = 2R_1$ and $R_4 = 3R_3$, if $V_1 = 10$ V, find V_2, V_3, and V_4.

16. In Figure 8–20, $I_1 = 30$ mA. If $R_2 = R_3 = R_4 = 2R_1$, find I_3 and I_T.

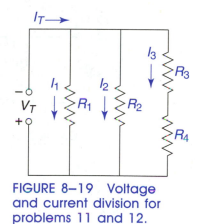

FIGURE 8–19   Voltage and current division for problems 11 and 12.

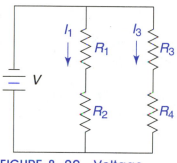

FIGURE 8–20   Voltage and current division for problems 15 and 16.

≡ 8.3
SUPERPOSITION THEOREM

The superposition theorem helps you analyze a circuit having two or more voltage sources by considering the effect of only one voltage source at a time. If there are two voltage sources, you simplify the circuit by redrawing it as two separate circuits, each with only one voltage source and the other voltage source shorted out. Then you combine algebraically, or superimpose, the two results to produce the currents or voltages of the original circuit. Study the following two examples, which show you how the theorem works.

EXAMPLE 8.7

Consider the series circuit in Figure 8–21(a) containing two voltage sources $V_1 = 12$ V and $V_2 = 18$ V. Find the total current I_T in the circuit and the voltage drop across R_1.

Solution   This circuit is not complex and can be solved directly or by using the superposition theorem. Both methods will be shown to illustrate how the theorem works and that the results agree.

EXAMPLE 8.7 (Cont.)

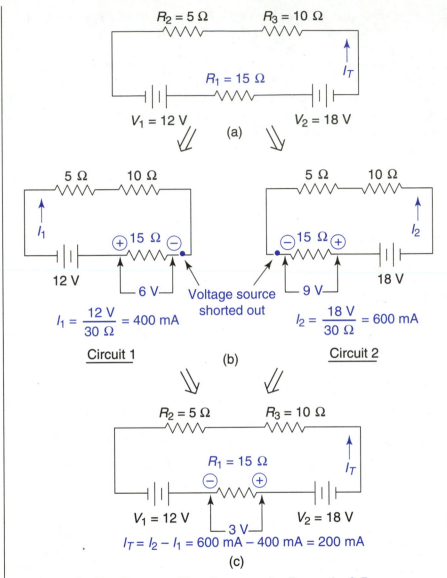

FIGURE 8–21   Superposition theorem for Example 8.7.

To apply the superposition theorem, redraw the circuit as two separate circuits. In each circuit, short out one of the voltage sources, leaving only the other voltage source as shown in Figure 8–21(b). The current in circuit 1 is

$$I_1 = \frac{V_1}{R_T} = \frac{12\,\text{V}}{10\,\Omega + 5\,\Omega + 15\,\Omega} = 0.40\,\text{A} = 400\,\text{mA}$$

and the voltage across R_1 is

$$I_1 R_1 = (0.40\,\text{A})(15\,\Omega) = 6.0\,\text{V}$$

EXAMPLE 8.7 (Cont.)

The current in circuit 2 and the voltage across R_1 is

$$I_2 = \frac{18 \text{ V}}{30 \, \Omega} = 0.60 \text{ A} = 600 \text{ mA}$$

$$I_2 R_1 = (0.60 \text{ A})(15 \, \Omega) = 9.0 \text{ V}$$

Note that the direction of the current in circuit 2 is opposite to the direction in circuit 1. Now combine the results algebraically as shown in Figure 8–21(c). Since I_2 flows opposite to I_1, the total current is the difference of the two currents:

$$I_T = I_2 - I_1 = 600 \text{ mA} - 400 \text{ mA} = 200 \text{ mA}$$

Also, since I_2 is greater than I_1, the direction of I_T is the same as the direction of I_2.

Similarly, the voltage across R_1 is the difference of the two voltage drops:

$$V_1 = 9.0 \text{ V} - 6.0 \text{ V} = 3.0 \text{ V}$$

The polarities across R_1 also agree with those in circuit 2 since the voltage drop is greater in this circuit.

To verify the results, the total current can also be found without using the superposition theorem by noting that the batteries are series-opposing. The total voltage is then the difference of the two voltages:

$$V_T = V_2 - V_1 = 18 \text{ V} - 12 \text{ V} = 6.0 \text{ V}$$

The total current and the voltage drop across R_1 are then

$$I_T = \frac{6.0 \text{ V}}{30 \, \Omega} = 0.20 \text{ A} = 200 \text{ mA}$$

$$I_T R_1 = (0.20 \text{ A})(15 \text{ V}) = 3.0 \text{ V}$$

These results agree with the results obtained using the superposition theorem. ▪

Study the next example, which shows a more complex circuit where the superposition theorem is very useful in finding the currents in the circuit.

EXAMPLE 8.8

Consider the circuit in Figure 8–22(a) with two voltage sources and three resistances. Find the currents I_1, I_2 and I_3.

Solution This is the same circuit as in Example 8.1 where the currents are found using Kirchhoff's laws and a system of three linear equations. The solution given here uses the superposition theorem, and the results will be the same as those of Example 8.1.

First redraw the circuit as two circuits, each with one voltage source. In each circuit, short out the other voltage source as shown in Figure 8–22(b). Show the current flow for the three currents in each circuit. Note that I_3 flows in the same direction in both circuits but that I_1 and I_2 flow in opposite directions. Therefore, when the results of the two circuits are superimposed, I_3 will be the sum of the currents in circuit 1 and circuit 2, whereas I_1 and I_2 will each be the difference of

EXAMPLE 8.8 (Cont.)

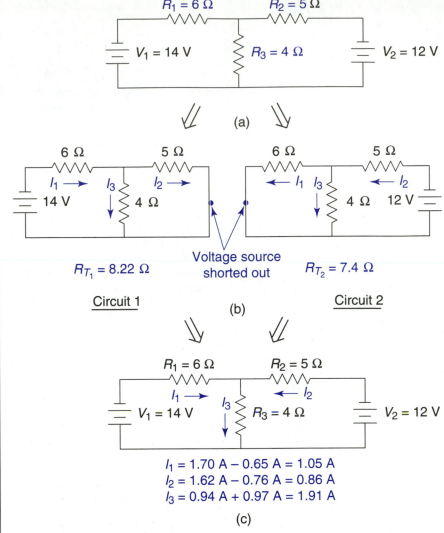

FIGURE 8–22 Superposition theorem for Example 8.8.

the currents in the two circuits. The calculations for each circuit are similar as follows:

Circuit 1. First find the total resistance of the circuit. In circuit 1, 5 Ω and 4 Ω are in parallel, and this parallel branch is in series with 6 Ω. The total resistance is then:

$$R_{T_1} = \frac{(5)(4)}{5+4} + 6 = \frac{20\ \Omega}{9} + 6\ \Omega = 8.22\ \Omega$$

Then find the total current through the battery which is I_1. Using Ohm's law,

$$I_1 = \frac{V_1}{R_{T_1}} = \frac{14\ \text{V}}{8.22\ \Omega} = 1.70\ \text{A}$$

EXAMPLE 8.8 (Cont.)

The other two currents I_2 and I_3 can be found using the current division formula for two parallel resistances (8.4):

$$I_2 = \frac{4}{5+4}(1.70\,\text{A}) = 0.76\,\text{A} = 760\,\text{mA}$$

$$I_3 = \frac{5}{5+4}(1.70\,\text{A}) = 0.94\,\text{A} = 940\,\text{mA}$$

Here, I_3 can also be found by just subtracting I_2 from I_1.

Circuit 2.   In circuit 2, 6 Ω and 4 Ω are in parallel, and this parallel branch is in series with 5 Ω. The total resistance of circuit 2 is then

$$R_{T_2} = \frac{(6)(4)}{6+4} + 5 = 2.4\,\Omega + 5\,\Omega = 7.4\,\Omega$$

The total current through the battery I_2 is

$$I_2 = \frac{12\,\text{V}}{7.4\,\Omega} = 1.62\,\text{A}$$

The other two currents are found using formula (8.4):

$$I_1 = \frac{4}{6+4}(1.62\,\text{A}) = 0.65\,\text{A} = 650\,\text{mA}$$

$$I_3 = \frac{6}{6+4}(1.62\,\text{A}) = 0.97\,\text{A} = 970\,\text{mA}$$

Now combine the results of the two circuits algebraically as shown in Figure 8–22(c). Since the direction of I_1 in circuit 1 is *opposite* to its direction in circuit 2, its value in the original circuit is the *difference* of the values in circuit 1 and circuit 2:

$$I_1 = 1.70\,\text{A} - 0.65\,\text{A} = 1.05\,\text{A} \approx 1.1\,\text{A}$$

Since I_1 in circuit 1 is *greater* than I_1 in circuit 2, its direction in the original circuit is the *same* as that in circuit 1.

Similarly, I_2 is the difference of the values in circuit 1 and circuit 2:

$$I_2 = 1.62\,\text{A} - 0.76\,\text{A} = 0.86\,\text{A} = 860\,\text{mA}$$

and its direction is the same as that in circuit 2. Since I_3 flows in the same direction in circuit 1 and circuit 2, its value is the sum of the currents in the two circuits.

$$I_3 = 0.94\,\text{A} + 0.97\,\text{A} = 1.91\,\text{A} \approx 1.9\,\text{A}$$

and its direction in the original circuit is the same as in circuit 1 or circuit 2.

The superposition method can be used for networks that contain components that are linear and bilateral. *Linear* means that the current is proportional to the voltage for each component. *Bilateral* means that the currents are the same when the polarities are reversed. Resistors, capacitors, and air-core inductors are generally considered to be linear and bilateral components.

EXERCISE 8.3

In problems 1 through 8, use the superposition theorem to find the currents or voltages. Round answers to two significant digits.

1. Given the series circuit in Figure 8–23, find the total current and the voltage drop across R_1.

2. Given the parallel circuit in Figure 8–24, where the two voltage sources are connected to a negative ground, find the current through R_1 and through R_2.

3. In the series-parallel circuit in Figure 8–25, $V_1 = 20$ V, $V_2 = 24$ V, $R_1 = 200$ Ω, and $R_2 = 300$ Ω. Find I_1 and I_2. (Note: A short across a resistance results in a resistance of zero.)

4. In the series-parallel circuit in Figure 8–25, $V_1 = 20$ V, $V_2 = 24$ V, $R_1 = 100$ Ω, and $R_2 = 50$ Ω. Find I_1 and I_2. (See Note in problem 3.)

5. In the series-parallel circuit in Figure 8–26, $V_1 = 12$ V, $V_2 = 18$ V, $R_1 = 10$ Ω, $R_2 = 20$ Ω, and $R_3 = 5$ Ω. Find I_1, I_2, and I_3.

6. In the series-parallel circuit in Figure 8–26, $V_1 = 15$ V, $V_2 = 30$ V, $R_1 = 10$ Ω, $R_2 = 20$ Ω, and $R_3 = 10$ Ω. Find I_1, I_2, and I_3.

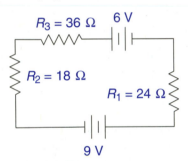

FIGURE 8–23 Series circuit with two voltage sources for problem 1.

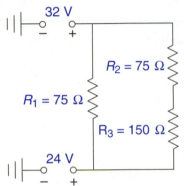

FIGURE 8–24 Parallel circuit with two voltage sources for problem 2.

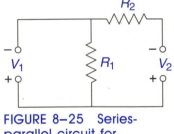

FIGURE 8–25 Series-parallel circuit for problems 3 and 4.

7. In the series-parallel circuit in Figure 8–27, $V_1 = 10$ V, $V_2 = 12$ V, $R_1 = 2.0$ kΩ, $R_2 = 3.0$ kΩ, and $R_3 = 2.0$ kΩ. Find I_1, I_2, and I_3. Note the polarity of the voltages.

8. In the series-parallel circuit in Figure 8–28, $V_1 = 18$ V, $V_2 = 14$ V, $R_1 = 20$ Ω, $R_2 = 30$ Ω, $R_3 = 40$ Ω, and $R_4 = 40$ Ω. Find I_1, I_2, I_3, and I_4.

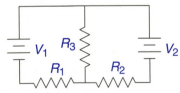

FIGURE 8–26 Series-parallel circuit for problems 5 and 6.

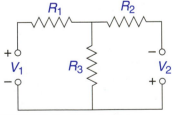

FIGURE 8–27 Series-parallel circuit for problem 7.

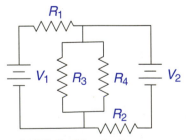

FIGURE 8–28 Series-parallel circuit for problem 8.

☰ 8.4
THEVENIN'S THEOREM

M. L. Thevenin, a French engineer, discovered one of the most useful theorems in network analysis. It is a simple but powerful idea that allows you to reduce a complex circuit to the simplest series circuit: a voltage source and a resistance.

Consider a load resistance R_L in some complex circuit represented by a black box as shown in Figure 8–29(a). No matter how complex the circuit, what happens in R_L is determined by two factors: the voltage experienced by the load and the equivalent resistance of the rest of the circuit. The voltage experienced by the load is called the Thevenin voltage V_{TH}. The equivalent resistance is called the Thevenin resistance R_{TH}.

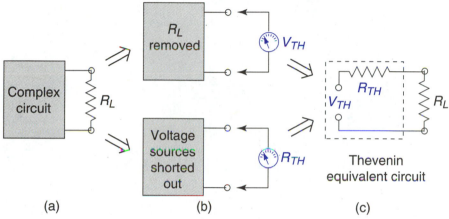

(a)   **(b)**   **(c)**

FIGURE 8–29   Thevenizing a circuit.

Given a circuit with a load resistance R_L, you calculate the Thevenin voltage V_{TH} as follows: Remove the load resistance and calculate the open circuit voltage across the terminals of R_L. You calculate the Thevenin resistance R_{TH} as follows: Short out the voltage source, or sources, and calculate the equivalent resistance across the open terminals of R_L. See Figure 8–29(b). In a laboratory, with an actual circuit, you can measure V_{TH} and R_{TH} with a multimeter.

Thevenin's theorem now says that the entire circuit without the load can be represented by a series circuit having a source voltage V_{TH} and a series resistance R_{TH}, as shown in Figure 8–29(c). The process of finding V_{TH} and R_{TH} is known as *Thevenizing a circuit*.

The following three examples illustrate how to Thevenize three different circuits. The first example is a series-parallel circuit with one voltage source.

EXAMPLE 8.9

Thevenize the circuit of Figure 8–30(a). Find the Thevenin voltage and the Thevenin resistance for the load R_L. Then draw the Thevenin equivalent circuit and find I_L (the current through R_L) and V_L (the voltage across R_L).

Solution   First remove R_L as shown in Figure 8–30(b). Now observe that since V_{TH} is an open circuit voltage, it is the same as V_3, the series voltage across R_3, because there is no current through R_2. Then V_{TH} is calculated as V_3, using the proportional voltage formula (8.3):

$$V_{TH} = V_3 = \frac{6}{4+6}(12\text{ V}) = 7.2\text{ V}$$

EXAMPLE 8.9 (Cont.)

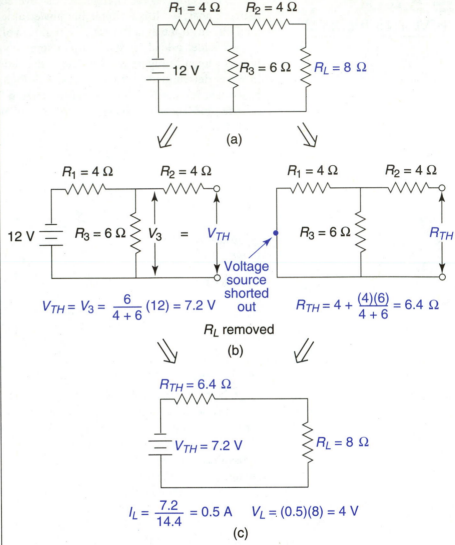

$$V_{TH} = V_3 = \frac{6}{4+6}(12) = 7.2\ V$$

$$R_{TH} = 4 + \frac{(4)(6)}{4+6} = 6.4\ \Omega$$

$$I_L = \frac{7.2}{14.4} = 0.5\ A \quad V_L = (0.5)(8) = 4\ V$$

FIGURE 8–30   Thevenizing a series-parallel circuit for Example 8.9.

To find R_{TH}, short out the voltage source, and compute the equivalent resistance of the series-parallel circuit shown in Figure 8–30(b):

$$R_{TH} = 4 + \frac{(4)(6)}{4+6} = 4\ \Omega + 2.4\ \Omega = 6.4\ \Omega$$

Now draw the Thevenin equivalent circuit as shown in Figure 8–30(c). Find I_L and V_L by applying Ohm's law to this series circuit:

$$I_L = \frac{7.2\ V}{8\ \Omega + 6.4\ \Omega} = 0.5\ A = 500\ mA$$

$$V_L = (0.5\ A)(8\ \Omega) = 4\ V$$

EXAMPLE 8.9 (Cont.)

You can also use the voltage divider rule to find V_L:

$$V_L = \frac{8}{8 + 6.4}(7.2\text{ V}) = 4\text{ V}$$

To verify the results in Example 8.9, you can find V_L and I_L directly from the original series-parallel circuit. The total resistance and the total current are

$$R_T = 4 + \frac{(12)(6)}{6 + 12} = 8\ \Omega$$

$$I_T = \frac{12\text{ V}}{8\ \Omega} = 1.5\text{ A}$$

Using the parallel current formula (8.4), I_L and V_L are

$$I_L = \frac{6}{6 + 12}(1.5\text{ V}) = 0.5\text{ A} = 500\text{ mA}$$

$$V_L = (0.5\text{ A})(8\ \Omega) = 4\text{ V}$$

These results agree with those in Example 8.9.

The next example shows how to Thevenize a circuit with two voltage sources.

EXAMPLE 8.10

Thevenize the circuit in Figure 8–31(a), and find I_L and V_L.

Solution A circuit of this type is solved using Kirchhoff's laws in Section 8.1. Using Thevenin's theorem, the results can be found more readily. To find V_{TH}, remove R_L as shown in Figure 8–31(b). The circuit is now a series circuit with two series-opposing voltages. The total voltage across R_1 and R_2 is then

$$V_T = 30\text{ V} - 24\text{ V} = 6\text{ V}$$

The open circuit voltage V_{TH} is equal to the voltage across R_2 combined with the voltage of the source V_2, as shown in Figure 8–31(c). Observe that the polarity across R_2 is such that it *adds* to the source voltage to produce V_{TH}:

$$V_{TH} = \frac{50}{150}(6\text{ V}) + 24\text{ V} = 26\text{ V}$$

Here, V_{TH} can also be found by *subtracting* the voltage across R_1 from 30 V, or by using the superposition theorem shown in Section 8-3.

To find R_{TH}, remove R_L and short out both voltages as shown in Figure 8–31(b). Redraw the circuit to show that R_{TH} is equivalent to R_1 and R_2 in parallel as shown in Figure 8–31(c):

$$R_{TH} = \frac{(50)(100)}{50 + 100} = \frac{5000}{150} = 33\ \Omega$$

The Thevenin equivalent circuit is now shown in Figure 8–31(d). Here, I_L and V_L are rounded to two significant digits.

$$I_L = \frac{26\text{ V}}{133\ \Omega} = 195\text{ mA} \approx 200\text{ mA}$$

$$V_L = (0.195\text{ A})(100\ \Omega) = 19.5\text{ V} \approx 20\text{ V}$$

EXAMPLE 8.10 (Cont.)

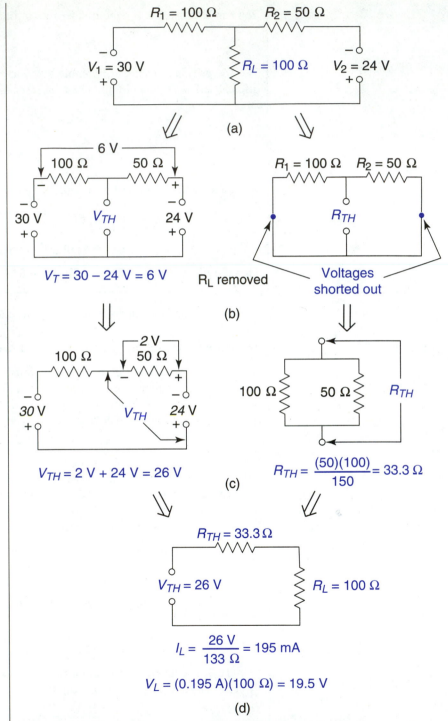

FIGURE 8–31   Thevenizing a circuit with two voltage sources for Example 8.10.

The last example shows how to Thevenize a bridge circuit. Such a circuit would be very difficult to solve without Thevenin's theorem.

EXAMPLE 8.11

Find the Thevenin equivalent circuit for R_L in the bridge circuit of Figure 8–32(a).

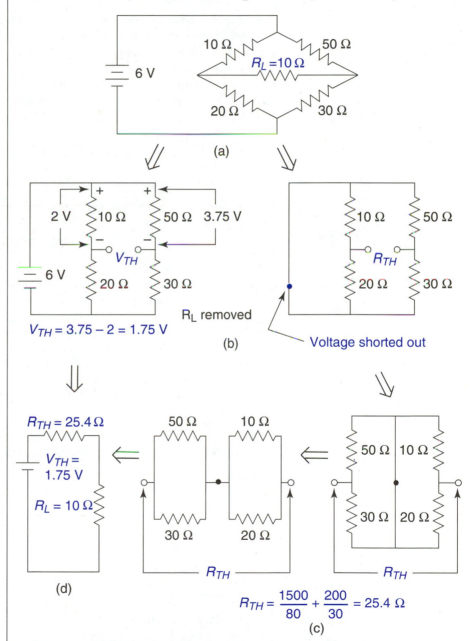

FIGURE 8–32 Thevenizing a bridge circuit for Example 8.11.

Solution Remove the load R_L, and redraw the circuit as shown in Figure 8–32(b). Apply Kirchhoff's voltage law to the loop containing V_{TH}, the 10-Ω and

EXAMPLE 8.11 (Cont.)

the 50-Ω resistances. Then V_{TH} is the difference in the voltage drops across the 10-Ω and the 50-Ω resistances:

$$V_{TH} = \frac{50}{80}(6\text{ V}) - \frac{10}{30}(6\text{ V}) = 3.75\text{ V} - 2\text{ V} = 1.75\text{ V} \approx 1.8\text{ V}$$

To find R_{TH}, short out the voltage source, and observe that the circuit can be redrawn to show it as two parallel banks in series. Figure 8–32(c) illustrates this process. The line containing the shorted voltage source is first placed between the resistances and the terminals for R_{TH} shown on the outside. The resistances are then separated into two parallel banks with the shorted voltage line connecting them. The Thevenin resistance is then equal to the sum of the equivalent resistances of each bank:

$$R_{TH} = \left[\frac{(50)(30)}{50+30}\right] + \left[\frac{(10)(20)}{10+20}\right] = 18.75\,\Omega + 6.67\,\Omega = 25.4\,\Omega \approx 25\,\Omega$$

The Thevenin equivalent circuit is shown in Figure 8–32(d).

ERROR BOX

A common error to watch out for in network analysis is incorrectly finding an open circuit voltage or short circuit current in certain situations.

Consider first the open circuit voltage. If there is no current flowing in a resistance, then there is *no* voltage drop across the resistance. The voltage at *each* end of the resistance will be the *same* with respect to any other circuit point.

If the two ends of a resistance are short circuited, then there will be no current flowing through the resistance. The resistance can be removed from the circuit, and it will not change the circuit.

See if you can correctly find the voltage or current in each of the practice problems.

Practice Problems: Find the open circuit voltage V in circuits a and b in Figure 8–33. Find the short circuit current I in circuits c and d in Figure 8–33.

Answers: a. 3 V b. 20 V c. 1.5 A d. 2.0 A

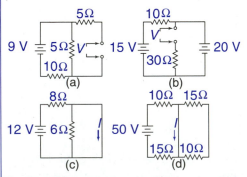

FIGURE 8–33 Open-circuit voltage and short-circuit current.

EXERCISE 8.4

In problems 1 through 12 solve each problem rounding answers to two significant digits.

1. Given the series-parallel circuit in Figure 8–34 with $V_T = 9$ V, $R_1 = 5$ Ω, $R_2 = 3$ Ω, $R_3 = 6$ Ω, and $R_L = 9$ Ω, Thevenize this circuit for R_L by first finding V_{TH} and R_{TH} and then I_L and V_L. See Example 8.9.

2. As in problem 1, Thevenize the circuit in Figure 8–34 when $V_T = 60$ V, $R_1 = 2.0$ kΩ, $R_2 = 1.0$ kΩ, $R_3 = 2.0$ kΩ, and $R_L = 1.0$ kΩ.

3. Given the series-parallel circuit in Figure 8–35 with $V_T = 120$ V, $R_1 = 1.0$ kΩ, $R_2 = 2.0$ kΩ, and $R_L = 3.0$ kΩ, find I_L and V_L two ways:
   (a) Using Thevenin's theorem.
   (b) Solving the circuit directly by first finding R_T and I_T.
   Check that the two sets of results agree. See Example 8.9.

4. As in problem 3, find I_L and V_L two ways for the circuit in Figure 8–35 when $V_T = 36$ V, $R_1 = 100$ Ω, $R_2 = 100$ Ω, and $R_L = 50$ Ω.

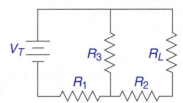

FIGURE 8–34 Series-parallel circuit for problems 1 and 2.

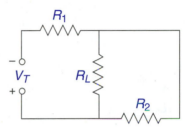

FIGURE 8–35 Series-parallel circuit for problems 3 and 4.

5. Given the circuit in Figure 8–36 with the two voltage sources $V_1 = 12$ V and $V_2 = 15$ V, Thevenize this circuit when $R_L = 6$ Ω, $R_1 = 4$ Ω, and $R_2 = 5$ Ω. Find V_{TH}, R_{TH}, I_L, and V_L. See Example 8.10.

6. As in problem 5, Thevenize the circuit in Figure 8–36 when $V_1 = 80$ V, $V_2 = 65$ V, $R_L = 100$ Ω, $R_1 = 100$ Ω, and $R_2 = 200$ Ω.

7. Given the circuit in Figure 8–37 with the two voltage sources $V_1 = 22$ V and $V_2 = 24$ V, Thevenize this circuit when $R_L = 1.5$ kΩ, $R_1 = 3.0$ kΩ, and $R_2 = 2.0$ kΩ. Find V_{TH}, R_{TH}, I_L, and V_L. Note the polarity of the voltage sources and the location of R_L. See Example 8.10.

8. As in problem 7, Thevenize the circuit in Figure 8–37 when $V_1 = 12$ V, $V_2 = 9.0$ V, $R_L = 50$ Ω, $R_1 = 200$ Ω, and $R_2 = 100$ Ω.

9. Given the bridge circuit in Figure 8–38 with $V_T = 15$ V, $R_1 = 20$ Ω, $R_2 = 30$ Ω, $R_3 = 30$ Ω, and $R_4 = 60$ Ω, find V_{TH} and R_{TH} for the load R_L. See Example 8.11.

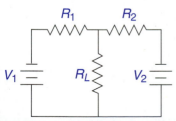

FIGURE 8–36 Circuit with two voltage sources for problems 5 and 6.

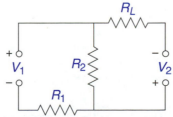

FIGURE 8–37 Circuit with two voltage sources for problems 7 and 8.

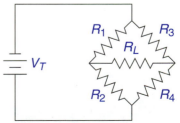

FIGURE 8–38 Bridge circuit for problems 9 and 10.

10. Given the bridge circuit in Figure 8–38 with $V_T = 60$ V, $R_1 = 200$ Ω, $R_2 = 50$ Ω, $R_3 = 100$ Ω, and $R_4 = 100$ Ω, find V_{TH} and R_{TH} for the load R_L. See Example 8.11.

11. Given the circuit in Figure 8–39 with $V_1 = 6.0$ V, $V_2 = 8.0$ V, $R_1 = 20$ Ω, $R_2 = 30$ Ω, and $R_L = 25$ Ω, find I_L and V_L two ways:
    (a) Applying Thevenin's theorem.
    (b) Applying the superposition theorem.
    Check that the two sets of results agree.

12. Given the circuit in Figure 8–39 with $V_1 = 28$ V, $V_2 = 20$ V, $R_1 = 60$ Ω, $R_2 = 40$ Ω, and $R_L = 100$ Ω, find I_L and V_L two ways:
    (a) Applying Thevenin's theorem.
    (b) Applying the superposition theorem.
    Check that the two sets of results agree.

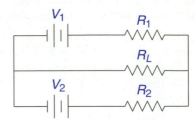

FIGURE 8–39 Thevenin's theorem and superposition theorem for problems 11 and 12.

≡ 8.5
NORTON'S THEOREM

E. L. Norton, an American scientist, devised another way to simplify a circuit that is similar to Thevenin's idea but that uses a current source instead of a voltage source. An *ideal current source* is an electrical component that supplies a constant current no matter what the load, in the same way that an ideal voltage source supplies a constant voltage. It is represented in a schematic as a circle with a current arrow as shown in the Norton circuit in Figure 8–40.

Voltage and Current Sources

A Thevenin circuit consisting of a voltage source and a series resistance can be shown to be equivalent to a circuit containing a current source I_N and a *parallel* resistance R_N, called a Norton circuit.

Consider the Thevenin circuit and the Norton circuit in Figure 8–40(a). Suppose $V_{TH} = 12$ V and $R_{TH} = 6$ Ω. The Norton equivalent circuit is then

$$I_N = \frac{V_{TH}}{R_{TH}} = \frac{12}{6} = 2 \text{ A}$$

$$R_N = R_{TH} = 6 \, \Omega$$

To show these circuits are equivalent, consider when the load terminals A and B are open as shown in Figure 8–40(a). The voltage across these terminals will be the same for both circuits, 12 V. When the load terminals are short circuited, as shown in Figure 8–40(b), the current through these terminals will be the same in both circuits, 2 A. Since the open voltages and short-circuit currents are the same in the Norton and Thevenin circuits, each circuit will have the same effect on a load resistance connected across AB and are, therefore, equivalent.

Nortonizing a Circuit

Given a circuit with a load resistance R_L, you calculate the Norton current I_N as follows: Remove the load resistance and short circuit the terminals of R_L. Then

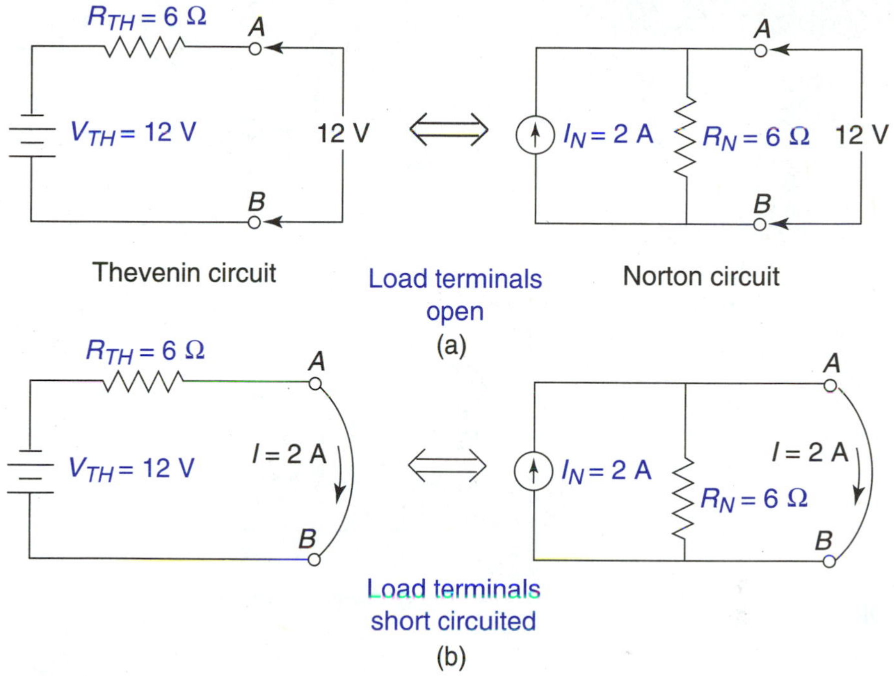

FIGURE 8–40 Equivalence of Thevenin circuit and Norton circuit.

calculate the short-circuit current through these terminals. You calculate the Norton resistance R_N in the same way as R_{TH}: Short out the source, or sources, and calculate the equivalent resistance across the open terminals of R_L.

The entire circuit without the load can then be represented by a parallel circuit having a current source I_N and a parallel resistance R_N. The process of finding I_N and R_N is known as *Nortonizing a circuit*. The following example illustrates how to apply Norton's theorem.

EXAMPLE 8.12

Nortonize the circuit of Figure 8–41(a): Find the Norton current and the Norton resistance for the load R_L. Then draw the Norton equivalent circuit and find I_L and V_L.

Solution   First remove R_L as shown in Figure 8–41(b). To find I_N, short circuit the load terminals, and find the current through these terminals. In this circuit, 3 Ω is in series with the parallel bank of 4 Ω and 6 Ω. The total resistance is then

$$R_T = 3 + \frac{(4)(6)}{4 + 6} = 3\,\Omega + 2.4\,\Omega = 5.4\,\Omega$$

The total current is

$$I_T = \frac{12\text{ V}}{5.4\,\Omega} = 2.22\text{ A}$$

Using the parallel current formula (8.4), I_N is

$$I_N = \frac{6}{6 + 4}\,(2.22\text{ A}) = 1.33\text{ A} \approx 1.3\text{ A}$$

EXAMPLE 8.12 (Cont.)

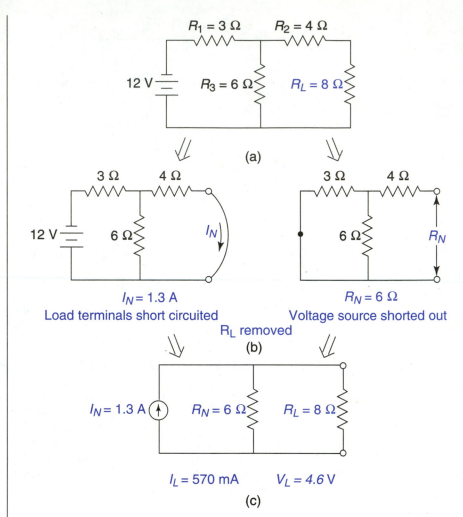

FIGURE 8–41 Nortonizing a circuit for Example 8.12.

To find R_N, short out the voltage source and compute the equivalent resistance across the open terminals of R_L as shown in Figure 8–41(b). In this circuit, 4 Ω is in series with the parallel bank of 3 Ω and 6 Ω:

$$R_N = 4 + \frac{(3)(6)}{3 + 6} = 4\,\Omega + 2\,\Omega = 6\,\Omega$$

The Norton equivalent circuit is shown in Figure 8–41(c) with a current source of 1.3 A and a resistance of 6 Ω parallel to the load. Then I_L and V_L are

$$I_L = \frac{6}{8 + 6}\,(1.33\,\text{A}) = 0.57\,\text{A} = 570\,\text{mA}$$

$$V_L = (0.5\,7\,\text{A})(8\,\Omega) = 4.56\,\text{V} \approx 4.6\,\text{V}$$

EXERCISE 8.5

In problems 1 through 6 solve each problem rounding answers to two significant digits.

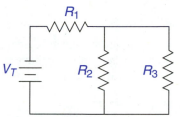

1. Given the series-parallel circuit in Figure 8–42 with $V_T = 14$ V, $R_1 = 6$ Ω, $R_2 = 8$ Ω and $R_3 = 10$ Ω,
   (a) Nortonize this circuit for R_3 as the load resistance by first finding I_N and R_N and then I_L and V_L.
   (b) Check your results by finding I_3 and V_3 by another method.

FIGURE 8–42 Series-parallel circuit for problems 1 and 2.

2. Given the series-parallel circuit in Figure 8–42 with $V_T = 24$ V, $R_1 = 50$ Ω, $R_2 = 100$ Ω, and $R_3 = 300$ Ω,
   (a) Nortonize this circuit for R_1 as the load resistance by first finding I_N and R_N and then I_L and V_L.
   (b) Check your results by finding I_1 and V_1 by another method.

3. Given the circuit in Figure 8–43 with $V_T = 50$ V, $R_1 = 200$ Ω, $R_2 = 100$ Ω, $R_3 = 100$ Ω, and $R_L = 300$ Ω,
   (a) Find I_L and V_L by applying Norton's theorem.
   (b) Show that you get the same results for I_L and V_L by applying Thevenin's theorem.

4. Given the circuit in Figure 8–43 with $V_T = 12$ V, $R_1 = 5$ Ω, $R_2 = 3$ Ω, $R_3 = 7$ Ω, and $R_L = 5$ Ω,
   (a) Find I_L and V_L by applying Norton's theorem.
   (b) Show that you get the same results for I_L and V_L by applying Thevenin's theorem.

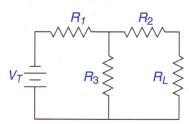

5. Given the Thevenin circuit and the equivalent Norton circuit in Figure 8–44 where $V_{TH} = 10$ V, $R_{TH} = R_N = 20$ Ω, $R_L = 5$ Ω, and

$$I_N = \frac{V_{TH}}{R_{TH}} = \frac{10}{20} = 0.5\text{A} = 500 \text{ mA}$$

FIGURE 8–43 Series-parallel circuit for problems 3 and 4.

show that I_L and V_L are the same for each circuit.

6. For the Thevenin circuit in Figure 8–44, V_L is given by the proportional voltage formula (8.3):

$$V_L = \frac{R_L}{R_{TH} + R_L}(V_{TH})$$

For the equivalent Norton circuit, I_L is given by the parallel current division formula (8.4):

$$I_L = \frac{R_N}{R_N + R_L}(I_N)$$

Show that if

$$I_N = \frac{V_{TH}}{R_{TH}}$$

$$R_N = R_{TH}$$

Then

$$R_L = \frac{V_L}{I_L}$$

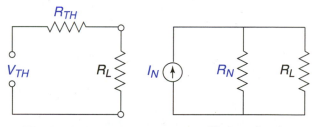

Thevenin circuit   Norton circuit

$$I_N = \frac{V_{TH}}{R_{TH}} \qquad R_N = R_{TH}$$

FIGURE 8–44 Equivalent Thevenin and Norton circuits for problems 5 and 6.

☰ CHAPTER HIGHLIGHTS

8.1 KIRCHHOFF'S LAWS

Electron current is the movement of electrons from (−) to (+). Electron current is used throughout the text. *Conventional current* flows from (+) to (−).

Kirchhoff's Current Law	The algebraic sum of all currents entering and leaving any point in a circuit equals zero.　　　(8.1)

Assign a positive value to a current entering a point and a negative value to a current leaving the point.

Kirchhoff's Voltage Law	The algebraic sum of the voltages around any closed path equals zero.　　　(8.2)

As you move around a closed path, assign a positive value to a voltage whose positive terminal comes first and a negative sign to a voltage whose negative terminal comes first. Study Examples 8.1 and 8.2 to better understand how Kirchhoff's laws are used to find the currents in a circuit.

8.2 VOLTAGE AND CURRENT DIVIDERS

Series Voltage Dividers.
For any resistance R_1 in a series circuit,

$$V_1 = \frac{R_1}{R_T}(V_T)$$

(8.3)

where V_T is the total voltage across all the series resistances.

Parallel Current Dividers.
For two resistances R_1 and R_2 in parallel,

$$I_1 = \frac{R_2}{R_1 + R_2}(I_T)$$

(8.4)

where I_T is the total current through both resistances.

For two or more resistances in parallel,

$$I_1 = \frac{G_1}{G_T}(I_T)$$

(8.5)

where $G_1 = \frac{1}{R_1}$ is the conductance of R_1 and G_T is the sum of all the parallel conductances.

8.3 SUPERPOSITION THEOREM

To analyze a circuit with two or more voltage sources, redraw it as two or more circuits each having only one voltage source with the other voltage sources shorted out. Then superimpose the results by combining them algebraically to find the currents or voltages of the original circuit. Study Example 8.8.

8.4 THEVENIN'S THEOREM

Thevenin's theorem reduces a complex circuit to an equivalent series circuit containing a Thevenin voltage V_{TH} and a Thevenin resistance R_{TH}. For a circuit with a load resistance R_L,

1. To find V_{TH}, remove R_L and calculate the open circuit voltage across the terminals of R_L.
2. To find R_{TH}, short out the voltage sources and calculate the equivalent resistance across the open terminals of R_L. Study Examples 8.9 and 8.10.

8.5 NORTON'S THEOREM

Norton's theorem, like Thevenin's theorem, reduces a complex circuit to an equivalent parallel circuit containing a Norton current source I_N and a Norton resistance R_N. For a circuit with a load resistance R_L,

1. To find I_N, remove R_L and short circuit the terminals of R_L. Then calculate the short circuit current through these terminals.
2. To find R_N, which is equal to R_{TH}, short out the voltage sources and calculate the equivalent resistance across the open terminals of R_L. Study Example 8.12.

≣ REVIEW QUESTIONS

In problems 1 through 12 solve each problem rounding answers to two significant digits.

1. Given the circuit in Figure 8–45 with $V_1 = 6$ V, V_2 = 6 V, $R_1 = 2$ Ω, $R_2 = 2$ Ω, and $R_3 = 4$ Ω, find the three currents I_1, I_2, and I_3 using Kirchhoff's laws.

2. Given the circuit in Figure 8–45, with $V_1 = 12$ V, $V_2 = 9$ V, $R_1 = 40$ Ω, $R_2 = 40$ Ω, and $R_3 = 60$ Ω, find the three currents I_1, I_2 and I_3 using Kirchhoff's laws.

3. Given the circuit in Figure 8–46, with $V_1 = 12$ V, $V_2 = 6$ V, $R_1 = 6$ Ω, $R_2 = 6$ Ω, $R_3 = 4$ Ω and $R_4 = 4$ Ω, find the current through each resistor using Kirchhoff's laws.

4. Given the circuit in Figure 8–46, with $V_1 = 120$ V, $V_2 = 120$ V, $R_1 = 80$ Ω, $R_2 = 100$ Ω, $R_3 = 100$ Ω and $R_4 = 80$ Ω, find the current through each resistor using Kirchhoff's laws.

5. For the series-parallel circuit in Figure 8–47,
   (a) Given only that $V_T = 26$ V, find V_1, V_2, and V_3.
   (b) Given only that $I_T = 92$ mA, find I_1 and I_2.

6. For the series-parallel circuit in Figure 8–48,
   (a) Given only that $V_T = 100$ V, find V_3 and V_4.
   (b) Given only that $I_T = 240$ mA, find I_1, I_2, and I_3.

7. Given the circuit in Figure 8–49, find I_1, I_2, and I_3 by applying the superposition theorem.

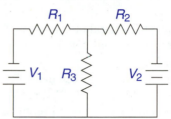

FIGURE 8–45   Kirchhoff's laws for problems 1 and 2.

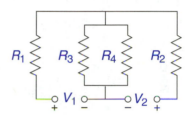

FIGURE 8–46   Kirchhoff's laws for problems 3 and 4.

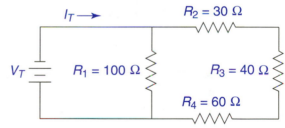

FIGURE 8–47   Series-parallel circuit for problem 5.

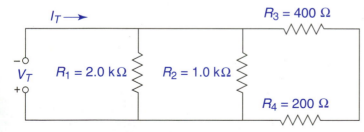

FIGURE 8–48   Series-parallel circuit for problem 6.

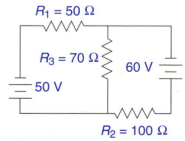

FIGURE 8–49   Superposition theorem for problem 7 and Thevenin's theorem for problem 8.

8. Given the circuit in Figure 8–49 with R_3 the load resistance, find the equivalent Thevenin voltage V_{TH} and Thevenin resistance R_{TH}. Then find I_3 and V_3. Check that I_3 is the same value as that found in problem 7.

9. Given the series-parallel circuit in Figure 8–50, Thevenize this circuit for the load resistance R_L. Find V_{TH}, R_{TH}, I_L, and V_L.

10. Nortonize the circuit in Figure 8–50 for the load resistance R_L. Find I_N, R_N, I_L and V_L. Check that the results agree with those in problem 9.

11. Given the circuit in Figure 8–51, find I_3 and V_3.

12. Given the bridge circuit in Figure 8–52, find V_{TH} and R_{TH} for the load R_L.

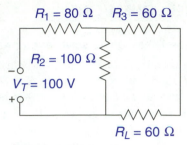

$R_1 = 80\ \Omega$ $R_3 = 60\ \Omega$

$R_2 = 100\ \Omega$

$V_T = 100\ V$

$R_L = 60\ \Omega$

FIGURE 8–50 Thevenin's theorem for problem 9 and Norton's theorem for problem 10.

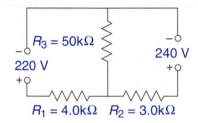

$R_3 = 50k\Omega$

220 V

240 V

$R_1 = 4.0k\Omega$ $R_2 = 3.0k\Omega$

FIGURE 8–51 Series-parallel circuit for problem 11.

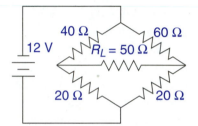

$R_L = 50\ \Omega$

40 Ω   60 Ω

12 V

20 Ω   20 Ω

FIGURE 8–52 Bridge circuit for problem 12.

Trigonometry

Courtesy of Motorola Computer Group, Tempe, Arizona

Technicians inspect and check integrated circuits.

This chapter reviews some basic geometry and studies the important concepts of trigonometry necessary for an understanding of alternating-current electricity. Trigonometry begins with angles and the right triangle, which is the building block for all geometric figures. It then leads to the study of angles in a circle, which is more natural since angles are a measure of rotation, or movement in a circle. Vectors and phasors, an important application of trigonometry, especially in electronics, are then studied followed by two laws that help to solve problems with vectors and phasors: the law of sines and the law of cosines.

Chapter Objectives

In this chapter, you will learn:

- How angles are measured in radians.
- How to change from degrees to radians and vice versa.
- The Pythagorean theorem and how to apply it.
- How to compute the sides of similar triangles using proportions.
- The definitions of the following trigonometric functions: sine, cosine, and tangent.
- How to find the trigonometric functions and their inverses for angles in a circle.
- The definitions of vectors and phasors and how to solve problems with vectors and phasors.
- The law of sines and the law of cosines and how to apply these laws to solve triangles and problems with vectors and phasors.

☰ 9.1
ANGLES AND TRIANGLES

Angles and Radian Measure

An *angle* is a measure of rotation between two radii in a circle. The center of the circle where the radii meet is called the *vertex* of the angle. See Figure 9–1. One complete rotation is defined as 360°. One complete rotation is also defined as 2π radians or 2π rad. Therefore, 2π rad = 360°. The abbreviation for radians, rad, is usually understood when there is no degree (°) symbol.

Dividing the equation 2π rad = 360° by 2 gives you the basic formula:

$$\pi \text{ rad } = 180° \text{ or } \pi = 180° \tag{9.1}$$

Remember this relationship to help you when working with radians. It leads to the following radian equivalents for common angles that you should be familiar with:

$$\frac{\pi}{2} \text{ rad} = 90°$$

$$\frac{\pi}{4} \text{ rad} = 45°$$

$$\frac{\pi}{3} \text{ rad} = 60°$$

$$\frac{\pi}{6} \text{ rad} = 30°$$

$$\frac{3\pi}{2} \text{ rad} = 270°$$

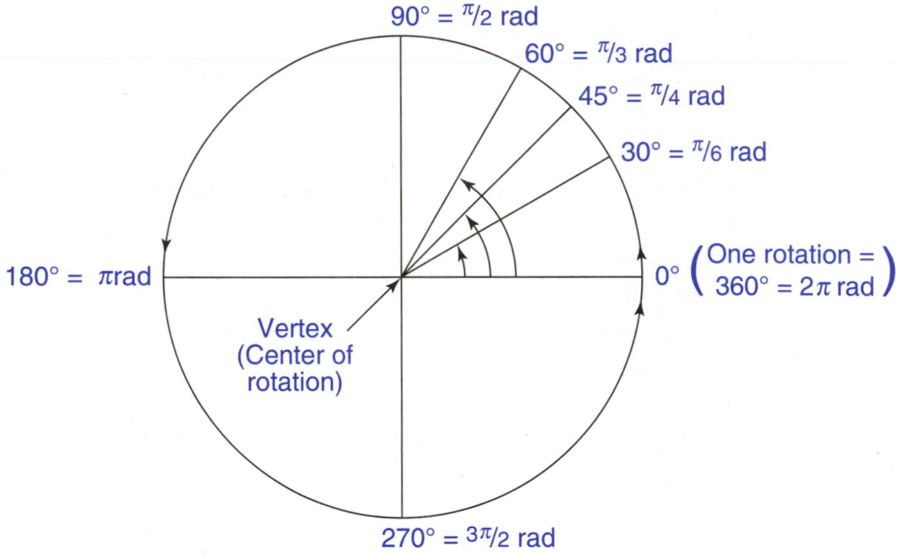

FIGURE 9–1   Basic angles.

Figure 9–1 shows these common angles in a circle. Rotation is counterclockwise starting from the right at 0°. Dividing both sides of (9.1) by π tells us that

$$1 \text{ rad} = \frac{180°}{\pi} \approx \frac{180°}{3.142} \approx 57.3° \tag{9.2}$$

This provides a conversion factor as follows.

Rule   To change from radians to degrees, multiply by $\frac{180°}{\pi}$ or 57.3°.

Rule   To change from degrees to radians, multiply by $\frac{\pi}{180°}$ or divide by 57.3°.

Study the next two examples, which show how to change from radians to degrees and vice versa.

EXAMPLE 9.1

Change to degrees:

1. $\frac{2\pi}{3}$ rad
2. 1.73 rad

Solution

1. To change $\frac{2\pi}{3}$ rad to degrees, multiply by $\frac{180°}{\pi}$ to divide out the π:

$$\frac{2\pi}{3}\,\text{rad} = \frac{2\pi}{3}\left(\frac{180°}{\pi}\right) = 120°$$

2. To change 1.73 rad to degrees, multiply by 57.3°:

$$1.73 \ \text{rad} = 1.73 \ (57.3°) = 99.1°$$

EXAMPLE 9.2

Change to radians

1. 135°
2. 46°

Solution

1. To change 135° to radians, multiply by $\frac{\pi}{180°}$ to express the angle in terms of π:

$$135° = 135°\left(\frac{\pi}{180°}\right) = \frac{3\pi}{4}$$

or divide by 57.3°:

$$\frac{135°}{57.3°} = 2.36\,\text{rad}$$

2. To change 46° to radians, multiply by $\frac{\pi}{180°}$ or divide by 57.3°:

$$46° = \frac{46°}{57.3°} = 0.80\,\text{rad}$$

The name "radian" is used because an angle of one radian cuts off an arc on the circle equal to one *radius* as shown in Figure 9–2.

Figure 9–3 shows various types of angles as explained here:

▪ A *right angle* is 90°. Angle C is a right angle.

▪ A *straight angle* is 180°.

▪ An *acute angle* is greater than zero and less than 90°. Angle A is an acute angle.

▪ *Perpendicular lines* meet at right angles denoted by a square at the vertex.

▪ An *obtuse angle* is greater than a right angle and less than 180°. Angle E is an obtuse angle.

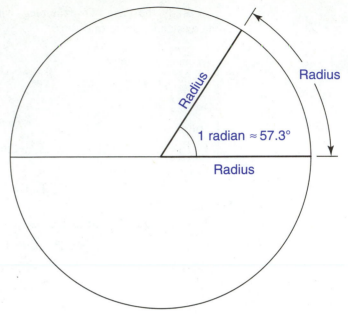

FIGURE 9–2 Radian measure.

- *Complementary angles* are two angles that add up to 90°. Angles *A* and *B* are complementary.
- *Supplementary angles* are two angles that add up to 180°. Angles *D* and *E* are supplementary, and angles *A* and *E* are supplementary.
- *Vertical angles* are opposite angles formed by two intersecting lines and are equal. Angles *A* and *D* are vertical angles and $\measuredangle A = \measuredangle D$.

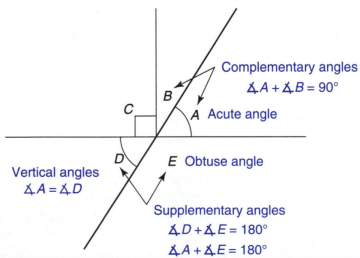

FIGURE 9–3 Types of angles for Example 9.3.

EXAMPLE 9.3

In Figure 9-3, given $\angle E = 110°$, find $\angle A$, $\angle B$, and $\angle D$.

Solution   Observe $\angle D$ is the supplement of $\angle E$ and $\angle A = \angle D$ because they are vertical angles. Therefore,

$$\angle A = \angle D = 180° - 110° = 70°$$

Note $\angle B$ is the complement of $\angle A$. Therefore,

$$\angle B = 90° - 70° = 20°$$

▪

Pythagorean Theorem

A *polygon* is a plane figure bounded by straight lines. A *triangle* is a three-sided polygon, a *rectangle* is a four-sided polygon, a *pentagon* is a five-sided polygon, and so on. Every polygon can be divided into triangles, and any triangle can be divided into two right triangles. The right triangle is therefore a basic building block for many figures. The study of the right triangle is called *trigonometry*. The longest side of any right triangle, which is opposite the right angle, is called the *hypotenuse* and labeled *c*. See Figure 9–4. The other two sides (also called the legs) are labeled *a* and *b*. The angles opposite these sides are labeled with corresponding capital letters. Angle *A* is opposite side *a*, angle *B* is opposite side *b*, and the right angle *C* is opposite the hypotenuse *c*.

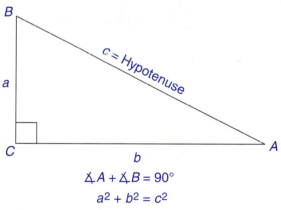

$$\angle A + \angle B = 90°$$
$$a^2 + b^2 = c^2$$

FIGURE 9–4 Right triangle and Pythagorean theorem.

An important theorem that applies to any triangle follows.

Rule   The three angles of any triangle add up to 180°.

Therefore, angles *A* and *B* in a right triangle add up to 90° and are complementary:

$$\angle A + \angle B = 90°$$

One of the most important and useful theorems in all of geometry, and mathematics, is the Pythagorean theorem. Although Babylonian and Egyptian surveyors used the theorem more than 3000 years ago, the earliest record of its formal proof was left by the Greek Pythagoras around 520 B.C.

> **Pythagorean Theorem**   In a right triangle, the sum of the squares of the sides equals the square of the hypotenuse.

This leads to the formula for the right triangle in Figure 9–4:

$$a^2 + b^2 = c^2 \qquad (9.3)$$

For example, if $a = 6$ and $b = 8$, then

$$6^2 + 8^2 = c^2$$

$$c^2 = 36 + 64 = 100$$

$$c = \sqrt{100} = 10$$

The following example shows how to apply the Pythagorean theorem to a practical problem to find the distance between two points.

EXAMPLE 9.4

A boat travels 5 mi south and then turns and travels east for 12 mi. How far is the boat from the starting point?

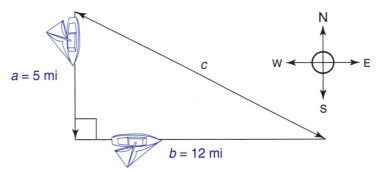

FIGURE 9– 5   Boat trip for Example 9.4.

Solution   Figure 9–5 shows the route that the boat travels. The direction of east is perpendicular to the direction of south. The distance of the boat from its starting point is the hypotenuse of a right triangle. You know $a = 5$ mi and $b = 12$ mi, and you need to find c. Apply the Pythagorean theorem:

$$a^2 + b^2 = c^2$$

$$(5)^2 + (12)^2 = c^2$$

EXAMPLE 9.4 (Cont.)

This is a quadratic equation containing only the square term. Solve for c^2, and then take the square root:

$$25 + 144 = c^2$$

$$c^2 = 169$$

$$c = \sqrt{169} = 13 \text{ mi}$$

The negative answer, -13, does not apply. Only the positive answer applies.

Proof of the Pythagorean Theorem

There are more than 400 proofs of the Pythagorean theorem, including one by a president of the United States, James Garfield. The following geometric proof may be similar to the one used by Pythagoras.

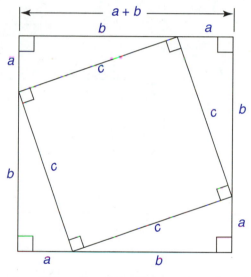

$$(a + b)^2 = c^2 + 4(1/2ab) \implies a^2 + b^2 = c^2$$

FIGURE 9–6 Proof of the Pythagorean theorem.

Figure 9–6 translates into the following verbal equation:

(Area of large square) = (Area of small square) + 4 × (Area of one right triangle)

This leads to

$$(a + b)^2 = c^2 + 4\left(\frac{1}{2}ab\right)$$

where the area of one right triangle $= \frac{1}{2}ab$ since one leg is the base and the other leg is the height. Multiplying out each side you obtain

$$a^2 + 2ab + b^2 = c^2 + 2ab$$

Cancellation of $2ab$ leads to the theorem:

$$a^2 + b^2 = c^2$$

Study the next example that closes the circuit and shows an important application of the Pythagorean theorem to a problem in electronics.

EXAMPLE 9.5

Close the Circuit

Figure 9–7(a) shows an ac circuit containing a resistance R and a capacitance C in parallel (RC parallel circuit). The total current I_T is related to the current through the capacitor I_C and the current through the resistance I_R by the current triangle shown in Figure 9–7(b). If $I_R = 45$ mA and $I_T = 80$ mA, find I_C.

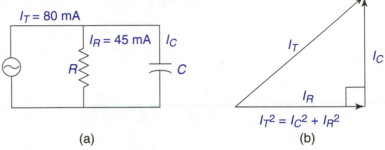

(a)   (b)

FIGURE 9– 7 *RC* parallel circuit for Example 9.5.

Solution The current diagram is a right triangle with sides I_C and I_R and hypotenuse I_T. The three currents are therefore related by the Pythagorean theorem:

$$I_C{}^2 + I_R{}^2 = I_T{}^2$$

Substitute the values in this Pythagorean formula using $I_R = 45$ mA and $I_T = 80$ mA:

$$I_C{}^2 + (45)^2 = (80)^2$$

Then

$$I_C{}^2 = (80)^2 - (45)^2 = 6400 - 2025 = 4375$$

Take the square root of both sides. The answer to two significant digits is then:

$$I_C = \sqrt{4375} = 66 \, \text{mA}$$

Again, only the positive answer applies as in Example 9.4.

Similar Polygons and Triangles

An important relationship between polygons is that of similarity. *Similar polygons* are two polygons that have the same shape but not necessarily the same size. The angles of a polygon determine the shape of a polygon. Therefore, if all the angles of one polygon are equal to the corresponding angles of another polygon, then the polygons are similar. Figure 9–8 shows two similar right triangles ABC and $A'B'C'$. The corresponding angles are equal:

$$\angle A = \angle A'$$
$$\angle B = \angle B'$$
$$\angle C = \angle C'$$

Polygons and particularly triangles whose angles are the same and are similar have their *corresponding sides in proportion*. A proportion is an equality between

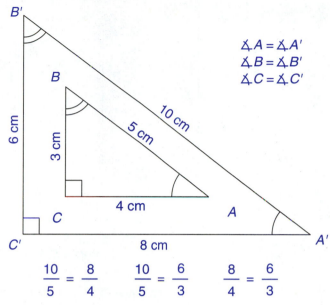

$$\angle A = \angle A'$$
$$\angle B = \angle B'$$
$$\angle C = \angle C'$$

$$\frac{10}{5} = \frac{8}{4} \qquad \frac{10}{5} = \frac{6}{3} \qquad \frac{8}{4} = \frac{6}{3}$$

FIGURE 9–8   Similar triangles and proportions.

two ratios. Consider the two right triangles in Figure 9–8. The sides of the large triangle are *twice* the sides of the small triangle. Therefore, you can write the following three proportions using the ratios of the corresponding sides:

$$\frac{10}{5} = \frac{8}{4}$$

$$\frac{10}{5} = \frac{6}{3}$$

$$\frac{8}{4} = \frac{6}{3}$$

These proportions can also be written from the small triangle to the large triangle:

$$\frac{5}{10} = \frac{4}{8}$$

$$\frac{5}{10} = \frac{3}{6}$$

$$\frac{4}{8} = \frac{3}{6}$$

Study the next example, which shows how to identify similar triangles and set up a proportion to find unknown sides.

EXAMPLE 9.6

Given right triangles ABC and BEF in Figure 9–9 with $CE = 5$, $BE = 10$, and $FE = 6$, find side AC and the hypotenuse AB.

Solution   The two right triangles are similar for the following reasons: The right angles are equal, and angle B is the same in both. When two angles of a triangle are equal to two corresponding angles in another triangle, the third angles must

EXAMPLE 9.6 (Cont.)

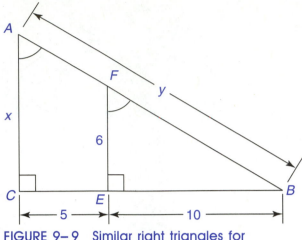

FIGURE 9–9 Similar right triangles for Example 9.6.

also be equal. This is because the three angles of any triangle add up to 180°. Therefore, $\angle A = \angle BFE$. Now identify the corresponding sides. *Corresponding sides are opposite equal angles.* Side *AC* corresponds to side *FE*, and side *BC* corresponds to side *BE*. Therefore, you can write the following proportion going from the large triangle to the small triangle:

$$\frac{AC}{FE} = \frac{BC}{BE}$$

Let *x* = side *AC* and note that side *BC* = 10 + 5 = 15. Then, substituting these values and *FE* = 6 in the proportion,

$$\frac{x}{6} = \frac{15}{10}$$

To solve a proportion you can *cross multiply*. This means multiplying the numerator of one fraction by the denominator of the other fraction and vice versa:

$$(x)(10) = (15)(6)$$

$$10x = 90$$

$$x = 9$$

Cross multiplication is equivalent to multiplying both sides of the equation by the product of the denominators.

Let *y* = side *AB*. You can find *y* by applying the Pythagorean theorem to the large triangle:

$$9^2 + 15^2 = y^2$$

$$y^2 = 225 + 81 = 306$$

$$y = \sqrt{306} = 17 \quad \text{(to two significant digits.)}$$

Similar figures are encountered in many technical problems. For example, the computer design model of a microcircuit is constructed similar to that of the actual circuit. The corresponding angles are equal, and the corresponding sides are in proportion. The scale of the model gives the ratio of the model distances to the actual distances. A scale of 100:1 means that the distances in the computer model are 100 times the distances in the actual circuit. Maps, photographs, or video images of any object are all examples of similarity. The next example illustrates an application of similar figures in electronics.

EXAMPLE 9.7

A rectangular microprocessor chip is 3.3 mm × 9.4 mm. If the image of the chip under a microscope has a width equal to 4.2 cm, what is the length of the image, and what is the scale of the image?

Solution   The microscopic image is similar to the actual chip but is much larger. The sides of the image and the chip are in proportion. Let λ (lambda) = Length of the image. To find λ, you can write the following proportion going from the image to the object:

$$\frac{\lambda}{9.4\,\text{mm}} = \frac{4.2\,\text{cm}}{3.3\,\text{mm}}$$

The units do not necessarily all have to be the same. If the units for the numerators agree and the units for the denominators agree, then the ratios will be equal. Cross multiply to obtain the solution:

$$(\lambda)(3.3\,\text{mm}) = (4.2\,\text{mm})(9.4\,\text{cm})$$

$$\lambda = \frac{(4.2\,\text{mm})(9.4\,\text{cm})}{3.3\,\text{mm}} = 12\,\text{cm}$$

Note that the millimeter (mm) units cancel leaving centimeter (cm) units for the answer. To find the scale of the image, divide one of the sides of the image by the corresponding side of the actual chip. However, you must use the same units to find the scale:

$$\text{Scale} = \frac{9.4\,\text{cm}}{3.3\,\text{mm}} = 940\,\text{mm} : 3.3\,\text{mm} = 285 : 1$$

▪

EXERCISE 9.1

In problems 1 through 8, change each angle to radians.

1. 60°
2. 45°
3. 90°
4. 180°

5. 315°
6. 150°
7. 50°
8. 27°

In problems 9 through 16, change each angle to degrees.

9. $\frac{\pi}{6}$ rad

10. $\frac{\pi}{3}$ rad

11. $\dfrac{3\pi}{2}$ rad

12. π rad

13. 5.76 rad

14. 3.93 rad

15. 1.21 rad

16. 0.026 rad

In problems 17 through 24, find all the lettered angles.

17.

FIGURE 9–10

21.

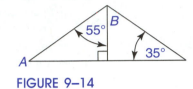

FIGURE 9–14

18.

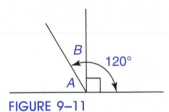

FIGURE 9–11

22.

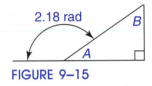

FIGURE 9–15

19.

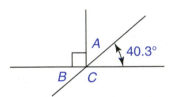

FIGURE 9–12

23.

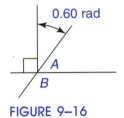

FIGURE 9–16

20.

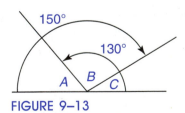

FIGURE 9–13

24.

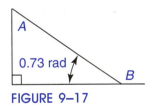

FIGURE 9–17

In problems 25 through 30, find the missing side of each right triangle (c = hypotenuse).

25. $a = 6$, $b = 8$, $c = ?$

26. $a = 8$, $b = 15$, $c = ?$

27. $a = ?$, $b = 2$, $c = 3$

28. $a = 90$, $b = ?$, $c = 150$

29. $a = 0.5$, $b = 0.7$, $c = ?$

30. $a = ?$, $b = 6.3$, $c = 9.2$

In problems 31 through 34, given that the triangles are similar, find x and y.

31.

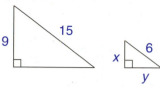

FIGURE 9–18

33.

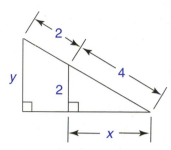

FIGURE 9–20

32.

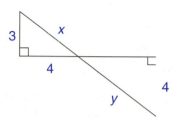

FIGURE 9–19

34.

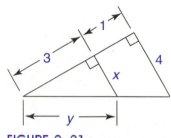

FIGURE 9–21

In problems 35 through 40, solve each applied problem. Round answers to two significant digits.

35. A ladder 8.0 ft long is placed 4.8 ft from a wall. How far up the wall does the ladder reach?

36. Two high-speed commuter trains leave a station at exactly 6 P.M. One train travels due north averaging 80 mi/h, and the other travels due west averaging 70 mi/h. How far apart are they at 6:30 P.M.?

37. Roxanne cycles 8 km south, 9 km east, and then 4 km further south. How far is she from her starting point? Hint: Draw one large right triangle.

38. A rectangular doorway is 1.4 m wide by 2.9 m high. Can a circular tabletop 3.3 m in diameter fit through the doorway?

39. The shadow of a building is 50 ft long. At the same time a tree 36 ft tall casts a shadow 20 ft long. How tall is the building? (Note that the triangles are similar.)

40. The shadow of a woman standing 10 m from a streetlight is 1.5 m long. If the woman is 180 cm tall, how high is the streetlight? (Note the different units and that the triangles are similar.)

Applications to Electronics   In problems 41 through 46 solve each applied problem. Round answers to two significant digits.

41. Figure 9–22(a) shows an ac circuit containing a resistance R and an inductance L in parallel (*RL* parallel circuit). The total current I_T is related to the current through the resistance I_R and the current through the inductance I_L by the current triangle shown in Figure 9–22(b). If $I_R = 180$ mA and $I_L = 140$ mA, find I_T. See Example 9.5.

42. In problem 41, find I_R given $I_T = 40$ mA and $I_L = 20$ mA.

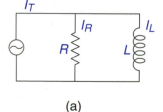

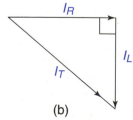

(a)   (b)

FIGURE 9–22   *RL* parallel circuit for problems 41 and 42.

43. Figure 9–23(a) shows an ac circuit containing a resistance R and an inductance L in series (RL series circuit). The impedance Z is related to the resistance R and the inductive reactance X_L by the impedance triangle shown in Figure 9–23(b). Find Z when $X_L = 25$ kΩ and $R = 15$ kΩ. See Example 9.5.

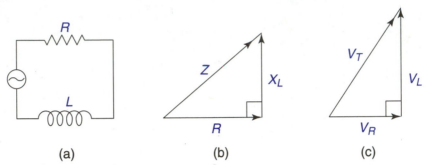

FIGURE 9–23 *RL* series circuit for problems 43 and 44.

44. In the RL series circuit in Figure 9–23, the total voltage V_T is related to the voltage across the inductance V_L and the voltage across the resistance V_R by the voltage triangle shown in Figure 9–23(c). Find V_L when $V_T = 20$ V and $V_R = 12$ V.

45. The dimensions of a rectangular computer screen are 27 cm × 20 cm. The screen size is to be scaled down proportionally for a notebook model. The width of the notebook screen is to be 125 mm. What should be the length of the notebook screen?

46. A circuit design is drawn to the scale of 35:1. If the length of a diode in the design is 2.5 cm, how long is the actual diode?

≡ 9.2
TRIGONOMETRY OF THE RIGHT TRIANGLE

Right triangles that have the same angles are similar. Therefore, they have the same ratios for any two corresponding sides. These ratios are very useful for finding sides and angles of right triangles when you know only some of the sides or angles. They are called the trigonometric ratios, or trigonometric functions, and are shown next.

Given the right triangle in Figure 9–24, the three basic trigonometric functions for the acute angle A are

$$\text{sine } A \text{ or } \sin A = \frac{\text{Opposite side}}{\text{Hypotenuse}} = \frac{a}{c}$$

$$\text{cosine } A \text{ or } \cos A = \frac{\text{Adjacent side}}{\text{Hypotenuse}} = \frac{b}{c}$$

$$\text{tangent } A \text{ or } \tan A = \frac{\text{Opposite side}}{\text{Adjacent side}} = \frac{a}{b} \qquad (9.4)$$

An easy way to remember these definitions is by the mnemonic

"S-OH, C-AH, T-OA"   (sin-Opp/Hyp, cos-Adj/Hyp, tan-Opp/Adj)

EXAMPLE 9.8

Given the triangle in Figure 9–24 with $a = 9$ and $b = 12$, find the values of the sine, cosine, and tangent of angle A and angle B using formulas (9.4).

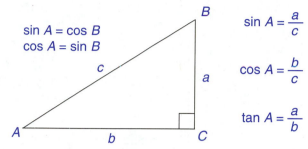

$$\sin A = \cos B$$
$$\cos A = \sin B$$

$$\sin A = \frac{a}{c}$$

$$\cos A = \frac{b}{c}$$

$$\tan A = \frac{a}{b}$$

FIGURE 9– 24   Trigonometric functions for Example 9.8.

Solution   First calculate the hypotenuse c using the Pythagorean theorem:

$$c^2 = a^2 + b^2$$
$$c^2 = 9^2 + 12^2$$
$$c^2 = 81 + 144 = 225$$
$$c = \sqrt{225} = 15$$

Then apply the definitions (9.4) above:

$$\sin A = \frac{a}{c} = \frac{9}{15} = 0.60 \qquad \sin B = \frac{b}{c} = \frac{12}{15} = 0.80$$

$$\cos A = \frac{b}{c} = \frac{12}{15} = 0.80 \qquad \cos B = \frac{a}{c} = \frac{9}{15} = 0.60$$

$$\tan A = \frac{a}{b} = \frac{9}{12} = 0.75 \qquad \tan B = \frac{b}{a} = \frac{12}{9} = 1.33$$

Observe that $\sin A = \cos B$ and $\cos A = \sin B$. This is always true when A and B are complementary since the side opposite A is the same side that is adjacent to B. Therefore, the sine of an angle equals the cosine of its complement. For example, $\sin 60° = \cos\ 30°$, $\sin 40° = \cos 50°$, $\sin 10.5° = \cos\ 79.5°$, and so on. Because of this relationship, sine and cosine are called cofunctions.

▪

EXAMPLE 9.9

Find the value of each trigonometric function using the calculator.

1. $\sin 50.3°$

2. $\tan 1.20$ rad

3. $\cos \dfrac{\pi}{5}$ rad

EXAMPLE 9.9 (Cont.)

Solution

1. To find the sin 50.3°, set the calculator for degrees by pressing the ⌈DRG⌉ or ⌈MODE⌉ key. The display should show "DEG" or "D." Some calculators are in the degree mode when you turn them on. Enter the angle, and press the trigo-nometric function key:

$$50.3 \; \boxed{\text{sin}} \; \rightarrow \; 0.769$$

2. To find the tan 1.20 rad, first set the calculator for radians using the ⌈DRG⌉ or ⌈MODE⌉ key. The display should show "RAD" or "R." Then enter the angle, and press the trigonometric function key:

$$\boxed{\text{DRG}} \; 1.20 \; \boxed{\text{tan}} \; \rightarrow \; 2.57$$

3. To find the cos $\dfrac{\pi}{5}$ rad, you can first change it to degrees and then find the co-sine, or set the calculator for radians and find it directly. Both methods are shown here:

Degrees:

$$\left(\frac{\pi}{5}\right)\left(\frac{180}{\pi}\right) = 36°$$

$$36 \; \boxed{\text{cos}} \; \rightarrow \; 0.809$$

Radians:

$$\boxed{\text{DRG}} \; \boxed{\pi} \; \boxed{\div} \; 5 \; \boxed{=} \; \boxed{\text{cos}} \; \rightarrow \; 0.809$$

When working with radians you can use the $\boxed{\pi}$ key on the calculator instead of entering the approximate value 3.142.

If any side and acute angle of a right triangle are known, or if any two sides of a right triangle are known, any other side or angle can be found using the trigono-metric functions. The next two examples show how to use the trigonometric func-tions to find the sides or angles of a right triangle.

EXAMPLE 9.10

Given a right triangle with ∡A = 36° and side c = 8, solve for the missing parts of the triangle.

Solution   You need to find ∡B and the two legs a and b of the triangle. Since ∡B is complementary to ∡A, ∡B = 90° − 36° = 54°. If you know the hypotenuse, you can use sine or cosine to find a and b. Using the sin A, you set up an equation to find a as follows:

$$\sin A = \frac{a}{c}$$

Then substitute the values for ∡A and c and solve for a:

$$\sin 36° = \frac{a}{8}$$

$$a = 8(\sin 36°) = 8(0.588) = 4.7$$

EXAMPLE 9.10 (Cont.)

Using the cos A, you solve for b as follows:

$$\cos A = \frac{b}{c}$$

$$\cos 36° = \frac{b}{8}$$

$$b = 8(\cos 36°) = 8(0.809) = 6.5$$

▪

EXAMPLE 9.11

Given a right triangle with $a = 6.2$ and $b = 5.5$, find $\angle A$.

Solution   Side a is opposite $\angle A$, and side b is adjacent to $\angle A$. Therefore, you can set up an equation to first find tan A:

$$\tan A = \frac{a}{b} = \frac{6.2}{5.5} = 1.13$$

Then, to find an angle knowing the value of a trigonometric function of the angle, you use the *inverse* trigonometric function. For tangent, this is $\boxed{\tan^{-1}}$ or $\boxed{\text{INV}}$ $\boxed{\tan}$ on the calculator. Enter the value and press the inverse function key:

$$1.13 \ \boxed{\text{INV}} \ \boxed{\tan} \ \rightarrow 48°$$

Some calculators use $\boxed{\text{2nd F}}$ or $\boxed{\text{SHIFT}}$ instead of $\boxed{\text{INV}}$.

▪

The next example shows how a surveyor can use trigonometry to find the height of a cliff by measuring an angle.

EXAMPLE 9.12

To find the height of a cliff, Jorge, a surveyor, chooses a point 100 m from the base of a cliff and sets up his transit (a telescope that can measure horizontal and vertical angles). He measures the angle of elevation of the top of the cliff to be 36.4°. The *angle of elevation* is the angle measured above the horizontal. See Figure 9–25. Ignoring Jorge's height and assuming the angle of elevation is measured from the ground, how high is the cliff?

Solution   You need to set up an equation using the side of the triangle you know, the side you are trying to find, and the trigonometric function of the angle that is the ratio of these two sides. The known side (100 m) is *adjacent* to the angle of 36.4°. The unknown side (the cliff height) is *opposite* to 36.4°. Therefore, choose the tangent function, which is opposite over adjacent. Letting h = height, the equation is

$$\tan 36.4° = \frac{\text{Opposite}}{\text{Adjacent}} = \frac{h}{100}$$

Then using the calculator,

$$h = 100\,(\tan 36.4°)$$

$$h = 100\,(0.737) = 73.7\,\text{m} \approx 74\,\text{m}$$

EXAMPLE 9.12 (Cont.)

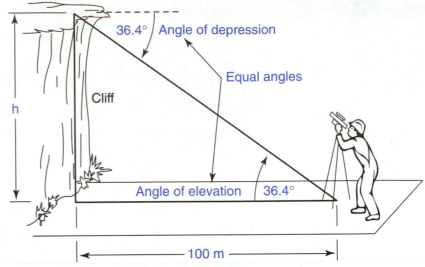

FIGURE 9– 25   Surveyor in Example 9.12.

For an observer on top of the cliff looking down at the surveyor, the *angle of depression* is the angle measured below the horizontal. The angle of depression is equal to the angle of elevation because the line of sight makes the same angle with the two parallel horizontal lines.

■

Study the next example which closes the circuit and shows an application to electronics using the trigonometric functions.

EXAMPLE 9.13

Close the Circuit

Figure 9–26(a) shows an ac circuit containing a resistance R and a capacitance C in series. The impedance Z is related to the resistance R and the capacitive reactance X_C by the impedance triangle shown in Figure 9–26(b). If $X_C = 10$ kΩ and $R = 5.6$ kΩ, find the angle θ between X_C and R and the impedance Z.

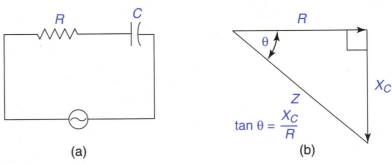

(a)   (b)

FIGURE 9– 26   *RC* series circuit for Example 9.13.

Solution   The trigonometric function of angle θ that is the ratio of X_C and R is the tangent. Therefore, you can write

$$\tan \theta = \frac{X_C}{R}$$

EXAMPLE 9.13 (Cont.)

Substitute the given values, and calculate the value of the tangent:

$$\tan \theta = \frac{10}{5.6} = 1.79$$

Now find the angle using [INV] [tan] or [tan⁻¹] on the calculator:

$$\theta = 1.79 \; [\text{INV}] \; [\text{tan}] \; \rightarrow 60.8° \approx 61°$$

Some calculators use [SHIFT] or [2nd F] instead of [INV].

To find Z, you can use the Pythagorean theorem, or you can use the sin θ (or cos θ):

$$\sin \theta = \frac{X_C}{Z}$$

$$\sin 60.8° = \frac{10}{Z}$$

When you use the sine or cosine function, the unknown Z appears in the denominator. You first need to multiply both sides by Z and then divide by sin 60.8°:

$$(Z) \sin 60.8° = \frac{10}{(Z)} (Z)$$

$$Z (\sin 60.8°) = 10$$

$$\frac{Z (\sin 60.8°)}{\sin 60.8°} = \frac{10}{\sin 60.8°}$$

$$Z = \frac{10}{\sin 60.8°} = \frac{10}{0.873} = 11 \text{ k}\Omega$$

The answers are rounded to two significant digits.

EXERCISE 9.2

In problems 1 through 10, find sin A, cos A, and tan A for each right triangle (c = hypotenuse).

1. $a = 8, b = 15, c = 17$
2. $a = 6, b = 4.5, c = 7.5$
3. $a = 10.5, b = 25$
4. $a = 11.5, b = 38$
5. $a = 10, c = 15$
6. $b = 4.5, c = 8.2$
7. $b = 0.30, c = 0.90$
8. $a = 6.6, c = 12.2$
9. $a = X, b = X$
10. $a = R, c = 2R$

In problems 11 through 18, find the value of each trigonometric function.

11. $\cos 60°$
12. $\tan 45°$
13. $\sin 80.6°$
14. $\cos 39.3°$
15. $\tan (0.568 \text{ rad})$
16. $\sin (1.035 \text{ rad})$
17. $\cos \left(\frac{\pi}{10} \text{ rad} \right)$
18. $\sin \left(\frac{\pi}{3} \text{ rad} \right)$

In problems 19 through 26, find the value of the given angle between 0° and 90°.

19. $\sin A = 0.317$

20. $\cos B = 0.0960$

21. $\tan \phi = 2.14$

22. $\tan \phi = 1.00$

23. $\tan \phi = 0.416$

24. $\tan \phi = 0.0989$

25. $\cos \theta = \dfrac{1}{3}$

26. $\sin \theta = \dfrac{3}{5}$

In problems 27 through 36, find all the missing sides and acute angles for each right triangle. In problems 33 and 34, give the angles in radians.

27. $A = 50°$, $c = 3.0$

28. $B = 17°$, $c = 8.2$

29. $B = 25.5°$, $a = 10.3$

30. $A = 53.2°$, $a = 0.45$

31. $a = 5$, $c = 8$

32. $a = 2.5$, $b = 6.2$

33. $B = 0.73$ rad, $a = 4.5$

34. $A = 1.1$ rad, $a = 3.3$

35. $\tan A = 1.0$, $a = 2.0$

36. $\sin B = 0.60$, $c = 10$

In problems 37 through 42, solve each applied problem. Round answers to two significant digits.

37. The angle of elevation of a building is 56° at a point 60 m from its base. How high is the building? See Example 9.12.

38. The angle of depression of a sailboat from the top of a lighthouse is 11°. If the top of the lighthouse is 150 ft above sea level, how far away from the base of the lighthouse is the sailboat? See Example 9.12.

39. A power cable is to be strung between the top of a building and the top of an electric utility pole. Gordon measures the angle of depression of the top of the pole from the top of the building to be 23°. If the building is 30 ft high and the pole is 60 ft high, how long is the distance from the top of the building to the top of the pole?

40. From a Rocky Mountain peak 9300 ft above sea level, Estela measures the angle of elevation of a nearby peak to be 15.5° (Figure 9-27). If the height of the other peak is 11,200 ft, how far is the *straight line air distance* between the peaks? (Neglect the Earth's curvature.)

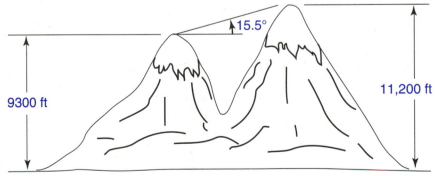

FIGURE 9–27 Distance between mountain peaks for problem 40.

41. After takeoff, the altimeter in a plane indicates a vertical rise of 10 mi/h at a constant velocity of 300 mi/h. Assuming the angle of ascent to be constant, find its value.

42. The safe angle ϕ for the bank of a highway curve is given by

$$\tan \phi = \frac{v^2}{gr}$$

where v = velocity, r = radius of curve, and g = gravitational acceleration. If $v = 24.6$ m/s (55 mi/h) and $r = 290$ m, find the value of the safe angle. Note $g = 9.81$ m/s^2.

Applications to Electronics In problems 43 through 50 solve each applied problem. Round answers to two significant digits.

43. The electrical force E acting on a body at an angle θ from the horizontal has a horizontal component E_x and a vertical component E_y. Figure 9–28 shows the force rectangle. If $E = 8.8$ N/C (newtons/coulomb) and $\theta = 47°$, find E_x and E_y.

44. In problem 43, if $E_x = 2.8$ N/C and $E_y = 3.6$ N/C, find E and the angle θ.

45. In Example 9.13, given $X_C = 2.6$ kΩ and $R = 3.7$ kΩ, find the angle θ and the impedance Z.

46. For the RC series circuit in Example 9.13, the voltage across the resistance V_R and the voltage across the capacitance V_C are related to the total voltage V_T by the voltage triangle shown in Figure 9–29. If $V_R = 8.0$ V and $V_C = 16$ V, find V_T and the angle θ.

47. In an ac circuit, the instantaneous voltage is given by

$$v = V_M \sin(\omega t)$$

where V_M = maximum voltage, ω = angular frequency in rad/s, and t = time in seconds. Find v when $V_M = 300$ V and $\omega t = 0.25\pi$ rad.

48. In problem 47, find the angle ωt between 0 and $\frac{\pi}{2}$ radians when $V_M = 300$ V and $v = 150$ V.

49. The true or active power of an ac circuit due to the resistance is given by

$$P = VI \cos \phi$$

where ϕ is called the phase angle. Find ϕ in degrees and radians when $V = 120$ V, $I = 1.5$ A, and $P = 100$ W.

50. The reactive power of an ac circuit due to the inductance and/or the capacitance is given by

$$P_X = VI \sin \phi$$

where ϕ is called the phase angle. Find ϕ in degrees and radians when $V = 240$ V, $I = 500$ mA, and $P_X = 100$ W.

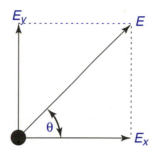

FIGURE 9–28
Electrical forces acting on a body for problems 43 and 44.

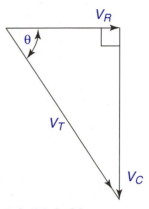

FIGURE 9–29
Voltage triangle in an *RC* circuit for problem 46.

≣ 9.3

TRIGONOMETRY OF THE CIRCLE

Since angles measure rotation or circular motion, the trigonometric functions can be defined for any angle, positive or negative, in a circle. This extends their definiton for an angle in a right triangle as follows.

Consider the circle in Figure 9–30 whose center is at the origin of a rectangular coordinate system. The direction of the radius is determined by the angle θ in the circle. The angle θ is measured from the positive x axis (initial side) to the radius (terminal side). Counterclockwise rotation is considered positive and clockwise rotation negative. The circle in Figure 9–30 contains four quadrants: $0° < I < 90°$, $< 270°$, $90° < II < 180°$, $180° < III$, $< 270°$, and $270° < IV < 360°$. The angle θ

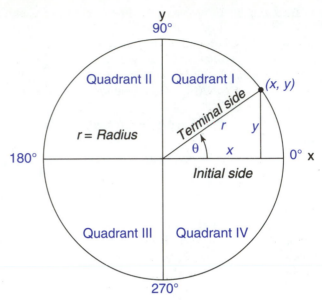

FIGURE 9–30 Trigonometry of the circle.

shown is a positive angle in the first quadrant. The *three trigonometric functions* of an angle θ whose terminal side passes through a point (x,y) on a circle of radius r are:

$$\sin\theta = \frac{y}{r}$$

$$\cos\theta = \frac{x}{r}$$

$$\tan\theta = \frac{y}{x} \tag{9.5}$$

Definitions (9.5) apply to *any* angle, positive or negative. The definitions for sin, cos, and tan are the same as those for a right triangle when the angle is in the first quadrant between 0° and 90°. The side x is the adjacent side, the side y is the opposite side, and the radius r is the hypotenuse. When the angle is in quadrant II, III, or IV, the definitions are applied using the coordinates of the point on the circle and the radius of the circle. Carefully study the following example, which shows how to apply the definitions for angles in any quadrant.

EXAMPLE 9.14

Draw each angle whose terminal side passes through the given point, and find the three trigonometric functions of the angle.

1. θ_1: (4,3)
2. θ_2: (−4,3)
3. θ_3: (−4,−3)
4. θ_4: (4,−3)

EXAMPLE 9.14 (Cont.)

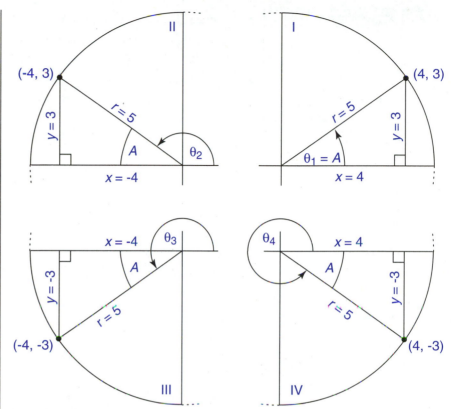

FIGURE 9–31 Angles in each quadrant for Example 9.14.

Solution   The four angles, one in each quadrant, are drawn in Figure 9–31. Study the diagram closely. For each case, a perpendicular is drawn from the point on the circle to the *x* axis forming a right triangle. The perpendicular *is always drawn to the x axis.* This is so side *y* is opposite the acute angle *A* shown in each triangle. The four right triangles are identical in size and shape. The sides of each triangle—*x, y,* and the radius *r*—are the same length in each quadrant.

For an angle in any quadrant, the Pythagorean theorem says

$$x^2 + y^2 = r^2 \text{ or } r = \sqrt{x^2 + y^2}$$

The radius r is always considered positive. Therefore, for all four cases, $r = +5$ since when $x = +4$ or -4, and $y = +3$ or -3,

$$r = \sqrt{x^2 + y^2} = \sqrt{(\pm 4)^2 + (\pm 3)^2} = \sqrt{16 + 9} = \sqrt{25} = 5$$

Now applying definitions (9.5), the three trigonometric functions for each case are

1. Quadrant I: $x = 4, y = 3, r = 5$

$$\sin \theta_1 = \frac{3}{5}$$

$$\cos \theta_1 = \frac{4}{5}$$

$$\tan \theta_1 = \frac{3}{4}$$

EXAMPLE 9.14 (Cont.)

2. Quadrant II: $x = -4$, $y = 3$, $r = 5$

$$\Rightarrow \sin \theta_2 = \frac{3}{5}$$

$$\cos \theta_2 = \frac{-4}{5}$$

$$\tan \theta_2 = \frac{3}{-4}$$

3. Quadrant III: $x = -4$, $y = -3$, $r = 5$

$$\sin \theta_3 = \frac{-3}{5}$$

$$\cos \theta_3 = \frac{-4}{5}$$

$$\Rightarrow \tan \theta_3 = \frac{3}{4}$$

4. Quadrant IV: $x = 4$, $y = -3$, $r = 5$

$$\sin \theta_4 = \frac{-3}{5}$$

$$\Rightarrow \cos \theta_4 = \frac{4}{5}$$

$$\tan \theta_4 = \frac{-3}{4}$$

Observe that the absolute values of the trigonometric functions are the same in each quadrant. They differ only in sign. The signs are determined by the point (x,y) on the terminal side. *All the functions are positive in the first quadrant, and only one function is positive in any other quadrant* (see the arrows). Angle A in each triangle is called the *reference angle* and is equal to θ_1 in each quadrant. Its importance is discussed after Example 9.16.

The positive functions for the angles in each quadrant are shown in Figure 9–32. One way to remember these positive functions is to memorize the following phrase: "All Star Trig Class." The first letters—A, S, T, C—are emphasized in Figure 9–32, starting from the first quadrant and reading counterclockwise.

Three other trigonometric functions are also defined for an angle in each quadrant. They are called the *reciprocal functions* since each is the reciprocal of one of the three basic functions:

$$\csc \text{ (cosecant) } \theta = \frac{1}{\sin \theta} = \frac{r}{y}$$

$$\sec \text{ (secant) } \theta = \frac{1}{\cos \theta} = \frac{r}{x}$$

$$\cot \text{ (cotangent) } \theta = \frac{1}{\tan \theta} = \frac{x}{y}$$

The reciprocal functions are not as important as sin, cos, and tan. They are used in trigonometric formulas in calculus.

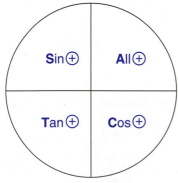

FIGURE 9–32 Positive trigonometric functions: "All Star Trig Class."

The next example shows how to find the values of the trig functions when given the value of just one trig function.

EXAMPLE 9.15

Given that $\tan \theta = -1$ and $\cos \theta$ is positive,

1. Draw θ, showing the values for x, y, and r.
2. Find the values of $\sin \theta$ and $\cos \theta$ applying formulas (9.5).

Solution

1. Figure 9–32 shows that $\cos \theta$ is positive in the first and fourth quadrant. However, since it is also given that $\tan \theta$ is negative, θ must be in the fourth quadrant. The sketch of the angle is shown in Figure 9–33. You can choose any positive value for x and negative value for y as long as $\tan \theta = \frac{y}{x} = -1$. For example, two sets of values are $x = 1$ and $y = -1$, or $x = 2$ and $y = -2$. The simplest set is $x = 1$ and $y = -1$. Then

$$r = \sqrt{x^2 + y^2} = \sqrt{(1)^2 + (-1)^2} = \sqrt{2} = 1.41$$

In Figure 9–33 the triangle is called the reference triangle. It is always drawn to the x axis so y is opposite the reference angle A. Reference triangles are discussed further with reference angles after Example 9.16.

2. To find the values of $\sin \theta$ and $\cos \theta$, use the values of x, y, and r from the reference triangle in Figure 9–33:

$$\sin \theta = \frac{y}{r} = \frac{-1}{\sqrt{2}} = -0.707$$

$$\cos \theta = \frac{x}{r} = \frac{1}{\sqrt{2}} = 0.707$$

Observe in Figure 9–33 that the legs of the reference triangle are equal. Therefore, the reference angle $A = 45°$ and $\theta = 360° - 45° = 315°$. You can therefore check these answers with the calculator using $\theta = 315°$. For example,

$$315° \boxed{\sin} \rightarrow -0.707$$

■

FIGURE 9– 33 Fourth quadrant angle and reference triangle for Example 9.15.

y

$\tan \theta = -1$
$\cos \theta$ is $\oplus$

θ

$x = 1$

A

$r = \sqrt{2} = 1.41$

$y = -1$

x

Study the next example that closes the circuit and shows an application of these ideas to a problem in electricity.

EXAMPLE 9.16

Close the Circuit

Figure 9–34(a) shows an ac circuit containing a resistance R and a capacitance C in series. The voltage across the resistance V_R and the voltage across the capacitance V_C are related to the total voltage V_T as shown in the voltage triangle in Figure 9–34(b). *The angle θ is considered a negative angle in the fourth quadrant.* If $V_R = 12$ V and $V_C = -15$ V, find V_T and $\tan \theta$.

Solution In this application, $x = V_R$, $y = V_C$, and $r = V_T$. Apply the Pythagorean theorem to find V_T:

$$V_T = \sqrt{V_R{}^2 + V_C{}^2} = \sqrt{(12)^2 + (-15)^2} = \sqrt{144 + 225} = \sqrt{369} = 19.2 \text{ V}$$

EXAMPLE 9.16 (Cont.)

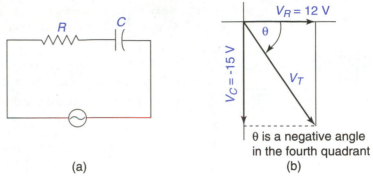

(a)   (b)

FIGURE 9–34 *RC series circuit and voltage triangle for Example 9.16.*

To find tan θ, apply definition (9.5):

$$\tan \theta = \frac{V_C}{V_R} = \frac{-15}{12} = -1.25$$

Reference Angles and Reference Triangles

In Figure 9–31 the triangle in each quadrant is called the reference triangle. It is always drawn to the *x* axis so that the reference angle *A* is always opposite side *y*. The following rule then applies.

> **Rule**   The absolute, or positive, value of the trigonometric function of an angle in any quadrant is equal to the trigonometric function of its reference angle.   (9.6)

Rule (9.6) means that the value of the trigonometric function of an angle will either be the same as the trigonometric function of its reference angle (positive) or will be opposite in sign (negative), depending on the quadrant. Figure 9–35 shows how the reference angle A is related to an angle θ in quadrants II, III, and IV. Examples for each quadrant are:

II: $\sin 150° = +\sin (180° - 150°) = +\sin 30° = 0.500$

III: $\cos 230° = -\cos (230° - 180°) = -\cos 50° = -0.643$

IV: $\tan 300° = -\tan (360° - 300°) = -\tan 60° = -1.73$

Although the value of the trigonometric function of any angle can be found directly on the calculator, it is necessary to understand how to use reference angles in many problems, especially when working with inverse trigonometric functions.

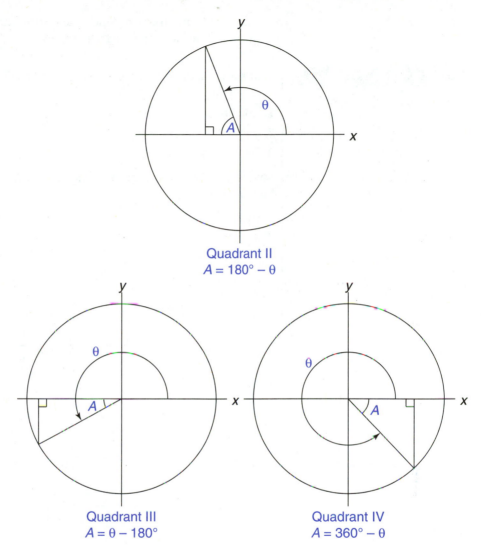

FIGURE 9–35 Reference angles and reference triangles.

Quadrant Angles

The quadrant angles are 0°, 90°, 180° and 270°. It is not possible to draw a reference triangle for these angles. The values of their sine, cosine, and tangent can be found directly with the calculator. The exceptions are tan 90° and tan 270°, which are infinitely large and have no defined value.

Negative Angles and Angles Greater Than 360°

To find the reference angle for negative angles and angles greater than 360°, you first find the coterminal angle between 0° and 360°. The coterminal angle is the angle having the same terminal side as the given angle and, therefore, the same reference angle and triangle. Study the next example, which shows how to work with angles of any size and reference angles.

EXAMPLE 9.17

Find the value of each of the trigonometric functions by using a reference angle or quadrant angle.

1. cos 110°

2. sin 450°

3. tan (−47°)

Solution Each of the given angles and their reference angles are shown in Figure 9–36.

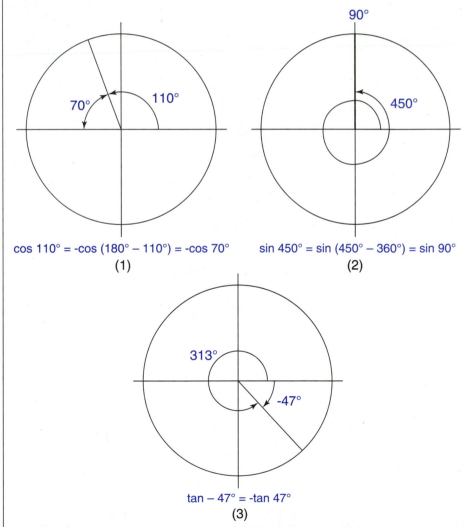

cos 110° = -cos (180° − 110°) = -cos 70°
(1)

sin 450° = sin (450° − 360°) = sin 90°
(2)

tan − 47° = -tan 47°
(3)

FIGURE 9– 36 Reference angles and quadrant angles for Example 9.17.

EXAMPLE 9.17 (Cont.)

1. To find cos 110°, first find the reference angle by subtracting from 180°:

$$\text{Reference angle} = 180° - 110° = 70°$$

The value of any trig function of 110° is then related to the same trig function of 70°. It is either equal to it or opposite in sign. Figure 9–32 shows that the cosine is negative in quadrant II. Therefore,

$$\cos 110° = -\cos 70° = -0.342$$

Check this result on your calculator by finding cos 110°.

2. To find sin 450°, first find the coterminal angle less than 360° by subtracting 360° (or multiples of 360°, if necessary): 450° − 360° = 90°. The value of any trig function of 450° is then equal to the same trig function of 90°. Therefore,

$$\sin 450° = \sin 90° = 1$$

Check this result on your calculator by finding sin 450°.

3. To find tan(−47°), you can find the reference angle by first adding 360° (or multiples of 360°, if necessary) to obtain a positive coterminal angle. For −47° the coterminal angle is −47° + 360° = 313°. The reference angle is then 360° − 313° = 47°. However, *for a negative angle in the fourth quadrant, the reference angle is just equal to the positive value of the angle*. Since tangent is negative in the fourth quadrant,

$$\tan(-47°) = -\tan 47° = -1.07$$

Check this result on your calculator by finding tan (−47°).

EXERCISE 9.3

In problems 1 through 12, draw the angle whose terminal side passes through the given point (x, y), and find the three trigonometric functions of the angle applying formulas (9.5).

1. (3,4)
2. (−8,6)
3. (5,−12)
4. (−15,−8)
5. (−1,−1)
6. (1.0,1.7)

7. (−3.9,5.2)
8. (17,−7.5)
9. (−0.40,−0.70)
10. (0.50,0.50)
11. (2.8,2.1)
12. (−3.6,1.5)

In problems 13 through 22, draw the angle and find the values of the other two trigonometric functions applying formulas (9.5).

13. $\sin \theta = -\dfrac{3}{5}$, tan θ positive

14. $\cos \theta = \dfrac{12}{13}$, sin θ negative

15. tan θ = 2.4, sin θ positive
16. tan θ = −0.75, cos θ positive

17. $\cos 2\pi ft = -0.60$, $\sin 2\pi ft$ negative
18. $\sin 2\pi ft = 0.50$, $\cos 2\pi ft$ negative
19. $\sin \omega t = 0.40$, ωt in quadrant I
20. $\cos \omega t = -0.707$, ωt in quadrant III
21. $\tan \phi_Y = -1.00$, ϕ_Y in quadrant II
22. $\tan \phi_Z = -1.33$, ϕ_Z in quadrant IV

In problems 23 through 40, express each function as a function of its reference angle and find the value of the function.

23. $\sin 130°$

24. $\cos 120°$

25. $\tan 150°$

26. $\tan 230°$

27. $\cos 335°$

28. $\sin 305°$

29. $\sin 540°$

30. $\cos 450°$

31. $\tan -53°$

32. $\tan -6.7°$

33. $\tan 1.10$ rad

34. $\tan 0.85$ rad

35. $\sin -200°$

36. $\cos -150°$

37. $\cos 720°$

38. $\sin -450°$

39. $\tan -\dfrac{\pi}{8}$ rad

40. $\tan -\dfrac{\pi}{6}$ rad

In problems 41 through 44, solve each applied problem. Round answers to three significant digits.

41. A force vector F in the second quadrant has components $F_x = -7.50$ N (N = newtons) and $F_y = 3.50$ N, as shown in Figure 9–37. Find $\sin \theta$, $\cos \theta$, and $\tan \theta$.

42. A formula for the area of a parallelogram with sides a and b and included angle θ is $A = ab \sin \theta$. Find the area of a structure in the shape of a parallelogram if $a = 5.30$ cm, $b = 6.20$ cm, and the included angle $\theta = 145°$.

43. The resultant velocity of a ship in knots (nautical mph) is given by
$$v = \sqrt{66.8 - 28.4 \cos 120°}$$
Calculate v.

44. The range of a projectile fired with velocity v_0 at an angle of elevation ϕ is given by
$$R = \frac{v_0^2 \sin 2\phi}{g}$$
where the acceleration due to gravity $g = 9.81$ m/s^2. Find R when $v_0 = 100$ m/s and
(a) $\phi = 45°$. **(b)** $\phi = 75°$.

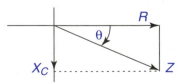

FIGURE 9–37 Force vector for problem 41.

F

$F_y = 3.50$ N

θ

$F_x = -7.50$ N

Applications to Electronics In problems 45 through 52, solve each applied problem. Round answers to three significant digits.

45. In Example 9.16, find V_T and $\tan \theta$ when $V_R = 14$ V and $V_C = -12$ V.

46. In Example 9.16, find V_R and $\tan \theta$ when $V_C = -5.5$ V and $V_T = 10$ V.

47. In the RC series circuit in Example 9.16 the resistance R and the capacitive reactance X_C are related to the impedance Z by the impedance triangle shown in Figure 9–38. If $\theta = -23°$ and the resistance $R = 4.6$ kΩ, find $\tan \theta$ and the capacitive reactance X_C. Note: X_C is usually expressed as a positive value.

R

θ

X_C

Z

FIGURE 9–38 RC series circuit impedance triangle for problems 47 and 48.

48. In problem 47, if $\theta = -56°$ and $Z = 3.80$ kΩ, find tan θ, X_C, and R.

49. In an ac circuit, the instantaneous current $i = I_M \sin \omega t$ where I_M = maximum current and ωt = angle of rotation of a generator. Find i when $I_M = 850$ mA and ωt is
(a) 90°.
(b) 120°.
(c) 5π/4 rad.

50. In the ac circuit of problem 49, the instantaneous voltage $v = V_M \cos \omega t$ where V_M = maximum voltage. Find v when $V_M = 10$ V and ωt is
(a) 180°.
(b) 400°.
(c) 3.5 rad.

51. In an ac circuit containing a capacitance, the instantaneous voltage is given by

$$v = V_M \sin (2\pi f t - \pi/2)$$

where V_M = maximum voltage, f = frequency in hertz, t = time in *seconds*, and the angle $(2\pi f t - \pi/2)$ is in *radians*. Find v when $V_M = 120$ V, $f = 50$ Hz, and
(a) $t = 20$ ms.
(b) $t = 18$ ms.

52. In an ac circuit, tan $\phi_Z = X/R$, where ϕ_Z = phase angle of the impedance and tan $\phi_Y = -X/R$, where ϕ_Y = phase angle of the admittance. If $X^2 + R^2 = Z^2$, find tan ϕ_Z and tan ϕ_Y when $X = 4.5$ kΩ and $Z = 5.5$ kΩ.

Courtesy of Hewlett-Packard Company.

An optical time-domain reflectometer allows telephone technicians to perform measurements along fiber links.

☰ 9.4
INVERSE FUNCTIONS

Example 9.11 introduced the inverse tangent function that is used to find an angle when the value of the tangent is known. This section studies all three inverse trigonometric functions and shows how to find the values of the angle in any quadrant.

Study the first example carefully. It illustrates what you need to consider when using inverse trigonometric functions.

EXAMPLE 9.18

Given $\cos \theta = -0.250$, find angle θ such that $0° \leq \theta < 360°$ (θ greater than or equal to 0° and less than 360°).

Solution   Since cosine is negative in the second and the third quadrant, there are two values for the angle θ that satisfy $\cos \theta = -0.250$. It is helpful to first find the reference angle and then determine the two angles. *If you enter the positive value of any trigonometric function in the calculator, the inverse function will always be an angle in the first quadrant,* which is the reference angle. Enter 0.25 and press the inverse cosine key $\boxed{\cos^{-1}}$ or $\boxed{\text{INV}}$ $\boxed{\cos}$ to obtain the reference angle:

$$0.25 \;\; \boxed{\text{INV}} \;\; \boxed{\cos} \;\; \rightarrow \;\; 75.5°$$

The reference angle for θ is then 75.5°. This is expressed mathematically as

$$\cos^{-1} 0.25 = 75.5°$$

and means "the angle whose cosine is 0.25 = 75.5°."

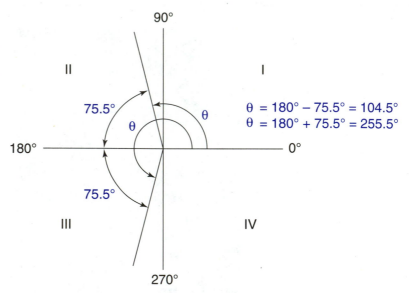

FIGURE 9–39   Inverse functions for Example 9.18.

Next find the two values of θ whose reference angle is 75.5° in the second and third quadrants. These are shown in Figure 9–39 and are found as follows:

$$\text{Quadrant II:} \quad \theta = 180° - 75.5° = 104.5°$$

$$\text{Quadrant III:} \quad \theta = 180° + 75.5° = 255.5°$$

EXAMPLE 9.18 (Cont.)

To find the angle in the second quadrant, you subtract from 180°. For the third quadrant, you add 180°, and for the fourth quadrant you subtract from 360°.

If you were to enter the negative value −0.25 in the calculator and press $\boxed{\cos^{-1}}$, you would get one of these answers, 104.5°, in the second quadrant. However, you still need to find the other answer.

▪

For each value of an inverse trigonometric function you enter, the calculator is programmed to give you only *one* defined value for the angle. That value is defined within the following specific range of values for each function:

$$-90° \le \sin^{-1}x \le 90°$$

$$0° \le \cos^{-1}x \le 180°$$

$$-90° < \tan^{-1}x < 90° \tag{9.7}$$

Note that the inverse functions will always give you angles in the first quadrant for *positive* values of *x*. Sin⁻¹ and tan⁻¹ give you *negative* angles in the fourth quadrant for negative values of *x,* and cos⁻¹ gives you angles in the second quadrant for negative values of *x*. The reason for these definitions is to provide a *continous* range of angles for each inverse function that includes all possible values. It follows that when you have a negative value for a trigonometric function, such as tan θ = −0.50, you can enter the absolute or positive value 0.50 in the calculator and you will get the reference angle for θ.

Study the next two examples, which show further how to use the inverse trigonometric functions.

EXAMPLE 9.19

Find tan⁻¹ −0.6783.

Solution When the inverse function tan⁻¹ (or sin⁻¹ or cos⁻¹) is used, it means only the one defined value for the angle given by (9.6). This is also called the principal value which is the calculator value. Enter the negative number, and press the inverse function key:

$$0.6783 \;\boxed{\pm}\; \boxed{\text{INV}}\; \boxed{\text{tan}}\; \to\; -34.1°$$

Therefore, tan⁻¹ −0.6783 = −34.1°.

▪

EXAMPLE 9.20

Given 3 sin ϕ + 1 = 0. Find ϕ in degrees and radians for $0 \le \phi < 2\pi$.

Solution This is a basic trigonometric equation. First solve for sin ϕ and then for ϕ:

$$3\sin\phi + 1 = 0$$

$$3\sin\phi = -1$$

$$\sin\phi = -\frac{1}{3} = -0.333$$

EXAMPLE 9.20 (Cont.)

The reference angle A = $\sin^{-1} 0.333 = 19.5°$. Sine is negative in the third and fourth quadrant. The two possible values for ϕ in degrees and radians are then

Quadrant III: $\phi = 180° + 19.5° = 199.5° = \dfrac{199.5}{57.3} = 3.48$ rad

Quadrant IV: $\phi = 360° - 19.5° = 340.5° = \dfrac{340.5}{57.3} = 5.94$ rad

---

Study the last example, which closes the circuit and is an application of inverse trigonometric functions to a problem in electricity.

EXAMPLE 9.21

Close the Circuit

The instantaneous voltage of ordinary house current is given by $v = 170 \sin 2\pi ft$, where the frequency $f = 60$ Hz and $t =$ time in seconds. Given $v = 120$ V, find

1. The angle $2\pi ft$ in degrees and radians for $0 \le 2\pi ft < 2\pi$.

2. The time t.

Solution

1. To find the angle $2\pi ft$, substitute 120 for v to yield the following equation:

$$120 = 170 \sin 2\pi ft$$

Next solve for $\sin 2\pi ft$:

$$\sin 2\pi ft = \frac{120}{170} = 0.7059$$

Then find the angle $2\pi ft$. The reference angle in degrees and radians is

$$\sin^{-1} 0.7059 = 44.9° = \frac{44.9}{57.3} = 0.784 \text{ rad}$$

Sine is positive in the first and second quadrants. The two solutions in degrees and radians are then

Quadrant I: $2\pi ft = 44.9° = 0.784$ rad

Quadrant II: $2\pi ft = 180° - 44.9° = 135.1° = \dfrac{135.1}{57.3} = 2.36$ rad

2. To find the time t, note that the angle $2\pi ft$ must be in radians because 2π is used instead of 360°. Let $f = 60$ Hz and solve for t using each of the preceding angles in radians:

Quadrant I: $2\pi(60)t = 0.784$

$$t = \frac{0.784}{120\pi} = 0.00208 \text{ s} = 2.08 \text{ ms}$$

Quadrant II: $2\pi(60)ft = 2.36$

$$t = \frac{2.36}{120\pi} = 0.00626 \text{ s} = 6.26 \text{ ms}$$

Radians are used in many technical applications instead of degrees because they lend to a better interpretation of the ideas.

EXERCISE 9.4

In problems 1 through 10, find the value of each inverse trigonometric function.

1. $\sin^{-1} 0.7071$
2. $\cos^{-1} 0.8660$
3. $\tan^{-1} 1.192$
4. $\tan^{-1} 2.456$
5. $\cos^{-1} -0.7865$

6. $\sin^{-1} -0.0997$
7. $\tan^{-1} -3.221$
8. $\tan^{-1} -0.2255$
9. $\tan (\sin^{-1} 0.50)$
10. $\sin (\tan^{-1} 1.0)$

In problems 11 through 18, find the angle to the nearest 0.1° for values between 0° and 360°.

11. $\cos \theta = 0.9063$
12. $\sin \theta = 0.6293$
13. $\tan \phi = 0.8451$
14. $\tan \phi = 1.245$

15. $\sin x = -0.5556$
16. $\cos x = -0.1334$
17. $\tan y = -0.5030$
18. $\tan y = -2.003$

In problems 19 through 26, find the angle in degrees and radians between 0 and 2π to three significant digits.

19. $\cos x = 0.5324$
20. $\sin x = -0.6316$
21. $3 \tan \phi = 3$
22. $\tan \phi + 1 = 0$
23. $2 \sin \theta + 1 = 0$

24. $3 \cos \theta - 1 = 0$
25. $\sin^2 \theta = \dfrac{1}{2}$
26. $\tan^2 \theta = 3$

In problems 27 and 28 solve each applied problem to three significant digits.

27. The components F_x and F_y of a resultant force F are shown in Figure 9–40. If $F_x = -6.20$ N (newtons) and $F_y = 8.70$ N, find the angle θ.

28. The difference between the true heading and the compass heading of an airplane is given by the following expression:

$$H = 60° - \tan^{-1} (1.40 \ \sin \ 65°)$$

Find the value of H.

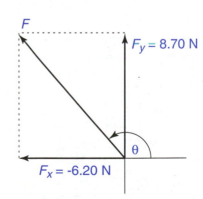

FIGURE 9–40 Force components for problem 27.

Applications to Electronics   In problems 29 through 34, solve each applied problem to three significant digits.

29. Figure 9–41(a) shows an ac circuit containing a resistance R and an inductance L in parallel (*RL* parallel circuit). The current through the resistance I_R and the current through the inductance I_L are related to the total current I_T by the current diagram shown in Figure 9–41(b). The angle θ is a negative angle in the fourth quadrant. If $I_R = 20$ mA and $I_L = 30$ mA, find θ.

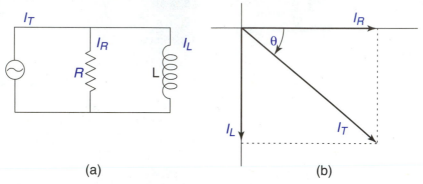

(a)   (b)

FIGURE 9–41   *RL* parallel circuit for problems 29 and 30.

30. In problem 29, find θ if $I_T = 380$ μA when $I_L = 150$ μA.

31. An electric field vector has components $E_x = 3.99$ N/C (newtons/coulomb) and $E_y = -2.03$ N/C as shown in Figure 9–42. Find the angle α of the resultant E.

32. In a problem involving an electric field the following expression arises: $\alpha = \tan^{-1} x + \tan^{-1}(1/x)$. Find α in radians when x is

(a) 0.50.

(b) 1.5.

Can you show that $\alpha = \pi/2$ rad for any value of x? (Hint: Draw a diagram.)

33. In Example 9.21, given that $v = 100$ V, find

(a) The angle $2\pi ft$ in radians for $0 \le 2\pi ft < 2\pi$.

(b) The time t.

34. In an ac circuit the instantaneous current is given by $i = 2.5 \cos \omega t$. If $i = 1.5$ mA, find

(a) The angle ωt in radians for $0 \le \omega t < 2\pi$.

(b) The time t if $\omega = 400$ rad/s.

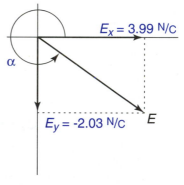

FIGURE 9–42   Electric field vectors for problem 31.

9.5

VECTORS AND PHASORS

Vectors and phasors are quantities that need two numbers to describe them. For example, force is a vector quantity and ac voltage is a phasor quantity. Both force and ac voltage can be represented on a graph as an arrow starting at the origin and ending at a point (x,y). Each can then be described in terms of the two numbers x and y that are the rectangular coordinates of the point. This is called rectangular form. They can also be described in terms of their *length* and the *angle* that they make with the *x* axis. This is called polar form. Polar form is more common for vectors and phasors. Therefore, a vector or phasor is a quantity that has both *magnitude* and *direction*.

The difference between a phasor and a vector is in the measurement of the direction. A vector's direction is an angle in two- or three-dimensional space. A phasor's direction is based on time. The angle of a phasor represents a time difference between two quantities, such as voltage and current, in an ac circuit. Vectors and phasors are treated the same way mathematically.

You are already familiar with some vector and phasor quantities. Some important examples are

1. **Velocity vector:** The velocity of the wind is 20 mi/h from the SE (halfway between south and east).
2. **Electromagnetic force vector:** A force exerted by an electromagnetic field on a particle is 2×10^{-6} N at an angle $\theta = 30°$ with the horizontal.
3. **Voltage phasor:** The voltage in an *RL* series circuit is 14 V with a phase angle $\theta = 47°$. This is written in polar form as 120 $\angle 47°$ V.
4. **Current phasor:** The total current in an *RL* parallel circuit is 50 mA with a phase angle of $\theta = -15°$. This is written in polar form as 50 $\angle -15°$ mA.

Most quantities you are familiar with can be described with one number. They are called *scalars*. For example, height, weight, and distance are scalars. The *magnitude* of a vector or phasor is also considered a scalar.

Adding Vectors and Phasors

Vectors and phasors are drawn as arrows and are usually identified with capital or bold letters. The length of the arrow represents the magnitude and the arrowhead shows the direction. Figure 9–43 shows how you add two vectors or phasors graphically. The procedure is as follows:

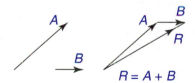

FIGURE 9–43 Addition of vectors or phasors.

> **Rule** To add two vectors or phasors *A* and *B*, place the tail of *B* at the head of *A* without changing its direction. The resultant $R = A + B$ goes from the tail of *A* to the head of *B*.

The first example shows how you can apply this principle to add two vectors or phasors mathematically.

EXAMPLE 9.22

Figure 9–44 shows a body acted on by a horizontal force $F_x = 18$ N at an angle $\theta_x = 180°$ and a vertical force $F_y = 10$ N at an angle $\theta_y = 90°$. Find the resultant force *F* acting on the body.

Solution   Add F_y to F_x by moving it parallel to itself so that its tail is on the head of F_x. The resultant *F* is the diagonal of the rectangle whose sides are F_x and F_y. You can also add F_x to F_y by moving it parallel to itself so that its tail is on the head of F_y. This produces the same resultant.

EXAMPLE 9.22 (Cont.)

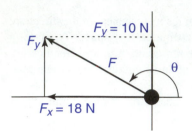

FIGURE 9–44 Force vectors for Example 9.22.

Since force is a vector, you need to find its magnitude *and* its direction. To find the magnitude of F apply the Pythagorean theorem to the right triangle:

$$F = \sqrt{F_x{}^2 + F_y{}^2} = \sqrt{18^2 + 10^2}$$

$$F = \sqrt{324 + 100} = \sqrt{424} = 20.6\,\text{N} \approx 21\,\text{N}$$

To find the direction, you need to find the angle θ in the second quadrant. First find the reference angle and then θ:

$$\text{Reference angle} = \tan^{-1}\left(\frac{F_y}{F_x}\right) = \tan^{-1}\left(\frac{10}{18}\right) = 29.1°$$

Then

$$\theta = 180° - 29.1° = 150.9° \approx 151°$$

Study the next example, which closes the circuit and shows an application of phasors in an ac circuit.

EXAMPLE 9.23

Close the Circuit

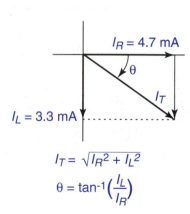

$$I_T = \sqrt{I_R{}^2 + I_L{}^2}$$

$$\theta = \tan^{-1}\left(\frac{I_L}{I_R}\right)$$

FIGURE 9–45 Current phasors for Example 9.23.

Figure 9–45 shows the phasor diagram for the currents in an *RL* parallel circuit. The current through the resistance $I_R = 4.7\ \angle 0°$ mA and the current through the inductance $I_L = 3.3\ \angle{-90°}$ mA. The negative angle indicates that I_L *lags* I_R by 90°. Find the total current I_T, which is the sum of the two phasors.

Solution For a phasor you must find the magnitude *and* the angle. Apply the Pythagorean theorem to find the magnitude of I_T:

$$I_T = \sqrt{I_R{}^2 + I_L{}^2} = \sqrt{4.7^2 + 3.3^2}$$

$$I_T = \sqrt{22.09 + 10.89} = \sqrt{32.98} = 5.7\,\text{mA}$$

To show the time difference, a negative angle in the fourth quadrant is always used to represent the phase angle θ rather than a positive angle. Therefore, one direct way to find θ is to use the inverse tangent function and attach a negative sign to I_L:

$$\theta = \tan^{-1}\left(\frac{-I_L}{I_R}\right) = \tan^{-1}\left(\frac{-3.3}{4.7}\right) = -35.1° \approx -35°$$

The solution in polar form is then $I_T = 5.7\ \angle{-35°}$ mA. You can also leave I_L positive and attach a negative sign to the angle.

When two vectors or phasors directly oppose each other—that is, they are 180° apart—they are easily added since the smaller one cancels part of the larger. The resultant magnitude is the *difference* in the magnitudes, and the resultant direction is that of the vector or phasor with the greater magnitude.

Study the next example, which closes the circuit and applies these ideas to phasors in an ac circuit.

ERROR BOX

A common error when working with vectors and phasors is getting the incorrect angle. For a *vector* in the fourth quadrant we often use a positive angle; for a phasor we use a *negative* angle. The calculator will give you a negative angle for $\tan^{-1}$ when you enter a negative number. For example, $\tan^{-1}(-2.14) = -65°$. This is the same direction as the positive angle $360° - 65° = 295°$. Apply this idea and see if you can get the correct angle for each vector or phasor sum.

Practice Problems: For each pair of vectors or phasors find the *angle* of the resultant.

1. Force vectors:
$$F_x = 0.25 \text{ N}, \ \theta_x = 0°; \ F_y = 0.15 \text{ N}, \ \theta_y = 270°$$

2. Acceleration vectors:
$$A_x = 3.6 \text{ m/s}^2, \ \theta_x = 0°; \ A_y = 8.4 \text{ m/s}^2, \ \theta_y = 270°$$

3. Voltage phasors:
$$V_R = 60 \ \angle 0° \text{ V}; \ V_C = 20 \ \angle{-90°} \text{ V}$$

4. Current phasors:
$$I_R = 22 \ \angle 0° \text{ mA}; \ I_L = 33 \ \angle{-90°} \text{ mA}$$

Answers: 1. 329° 2. 293° 3. −18° 4. −56°

EXAMPLE 9.24

Close the Circuit

Figure 9–46(a) shows the phasor diagram for an ac circuit containing a resistance, an inductance, and a capacitance in series (*RLC* series circuit). Given the resistance $R = 2.0 \ \angle 0° \text{ k}\Omega$, the capacitive reactance $X_C = 3.0 \ \angle{-90°} \text{ k}\Omega$, and the inductive reactance $X_L = 4.0 \ \angle 90° \text{ k}\Omega$, find the circuit impedance Z, which is the sum of these three phasors.

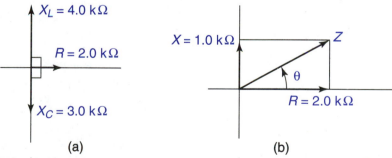

(a)   (b)

FIGURE 9–46 *RLC* series circuit impedance phasors for Example 9.24.

Solution The phasors X_C and X_L oppose each other in an ac circuit. Add them together first. Since they are opposite in direction, the resultant magnitude, which is the net reactance X, is their difference:

$$X = X_L - X_C = 4.0 - 3.0 = 1.0 \text{ k}\Omega$$

EXAMPLE 9.24 (Cont.)

The phase angle of X is that of X_L, $90°$, since X_L is larger than X_C. Therefore, $X = 1.0 \angle 90°$ kΩ.

The phasor diagram can then be reduced to two phasors as shown in Figure 9–46(b). You can now add these two perpendicular phasors to give the circuit impedance Z. The magnitude of Z is

$$Z = \sqrt{1.0^2 + 2.0^2} = \sqrt{5.0} = 2.2 \text{ k}\Omega$$

The phase angle of Z is

$$\theta = \tan^{-1}\left(\frac{X}{R}\right) = \tan^{-1}\left(\frac{1.0}{2.0}\right) = 27°$$

The circuit impedance in polar form is then $Z = 2.2 \angle 27°$ kΩ.

Components

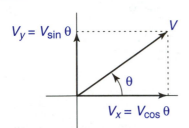

$V_y = V_{\sin\theta}$

V

θ

$V_x = V_{\cos\theta}$

FIGURE 9–47   Vector or phasor components.

In the preceding examples, two perpendicular vectors or phasors are combined into one resultant. The reverse of this process is also very useful when adding vectors or phasors that are not perpendicular. When a vector or phasor is placed on a rectangular coordinate system with its tail at the origin, it can be resolved into *components* along the x and y axes. In Figure 9–44, F_x is the x component of F, and F_y is the y component. In Figure 9–45 I_R is the x component of I_T and I_L is the y component. The relationship between a vector or phasor V and its components V_x and V_y is shown in Figure 9–47. The definitions of sine and cosine yield the formulas for the components:

$$\cos\theta = \frac{V_x}{V} \Rightarrow V_x = V\cos\theta \quad (x \text{ component})$$

$$\sin\theta = \frac{V_y}{V} \Rightarrow V_y = V\sin\theta \quad (y \text{ component}) \tag{9.8}$$

These formulas apply for any angle θ in any quadrant.

When you need to add two or more vectors or phasors that are not perpendicular, you first resolve each into x and y components. Then you add the components to give a resultant x component and a resultant y component. Finally, you combine the x and y components to produce the resultant. Study the next example that closes the circuit and illustrates this process in an application with impedance phasors.

EXAMPLE 9.25

Close the Circuit

Given the two impedance phasors $Z_1 = 7.3 \angle 25°$ kΩ and $Z_2 = 5.5 \angle -55°$ kΩ, find the resultant phasor Z.

Solution   Draw the phasors on a coordinate graph as shown in Figure 9–48(a). Set up a table for the components as follows by applying formula (9.8). Then add the columns to produce the x and y components of the resultant.

EXAMPLE 9.25 (Cont.)

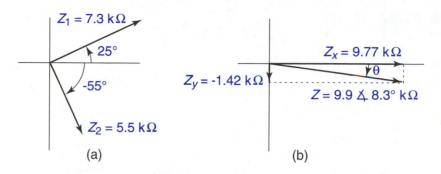

FIGURE 9–48   Adding impedance phasors for Example 9.25.

		x components			y components	
Z_1	7.3 cos 25°	= 6.62 kΩ		7.3 sin 25°	=	3.09 kΩ
Z_2	5.5 cos (–55°)	= 3.15 kΩ		5.5 sin (–55°)	=	– 4.51 kΩ
Resultant		Z_x = 9.77 kΩ			Z_y	= – 1.42 kΩ

Note that the formulas for the components work for all angles and produce the correct sign that indicates positive or negative direction. Figure 9–48(b) shows the components of the resultant.

The magnitude of Z is then

$$Z = \sqrt{9.77^2 + (-1.42)^2} = \sqrt{97.5} = 9.9\,\text{k}\Omega$$

and the negative phase angle is

$$\theta = \tan^{-1}\left(\frac{Z_x}{Z_y}\right) = \tan^{-1}\left(\frac{-1.42}{9.77}\right) = -8.3°$$

Observe that when you use the negative component in the inverse tangent formula you obtain the correct negative angle. The resultant phasor is then $Z = 9.9 \angle -8.3°\,\text{k}\Omega$. ▪

EXERCISE 9.5

In problems 1 through 6, find the resultant of each pair of vectors with an angle between 0° and 360°.

1. $F_x = 10$ N, $\theta_x = 0°$
   $F_y = 24$ N, $\theta_y = 90°$

2. $F_x = 44$ N, $\theta_x = 180°$
   $F_y = 26$ N, $\theta_y = 90°$

3. $V_1 = 53$ ft/s, $\theta_1 = 180°$
   $V_2 = 26$ ft/s, $\theta_2 = 90°$

4. $A_1 = 12$ m/s^2, $\theta_1 = 0°$
   $A_2 = 9.8$ m/s^2, $\theta_2 = 270°$

5. $F_1 = 4 \times 10^{-6}$ N, $\theta_1 = 180°$
   $F_2 = 5 \times 10^{-6}$ N, $\theta_2 = 270°$

6. $E_1 = 400$ N/C, $\theta_1 = 180°$
   $E_2 = 300$ N/C, $\theta_2 = 270°$

In problems 7 through 16, find the resultant of each pair of phasors with an angle between – 90° and 90°.

7. $R = 33 \angle 0°$ kΩ
   $X_L = 22 \angle 90°$ kΩ

8. $R = 1.2 \angle 0°$ kΩ
   $X_L = 1.5 \angle 90°$ kΩ

9. $R = 2.4 \angle 0° \text{ k}\Omega$
   $X_C = 4.7 \angle -90° \text{ k}\Omega$

10. $R = 18 \angle 0° \text{ k}\Omega$
    $X_C = 8.2 \angle -90° \text{ k}\Omega$

11. $V_L = 24 \angle 90° \text{ V}$
    $V_C = 12 \angle -90° \text{ V}$

12. $I_C = 340 \angle 90° \text{ mA}$
    $I_L = 770 \angle -90° \text{ mA}$

13. $R = 1.0 \angle 0° \text{ k}\Omega$
    $X = 820 \angle -90° \Omega$

14. $R = 910 \angle 0° \Omega$
    $X = 1.6 \angle 90° \text{ k}\Omega$

15. $I_R = 160 \angle 0° \text{ μA}$
    $I_C = 250 \angle 90° \text{ μA}$

16. $V_R = 28 \angle 0° \text{ V}$
    $V_C = 18 \angle -90° \text{ V}$

In problems 17 through 20, solve each applied problem. Round answers to two significant digits.

17. A horizontal force of 23 N acts to the right on a body, and a vertical force of 40 N acts downward on it. Find the magnitude and direction of the resultant force.

18. A taxi driver traveling at 15 mi/h throws a cigarette out of the window perpendicular to his direction and at a speed of 5 mi/h. What is the velocity of the cigarette as it leaves his hand? Find the magnitude and the angle with respect to the taxi's direction. Let the taxi's direction be *x* and the cigarette's direction be *y*.

19. Walt draws his boat's velocity vector on a coordinate graph as 5.5 mi/h at an angle of 0°. On the same graph he draws a current vector of 2.0 mi/h acting on his boat at an angle of 40°. Find the velocity of the boat with respect to the ground, which is the resultant of the two velocity vectors. (Hint: Add components as in Example 9.25.)

20. In Figure 9–49, Doug is flying a light airplane due north at 130 kn (knots = nautical miles per hour). The plane is experiencing a 40-kn headwind from the northwest (45° west of north). What is the resultant velocity of Doug's plane? Find the magnitude and the angle measured from due north. (Place the tail of the wind vector at the origin and add components as in Example 9.25.)

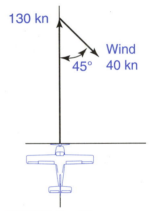

130 kn

Wind
45°   40 kn

FIGURE 9– 49
Airplane velocity for
problem 20.

Applications to Electronics In problems 21 through 30, solve each applied problem. Round answers to two significant digits.

21. Figure 9–50 shows the phasor diagram for the voltages in a series *RC* circuit. The voltage through the capacitance lags the voltage through the resistance by 90°. If the magnitude of $V_R = 33$ V and the magnitude of $V_C = 63$ V, find the total voltage V_T, which is the phasor sum of the voltages.

22. In problem 21, find V_T if the magnitude of $V_R = 4.5$ V and the magnitude of $V_C = 2.6$ V.

23. In Example 9.23, find I_T when $I_R = 50 \angle 0° \text{ mA}$ and $I_L = 70 \angle -90° \text{ mA}$.

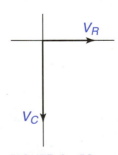

V_R

V_C

FIGURE 9– 50
Voltages in a
series *RC* circuit
for problem 21.

24. In a parallel RC circuit, the current through the capacitance I_C leads the current through the resistance I_R by 90°. Find I_T when $I_R = 90 \angle 0°$ μA and $I_C = 200 \angle 90°$ μA.

25. In Example 9.24, find the circuit impedance Z when $R = 15 \angle 0°$ kΩ, $X_L = 30 \angle 90°$ kΩ, and $X_C = 20 \angle -90°$ kΩ.

26. In Example 9.24, find the circuit impedance Z when $R = 4.3 \angle 0°$ kΩ, $X_L = 3.6 \angle 90°$ kΩ, and $X_C = 7.5 \angle -90°$ kΩ. (Note: X_C is larger than X_L.)

27. Given the two impedance phasors $Z_1 = 44 \angle -56°$ kΩ and $Z_2 = 22 \angle 86°$ kΩ, find the resultant phasor. See Example 9.25.

28. Given the two impedance phasors $Z_1 = 5.2 \angle 12°$ kΩ and $Z_2 = 6.7 \angle -15°$ kΩ, find the resultant phasor. See Example 9.25.

29. At a point in an electric field, two electric charges produce the intensity vectors: $E_1 = 1.9 \times 10^3$ N/C with angle $\theta_1 = 130°$ and $E_2 = 2.3 \times 10^3$ N/C with angle $\theta_2 = 200°$. Find the resultant vector. See Example 9.25.

30. At a point in a magnetic field, the magnetic flux density B_1 due to current in one wire is 4.0×10^{-6} T (tesla) and the flux density B_2 due to current in another wire is 6.0×10^{-6} T. If vectors B_1 and B_2 placed with their tails together make an angle of 45°, what is the resultant flux density? Find the angle measured from B_1. See Example 9.25.

≡ 9.6
LAW OF SINES AND LAW OF COSINES

The trigonometry studied so far applies to right triangles. There are two important laws, however, that can be used to find the sides or angles of any triangle. They are the law of sines and the law of cosines.

Law of Sines

Given any triangle ABC (Figure 9–51), the *law of sines* states

$$\frac{a}{\sin A} = \frac{b}{\sin B} = \frac{c}{\sin C} \tag{9.9}$$

This expression represents three different formulas. Each one is between two different sides and angles of the triangle. The law of sines says that the side of a triangle is proportional to the sine of the opposite angle. For example, sin 70° = 0.94 and is approximately twice sin 28° = 0.47. If angle A in Figure 9–51 increases from 28° to 70° and side b and angle B do not change, side a will approximately double in length.

The law of sines can be used to find the missing parts of a triangle when either of the following is true:

1. Two angles and a side opposite one of the angles are known.
2. Two sides and an angle opposite one of the sides are known.

The following example shows how to apply the law of sines.

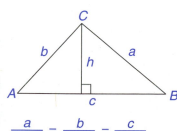

$$\frac{a}{\sin A} = \frac{b}{\sin B} = \frac{c}{\sin C}$$

FIGURE 9–51 Law of sines for Example 9.26.

EXAMPLE 9.26

In Figure 9–51, given $\angle A = 40°$, $\angle B = 60°$, and $c = 5.0$ cm, find the three missing parts of triangle ABC.

Solution   You know two angles and side c but you need $\angle C$ to use the law of sines. Subtract the two angles from 180° to find $\angle C$:

$$\angle C = 180° - \angle A - \angle B = 180° - 40° - 60° = 80°$$

You now have side c and $\angle C$ and can use the law of sines to find a and b. To find a use (9.9) with a and c:

$$\frac{a}{\sin A} = \frac{c}{\sin C}$$

Solve for a and substitute:

$$a = \frac{c(\sin A)}{\sin C} = \frac{(5.0)(\sin 40°)}{\sin 80°}$$

$$a = \frac{(5.0)(0.643)}{(0.985)} = 3.3 \text{ cm}$$

To find b use (9.9) with b and c:

$$\frac{b}{\sin B} = \frac{c}{\sin C}$$

Solve for b and substitute:

$$b = \frac{c(\sin B)}{\sin C} = \frac{(5.0)(\sin 60°)}{\sin 80°}$$

$$b = \frac{(5.0)(0.866)}{(0.985)} = 4.4 \text{ cm}$$

Proof of the Law of Sines

The following proof of the law of sines applies to acute triangle ABC in Figure 9–51. The proof for an obtuse or right triangle is similar. In triangle ABC,

$$\frac{h}{b} = \sin A \Rightarrow h = b(\sin A)$$

$$\frac{h}{a} = \sin B \Rightarrow h = a(\sin B)$$

Equate the two expressions for h:

$$a(\sin B) = b(\sin A)$$

Divide by $(\sin A)(\sin B)$ to obtain

$$\frac{a \, \cancel{(\sin B)}}{\sin A \, \cancel{(\sin B)}} = \frac{b \, \cancel{(\sin A)}}{\sin B \, \cancel{(\sin A)}}$$

Similarly, it can be shown that

$$\frac{b}{\sin B} = \frac{c}{\sin C}$$

Therefore, any two of the three ratios in (9.9) are equal to each other.

Sometimes the law of sines provides two possible solutions, but only one of them applies.

Study the next example, which closes the circuit and illustrates this situation in an electrical problem with vectors.

EXAMPLE 9.27

Close the Circuit

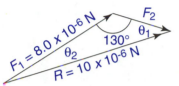

FIGURE 9–52 Electrical forces for Example 9.27.

Two electrical forces F_1 and F_2 act on a particle in an electrical field at an angle of 130° when added head to tail. If $F_1 = 8.0 \times 10^{-6}$ N and the resultant R is measured to be 10×10^{-6} N, find the magnitude of F_2.

Solution   Figure 9–52 shows the vector diagram with F_2 added to F_1. You know two sides of the triangle and the angle opposite one of them. It is necessary to find the angle θ_1 between R and F_2 first, using the law of sines. Then you can find θ_2 and F_2. Apply (9.9) to sides F_1 and R:

$$\frac{F_1}{\sin \theta_1} = \frac{R}{\sin 130°}$$

Invert both fractions in the proportion and solve for $\sin \theta$:

$$\frac{\sin \theta}{F_1} = \frac{\sin 130°}{R}$$

$$\sin \theta_1 = \frac{F_1 (\sin 130°)}{R} = \frac{8.0 \times 10^{-6}(0.7660)}{10 \times 10^{-6}} = 0.6128$$

At this point, you need to find the inverse sine of 0.6128. However, there are two possible angles that could apply—one in the first quadrant and one in the second quadrant. The first quadrant angle is

$$\theta_1 = \sin^{-1} (0.6128) = 37.8°$$

The second quadrant angle is

$$\theta_1 = 180° - 37.8° = 142.2°$$

However, 142.2° cannot be a solution to the problem because the triangle of forces can only have one obtuse angle. Therefore,

$$\theta_1 = 37.8° \text{ and } \theta_2 = 180° - 130° - 37.8° = 12.2°.$$

Now find F_2 by using the law of sines between F_2 and R:

$$F_2 = \frac{R(\sin 12.2°)}{\sin 130°} = \frac{10 \times 10^{-6}(0.211)}{0.766} = 2.8 \times 10^{-6} \text{ N (two significant digits)}$$

Law of Cosines

Given any triangle ABC, the *law of cosines* states

$$c^2 = a^2 + b^2 - 2ab(\cos C) \qquad (9.10)$$

The law of cosines is a generalization of the Pythagorean theorem. When $\angle C$ = 90°, cos 90° = 0, and (9.10) becomes $c^2 = a^2 + b^2$.

The law of cosines can be used to find the parts of a triangle when the law of sines does not apply. This is represented by either of the following two cases:

1. Two sides and the included angle are known.
2. Three sides are known.

EXAMPLE 9.28

Given triangle ABC in Figure 9–53 with $\angle C = 55°$, a = 12 mm, and b = 15 mm, find the three missing parts of triangle ABC.

Solution Since you are given two sides and the included angle, use the law of cosines first to find side c:

$$c^2 = a^2 + b^2 - 2ab(\cos C)$$

$$c^2 = (12)^2 + (15)^2 - 2(12)(15)(\cos 55°)$$

$$c = 144 + 225 - 360(0.5736) = 162.5$$

Then

$$c = \sqrt{162.5} = 12.75 \text{ mm}$$

Now you can use the law of sines to find A (or B):

$$\sin A = \frac{a(\sin C)}{c} = \frac{(12)(\sin 55°)}{12.75} = 0.7711$$

Then

$$\angle A = \sin^{-1}(0.711) = 50°$$

$$\angle B = 180° - 55° - 50° = 75°$$

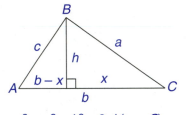

$$c^2 = a^2 + b^2 - 2ab(\cos C)$$

FIGURE 9–53 Law of cosines for Example 9.28.

Proof of the Law of Cosines

The following proof of the law of cosines applies to triangle ABC in Figure 9–53. The proof for an obtuse triangle is similar. Apply the Pythagorean theorem to each of the right triangles in triangle ABC:

$$c^2 = (b - x)^2 + h^2$$

$$a^2 = x^2 + h^2$$

Subtract the two equations canceling out h^2:

$$c^2 - a^2 = (b - x)^2 - x^2$$

Expand $(b - x)^2$ canceling out x^2 and solve for c^2:

$$c^2 - a^2 = b^2 - 2bx + x^2 - x^2$$

$$c^2 = a^2 + b^2 - 2bx$$

Now

$$\frac{x}{a} = \cos C \implies x = a(\cos C)$$

Substitute $a(\cos C)$ for x to produce the law of cosines:

$$c^2 = a^2 + b^2 - 2ab(\cos C)$$

By switching the letters in the triangle in Figure 9–53, the law of cosines can be written two other ways:

$$b^2 = a^2 + c^2 - 2ac(\cos\ B)$$

$$a^2 = b^2 + c^2 - 2bc(\cos\ A)$$

However, it is only necessary to memorize (9.10). By switching the letters or labeling the triangle accordingly, the angle in the formula can be made to correspond to the given angle, or the angle to be found, as the next example shows.

EXAMPLE 9.29

Given triangle ABC with $\angle B = 20°$, $a = 0.45$ in., and $c = 0.68$ in., find the three missing parts of the triangle.

Solution   In (9.9) switch b with c, and B with C, to obtain

$$b^2 = a^2 + c^2 - 2ac(\cos\ B)$$

Then

$$b^2 = (0.45)^2 + (0.68)^2 - 2(0.45)(0.68)(\cos\ 20°)$$

$$b^2 = 0.2025 + 0.4624 - 0.5751 = 0.0898$$

$$b = \sqrt{0.0898} = 0.300 \quad (\text{three significant digits})$$

Use the law of sines to find $\angle C$:

$$\sin\ C = \frac{c(\sin\ B)}{b} = \frac{(0.68)(0.342)}{0.300} = 0.775$$

At this point, a word of caution is necessary: the reference angle = $\sin^{-1} 0.775$ and $\angle C$ can be 51° or 129°. If you choose $\angle C = 51°$, then $\angle A = 180° - 51° - 20° = 109°$. This is not possible because $\angle C$ *must be the largest angle since it is opposite the largest side.* Hence the only solution is

$$\angle C = 129°$$

$$\angle A = 180° - 129° - 20° = 31°$$

You can avoid this confusion by always finding the smaller angle first if you have two unknown angles. When you have three unknown angles, such as in the next example, find the largest angle first.

◾

Study the last example, which closes the circuit and shows an application of the law of cosines to a problem in electronics.

EXAMPLE 9.30

Close the Circuit

Figure 9–54 shows the phasor diagram for the currents in a three-phase alternator. If the phase current $I_P = 14$ A, find the line current I_T.

Solution   The phasor diagram is an isosceles triangle with two of the sides equal to I_P. Figure 9–54 shows the exterior angle at the vertex of the equal sides to be 60°. Therefore, the interior angle formed by the equal sides is $180° - 60° = 120°$.

EXAMPLE 9.30 (Cont.)

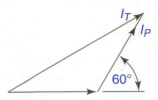

FIGURE 9–54 Three-phase alternator currents for Example 9.30.

Let I_T = side c in the law of cosines (9.10). Then $\angle C = 120°$, $a = b = I_P = 14$ A, and you have

$$c^2 = a^2 + b^2 - 2ab(\cos C)$$

$$I_T{}^2 = (14)^2 + (14)^2 - 2(14)(14) \cos (120°)$$

$$I_T{}^2 = 196 + 196 - 392(-0.5) = 588$$

$$I_T = \sqrt{588} = 24 \text{ A}$$

■

EXERCISE 9.6

In problems 1 through 16, find the missing parts of each triangle. Round answers to two significant digits.

1. $A = 65°$, $B = 75°$, $b = 8$
2. $A = 100°$, $C = 35°$, $a = 6$
3. $B = 98°$, $C = 40°$, $a = 14$
4. $A = 57°$, $B = 48°$, $c = 2.0$
5. $a = 5.0$, $b = 4.0$, $A = 67°$
6. $b = 860$, $c = 620$, $B = 87°$
7. $a = 0.46$, $c = 0.57$, $C = 105°$
8. $a = 11$, $b = 16$, $B = 151°$

9. $C = 70°$, $a = 5$, $b = 7$
10. $C = 55°$, $a = 12$, $b = 18$
11. $B = 145°$, $a = 0.13$, $c = 0.31$
12. $A = 110°$, $b = 180$, $c = 230$
13. $a = 3.1$, $b = 4.2$, $c = 4.9$
14. $a = 39$, $b = 29$, $c = 40$
15. $a = 10$, $b = 11$, $c = 19$
16. $a = 420$, $b = 220$, $c = 270$

In problems 17 through 22, solve each applied problem. Round answers to two significant digits.

17. Two power lines are to be strung across a river from tower A to towers B and C, which are 840 ft apart, on the other side of the river. A surveyor measures angle ABC to be 51° and angle BCA to be 96°. What are the distances AB and AC?

18. Two fire lookout towers A and B are 17 km apart. Tower A spots a fire at C and measures angle CAB to be 45°, whereas tower B measures angle ABC to be 38°. Which tower is closer to the fire, and by how much?

19. Each of the sides of the Pentagon in Virginia is 280 m long. How far is it in meters from one corner of the building to a nonadjacent corner? (Note: The interior angle of a regular pentagon is 108°.)

20. Ed hikes into the woods in a northeast direction (45° east of north) for 5 mi and then turns due east and hikes 6 mi. How far is he from his starting point?

21. Christine, a pilot, wishes to fly southwest. If her airspeed is 220 km/h and there is an 80 km/h wind from due north, how many degrees south of west should she fly so that her resultant direction will be southwest? Find θ in Figure 9–55.

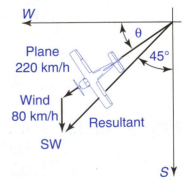

FIGURE 9–55 Flight direction for problem 21.

22. A tugboat is towing a barge as shown in Figure 9–56. The tension in each cable is 1000 lb. What is the resultant force exerted by the tug?

Applications to Electronics   In problems 23 through 28, solve each applied problem. Round answers to two significant digits.

23. Two electrical forces F_1 and F_2 act on a particle in an electrical field at an angle θ as shown in Figure 9–57. If $F_1 = 9.0 \times 10^{-6}$ N, $\theta = 60°$, and $R = 13 \times 10^{-6}$ N, find the magnitude of F_2.

24. In Figure 9–57, find the magnitude of F_1 if $F_2 = 1.5 \times 10^{-6}$ N, $\theta = 100°$, and $R = 2.5 \times 10^{-6}$ N.

25. In Example 9.30, find the line current I_T when the phase current $I_P = 25$ A.

26. In Example 9.30,
   (a) Find the phase current I_P when the line current $I_T = 40$ A.
   (b) Find a formula for I_T in terms of I_P.

27. Two phasors when added together produce the parallelogram shown in Figure 9–58, where the resultant is the diagonal of the parallelogram. If the magnitude of $Z_1 = 7.5$ kΩ, the magnitude of $Z_2 = 11$ kΩ, and $\alpha = 50°$, use the law of cosines to find the magnitude of Z_T and the angle θ.

28. In Figure 9–58, find Z_T and θ when the magnitude of $Z_1 = 2.5$ kΩ, the magnitude of $Z_2 = 4.5$ kΩ, and $\alpha = 45°$.

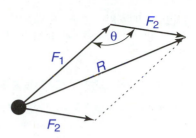

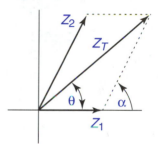

FIGURE 9–57   Electrical forces for problems 23 and 24.

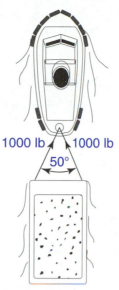

FIGURE 9–56 Tug pulling barge for problem 22.

FIGURE 9–58 Addition of phasors for problems 27 and 28.

≡ CHAPTER HIGHLIGHTS

9.1 ANGLES AND TRIANGLES

The basic relationship between radians and degrees is given by

$$\pi \text{ rad } = 180° \qquad (9.1)$$

This leads to the radian equivalents:

$$\frac{\pi}{2} \text{ rad} = 90°$$

$$\frac{\pi}{4} \text{ rad} = 45°$$

$$\frac{\pi}{3} \text{ rad} = 60°$$

$$\frac{\pi}{6} \text{ rad} = 30°$$

$$\frac{3\pi}{2} \text{ rad} = 270°$$

To change from radians to degrees, multiply by $\frac{180°}{\pi}$ or 57.3°. To change from degrees to radians, multiply by $\frac{\pi}{180°}$ or divide by 57.3°.

The sum of the angles of any triangle add up to 180°.

For a right triangle with hypotenuse c and sides a and b, the Pythagorean theorem states:

$$a^2 + b^2 = c^2$$

Similar triangles have corresponding angles equal and corresponding sides in proportion. Study examples 9.6 and 9.7.

9.2 TRIGONOMETRY OF THE RIGHT TRIANGLE

For an acute angle A in a right triangle with hypotenuse c and sides a and b, the three basic trigonometric functions are

$$\sin A = \frac{\text{Opposite side}}{\text{Hypotenuse}} = \frac{a}{c}$$

$$\cos A = \frac{\text{Adjacent side}}{\text{Hypotenuse}} = \frac{b}{c}$$

$$\tan A = \frac{\text{Opposite side}}{\text{Adjacent side}} = \frac{a}{c} \qquad (9.4)$$

Study Examples 9.10 and 9.13 to understand how to use and apply the basic trigonometric functions.

9.3 TRIGONOMETRY OF THE CIRCLE

The three trigonometric functions for an angle θ whose terminal side passes through a point (x,y) on a circle of radius r are

$$\sin \theta = \frac{y}{r}$$

$$\cos \theta = \frac{x}{r}$$

$$\tan \theta = \frac{y}{x} \qquad (9.5)$$

The functions are positive as follows: All in quadrant I, sin in II, tan in III, cos in IV. Remember "All Star Trig Class." See Figure 9–32.

Reference triangles are always drawn to the x axis with the reference angle between the terminal side of the angle and the x axis. See Figure 9–35. The absolute or positive value of a trigonometric function is equal to the function of its reference angle.

9.4 INVERSE FUNCTIONS

The range of values for the inverse trigonometric functions are

$$-90° \le \sin^{-1} x \le 90°$$

$$0° \le \cos^{-1} x \le 180°$$

$$-90° < \tan^{-1} x < 90° \qquad (9.7)$$

The inverse functions always give you the reference angle when you enter the absolute or positive value of x in the calculator.

9.5 VECTORS AND PHASORS

A vector or phasor is a quantity that has both magnitude and direction.

Add two vectors or phasors by placing the tail of one on the head of the other without changing its direction. See Figure 9–43.

Study Examples 9.23 and 9.24, which show how to add phasors in an ac circuit. The components of a vector or phasor V are

$$V_x = V \cos \theta \qquad (x \text{ component})$$

$$V_y = V \sin \theta \qquad (y \text{ component}) \qquad (9.8)$$

Phasors or vectors that are not perpendicular are added by combining their components as shown in Example 9.25.

9.6 LAW OF SINES AND LAW OF COSINES

For any triangle ABC, the law of sines states

$$\frac{a}{\sin A} = \frac{b}{\sin B} = \frac{c}{\sin C} \qquad (9.9)$$

Use the law of sines to find the parts of any triangle when

1. Two angles and a side opposite one of the angles are known.

2. Two sides and an angle opposite one of the sides are known.

For any triangle ABC, the law of cosines states

$$c^2 = a^2 + b^2 - 2ab(\cos C) \qquad (9.10)$$

Use the law of cosines when you cannot use the law of sines. That is, when

1. Two sides and the included angle are known.

2. Three sides are known.

≡ REVIEW QUESTIONS

In problems 1 through 4, change degrees to radians or vice versa.

1. 30°

2. 225°

3. $\frac{\pi}{3}$ rad

4. 1.5 rad

In problems 5 and 6, find all the lettered angles.

5. Figure 9–59

6. Figure 9–60

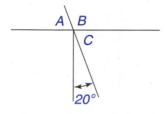

FIGURE 9–59

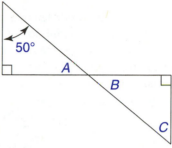

FIGURE 9–60

In problems 7 through 10, find sin A, cos A, and tan A for each right triangle.

7. $a = 8$, $b = 6$, $c = 10$

8. $a = 16$, $b = 30$, $c = 34$

9. $a = 10$, $b = 24$

10. $a = 2.1$, $c = 7.5$

In problems 11 through 14, find the three trigonometric functions of angle θ.

11. The terminal side of ∡ θ passes through (−5,12)

12. The terminal side of ∡ θ passes through (−0.80, −0.60)

13. sin θ = −0.30, cos θ positive

14. tan θ = 2.0, θ in quadrant III

In problems 15 through 18, find the angle in degrees and radians for 0° ≤ θ < 360°.

15. $\sin \theta = \frac{1}{2}$

16. $\tan \phi = -0.966$

17. $\tan x + 3.0 = 0$

18. $3 \cos y = 1$

In problems 19 through 22, find the resultant of the pair of vectors or phasors.

19. $F_X = 16$ N, $\theta_x = 0°$
   $F_Y = 12$ N, $\theta_Y = 90°$

20. $V_1 = 3.2$ m/s, $\theta_1 = 180°$
   $V_2 = 1.1$ m/s, $\theta_2 = 270°$

21. $R = 10 \ \angle 0° \ k\Omega$
   $X_C = 7.5 \ \angle{-90°} \ k\Omega$

22. $I_C = 750 \ \angle 90° \ mA$
   $I_L = 900 \ \angle{-90°} \ mA$

In problems 23 through 26, find the missing parts of each triangle.

23. ∡ C = 90°, $a = 3$, $b = 5$

24. ∡ A = 110°, $a = 18$, $b = 12$

25. ∡ C = 60°, $a = 10$, $b = 15$

26. ∡ A = 25°, $b = 40$, $c = 60$

In problems 27 through 32, solve each applied problem. Round answers to two significant digits.

27. A boat travels 20 mi south and 30 mi east. How far is the boat from its starting point?

28. A sketch of a rectangular computer cabinet is drawn to the scale of 2:5. If the dimensions in the sketch are 6.0 in by 8.0 in. by 2.0 in., what are the actual dimensions of the cabinet?

29. The angle of elevation of the top of a wind generator is 75° at a point 20 m from the base of the tower. How high is the generator?

30. A rectangular circuit chip measures 2.3 mm by 4.8 mm. What angle does the diagonal make with the longer side?

31. A body is acted on by a horizontal force of 8.8 lb acting to the left and a vertical force of 5.1 lb acting down. Find the magnitude and direction of the resultant force.

32. David measures the three sides of his triangular garden to be 9.0 m, 10 m, and 13 m. He believes it to be a right triangle. Is he correct? Find the three angles of the triangle.

Applications to Electronics In problems 33 through 40, solve each applied problem. Round answers to two significant digits.

33. Figure 9–61 shows the impedance triangle for an RL series circuit. If the magnitude of $R = 8.2$ kΩ and $X_L = 2.4$ kΩ, find the magnitude and phase angle θ of the impedance Z.

34. In computing the rotational velocity of an electrical generator the following expression arises:

$$\omega = (10^3 \sin \theta)\left(1 + \frac{\cos \theta}{3}\right)$$

Find ω in rads/s when θ is

(a) $\dfrac{3\pi}{4}$ rad

(b) 4.19 rad.

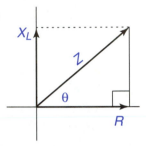

FIGURE 9– 61
RL series circuit
impedance triangle
for problem 33.

35. The angle α in degrees of a laser beam is expressed by

$$\alpha = 2\tan^{-1}\left(\frac{w}{2d}\right)$$

where w = width of the beam and d = distance from the source. Find α when $w = 1.0$ m and $d = 1000$ m.

36. In an ac circuit, the current at any time t is given by

$$i = I_M \cos \omega t$$

where I_M = maximum current, ω = angular velocity in rad/s, and t = time in seconds. Find i when $I_M = 4.0$ A, $\omega = 120\pi$ rad/s, and t is

(a) 25 ms.   (b) 22 ms.

37. In problem 36, find the angle ωt in radians for $0 \le \omega t < 2\pi$ and the time t when $i = 3.0$ A.

38. For each set of phasors for an ac series circuit find the resultant of the two phasors:

(a) *RL* series circuit: $R = 16$ $\angle 0°$ kΩ,
   $X_L = 20$ $\angle 90°$ kΩ.

(b) *RC* series circuit: $V_R = 5$ $\angle 0°$ V,
   $V_C = 5$ $\angle -90°$ V.

39. Given the two impedance phasors $Z_1 = 22$ $\angle 20°$ kΩ and $Z_2 = 13$ $\angle -30°$ kΩ, find the resultant phasor.

40. Figure 9–62 shows two phasors added together with an angle of 25° between them. If the magnitude of $Z_1 = 1.0$ kΩ and $Z_2 = 2.0$ kΩ, find the magnitude of the resultant Z and the angle θ it makes with Z_1. Find θ to the nearest degree.

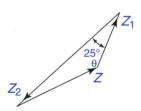

FIGURE 9– 62
Impedance phasors
for problem 40.

Alternating Current

Courtesy of DeVry Institutes.

Students in an AC circuits laboratory.

This chapter, like Chapter 5 on dc circuits, closes the circuit and shows the basic connection between the sine and cosine functions and the theory of alternating current. The sine and cosine functions are periodic functions; that is, they repeat their values after a certain number of degrees known as their period or cycle. They are essentially the same function, and their graphs are called sinusoidal curves. Alternating current and voltage are also periodic in nature. The graphs of alternating current and voltage can be represented as sinusoidal curves or waves. This chapter studies sinusoidal waves and the fundamentals of ac electricity, including peak values, effective or root mean square values, frequency, period, and phase angles.

Close the Circuit

Chapter Objectives

In this chapter, you will learn:

- How to graph a sine and cosine curve.
- How to graph alternating-current and voltage waves and find instantaneous values.
- How to find peak and effective, or root mean square, values of voltage and current.
- How to find the period and frequency of an ac wave.
- How to determine the phase angle of an ac wave.

≡ 10.1
SINUSOIDAL CURVES AND ALTERNATING CURRENT

Sine Curve

The graphs of the sine and cosine functions are the same except for their position. They are generated by the movement of a point around a circle. They are also generated by the alternating current electricity produced when a coil of wire rotates in a magnetic field under certain conditions. The first example illustrates the basic sine curve using angles in radians and degrees. However, radians, which were introduced in Section 9.1, are used instead of degrees in most applications. Study the example well as it is the basis for many of the ideas in this chapter.

EXAMPLE 10.1

Graph $y = \sin \theta$ from $\theta = 0$ to $\theta = 2\pi$ radians (360°).

Solution   To graph the basic sine function, you need to construct a table of values of x and $y = \sin \theta$ using a convenient interval. Here, $\frac{\pi}{6}$ rad or 30° is used as the interval for the table that follows because it is a common angle and provides enough points to sketch the graph. Remember π rad = 180°. Using a calculator,

EXAMPLE 10.1 (Cont.)

find the values of sin θ for each angle. For example, to calculate the $\sin \frac{7\pi}{6}$ (210°), first set your calculator for radians using the ⎡DRG⎤ key or a similar key as follows:

$$\boxed{\text{DRG}} \; 7 \; \boxed{\times} \; \boxed{\pi} \; \boxed{\div} \; 6 \; \boxed{=} \; \boxed{\sin} \; \rightarrow \; -0.50$$

In the following table, angles are shown in both degrees and radians to help you become familiar with the radian values.

Degrees	0	30	60	90	120	150	180	210
Radians	0	$\frac{\pi}{6}$	$\frac{\pi}{3}$	$\frac{\pi}{2}$	$\frac{2\pi}{3}$	$\frac{5\pi}{6}$	π	$\frac{7\pi}{6}$
$y = \sin\theta$	0	0.50	0.87	1	0.87	0.50	0	-0.50

Degrees	240	270	300	330	360
Radians	$\frac{4\pi}{3}$	$\frac{3\pi}{2}$	$\frac{5\pi}{3}$	$\frac{11\pi}{6}$	2p
$y = \sin\theta$	-0.87	-1	-0.87	-0.50	0

The graph of the sine function is shown in Figure 10–1. It illustrates how the sine curve is generated by motion in a circle. If the radius of the circle is equal to 1, then $\sin\theta = \frac{y}{1}$ or y = sin θ. Therefore, the height of the point, which is the end of the radius, is equal to the height of the sine curve. By projecting these heights to the graph on the right as the radius revolves in a circle, you can see how the movement of the point around the circle produces the sine curve.

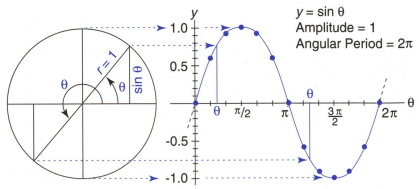

FIGURE 10–1   Sine curve showing projections of points on the unit circle for Example 10.1.

Observe that the sine curve begins at zero, reaches a maximum of 1 at $\frac{\pi}{2}$, and returns to zero at π. The maximum value is called the *amplitude* of the curve. The sine curve then becomes negative and reaches a minimum of –1 at $\frac{3\pi}{2}$ and returns

EXAMPLE 10.1 (Cont.)

to zero at 2π. The curve then starts the same cycle after 2π and repeats itself. The sine curve goes through one complete cycle every 2π radians. The number of radians in one cycle is called the *angular period* of the curve. The amplitude of the sine curve is then 1, and its angular period is 2π.

▪

Cosine Curve

The cosine curve is essentially the same as the sine curve except for its position. Instead of beginning at zero, it begins at the maximum value 1 and ends at 1. Study the next example, which illustrates the cosine curve.

EXAMPLE 10.2

Graph $y = \cos \theta$ from $\theta = 0$ to $\theta = 2\pi$ rad.

Solution   The table of values and the graph (Figure 10–2) are shown here.

Degrees	0	30	60	90	120	150	180	210
Radians	0	$\dfrac{\pi}{6}$	$\dfrac{\pi}{3}$	$\dfrac{\pi}{2}$	$\dfrac{2\pi}{3}$	$\dfrac{5\pi}{6}$	π	$\dfrac{7\pi}{6}$
$y = \cos \theta$	1	0.87	0.50	0	-0.50	-0.87	-1	-0.87

Degrees	240	270	300	330	360
Radians	$\dfrac{4\pi}{3}$	$\dfrac{3\pi}{2}$	$\dfrac{5\pi}{3}$	$\dfrac{11\pi}{6}$	2p
$y = \cos \theta$	-0.50	0	0.50	0.87	1

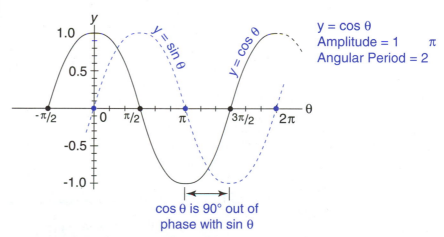

$y = \cos \theta$
Amplitude = 1
Angular Period = 2π

cos θ is 90° out of phase with sin θ

FIGURE 10–2   Cosine curve with sine curve compared for Example 10.2.

EXAMPLE 10.2 (Cont.)

Figure 10–2 also shows the sine curve next to the cosine curve, which is extended to $\frac{-\pi}{2}$ for comparison. Both curves have an amplitude equal to 1 and an angular period equal to 2π. The difference is *cos θ is 90° out of phase with sin θ*.

If you move sin θ to the left 90° (or cos θ to the right 90°), both curves will match exactly. Hence, sine and cosine curves are both called sinusoidal curves.

The amplitude of a sinusoidal curve can be changed by a coefficient in front of the function as follows:

$$\text{Given } y = a \sin θ \text{ or } y = a \cos θ, \text{ the amplitude} = |a| \quad (10.1)$$

For example, the graph of $y = 3 \sin θ$ looks the same as $y = \sin θ$ except that it reaches a maximum of 3 and a minimum of –3. See Figure 10–3. If a is negative such as $y = -10 \cos θ$, the amplitude is still positive: $|-10| = 10$.

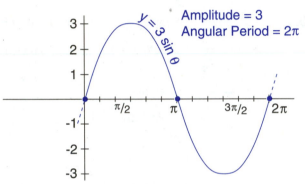

FIGURE 10–3 Graph of y = 3 sin θ.

Graphs of the sine and cosine curves can be done on graphing calculators such as Texas Instruments' calculators TI-81, TI-82, or TI-85. For example, to draw the graph of $y = 3 \sin θ$ on the TI-82 or TI-85, first press [MODE] and the down arrow to set the calculator for radians. Then one set of keystrokes for the TI-82 is

[ZOOM] 7 [Y=] [CLEAR] [SIN] [X, T, Θ] [GRAPH]

where [ZOOM] 7 resets the scale on the x axis from $x = -2\pi$ to $x = 2\pi$ and the scale on the y axis from –3 to 3. One set of keystrokes for the TI-85 is

[GRAPH] [F3] [MORE] [F3] [2nd] [F1] [CLEAR] [3] [×] [SIN] [x-VAR] [2nd] [F5]

where [F3] [MORE] [F3] sets the scale on the x axis from $x = -8.25$ rad to $x = 8.25$ rad and the scale on the y axis from -4 to $+4$. Appendix A shows some basic graphing functions on the TI-82.

Alternating Current

Alternating current behaves in the same way as a sinusoidal curve. It is produced in a coil of wire that rotates in a circle in a magnetic field. See Figure 10–4. When the coil is in position (1), there is no induced voltage and no current flow in the wire.

As the coil turns, there is an induced voltage that increases, and the current begins to flow in one direction. The voltage and current reach a maximum in position (2) when the coil has turned $\frac{\pi}{2}$ radians (90°). As the coil turns more, the induced voltage decreases, and the current decreases. The voltage and the current reach zero at position (3) after π radians (180°). As the coil turns more, there is an induced voltage of opposite polarity, and the current begins to move in the other direction. The voltage and current reach a maximum in the other direction at position (4), which is $\frac{3\pi}{2}$ radians (270°). As the coil continues to turn the induced voltage again decreases, and the current decreases and returns to zero at 2π radians (360°), which is back to position (1). The cycle then repeats itself.

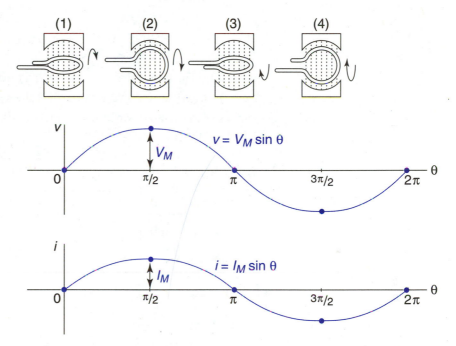

FIGURE 10–4 Induced voltage and current in a coil of wire.

The graph of the induced voltage and the current are both sinusoidal waves expressed by the following equations:

$$v = V_M \sin \theta$$

$$i = I_M \sin \theta$$

where v and i represent instantaneous values of the voltage and current. The maximum value is called the *peak value* and is equal to V_M for the voltage and I_M for the current. Also, V_P and I_P are used to represent peak values. The number of radians in one cycle is 2π.

Study the next example, which closes the circuit and shows how to graph an ac wave.

EXAMPLE 10.3

Close the Circuit

Graph the sine wave voltage $v = 170 \sin \theta$.

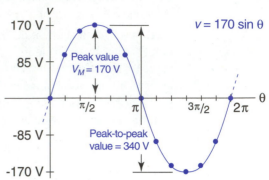

FIGURE 10–5   Sine wave voltage for Example 10.3.

Solution   The peak value $V_M = 170$ V, and the number of radians in one cycle is 2π. Construct the following table of values from 0 to 2π for angles every $\frac{\pi}{6}$ radians. For example, when $\theta = \frac{4\pi}{3}$ rad (240°), the value of v is found on the calculator using radians as follows:

$$4 \;\boxed{\times}\; \boxed{\pi} \;\boxed{\div}\; 3 \;\boxed{=}\; \boxed{\sin} \;\boxed{\times}\; 170 \rightarrow -147 \text{ V}$$

The table of values is then

Degrees	0	30	60	90	120	150	180	210
Radians	0	$\frac{\pi}{6}$	$\frac{\pi}{3}$	$\frac{\pi}{2}$	$\frac{2\pi}{3}$	$\frac{5\pi}{6}$	π	$\frac{7\pi}{6}$
$v = 120 \sin \theta$	0	85	147	170	147	85	0	-85

Degrees	240	270	300	330	360
Radians	$\frac{4\pi}{3}$	$\frac{3\pi}{2}$	$\frac{5\pi}{3}$	$\frac{11\pi}{6}$	$2p$
$v = 120 \sin \theta$	-147	-170	-147	-85	0

Carefully set up a scale, and plot the points to show the sine wave voltage (Figure 10–5). The peak-to-peak value, which is the difference between the maximum and minimum values, is equal to

$$\text{Peak-to-peak value} = 170 \text{ V} - (-170 \text{ V}) = 340 \text{ V}$$

This is the same as twice the peak value:

$$\text{Peak-to-peak value} = 2V_M = 2(170 \text{ V}) = 340 \text{ V}$$

To do example 10.3 on the TI-82 graphing calculator, press $\boxed{\text{WINDOW}}$ to enter the range of values for x and y. Set x from 0 to 2π (6.28) and y from -170 to 170. (See Appendix A.) The press the following:

$$\boxed{\text{Y=}} \;\boxed{\text{clear}}\; 170 \;\boxed{\times}\; \boxed{\text{SIN}} \;\boxed{\text{X, T, }\Theta}\; \boxed{\text{graph}}$$

In an ac circuit containing only a resistance, the sine wave current is *in phase* with the sine wave voltage, as shown in Figure 10–4. Both waves start at 0 and end at 2π. The sine wave current is graphed in the same way as the sine wave voltage. There are many other technical applications of sinusoidal curves such as sound waves, light waves, and radio waves, and the mechanical vibration of a wire, spring, and beam. You may well encounter some of these applications in other courses.

EXERCISE 10.1

In problems 1 through 6, graph each sinusoidal curve from $\theta = 0$ to $\theta = 2\pi$. Give the amplitude and the angular period of the curve.

1. $y = 50 \sin \theta$

2. $y = 100 \sin \theta$

3. $y = 20 \cos \theta$

4. $y = 10 \cos \theta$

5. $y = 2.5 \sin \theta$

6. $y = 1.5 \cos \theta$

Applications to Electronics In problems 7 through 14, graph each ac voltage or current wave from $\theta = 0$ to $\theta = 2\pi$. Give the peak value, the peak-to-peak value, and the number of radians in one cycle.

7. $v = 100 \sin \theta$

8. $v = 60 \sin \theta$

9. $i = 5 \cos \theta$

10. $i = 3 \cos \theta$

11. $v = 300 \sin \theta$

12. $v = 150 \cos \theta$

13. $i = 1.2 \cos \theta$

14. $i = 0.50 \sin \theta$

In problems 15 through 18 solve each applied problem. Round answers to two significant digits.

15. The current in an ac circuit containing an inductance is given by $i = 2.0 \sin \theta$, while the voltage is given by $v = 110 \cos \theta$. Sketch these two ac waves on the same graph to show that the voltage and the current are 90° out of phase. Use different scales for current and voltage.

16. The voltage in an ac circuit containing a capacitance is given by $v = 200 \sin \theta$, while the current is given by $i = 4.0 \cos \theta$. Sketch these two ac waves on the same graph to show that the voltage and the current are 90° out of phase. Use different scales for voltage and current.

17. The voltage in an ac circuit containing a resistance is given by $v = 100 \sin \omega t$ where ω = angular velocity, t = time in seconds, and ωt = angle in radians. If the angular velocity $\omega = 400$ rad/s,
  (a) Find v when $t = 1.0$ ms.
  (b) Find ωt in radians between 0 and 2π when $v = 50$ V.

18. The current in amps in an ac circuit containing a resistance is given by $i = 1.5 \sin \omega t$ where ω = angular velocity, t = time in seconds, and ωt = angle in radians. If the angular velocity $\omega = 300$ rad/s,
  (a) Find i when $t = 2.0$ ms.
  (b) Find the angle in radians between 0 and 2π when $i = 500$ mA.

10.2
ROOT MEAN SQUARE VALUES

The current and voltage in an ac circuit are continuously changing and assume different values every instant. However, we assign only one value, such as 120 V for ordinary household voltage, or 15 A for the current rating of a household fuse. These values are a type of mathematical "average" and are calculated over one cycle of the ac sine wave, called the *root mean square* or effective values. The root mean square (rms) value of an ac wave represents the value of a dc voltage or current that produces the same electrical energy as the ac voltage or current. For example, an ac sine wave voltage whose peak value is 170 V has an rms value of 120 V. When this ac source is connected to a certain resistance, it will dissipate the same amount of electrical energy as a dc voltage of 120 V connected to the same resistance. The name "root mean square" is used because of the way it is calculated. First, the instantaneous values of the voltage or current are squared. Second, the average or mean of these squared values is calculated. Third, the square root of this mean is computed. When this is done precisely, using the methods of calculus on a sine wave, it yields the following formulas:

$$V_{rms} = \frac{V_M}{\sqrt{2}} = 0.7071(V_M)$$

$$I_{rms} = \frac{I_M}{\sqrt{2}} = 0.7071(I_M)$$

(10.2)

For example, when $V_M = 170$ V: $V_{rms} = 0.7071(170$ V$) = 120$ V, and when $I_M = 4.0$ A: $I_{rms} = 0.7071(4.0$ A$) = 2.8$ A.

An approximate calculation for the rms value of the ac voltage v = 170 sin θ is shown in the following table for intervals of θ every $\frac{\pi}{12}$ rad (15°) from $\theta = \frac{\pi}{12}$ to θ = π rad.

θ	sin θ	170 sin θ	(170 sin θ)²
$\frac{\pi}{12}$	0.2588	44.0	1,936
$\frac{\pi}{6}$	0.5000	85.0	7,225
$\frac{\pi}{4}$	0.7071	120.2	14,450
$\frac{\pi}{3}$	0.8660	147.2	21,675
$\frac{5\pi}{12}$	0.9659	164.2	26,964
$\frac{\pi}{2}$	1.000	170.0	28,900
$\frac{7\pi}{12}$	0.9659	164.2	26,964
$\frac{2\pi}{3}$	0.8660	147.2	21,675
$\frac{3\pi}{4}$	0.7071	120.2	14,450
$\frac{5\pi}{6}$	0.5000	85.2	7,225
$\frac{11\pi}{12}$	0.2588	44.0	1,936
π	0.0000	0.0	0
		TOTAL	173,400

The last column gives the squares of the instantaneous voltages, and its sum is 173,400. The mean square or average of this column is

$$\text{Mean square} = \frac{173{,}400}{12} = 14{,}450$$

The root mean square of the voltage is then approximately:

$$V_{rms} \approx \sqrt{14{,}450} = 120.2 \text{ V}$$

This value agrees to four significant digits with that obtained using formula (10.2):

$$V_{rms} = 0.7071(170 \text{ V}) = 120.2 \text{ V}$$

To find the peak values from the rms values, use the reverse of formulas (10.2):

$$V_M = \sqrt{2}(V_{rms}) = 1.414(V_{rms})$$

$$I_M = \sqrt{2}(I_{rms}) = 1.414(I_{rms}) \tag{10.3}$$

For example, when $V_{rms} = 50$ V: $V_M = 1.414(50 \text{ V}) = 71$ V, and when $I_{rms} = 1.5$ A: $I_M = 1.414(1.5 \text{ A}) = 2.1$ A.

Whenever voltage and current values are given for an ac circuit, they are always rms or effective values unless stated otherwise. Since the rms or effective value of an ac source provides the same power to a resistance as a dc source of that value, Ohm's law and the power law can be applied to rms values in an ac circuit containing only resistances:

$$I_{rms} = \frac{V_{rms}}{R}$$

$$P = (V_{rms})(I_{rms})$$

Study the next example, which applies these ideas.

| EXAMPLE 10.4 | Given an ac circuit with a resistance $R = 100\ \Omega$ and source voltage $V = 240$ V, find I_{rms}, I_M, and the power P dissipated in the resistance. |

Solution Since it is not indicated otherwise, the ac voltage given is the effective or rms value: $V_{rms} = 240$ V. To find I_{rms}, apply Ohm's law:

$$I_{rms} = \frac{V_{rms}}{R} = \frac{240 \text{ V}}{100\ \Omega} = 2.4 \text{ A}$$

Apply (10.3) to find I_M:

$$I_M = 1.414(2.4 \text{ A}) = 3.4 \text{ A}$$

To find the power P, apply the power law using the effective values:

$$P = V_{rms}I_{rms} = (240 \text{ V})(2.4 \text{ A}) = 576 \text{ W}$$

EXERCISE 10.2

In all problems, round answers to two significant digits.

In problems 1 through 6, find the effective value of the ac voltage or current.

1. $V_M = 340$ V

2. $V_M = 160$ V

3. $I_M = 2.2$ A

4. $I_M = 150$ mA

5. $V_M = 56$ V

6. $I_M = 5.2$ mA

In problems 7 through 12, find the peak value of the ac voltage or current.

7. $V = 110$ V

8. $V = 230$ V

9. $I = 1.2$ A

10. $I = 600$ mA

11. $I = 25$ mA

12. $V = 75$ V

In problems 13 through 18 solve each applied problem.

13. An ac circuit contains a resistance $R = 50$ Ω connected to a voltage $V = 110$ V. Find I_{rms}, I_M, and the power P dissipated in the resistance.

14. An ac circuit contains a resistance $R = 5.1$ kΩ connected to a voltage source.
If $I = 22$ mA, find V_{rms}, V_M, and the power P dissipated in the resistance.

15. Given the ac series circuit in Figure 10–6 with $R_1 = 20$ Ω, $R_2 = 30$ Ω, $R_3 = 10$ Ω, and the current $I = 55$ mA, find the source voltage V and P_1.

16. Given the ac series circuit in Figure 10–6 with $R_1 = 750$ Ω, $R_2 = 1.2$ kΩ, $R_3 = 1.6$ kΩ, and the source voltage $V = 100$ V, find I and P_1.

17. Given the ac parallel circuit in Figure 10–7 with $R_1 = 1.2$ kΩ, $R_2 = 1.5$ kΩ, and $V = 80$ V, find I_1 and P_2.

18. Given the ac parallel circuit in Figure 10–7 with $R_1 = 620$ Ω, $R_2 = 510$ Ω, and $I_1 = 450$ mA, find V and I_2.

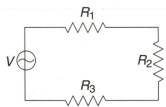

FIGURE 10–6 AC series circuit for problems 15 and 16.

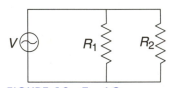

FIGURE 10–7 AC parallel circuit for problems 17 and 18.

☰ 10.3
FREQUENCY AND PERIOD

When alternating current is generated by revolving a coil of wire in a magnetic field, the speed of rotation is an important factor in the sine wave voltage and current. The speed of rotation ω (omega) is called the *angular velocity*. It is the rate of change of the angle θ with respect to time t and is measured in radians per second (rad/s).

$$\omega = \frac{\theta}{t} \text{ or } \theta = \omega t$$

(10.4)

For example, suppose $\omega = 120\pi$ rad/s. Since 2π rad $= 1$ rotation, this means

$$\text{Rotations per second} = \frac{120\,\pi}{2\,\pi} = 60$$

Rotations per second is the same as cycles per second and is called the *frequency f*. The units for f are hertz (Hz). Therefore you have

$$\omega = 120\pi \text{ rad/s} \implies f = \frac{\omega}{2\pi} = \frac{120\,\pi}{2\pi} = 60 \text{ Hz}$$

The relationship between frequency and angular velocity is then:

$$f = \frac{\omega}{2\pi} \text{ or } \omega = 2\pi f \tag{10.5}$$

Combining (10.4) and (10.5) gives you

$$\theta = \omega t = 2\pi f t$$

The equations for the sine wave voltage and current can then be written in terms of angular velocity ω or frequency f:

$$v = V_M \sin \omega t = V_M \sin 2\pi f t$$

$$i = I_M \sin \omega t = I_M \sin 2\pi f t \tag{10.6}$$

The *period* of an ac wave is the *time T* for one cycle and is the reciprocal of the frequency:

$$T = \frac{1}{f} = \frac{2\pi}{\omega} \tag{10.7}$$

For example, if $f = 60$ Hz, then $T = \frac{1}{60}$s $= 17$ ms. This is the frequency and period for ordinary household current in the United States with $V_{rms} = 120$ V. In Europe and elsewhere, voltages of 220 V and frequencies of 50 Hz are used. Higher frequencies and smaller periods are encountered in circuits that transmit and receive radio waves. For example, a radio wave whose frequency $f = 50$ MHz has a period

$$T = \frac{1}{50 \text{ MHz}} = \frac{1}{50 \times 10^6 \text{ Hz}} = 0.020 \,\mu\text{s} \text{ or } 20 \text{ ns}$$

Study the next example, which combines the above ideas.

EXAMPLE 10.5

Given the ac sine wave current $i = 3.0 \sin 500\pi t$, find the angular velocity, frequency, and the period of the wave. Sketch the graph of current versus time for one cycle from $t = 0$.

Solution Applying formula (10.6), the angular velocity is the coefficient of t:

$$\omega = 500\pi$$

Apply formula (10.5) to find the frequency:

$$f = \frac{\omega}{2\pi} = \frac{500\pi}{2\pi} = 250 \text{ Hz}$$

EXAMPLE 10.5 (Cont.)

From (10.7), the period is the reciprocal of the frequency:

$$T = \frac{1}{f} = \frac{1}{250} s = 0.0040\,s = 4.0\,ms$$

Since the period of the wave is given in units of time, the values of the current are plotted against time rather than radians. To sketch the sine wave, you can plot the maximum, minimum, and zero values and then sketch the curve since you know its shape is sinusoidal. These values correspond to $\theta = 0, \frac{\pi}{2}, \pi, \frac{3\pi}{2}$, and 2π radians. For each of these values, calculate t in milliseconds using formula (10.4) solved for t:

$$t = \frac{\theta}{\omega} = \frac{\theta}{500\,\pi}$$

Then

θ (rad)	0	$\dfrac{\pi}{2}$	π	$\dfrac{3\pi}{2}$	2π
t (ms)	0	1.0	2.0	3.0	4.0

You can also divide the period $T = 4.0$ ms into four parts to obtain these points since each represents one quarter of the cycle. The peak value is $I_M = 3.0$ A and the table of values for i is then

t (ms)	0.0	1.0	2.0	3.0	4.0
i (A)	0.0	3.0	0.0	−3.0	0.0

The sketch is shown in Figure 10–8.

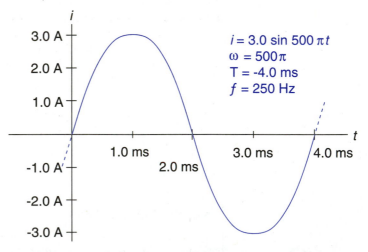

FIGURE 10– 8 Frequency and period of AC sine wave for Example 10.5.

ERROR BOX

A common error to watch out for when working with ac waves is not to confuse frequency, period, and the number of radians in one cycle. For any sinusoidal curve sin θ or cos θ, the number of radians in one cycle is 2π. This was also called the *angular* period in Section 10.1. However, for an ac wave V_M sin 2πft or I_M sin 2πft, the number of radians in one cycle is 2π, but the *period* is measured in units of time and is not equal to 2π. It is equal to the value of t that makes the angle 2πft equal to 2π. This value is $t = \dfrac{1}{f}$. See if you can do the practice problems correctly.

PRACTICE PROBLEMS   For each ac wave, find the number of radians in one cycle, the frequency, and the period.

1. 170 sin 120πt   2. 170 sin 377t   3. 3 sin 100πt   4. 3 sin 314t

Answers: 4. 2π, 50 Hz, 20 ms   3. 2π, 50 Hz, 20 ms

2. 2π, 60 Hz, 17 ms   1. 2π, 60 Hz, 17 ms

The next example shows how to find instantaneous values for an ac wave.

EXAMPLE 10.6

Given the ac sine wave voltage $v = 160 \sin 100\pi t$,

(1) Find the value of v when $t = 13$ ms.

(2) Find the value of t when $V = 50$ V.

Solution

1. To find the value of v, substitute $t = 13$ ms $= 13 \times 10^{-3}$ s into the equation of the curve:

$$v = 160 \sin 100\pi(13 \times 10^{-3}) = 160 \sin (4.08 \text{ rad }) = -130 \text{ V}$$

Note that the angle 4.08 is in radians and is in the third quadrant. This gives a negative value for the voltage. You need to set your calculator for radians to compute this correctly. You can also find v using degrees by letting π = 180°:

$$v = 160 \sin 100(180°)(13 \times 10^{-3}) = 160 \sin 234° = -130 \text{ V}$$

2. To find the value of t, you must use the inverse sine with radian measure. Substitute 50 V for V:

$$50 = 160 \sin 100\pi t$$

Then:

$$\sin 100\pi t = \frac{50}{160} = 0.313$$

The reference angle in radians is

$$\sin^{-1} 0.313 = 0.318 \text{ rad}$$

Since the value of the sin is positive, there are two solutions for the angle in the first and second quadrant:

I: θ = 0.318 rad

II: θ = π − 0.318 = 2.82 rad

EXAMPLE 10.6 (Cont.)

Each value of θ yields a value of t:

$$100\pi t = 0.318 \Rightarrow t = \frac{0.318}{100\pi} = 1.0 \text{ ms}$$

$$100\pi t = 2.82 \Rightarrow t = \frac{2.82}{100\pi} = 9.0 \text{ ms}$$

EXERCISE 10.3

For all problems round answers to two significant digits.

In problems 1 through 6, find the frequency f and the period T for each value of the angular velocity ω.

1. 110π rad/s
2. 220π rad/s
3. 1000π rad/s

4. 600π rad/s
5. 400 rad/s
6. 300 rad/s

In problems 7 through 12, find the period T for each value of the frequency f.

7. 60 Hz
8. 50 Hz
9. 100 kHz

10. 60 kHz
11. 70 MHz
12. 100 MHz

In problems 13 through 16, give the peak value, angular velocity ω, frequency f, and the period T for each ac wave.

13. $i = 4.0 \sin 120\pi t$

14. $i = 1.0 \sin 100\pi t$

15. $v = 310 \sin 300t$

16. $v = 170 \sin 400t$

In problems 17 and 18 find the angular velocity, the frequency, and the period for the given ac wave.

17. The voltage wave in Figure 10–9.

18. The current wave in Figure 10–10.

In problems 19 through 22, write the equation of the ac wave having the given conditions.

19. $V_M = 180$ V, $f = 60$ Hz
20. $I_M = 2.5$ A, $f = 50$ Hz
21. $I_M = 4.0$ A, $T = 20$ ms
22. $V_M = 60$ V, $T = 18$ ms

In problems 23 through 26, sketch each ac wave for one cycle. Give the angular velocity ω, the frequency f, and the period T.

23. $v = 150 \sin 100\pi t$

24. $v = 340 \sin 120\pi t$

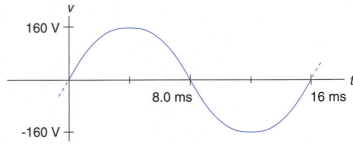

FIGURE 10–9   AC voltage wave for problem 17.

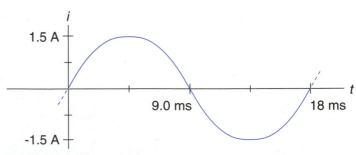

FIGURE 10–10   AC current wave for problem 18.

25. $i = 2.5 \sin 400\pi t$

26. $i = 1.0 \sin 200\pi t$

In problems 27 through 30 solve each applied problem.

27. Given the ac sine wave voltage $v = 160 \sin 120\pi t$,
   (a) Find v when $t = 7.0$ ms.
   (b) Find t when $v = 100$ V.

28. Given the ac sine wave voltage $v = 300 \sin 100\pi t$,
   (a) Find v when $t = 17$ ms.
   (b) Find t when $v = -100$ V.

29. Given the ac sine wave current $i = 2.0 \sin 100\pi t$,
   (a) Find i when $t = 8.0$ ms.
   (b) Find t when $i = -1.0$ A.

30. Given the ac sine wave current $i = 3.0 \sin 120\pi t$,
   (a) Find i when $t = 15$ ms.
   (b) Find t when $i = 2.0$ A.

10.4

PHASE ANGLE

Figure 10–2 in Section 10.1 compares the cosine curve with the sine curve. It shows that $\cos \theta$ is 90° out of phase with $\sin \theta$. We say that $\cos \theta$ *leads* $\sin \theta$ by 90°. This is because $\cos \theta$ is zero at –90° and therefore "starts" 90° before $\sin \theta$. $\cos \theta$ is already at the peak value when $\sin \theta$ is just starting at 0°. You can also say that $\sin \theta$ *lags* $\cos \theta$ by 90°. If you move $\sin \theta$ to the left 90°, it will match $\cos \theta$ exactly. Since $\sin \theta$ *lags* $\cos \theta$, this movement to the left is done by *adding a phase angle* of $\frac{\pi}{2}$ to $\sin \theta$:

$$\sin\left(\theta + \frac{\pi}{2}\right) = \cos \theta$$

When you compute values of $\sin\left(\theta + \frac{\pi}{2}\right)$, you will get the exact same values as $\cos \theta$. For example, when $\theta = \frac{3\pi}{4}$ rad (135°),

$$\sin\left(\frac{3\pi}{4} + \frac{\pi}{2}\right) = \sin\frac{5\pi}{4} = -0.707$$

$$\cos\frac{3\pi}{4} = -0.707$$

The concept of lead or lag for ac sine waves is important in the theory of alternating current. In an ac circuit containing only a resistance, the voltage wave is *in phase* with the current wave. The phase difference between the two waves is 0°. However, in an ac circuit containing only an inductance, the voltage wave *leads* the current wave by 90°. In an ac circuit containing only a capacitance, the voltage wave *lags* the current wave by 90°. In an ac circuit containing an inductance and a capacitance, the voltage can lead or lag the current by any phase difference between 0° and 90°.

The equations of the sine wave voltage or current with a phase angle ϕ (phi) take the form:

$$v = V_M \sin (\omega t + \phi)$$

$$i = I_M \sin (\omega t + \phi) \tag{10.8}$$

The phase angle ϕ tells you how many degrees the wave leads a sine wave with the same frequency and $\phi = 0$ rad. Phase differences can apply only to waves of the same frequency. Study the following example, which illustrates the concept of phase angle in an ac circuit.

EXAMPLE 10.7

Figure 10–11(a) shows an ac circuit containing an inductance L. In this circuit the current is given by $i = I_M \sin \omega t$ and the voltage across the inductance by

$$v = V_M \sin \left(\omega t + \frac{\pi}{2} \right)$$

Given $f = 60$ Hz, $I_M = 2.0$ A, and $V_M = 100$ V, sketch i versus t and v versus t on the same graph for one cycle from $t = 0$. Determine the time difference between the two waves.

Solution   Apply (10.5) to express ω in terms of f:

$$\omega = 2\pi f = 2\pi(60) = 120\pi$$

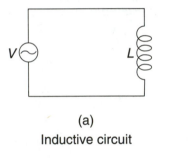

(a)
Inductive circuit

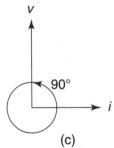

(c)
Voltage and current phasors

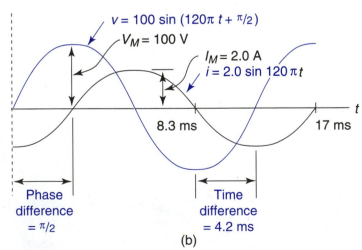

(b)

FIGURE 10– 11   Voltage leads current in an AC inductive circuit for Example 10.7.

EXAMPLE 10.7 (Cont.)

The current and voltage equations are then

$$i = 2.0 \sin 120\pi t$$

$$v = 100 \sin\left(120\pi t + \frac{\pi}{2}\right)$$

The period for both curves is

$$T = \frac{1}{f} = \frac{1}{60}\,\text{s} = 16.7\,\text{ms} \approx 17\,\text{ms}$$

The current wave has a phase angle $\phi = 0$ rad and, therefore, starts at $t = 0$. The voltage wave has a phase angle $\phi = \frac{\pi}{2}$ rad and, therefore, *leads* the current wave by $\frac{\pi}{2}$ radians. It starts at $\frac{-\pi}{2}$ radians and reaches peak value at $t = 0$. The sketch of both sine waves is shown in Figure 10–11(b) with separate scales for i and v.

To find the time difference between the two waves, you must convert the phase difference into time. Since $\theta = wt$, you have

$$t = \frac{\theta}{\omega}$$

Then a phase difference of $\frac{\pi}{2}$ represents a time difference of

$$t = \frac{\dfrac{\pi}{2}}{120\pi} = \frac{1}{240}\,\text{s} = 4.2\,\text{ms}$$

You can also observe that the phase difference is one-fourth of a cycle, which is $\frac{16.7}{4} = 4.2$ ms.

Phase differences in ac circuits are usually shown as angles between phasors. In Figure 10–11(c), the voltage phasor is shown at 90°, while the current phasor is at 0°. This represents the voltage and current at $t = 0$. As time increases, this phasor diagram would rotate counterclockwise with the phase difference of 90° remaining constant. However, since the phase difference does not change, the phasors only need to be drawn in one position.

■

EXERCISE 10.4

For all problems, round answers to two significant digits.

In problems 1 through 4, give the peak value, frequency, period, and phase angle for each ac wave.

1. $i = 1.5 \sin(100\pi t + \pi)$

2. $i = 2.3 \sin\left(120\pi t + \frac{\pi}{2}\right)$

3. $v = 170 \sin\left(110\pi t + \frac{\pi}{4}\right)$

4. $v = 300 \sin\left(200\pi t + \frac{\pi}{2}\right)$

In problems 5 through 10, write the equation of the ac sine wave having the given conditions.

5. $I_M = 3.0$ A, $\omega = 120\pi$, $\phi = \frac{\pi}{2}$ rad

6. $V_M = 400$ V, $\omega = 120\pi$, $\phi = 0$ rad

7. $V_M = 320$ V, $f = 50$ Hz, $\phi = \frac{\pi}{4}$ rad

8. $I_M = 1.2$ A, $f = 60$ Hz, $\phi = \frac{\pi}{2}$ rad

9. $V_M = 170$ V, $T = 18$ ms, $\phi = \pi$ rad

10. $I_M = 2.2$ A, $T = 25$ ms, $\phi = \frac{\pi}{8}$ rad

In problems 11 through 16 solve each applied problem.

11. Given an ac circuit containing a capacitance C, the voltage across the capacitance is given by $v = 200 \sin 120\pi t$ and the current in the circuit by

$$i = 1.0 \sin\left(120\pi t + \frac{\pi}{2}\right)$$

The voltage lags the current by $\frac{\pi}{2}$ rad $= 90°$. Sketch i versus t and v versus t on the same graph for one cycle from $t = 0$. Find the time difference between the two waves. See Example 10.7.

12. An ac circuit contains a resistance and an inductance in series (*RL* circuit). The voltage across the resistance is given by $v_R = 100 \sin 100\pi t$ and the voltage across the inductance by $v_L = 100 \sin\left(100\pi t + \frac{\pi}{2}\right)$. Here v_L leads v_R by $\frac{\pi}{2}$ rad $= 90°$. Sketch v_R versus t and v_L versus t on the same graph for one cycle from $t = 0$. Find the time difference between the two voltages. See Example 10.7.

13. For the ac waves in Figure 10–12, find the frequency, period, and time difference, and tell the angle that v leads or lags i.

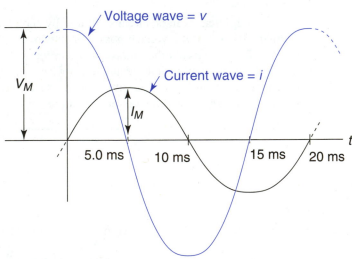

FIGURE 10–12 AC voltage and current waves for problem 13.

14. For the ac waves in Figure 10–13, find the frequency, period, and time difference, and tell the angle that v leads or lags i.

15. Given the current wave

$$i = 3.1 \sin\left(120\pi t + \frac{\pi}{2}\right)$$

Find i when
(a) $t = 6.0$ ms.
(b) $t = 14$ ms.

16. Given the voltage wave

$$v = 160 \sin\left(110\pi t + \frac{\pi}{4}\right)$$

Find v when
(a) $t = 5.0$ ms.
(b) 15 ms.

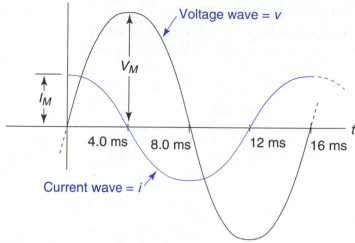

FIGURE 10–13   AC voltage and current waves for problem 14.

CHAPTER HIGHLIGHTS

10.1 SINUSOIDAL CURVES AND ALTERNATING CURRENT

The sine and cosine curves are called sinusoidal curves. They are the same curve, but the cosine leads the sine by $\frac{\pi}{2}$ rad or 90°. See Figure 10–2. The maximum value is called the amplitude, and the number of radians in one cycle is called the angular period. For the sinusoidal functions,

$$y = a \sin \theta$$
$$y = a \cos \theta \tag{10.1}$$

the amplitude = $|a|$ and the angular period = 2π.

Alternating current is generated in a coil of wire that rotates in a magnetic field. Alternating current and voltage are sinusoidal waves expressed by the following equations:

$$v = V_M \sin \theta$$
$$i = I_M \sin \theta$$

where v and i are the instantaneous values of the voltage and current. The maximum values V_M and I_M are called peak values. See Figure 10–5 for the graph of an ac voltage wave.

10.2 ROOT MEAN SQUARE VALUES

The root mean square or effective value of an ac voltage or ac current wave is the dc value that produces the same electrical energy. The rms values of voltage and current for a sine wave are

$$V_{rms} = \frac{V_M}{\sqrt{2}} = 0.7071(V_M)$$
$$I_{rms} = \frac{I_M}{\sqrt{2}} = 0.7071(I_M) \tag{10.2}$$

To find the peak values from the rms values, use the reverse of (10.2):

$$V_M = \sqrt{2}(V_{rms}) = 1.414(V_{rms})$$
$$I_M = \sqrt{2}(I_{rms}) = 1.414(I_{rms}) \tag{10.3}$$

The rms values are always understood to be the ones given for an ac circuit unless stated otherwise. In a purely resistive ac circuit, Ohm's law and the power law apply to the rms values of voltage and current.

10.3 FREQUENCY AND PERIOD

The equations for ac voltage and current in terms of the angular velocity ω in rad/s or the frequency f in Hz, and the time t in seconds are

$$v = V_M \sin \omega t = V_M \sin 2\pi f t$$

$$i = I_M \sin \omega t = I_M \sin 2\pi f t \qquad (10.6)$$

The angle $\theta = \omega t = 2\pi f t$ and $\omega = 2\pi f$.

The period of an ac wave is the time T for one cycle where

$$T = \frac{1}{f} \qquad (10.7)$$

See Figure 10–8 for the sketch of an ac wave showing frequency and period.

10.4 PHASE ANGLE

The equations of the sine wave voltage or current with a phase angle ϕ (phi) take the following form:

$$v = V_M \sin (\omega t + \phi)$$

$$i = I_M \sin (\omega t + \phi) \qquad (10.8)$$

The phase angle ϕ tells you how many degrees the wave leads a sine wave with the same frequency and $\phi = 0$ rad.

An inductance in an ac circuit causes the voltage wave to lead the current wave by 90°. A capacitance in an ac circuit causes the voltage wave to lag the current wave by 90°. Study Example 10.7, which illustrates phase and time difference.

≡ REVIEW QUESTIONS

In all problems, round answers to two significant digits.

In problems 1 through 4, give the peak value, effective value, angular velocity, frequency, period, and phase angle for each ac wave.

1. $v = 160 \sin 120\pi t$

2. $i = 3.0 \sin 200 \pi t$

3. $i = 1.2 \sin \left(240\pi t + \dfrac{\pi}{2} \right)$

4. $v = 320 \sin \left(400\pi t + \dfrac{\pi}{3} \right)$

In problems 5 through 8, write the equation of the ac wave having the given conditions.

5. $I_M = 2.4$ A, $f = 100$ Hz, $\phi = 0$ rad

6. $V_M = 50$ V, $f = 50$ Hz, $\phi = 0$ rad

7. $V_M = 300$ V, $f = 60$ Hz, $\phi = \dfrac{\pi}{2}$ rad

8. $I_M = 1.5$ A, $T = 40$ ms, $\phi = \dfrac{\pi}{6}$ rad

In problems 9 through 16 solve each applied problem.

9. Given an ac circuit containing only a resistance $R = 1.5$ kΩ, if the current $I = 50$ mA, find the voltage V, its peak value V_M, and the power P dissipated in the resistance.

10. A series circuit contains two resistances connected to an ac power supply of 120 V. If $R_1 = 30$ Ω and $R_2 = 20$ Ω, find the current I, its peak value I_M, and the voltage drop across R_1.

11. An ac wave takes 3 ms to go from 0 to peak value. What is the frequency and period of the wave?

12. Sketch the voltage wave $v = 90 \sin 150\pi t$ for one cycle from $t = 0$. Give the frequency and period of the wave.

13. In Figure 10–14 find the frequency, period, and time difference of the two ac waves.

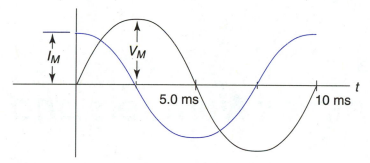

FIGURE 10–14 AC waves for problem 13.

14. An ac circuit contains a resistance and a capacitance in series (RC circuit). The voltage across the capacitance is given by $v_C = 200 \sin 120\pi t$. The voltage v_R across the resistance leads the voltage across the capacitance by 90° and has the same peak value.
   (a) Write the equation of the voltage wave v_R.
   (b) Sketch the two waves on the same graph for one cycle from $t = 0$, and find the time difference between the two waves.

15. Given the current wave $i = 0.50 \sin 150\pi t$,
   (a) Find i when $t = 2.0$ ms.
   (b) Find t when $i = 250$ mA.

16. Given the voltage wave $v = 400 \sin \left(200\pi t + \dfrac{\pi}{2}\right)$, *find v when*
   (a) $t = 2.0$ ms.
   (b) $t = 5.0$ ms.

Complex Numbers and Phasors

Photo courtesy of Hewlett-Packard Company.

A technician designs custom software using an automatic board testing program.

All the numbers that have been used so far are called "real" numbers. For example, 10, –1/2, 1.5, $\sqrt{2}$, and π are all real numbers. In Chapter 7, the square root of a negative number, $\sqrt{-4}$, appears as the root of a quadratic equation. The number $\sqrt{-4}$ is not a real number but is called an "imaginary number." When square roots of negative numbers first appeared about 500 years ago, they were found to have little application; hence, they were called imaginary. However, today they have a very important application in electronics and are as real as "real" numbers, but somewhat different rules apply to their operations.

When an imaginary number is combined with a real number, it forms a complex number. For example, $3 + \sqrt{-4}$ is a complex number. Complex numbers are the basis for the mathematics of ac networks. They are not difficult to work with and are not really as complex as their name implies.

Chapter Objectives

In this chapter, you will learn:

- The j operator and its basic operations.
- How to add, multiply, and divide complex numbers.
- Complex phasors and how to graph them.
- How to change a complex phasor from rectangular form to polar form and vice versa.
- How to add, multiply, and divide complex phasors in polar form.

11.1

THE j OPERATOR

The *imaginary unit* or j *operator* is defined as

$$j = \sqrt{-1}$$

or

$$j^2 = (\sqrt{-1})^2 = -1 \tag{11.1}$$

It is important to understand that j is *just a symbol* for $\sqrt{-1}$. We write j instead of $\sqrt{-1}$ because it is easier to work with. In mathematics, i is used to represent $\sqrt{-1}$. However, since i is used for instantaneous current in electricity, j is used for the imaginary unit in science and technology.

Square roots of negative numbers are called *imaginary* numbers. The j operator allows you to simplify imaginary numbers by separating the j as follows:

$$\sqrt{-4} = \sqrt{-1}\sqrt{4} = j\sqrt{4} = j\,2$$

Observe that the j is written first and then the coefficient. The negative sign under the radical becomes a j on the outside, and it is not necessary to write $\sqrt{-1}$. That is, for any positive number x,

$$\sqrt{-x} = j\sqrt{x}$$

For example,

$$\sqrt{-0.09} = j\sqrt{0.09} = j0.3$$

$$-3\sqrt{-25} = -3(j\sqrt{25}) = -3(j5) = -j15$$

Note that $-j$ is not the same as j. They are as different as -1 and 1. The multiplication and division rules for radicals in Section 7.2 do not apply to square roots of negative numbers. To multiply and divide imaginary numbers, you must *first separate the j operator*. Study the following examples that show how to work with the j operator.

EXAMPLE 11.1

Simplify:

$$(\sqrt{-9})(\sqrt{-16})$$

Solution   First separate the j operator:

$$(\sqrt{-9})(\sqrt{-16}) = (j\sqrt{9})(j\sqrt{16})$$

Then multiply the j's and simplify the radicals:

$$(j\sqrt{9})(j\sqrt{16}) = j^2(3)(4)$$

Now, *whenever j^2 appears you replace it with -1:*

$$j^2(3)(4) = (-1)(3)(4) = -12$$

The result is a real number and has no j operator. Note that if you tried to multiply under the radicals before separating the j's, you would get $\sqrt{(-9)(-16)} = \sqrt{144} = 12$, which is not the correct answer.

■

EXAMPLE 11.2

Simplify:

$$(-5\sqrt{-0.8})(3\sqrt{0.2})$$

Solution   First separate the j operator:

$$(-5\sqrt{-0.8})(3\sqrt{0.2}) = (-j5\sqrt{0.8})(3\sqrt{0.2})$$

You can now multiply under the radicals since the numbers under the radicals are both positive:

$$(-j5\sqrt{0.8})(3\sqrt{0.2}) = -j15\sqrt{0.16} = -j15(0.4) = -j6$$

■

Since j is a radical, it can be eliminated from the denominator of a fraction by multiplying the numerator and denominator by j:

$$\frac{3}{j} = \frac{3(j)}{j(j)} = \frac{j3}{j^2} = \frac{j3}{-1} = -j3$$

Multiplying the numerator and denominator of a fraction by j is how you divide imaginary numbers, as the next example shows.

EXAMPLE 11.3

Divide:

$$\frac{\sqrt{50}}{10\sqrt{-2}}$$

Solution   Separate the j first, then divide the radicals:

$$\frac{\sqrt{50}}{10\sqrt{-2}} = \frac{\sqrt{50}}{j10\sqrt{2}} = \frac{\sqrt{25}}{j10} = \frac{5}{j10}$$

Now to divide by j, multiply the numerator and denominator by j and simplify:

$$\frac{5}{j10} = \frac{5(j)}{j10(j)} = \frac{j5}{(j^2)10} = \frac{j5}{-10} = -\frac{j}{2} \text{ or } -j\,0.5$$

The imaginary unit is called the *j* operator because of its special properties. When you multiply repeatedly by the *j* operator, a cycle occurs that repeats after four multiplications. Observe the pattern that results by applying the rules for exponents to powers of j:

$$j^0 = +1$$
$$j^1 = j$$
$$j^2 = -1$$
$$j^3 = j^2(j) = (-1)(j) = -j$$
$$j^4 = (j^2)(j^2) = (-1)(-1) = +1$$
$$j^5 = (j^4)(j) = (+1)(j) = j \text{ etc.}$$

This is represented graphically where the horizontal axis is the real axis containing the numbers 1 and −1 and the vertical axis is the imaginary axis containing j and $-j$. See Figure 11–1.

> **Rule**   Each time you multiply by j it is the same as rotating counterclockwise by 90°.

The j operator "performs the operation" of 90° rotation each time you multiply by it. This is important when working with phasors in ac networks and is studied more in Section 11.3.

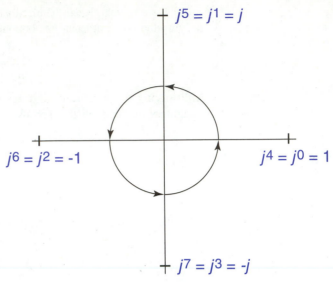

FIGURE 11–1   90° rotation of *j* operator.

ERROR BOX

A common error when working with the *j* operator is confusing *j*, −*j*, −1, and 1. Each of these are different numbers, and none of them can be changed to any of the others. The *j* operator is just a *symbol* for $\sqrt{-1}$. We write *j* *instead of* the radical symbol. Also, j^2 is equal to −1 and we *always change* j^2 to −1 when it occurs.

See if you can do the practice problems correctly.

Practice Problems   Simplify each of the following.

1. $(\sqrt{4})(\sqrt{-4})$
2. $(-\sqrt{4})(\sqrt{-4})$
3. $(\sqrt{-4})(\sqrt{-4})$
4. $(-\sqrt{-4})(\sqrt{-4})$
5. $(-\sqrt{-4})(-\sqrt{-4})$
6. $(-\sqrt{4})(-\sqrt{-4})$

Answers: 1. $j4$ 2. $-j4$ 3. -4 4. 4 5. -4 6. $j4$

EXERCISE 11.1

In problems 1 through 12, simplify and express in terms of the *j* operator.

1. $\sqrt{-9}$
2. $\sqrt{-49}$
3. $\sqrt{-0.04}$
4. $\sqrt{-1.21}$
5. $-\sqrt{-16}$
6. $-\sqrt{-100}$

7. $2\sqrt{-0.01}$

8. $-3\sqrt{-0.25}$

9. $\sqrt{-\dfrac{1}{4}}$

10. $\sqrt{-\dfrac{1}{100}}$

11. $\sqrt{-4 \times 10^{6}}$

12. $\sqrt{-25 \times 10^{-6}}$

In problems 13 through 26, multiply and simplify.

13. $(\sqrt{-36})(\sqrt{-4})$

14. $(\sqrt{-2})(\sqrt{-8})$

15. $(\sqrt{-0.1})(\sqrt{0.1})$

16. $(\sqrt{0.5})(\sqrt{-0.5})$

17. $(-\sqrt{-5})(\sqrt{-20})$

18. $(-\sqrt{-8})(\sqrt{2})$

19. $(-\sqrt{0.4})(-\sqrt{-0.9})$

20. $(-\sqrt{-10})(-\sqrt{-2.5})$

21. $(4\sqrt{-10})(-10\sqrt{40})$

22. $(-5\sqrt{-3})(-5\sqrt{-27})$

23. $(\sqrt{2 \times 10^{3}})(\sqrt{-8 \times 10^{3}})$

24. $(\sqrt{-1.5 \times 10^{-3}})(\sqrt{-13.5 \times 10^{-3}})$

25. $(-2j)(6j)(4j)$

26. $(3j)(-7j)(-j)$

In problems 27 through 34, divide and simplify.

27. $\dfrac{4}{\sqrt{-64}}$

28. $\dfrac{-2}{\sqrt{-25}}$

29. $\dfrac{\sqrt{8}}{\sqrt{-2}}$

30. $\dfrac{\sqrt{500}}{\sqrt{-5}}$

31. $\dfrac{1.2\sqrt{0.01}}{\sqrt{-0.04}}$

32. $\dfrac{5.5\sqrt{4}}{\sqrt{-0.25}}$

33. $\dfrac{10 \times 10^{3}}{\sqrt{-25 \times 10^{12}}}$

34. $\dfrac{-2 \times 10^{6}}{\sqrt{-9 \times 10^{6}}}$

Applications to Electronics   In problems 35 through 38 solve each applied problem to two significant digits.

35. The voltage in an ac circuit is given by Ohm's law $V = IZ$, where $I =$ current and $Z =$ impedance. Find V when $I = j10$ mA and $Z = -j3.5$ kΩ.

36. Find V in problem 35 when $I = j620$ µA and $Z = j15$ kΩ.

37. The impedance in an ac circuit is given by Ohm's law $Z = \dfrac{V}{I}$, where $V =$ voltage and $I =$ current. Find Z when $V = 5.5$ V and $I = j25$ mA.

38. Find I in problem 37 when $V = 12$ V and $Z = j1.6$ kΩ.

☰ 11.2
OPERATIONS WITH COMPLEX NUMBERS

When an imaginary number is combined with a real number, it is called a complex number. For example, $2.1 - j5.3$ is a complex number. Here, 2.1 is the real part, and $-j5.3$ is the imaginary part. All complex numbers have the form $x + jy$, where x and y are both real numbers. The group of complex numbers contain all the mathematical numbers needed for technical calculations as shown in Figure 11–2. When $x = 0$, you have a pure imaginary number jy. When $y = 0$, you have just the real number x.

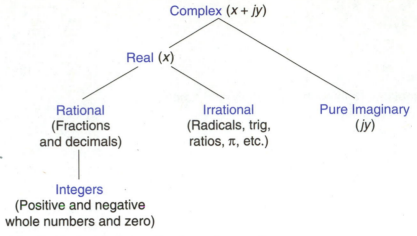

FIGURE 11–2 Mathematical number systems.

Addition of Complex Numbers

Addition of complex numbers is straightforward. To add two complex numbers, you add the real and imaginary parts separately. Treat the j operator like a variable in algebra as follows:

$$(2.1 - j5.3) + (3.4 + j4.2) = (2.1 + 3.4) + (-j5.3 + j4.2) = 5.5 - j1.1$$

EXAMPLE 11.4

Add the following complex numbers:

$$(1 - j2) - (3 - j5) + (-2 + j3) = 1 - j2 - 3 + j5 - 2 + j3$$

Solution   Add all the real parts and all the imaginary parts separately:

$$(1 - j2) - (3 - j5) + (-2 + j3)$$
$$= (1 - 3 - 2) + (- j2 + j5 + j3) = -4 + j6$$

Multiplication of Complex Numbers

Multiplication and division of complex numbers follow the rules of algebra with j being treated like a variable. To multiply two complex numbers you use the FOIL method—first, outside, inside, last—and multiply the four products separately (see Section 3.3). However, you must remember to replace j^2 by -1. For example, to multiply $3 - j4$ and $2 + j5$, you proceed as follows:

$$\begin{array}{cccc} \text{F} & \text{O} & \text{I} & \text{L} \end{array}$$

$$(3-j4)(2+j5) = (3)(2) + (3)(j5) + (-j4)(2) + (-j4)(j5)$$

$$= 6 + j15 - j8 - (j^2)(20) = 6 + j7 - (-1)(20)$$

$$= 26 + j7$$

Since j^2 is always replaced by -1, the product of two complex numbers will always be a complex number in the form $x + jy$. Study the next example, which illustrates the product of three numbers.

EXAMPLE 11.5

Multiply:
$$j2(2 - j)(1 + j)$$

Solution   Multiply the last two complex numbers first, and then multiply the result by $j2$:
$$j2(2 - j)(1 + j) = j2(2 + j2 - j - j^2) = j\,2(3 + j)$$
$$= j6 + j^2 2 = -2 + j6$$

Note that the real part of the complex number is written first.

Division of Complex Numbers

The *complex conjugate* of $x + jy$ is $x - jy$. When complex conjugates are multiplied together, the result is a real number since the outside and the inside products cancel:

$$(x + jy)(x - jy) = x^2 - jxy + jxy - j^2 y^2 = x^2 - (-1)y^2 = x^2 + y^2$$

Applying this idea, division of complex numbers is done by multiplying the numerator and denominator by the conjugate of the denominator as follows:

$$\frac{2 + j3}{5 - j4} = \frac{2 + j3(5 + j4)}{5 - j4(5 + j4)} = \frac{10 + j8 + j15 + j^2 12}{25 - j^2 16} = \frac{10 + j23 + (-1)12}{25 - (-1)16} = \frac{-2 + j23}{41}$$

The answer can be left as one fraction or separated and expressed as separate fractions or decimals:

$$\frac{-2 + j23}{41} = \frac{-2}{41} + j\frac{23}{41} \doteq -0.049 + j\,0.56$$

Study the next example, which closes the circuit and shows an application of complex numbers to an ac circuit problem.

EXAMPLE 11.6

Close the Circuit

The total impedance of an ac circuit containing two impedances Z_1 and Z_2 in parallel is given by

$$Z_T = \frac{Z_1 Z_2}{Z_1 + Z_2}$$

Find Z_T when $Z_1 = 1 + j$ kΩ and $Z_2 = 1 - j2$ kΩ

Solution   Substitute into the formula to obtain

$$Z_T = \frac{(1 + j)(1 - j2)}{(1 + j) + (1 - j2)}$$

EXAMPLE 11.6 (Cont.)

Multiply the numbers in the numerator, and add the numbers in the denominator:

$$Z_T = \frac{1 - j2 + j - j^2 2}{2 - j} = \frac{1 - j + 2}{2 - j} = \frac{3 - j}{2 - j}$$

Then divide the complex numbers by multiplying the numerator and the denominator by the conjugate of the denominator, which is $2 + j$:

$$Z_T = \frac{3 - j(2 + j)}{2 - j(2 + j)} = \frac{6 + j3 - j^2 - j^2}{4 - j^2} = \frac{7 + j}{5} \, k\Omega$$

The answer in decimal form is:

$$Z_T = \frac{7}{5} + j\frac{1}{5} \, k\Omega = 1.4 + j0.20 \, k\Omega$$

You can add, multiply, and divide complex numbers on your calculator if it has a $\boxed{\text{CPLX}}$ mode. You must first put it in the complex mode by pressing $\boxed{\text{MODE}}$ and the key that corresponds to $\boxed{\text{CPLX}}$. You then enter the real and imaginary parts separately, using the $\boxed{\text{i}}$ key or an equivalent key for the imaginary part, and perform the operation. For example, one way to do $\frac{3 - j}{2 - j}$ from Example 11.6 using a Casio fx or Radio Shack calculator that has complex mode is:

$\boxed{\text{MODE}}$ $\boxed{\text{EXP}}$ 3 $\boxed{+}$ 1 $\boxed{+/-}$ $\boxed{\text{i}}$ $\boxed{)}$ $\boxed{\div}$ $\boxed{(}$ 2 $\boxed{-}$ $\boxed{\text{i}}$ $\boxed{=}$ → 1.4 $\boxed{\text{Re↔Im}}$ → 0.20

The key $\boxed{\text{EXP}}$ corresponds to the $\boxed{\text{CPLX}}$ mode. The $\boxed{\text{i}}$ key is the imaginary key and is the same as the *j* operator. You need to close the parentheses after you enter the first complex number or press the $\boxed{=}$ key. Then you need to open the parentheses before you enter the second complex number, but it is not necessary to close the parentheses. The $\boxed{=}$ key does that. When you press $\boxed{=}$, the real part is displayed, and when you press $\boxed{\text{Re↔Im}}$, the imaginary part is displayed.

EXERCISE 11.2

In problems 1 through 36, perform the operations and express the answers in the form $x + jy$.

1. $(1 - j2) + (3 + j4)$
2. $(5 + j) + (3 - j2)$
3. $(3 + j2) - (3 - j3)$
4. $(4 - j2) - (6 - j2)$
5. $(5.3 + j9.1) + (3.1 - j6.4)$
6. $(1.0 - j1.3) - (0.9 + j0.7)$
7. $(1 - j) - (2 + j) + (1 - j2)$
8. $(5 + j3) - 2(5 - j3)$
9. $(3 - j2)(5 + j)$
10. $(2 - j5)(3 - j4)$
11. $(2 + j2)^2$
12. $(1 - j)^2$
13. $(1.4 - j7.0)(3.5 + j5.0)$
14. $(-0.80 + j0.60)(1.5 + j0.50)$
15. $(5 + j6)(5 - j6)$
16. $(-7 + j2)(-7 - j2)$
17. $3(5 - j2)(1 + j4)$
18. $j6(1 - j)(2 + j)$
19. $j2(1 + j)^2$
20. $(1 + j)(1 - j)^2$

21. $\dfrac{1}{2+j4}$

22. $\dfrac{1}{5-j5}$

23. $\dfrac{5+j5}{1-j2}$

24. $\dfrac{6-j4}{1-j}$

25. $\dfrac{j2}{-1+j}$

26. $\dfrac{j10}{3+j}$

27. $\dfrac{0.5}{0.1+j0.2}$

28. $\dfrac{1.1}{1.3-j1.9}$

29. $\dfrac{5+j}{1+j2}$

30. $\dfrac{4-j3}{2-j7}$

31. $\dfrac{0.10\,(50-j40)}{1.0-j}$

32. $\dfrac{30-j20}{(-j)(50+j30)}$

33. $\dfrac{(1+j)(2-j)}{2+j3}$

34. $\dfrac{3-j}{(2+j)(1+j)}$

35. $\dfrac{(2+j)(-2+j)}{(2+j)+(-2+j)}$

36. $\dfrac{(4-j)(1+j)}{(4-j)+(1+j)}$

In problems 37 and 38, perform the operations to solve the problem.

37. Any quadratic equation can be factored if the factors can be expressed as irrational or complex numbers. Show that the factors of $x^2 - 2x + 2 = 0$ are $[x - (1 + j)]\,[x - (1 - j)] = 0$ by multiplying out the factors. (Note: The roots are the negative of the factors $x = 1 + j$ and $x = 1 - j$.)

38. Prove $\sqrt{j} = \pm\dfrac{1+j}{\sqrt{2}}$ by showing that

$$\left(\pm\dfrac{1+j}{\sqrt{2}}\right)^2 = j$$

Applications to Electronics   In problems 39 through 44 solve each applied problem to two significant digits.

39. The total impedance of an ac circuit containing two impedances Z_1 and Z_2 in series is given by $Z_T = Z_1 + Z_2$. Find Z_T when $Z_1 = 5.5 + j6.7$ kΩ and $Z_2 = 8.2 - j3.5$ kΩ.

40. In problem 39, find Z_1 when $Z_T = 3.5 + j4.5$ kΩ and $Z_2 = 5.2 - j3.3$ kΩ.

41. In Example 11.6, find Z_T when $Z_1 = 1 - j$ kΩ and $Z_2 = 1 + j2$ kΩ.

42. In Example 11.6, find Z_T when $Z_1 = 10 + j10$ kΩ and $Z_2 = 10 - j20$ kΩ.

43. The voltage in an ac circuit is given by Ohm's law $V = IZ$ where I = current and Z = circuit impedance. Find V when $I = 20 + j10$ mA and $Z = 3 - j4$ kΩ.

44. In problem 43, find I when $V = 120$ V and $Z = 500 + j200$ Ω.

☰ 11.3
COMPLEX PHASORS AND POLAR FORM

Complex numbers can be represented as phasors with a magnitude and a direction angle. In phasor form they are very useful for applications to ac circuits. Given a complex number $x + jy$, the phasor is represented by the arrow drawn from the origin to the point (x, y) as shown in Figure 11–3. The x axis represents real numbers, and the y axis represents imaginary numbers. Figure 11–3 is called the *complex plane,* or Gaussian plane after Karl Gauss (1777–1855), the foremost German mathematician of the nineteenth century.

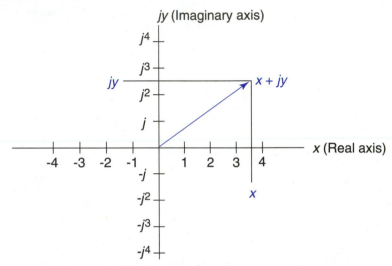

FIGURE 11–3 Complex plane or Gaussian plane.

Drawing complex phasors is similar to plotting points on the graph. Figure 11–4 shows the complex phasors $3 + j$, $2 - j3$, $-1 - j$, $-4 + j0$, and $0 + j2$. Note that a real number $x + j0$ is a horizontal phasor and a pure imaginary number $0 + jy$ is a vertical phasor.

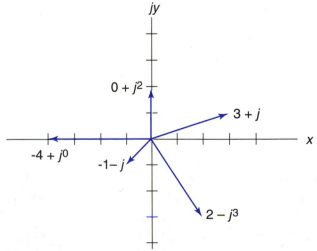

FIGURE 11–4 Complex phasors.

Complex phasors are added the same way as vectors and other phasors by adding components. The real part is the x component, and the imaginary part is the y component. Adding components is the same as adding complex numbers shown in Section 11.2. Study the next example.

EXAMPLE 11.7

Add the complex phasors $2 + j$ and $1 + j3$ algebraically and graphically.

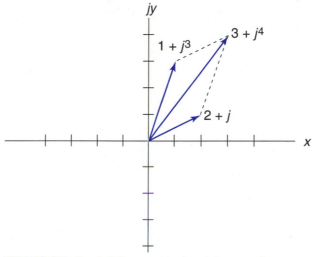

FIGURE 11–5 Adding complex phasors for Example 11.7.

Solution   The algebraic sum is $(2 + j) + (1 + j3) = 3 + j4$. This is shown graphically in Figure 11–5. The components of the resultant phasor $3 + j4$ consist of the sum of the x components (real numbers) and the sum of the y components (imaginary numbers). The resultant $3 + j4$ is the diagonal of the parallelogram shown.

■

Polar Form

The complex phasor $x + jy$ is in rectangular form. As with any vector or phasor it has magnitude, which is its length, and direction, which is the angle it makes with the positive x axis. Polar form expresses the phasor in terms of its length Z and angle θ as $Z\angle\theta$. The formulas to change from rectangular to polar form come from trigonometry as shown in Figure 11–6:

$$Z = \sqrt{x^2 + y^2}$$

$$\theta = \tan^{-1}\left(\frac{y}{x}\right) \quad (x \neq 0)$$

These can be combined and written in rectangular and polar form as:

$$x + jy = \sqrt{x^2 + y^2} \;\angle\tan^{-1}\left(\frac{y}{x}\right) = Z\angle\theta \quad x \neq 0$$

(11.2)

For example, to change the phasor $1 + j2$ to polar form, apply (11.2) where $x = 1$ and $y = 2$:

$$1 + j2 = \sqrt{1^2 + 2^2} \;\underline{/\tan^{-1}}\left(\frac{2}{1}\right) = \sqrt{5} \;\underline{/63°} = 2.2 \;\underline{/63°}$$

Observe that the inverse tangent function $\tan^{-1}$ gives the correct angle in the first quadrant when $\frac{y}{x}$ is positive and the correct negative angle in the fourth quadrant when $\frac{y}{x}$ is negative.

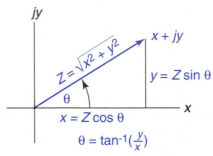

$$x + jy = Z\cos\theta + Z\sin\theta = Z\underline{/\theta}$$

FIGURE 11–6 Polar form of complex phasor.

Study the next two examples that show more conversions to polar form.

EXAMPLE 11.8

Change $4 - j3$ to polar form.

Solution Apply (11.2) with $x = 4$ and $y = -3$:

$$4 - j3 = \sqrt{4^2 + (-3)^2} \;\underline{/\tan^{-1}}\left(-\frac{3}{4}\right) = \sqrt{25} \;\underline{/\tan^{-1}}\;(-0.75) = 5 \;\underline{/-37°}$$

When you enter a negative number on the calculator and press $\boxed{\tan^{-1}}$, the calculator will correctly give you a negative angle in the fourth quadrant.

▪

EXAMPLE 11.9

Change $0 - j25$ to polar form.

Solution To change $0 - j25$ to polar form, observe that $x = 0$ and that the phasor lies along the negative y axis. The angle is therefore –90°. Note that *formula (11.2) for θ will not work when x = 0*. The magnitude is just the value of $y = 25$. The polar form is then $25 \;\underline{/-90°}$.

▪

To change from polar form to rectangular form, the formulas: $x = Z\cos\theta$ and $y = Z\sin\theta$ shown in Figure 11–6, yield:

$$Z\;\underline{/\theta} = Z\cos\theta + j(Z\sin\theta) = x + jy \qquad (11.3)$$

The formulas for x and y are the same as formula (9.7) for the components of a phasor that are shown in Section 9–5.

For example, to change $20 \angle 30°$ to rectangular form, apply (11.3) where $Z = 20$ and $\theta = 30°$:

$$20 \angle 30° = 20 \cos 30° + j(20 \sin 30°) = 17 + j10$$

Study the next two examples, which show more conversions to rectangular form.

EXAMPLE 11.10

Change $45 \angle -40°$ to rectangular form.

Solution   Apply formula (11.3) with $Z = 45$ and $\theta = -40°$:

$$45 \angle -40° = 45 \cos (-40°) + j45 \sin (-40°) = 34 - j29$$

▪

EXAMPLE 11.11

Change $3.5 \angle 0°$ to rectangular form.

Solution   Observe that for an angle $\theta = 0°$, the phasor lies along the positive x axis. Therefore, $y = 0$ and $x = 3.5$. In rectangular form, the phasor is then $3.5 + j0$.

You can also use formula (11.3), which will work for any angle θ.

▪

You can use your calculator to change from rectangular to polar form or vice versa if you have the keys [R→P] and [P→R]. Depending on your calculator, the key sequence may vary. For example, one way to change $4 - j3$ from Example 11.8 to polar form, using a Casio fx or Radio Shack calculator is

$$4 \boxed{\text{INV}} \boxed{\text{R→P}} 3 \boxed{+/-} \boxed{=} \rightarrow 5 \boxed{x↔y} \rightarrow -37°$$

The value of Z (or x) is displayed after you press $\boxed{=}$ and the value of θ (or y) is displayed after you press $\boxed{x↔y}$.

To change $2.8 \angle 50°$ to rectangular form, the keystrokes are:

$$2.8 \boxed{\text{INV}} \boxed{\text{P→R}} 50 \boxed{=} \rightarrow 1.8 \boxed{x↔y} \rightarrow 2.1$$

therefore $2.8 \angle 50° = 1.8 + j2.1$

Operations with Complex Phasors

You add phasors in polar form by first *changing to rectangular form and adding components* as shown in Example 11.7. You then change back to polar form. Study the next example.

EXAMPLE 11.12

Add the complex phasors:

$$40 \angle 25° + 50 \angle{-45°}$$

Solution   Change each phasor to rectangular form:

$$40 \angle 25° = 40 \cos 25° + j40 \sin 25° = 36.3 + j16.9$$

$$50 \angle{-45°} = 50 \cos (-45°) + j50 \sin (-45°) = 35.4 + -j35.4$$

Add components:

$$(36.3 + 35.4) + j(16.9 - 35.4) = 71.7 - j18.5$$

Change back to polar form:

$$71.7 - j18.5 = \sqrt{71.7^2 + (-18.5)^2} \angle \tan^{-1}\left(\frac{-18.5}{71.7}\right) = 74 \angle{-14°}$$

Phasors can be readily multiplied and divided in polar form using the following formulas:

$$(Z_1 \angle \theta_1)(Z_2 \angle \theta_2) = Z_1 Z_2 \angle (\theta_1 + \theta_2) \qquad (11.4)$$

$$\frac{Z_1 \angle \theta_1}{Z_2 \angle \theta_2} = \frac{Z_1}{Z_2} \angle (\theta_1 - \theta_2) \qquad (11.5)$$

Multiplying Complex Phasors (Polar Form)   *Multiply* the magnitudes and *add* the angles.

For example:

$$(2 \angle 30°)(3 \angle 40°) = (2)(3) \angle (30° + 40°) = 6 \angle 70°$$

Dividing Complex Phasors (Polar Form)   *Divide* the magnitudes and *subtract* the angles.

For example:

$$\frac{6 \angle 50°}{3 \angle 60°} = \frac{6}{3} \angle (50° - 60°) = 2 \angle{-10°}$$

It can be shown with formula (11.4) that multiplying any phasor by the j operator rotates that phasor by 90°. In polar form $j = 1 \angle 90°$. Therefore, for any phasor $Z \angle \theta$:

$$(Z \angle \theta)(j) = (Z \angle \theta)(1 \angle 90°) = (Z)(1) \angle (\theta + 90°) = Z \angle (\theta + 90°)$$

Study the next example, which illustrates many of these ideas.

EXAMPLE 11.13

Perform the operations, and express the answer in polar form:

$$\frac{(1+j)(1-j2)}{3+j}$$

Solution   One method is to first change each phasor to polar form:

$$1+j = \sqrt{1^2+1^2} \ \angle \tan^{-1}\left(\frac{1}{1}\right) = 1.41 \ \angle 45°$$

$$1-j2 = \sqrt{1^2+(-2)^2} \ \angle \tan^{-1}\left(\frac{-2}{1}\right) = 2.24 \ \angle -63.4°$$

$$3+j = \sqrt{3^2+1^2} \ \angle \tan^{-1}\left(\frac{1}{3}\right) = 3.16 \ \angle 18.4°$$

Then express the example in polar form, and multiply the numerator by applying (11.4):

$$\frac{(1+j)(1-j2)}{3+j} = \frac{(1.41 \ \angle 45°)(2.24 \ \angle -63.4°)}{3.16 \ \angle 18.4°}$$

$$= \frac{(1.41)(2.24) \ \angle[45°+(-63.4°)]}{3.16 \ \angle 18.4°} = \frac{3.16 \ \angle -18.4°}{3.16 \ \angle 18.4°}$$

Then divide by applying (11.5):

$$\frac{3.16 \ \angle -18.4°}{3.16 \ \angle 18.4°} = \frac{3.16}{3.16} \ \angle(-18.4°-18.4°) = 1.0 \ \angle -37°$$

Another method is to do the operations in rectangular form and then change to polar form:

$$\frac{(1+j)(1-j2)}{3+j} = \frac{1-j2+j-j^2 2}{3+j} = \frac{3-j}{3+j} = \frac{3-j(3-j)}{3+j(3-j)}$$

$$= \frac{8-j6}{10} = 0.80-j0.60$$

Then $0.80 - j0.60$ in polar form is

$$0.80-j0.60 = \sqrt{0.80^2+(-0.60)^2} \ \angle \tan^{-1}\left(\frac{-0.60}{0.80}\right)$$

$$= \sqrt{0.64+0.36} \ \angle \tan^{-1}(-0.75) = 1.0 \ \angle -37°$$

Observe that each method gives the same answer. However, if the phasors are given in polar form, it is easier to multiply and divide in that form.

Study the last example, which closes the circuit and shows an application of complex phasors to ac circuits.

EXAMPLE 11.14

Close the Circuit

In an ac circuit, Ohm's law $V = IZ$ applies when the voltage V, the current I, and the impedance Z are complex phasors. As with dc circuits, the units must be consistent. Given an ac circuit where $V = 20 \angle 45°$ V and $Z = 2.5 \angle -45°$ kΩ, find the current I.

Solution   To find I, apply Ohm's law in the form $I = \dfrac{V}{Z}$ and express Z in ohms:

$$I = \frac{V}{Z} = \frac{20 \angle 45° \text{ V}}{2500 \angle -45° \text{ Ω}}$$

Apply (11.5) and divide in polar form:

$$I = \frac{20}{2500} \angle [-45° - (-45°)] = 8.0 \angle 90° \text{ mA}$$

■

Complex phasors play an important part in the mathematics of ac circuits. Their importance is shown in the next two chapters on series ac circuits and parallel ac circuits.

EXERCISE 11.3

In problems 1 through 6, graph each complex phasor.

1. $1 + j2$
2. $4 - j3$
3. $0 - j$

4. $0 - j4$
5. $-4 + j0$
6. $5 + j0$

In problems 7 through 12, add the complex phasors algebraically and graphically.

7. $(1 + j2) + (3 + j)$
8. $(3 + j4) + (2 + j)$
9. $(2 + j3) + (2 - j3)$
10. $(3 - j2) + (3 + j2)$

11. $(0 + j2) + (5 + j0)$
12. $(0 - j3) + (3 + j0)$
13. $(0 + j3) + (0 - j2)$
14. $(0 - j6) + (0 + j4)$

In problems 15 through 26, change each complex phasor to polar form.

15. $2 + j2$
16. $1 - j$
17. $2 - j4$
18. $3 - j4$
19. $0 - j55$
20. $0 + j40$

21. $75 + j0$
22. $-100 + j0$
23. $15 + j25$
24. $8 + j15$
25. $0.5 - j1.2$
26. $1.2 - j0.9$

In problems 27 through 38, change each complex phasor to rectangular form.

27. $3 \angle 45°$

28. $5 \angle 60°$

29. $50 \angle -30°$

30. $45 \angle -20°$

31. $100 \angle 0°$

32. $60 \angle 90°$

33. $10 \angle -90°$

34. $75 \angle -90°$

35. $3.2 \angle -26°$

36. $4.3 \angle 82°$

37. $15 \angle \pi/2$

38. $90 \angle \pi/4$

In problems 39 through 44, add the complex phasors by first changing to rectangular form. Express answers in polar form. See Example 11.12.

39. $30 \angle 30° + 60 \angle 45°$

40. $25 \angle 60° + 45 \angle 30°$

41. $50 \angle 10° + 100 \angle -20°$

42. $3 \angle 75° + 4 \angle -40°$

43. $10 \angle 90° + 20 \angle -90°$

44. $120 \angle 0° + 120 \angle 90°$

In problems 45 through 56, perform the operations, and express the answer in polar form. Apply formulas (11.4) and (11.5).

45. $(2 \angle 40°)(4 \angle 20°)$

46. $(5 \angle 50°)(10 \angle 10°)$

47. $(10 \angle -30°)(20 \angle -20°)$

48. $(15 \angle -50°)(30 \angle 70°)$

49. $(30 \angle 90°)(6 \angle 0°)$

50. $(12 \angle -90°)(24 \angle 90°)$

51. $\dfrac{60 \angle 70°}{20 \angle 20°}$

52. $\dfrac{8 \angle 90°}{4 \angle 0°}$

53. $\dfrac{10 \angle -45°}{50 \angle 15°}$

54. $\dfrac{120 \angle -90°}{10 \angle 0°}$

55. $\dfrac{12 \angle 60°}{8 \angle -30°}$

56. $\dfrac{7.5 \angle -45°}{15 \angle -90°}$

In problems 57 through 62, perform the operations, and express the answer in polar form.

57. $(1 + j)(2 - j2)$

58. $(4 - j3)(3 + j4)$

59. $\dfrac{12 + j9}{6 - j8}$

60. $\dfrac{2 - j}{1 + j}$

61. $\dfrac{(1 + j2)(1 - j)}{1 + j}$

62. $\dfrac{(1 + j)(2 + j)}{1 - j}$

Applications to Electronics   In problems 63 through 70, solve each applied problem to two significant digits.

63. An impedance phasor in rectangular form is given by $Z = R + jX$, where R is the resistance and X is the reactance. If $R = 5.5$ kΩ and $X = 2.5$ kΩ, write Z in polar form.

64. An impedance phasor in polar form is given by $Z = 1.6 \angle -45°$ kΩ. The real component is the resistance R, and the imaginary component is the reactance X. Write Z in rectangular form.

65. The phasor current in one branch of a parallel ac circuit is given by $I_1 = 2 \angle 30°$ mA and in another branch by $I_2 = 3 \angle 60°$ mA. Find the total current in both branches in polar form.

66. In an ac series circuit, one impedance is given by $Z_1 = 40 \angle 0°$ kΩ and another by $Z_2 = 30 \angle -45°$ kΩ. If the total impedance $Z_T = Z_1 + Z_2$, find Z_T in polar form.

67. In an ac circuit, the voltage $V = 12 \angle 50°$ V, and the impedance $Z = 3.3 \angle -15°$ kΩ. Using Ohm's law, find the current I. See Example 11.14.

68. In an ac circuit, the voltage $V = 30 \angle -45°$ V, and the current $I = 12 \angle 30°$ mA. Using Ohm's law, find the impedance Z. See Example 11.14.

69. The total impedance of an ac circuit containing two impedances Z_1 and Z_2 in parallel is given by

$$Z_T = \frac{Z_1 Z_2}{Z_1 + Z_2}$$

Find Z_T in polar form when $Z_1 = 60 \angle 30°$ kΩ and $Z_2 = 50 \angle 90°$ kΩ.

70. In problem 69, find Z_T when $Z_1 = 100 \angle -90°$ kΩ and $Z_2 = 100 \angle 45°$ kΩ.

≡ CHAPTER HIGHLIGHTS

11.1 THE j OPERATOR

The *imaginary unit* or j *operator* is defined as

$$j = \sqrt{-1}$$

$$\text{or } j^2 = -1 \qquad (11.1)$$

To simplify the square root of a negative number the negative sign becomes a j on the outside. That is, for any positive number x,

$$\sqrt{-x} = j\sqrt{x}$$

To multiply or divide imaginary numbers, separate the j first. Treat j like an algebraic variable, but replace j^2 by –1. To divide by j, multiply numerator and denominator by j.

The j operator has the effect of 90° rotation when you multiply by it. See Figure 11–1.

11.2 OPERATIONS WITH COMPLEX NUMBERS

A complex number has the form $x + jy$. To add two complex numbers, you add the real (x) and imaginary (y) parts separately. To multiply two complex numbers, use the FOIL method to obtain the four products, treating j as a variable and replacing j^2 by –1. To divide complex numbers, multiply numerator and denominator by the complex conjugate of the denominator. Note that $x + jy$ and $x - jy$ are complex conjugates.

11.3 COMPLEX PHASORS AND POLAR FORM

A complex phasor $x + jy$ is represented by the arrow drawn from the origin to the point (x, y). See Figure 11–6.

The formula to change from rectangular form to polar form is

$$x + jy = \sqrt{x^2 + y^2} \; \angle \tan^{-1}\left(\frac{y}{x}\right) = Z \angle \theta$$

$$(11.2)$$

The formula to change from polar form to rectangular form is:

$$Z \angle \theta = Z \cos \theta + jZ \sin \theta = x + jy$$

$$(11.3)$$

To add complex phasors in polar form, change to rectangular form and add the real and imaginary components separately. Study Example 11.12. To multiply or divide complex phasors in polar form, apply the formulas:

$$(Z_1 \angle \theta_1)(Z_2 \angle \theta_2) = Z_1 Z_2 \angle (\theta_1 + \theta_2) \quad (11.4)$$

$$\frac{Z_1 \angle \theta_1}{Z_2 \angle \theta_2} = \frac{Z_1}{Z_2} \angle (\theta_1 - \theta_2) \qquad (11.5)$$

Study Examples 11.13 and 11.14.

≡ REVIEW QUESTIONS

In problems 1 through 12, perform the operations and express the result in the form $x + jy$.

1. $(\sqrt{-16})(\sqrt{-64})$

2. $(-\sqrt{3})(\sqrt{-27})$

3. $\dfrac{5}{\sqrt{-100}}$

4. $\dfrac{\sqrt{0.09}}{\sqrt{-0.25}}$

5. $(2 + j5) + (4 - j3)$

6. $2(1 - j) + (8 - j4)$

7. $(5 + j4)(2 - j3)$

8. $(1.2 - j1.5)(1.0 - j2.2)$

9. $\dfrac{1}{1 - j2}$

10. $\dfrac{3 + j}{1 + j2}$

11. $\dfrac{6 - j2}{1 + j}$

12. $\dfrac{4 + j3}{4 - j3}$

In problems 13 through 16, change the complex phasor from rectangular to polar form or vice versa.

13. $1 + j3$

14. $0 - j45$

15. $100 \angle 30°$

16. $5.5 \angle 90°$

In problems 17 through 22, perform the operations, and express the answer in polar form.

17. $20 \angle 45° + 40 \angle -60°$

18. $7.5 \angle 90° + 4.5 \angle -90°$

19. $(3.2 \angle 25°)(4.5 \angle 15°)$

20. $\dfrac{110 \angle 90°}{5 \angle 45°}$

21. $\dfrac{(1 + j)(1 - j)}{(1 + j) + (1 - j)}$

22. $\dfrac{(10 \angle 45°)(10 \angle -90°)}{10 \angle 45° + 10 \angle -90°}$

Applications to Electronics   In problems 23 through 26 solve each applied problem to two significant digits.

23. The voltages in a series circuit are given by the phasors $V_1 = 15 \angle 34°$ V and $V_2 = 10 \angle 90°$ V. Find the total voltage $V_T = V_1 + V_2$.

24. In an ac circuit, the current $I = 5 \angle 45°$ mA and the impedance $Z = 15 \angle -15°$ kΩ. Find the voltage V using Ohm's law $V = IZ$.

25. The admittance Y of an ac circuit is related to the impedance Z by $Y = \dfrac{1}{Z}$. It is measured in siemens (S). Find Y when $Z = 20 \angle -45°$ kΩ.

26. A parallel ac circuit contains the two impedances $Z_1 = 30 \angle 0°$ kΩ and $Z_2 = 10 \angle 60°$ kΩ. Find the total impedance Z_T using the following formula:

$$Z_T = \frac{Z_1 Z_2}{Z_1 + Z_2}$$

Series AC Circuits

Courtesy of Arthur Kramer and NYC Technical College.

Students working in a basic electronics laboratory.

In Chapter 10, alternating current was introduced, and ac circuits containing only resistance were discussed. However, there are two other basic types of circuit components found in ac circuits: inductances and capacitances. Each of these tend to oppose the flow of current but in a different way than a resistance. They are called *reactances*. Resistances and reactances are two types of *impedances* found in ac circuits.

Close the Circuit

This chapter studies three different combinations of impedances in a series ac circuit: a resistance-inductance or *RL* circuit, a resistance-capacitance or *RC* circuit, and a resistance-inductance-capacitance or *RLC* circuit. The mathematical relationships that apply to these circuits use complex phasors and the ideas presented in Chapter 11.

Chapter Objectives

In this chapter, you will learn:

- Inductive reactance and how to calculate it.
- Ohm's law for ac circuits.
- How to find the total impedance and voltage of a series *RL* circuit.
- Capacitive reactance and how to calculate it.
- How to find the total impedance and voltage of a series *RC* circuit.
- How to find the total impedance and voltage of a series *RLC* circuit.
- How to find the resonant frequency of a series *RLC* circuit.

≡ 12.1
INDUCTIVE REACTANCE AND *RL* CIRCUITS

Inductive Reactance

An inductor is a coil of wire that has a voltage induced in it by an ac current. The measure of the coil's capacity to produce voltage is called inductance L, which is measured in henrys (H). The inductive reactance X_L is a measure of the inductor's effect on the applied voltage and current. Reactance is measured in the same units as resistance, ohms. It depends on the inductance L and the frequency f given by the following formula:

$$X_L = 2\pi f L \qquad (12.1)$$

For example, an inductance $L = 100$ mH with a frequency $f = 2.0$ kHz has an inductive reactance of

$$X_L = 2(3.14)(2.0 \text{ kHz})(100 \text{ mH}) = 1.26 \text{ k}\Omega \approx 1.3 \text{ k}\Omega$$

If you need to find L or f, and are given the other quantities, you can solve formula (12.1) for either L or f. For example, an inductance $L = 70$ mH will have an inductive reactance $X_L = 500 \ \Omega$ at a frequency of

$$f = \frac{X_L}{2\pi L} = \frac{500 \ \Omega}{2(3.14)(70 \text{ mH})} = 1.14 \text{ kHz} \approx 1.1 \text{ kHz}$$

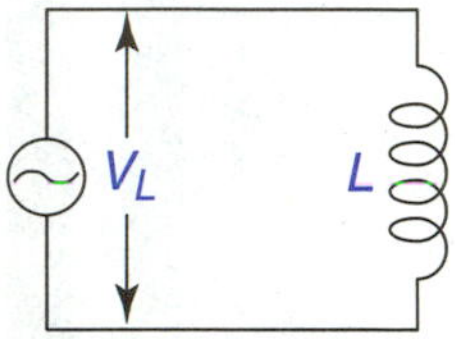

One of the reasons X_L is measured in ohms is because reactance is related to voltage and current in an ac circuit in the same way as resistance is related to voltage and current in a dc circuit. Ohm's law applies to the reactance X_L, the voltage across the inductance V_L, and the current through the inductance I:

$$I = \frac{V_L}{X_L}$$

For example, Figure 12–1 shows an ac source connected to an inductance. If $V_L = 10$ V and $X_L = 2.6$ kΩ, the current in the circuit is

$$I = \frac{V_L}{X_L} = \frac{10 \text{ V}}{2.6 \text{ k}\Omega} = 3.8 \text{ mA}$$

The following example further illustrates these ideas.

$$X_L = 2 \pi f_L$$

$$I = \frac{V_L}{X_L}$$

FIGURE 12–1
Inductive reactance for Example 12.1.

EXAMPLE 12.1

Given the inductive circuit in Figure 12–1 with $V_L = 55$ V, $f = 30$ kHz, and $I = 4.5$ mA, find X_L and L.

Solution   To find X_L use Ohm's law solved for X_L:

$$I = \frac{V_L}{X_L} \Rightarrow X_L = \frac{V_L}{I}$$

$$X_L = \frac{V_L}{I} = \frac{55 \text{ V}}{4.5 \text{ mA}} = 12 \text{ k}\Omega$$

Then, to find L, use (12.1) solved for L:

$$L = \frac{X_L}{2\pi f} = \frac{12 \text{ k}\Omega}{2(3.14)(30 \text{ kHz})} = 65 \text{ mH}$$

RL Circuits

When a resistance and an inductance are connected in series to an ac source, the sine wave voltage across the inductance V_L is out of phase with the sine wave voltage across the resistance V_R. There is an induced voltage in the inductance that tends to oppose the current flow and causes the current I to lag V_L by 90°. The current must be the same throughout a series circuit and is in phase with V_R. As a result, V_L *leads the current and* V_R *by 90°.*

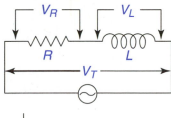

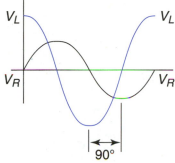

V_L leads V_R by 90°

FIGURE 12–2 Voltage waves in an *RL* series circuit for Examples 12.2 and 12.3.

Figure 12–2 shows the *RL* circuit and the two voltages. The voltage V_R is a sine wave with a phase angle of 0°, the voltage V_L is a sine wave with a phase angle of 90° and, therefore, leads V_R by 90°.

Figure 12–3a shows the phasor diagram of the voltages and the current in the *RL* series circuit. The voltage V_R is the real component, and the voltage V_L is the imaginary component of the resultant or total voltage V_T. The total voltage in rectangular and polar form is then given by:

$$V_T = V_R + jV_L = \sqrt{V_R^2 + V_L^2} \;\angle\tan^{-1}\left(\frac{V_L}{V_R}\right) \qquad (12.2)$$

The resistance and the reactance have the same phase relationship as the voltages, where R is the real component and X_L is the imaginary component of the total circuit impedance Z:

$$Z = R + jX_L = \sqrt{R^2 + X_L^2} \;\angle\tan^{-1}\left(\frac{X_L}{R}\right) \qquad (12.3)$$

Figure 12–3b shows the impedance triangle. Here, Z has the same phase angle θ as V_T since by Ohm's law,

$$\frac{V_L}{V_R} = \frac{IX_L}{IR} = \frac{X_L}{R}$$

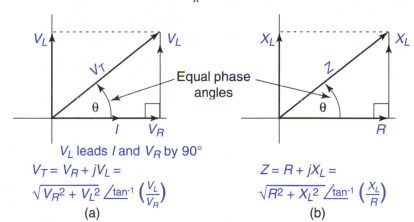

V_L leads *I* and V_R by 90°

$$V_T = V_R + jV_L =$$
$$\sqrt{V_R^2 + V_L^2}\;\angle\tan^{-1}\left(\frac{V_L}{V_R}\right)$$
(a)

$$Z = R + jX_L =$$
$$\sqrt{R^2 + X_L^2}\;\angle\tan^{-1}\left(\frac{X_L}{R}\right)$$
(b)

FIGURE 12–3 Phasor relationships in an *RL* series circuit.

Study the following examples that show how to apply these ideas.

EXAMPLE 12.2

Given the *RL* series circuit in Figure 12–2 with $R = 800\ \Omega$ and $X_L = 1.2\ \text{k}\Omega$, find the total impedance Z of the circuit in rectangular and polar form.

Solution Draw the impedance triangle with R as the real component and X_L as the imaginary component of the complex phasor Z. See Figure 12–4. Then apply (12.3):

$$Z = R + jX_L = 800 + j1200\ \Omega = \sqrt{800^2 + 1200^2}\;\angle\tan^{-1}\left(\frac{1200}{800}\right)$$

$$= \sqrt{2.08\times10^6}\;\angle\tan^{-1}(1.5) = 1.44\;\angle 56°\ \text{k}\Omega \approx 1.4\;\angle 56°\ \text{k}\Omega$$

EXAMPLE 12.2 (Cont.)

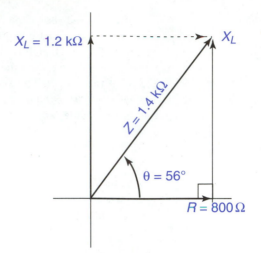

$$Z = 800 + j1200 = \sqrt{800^2 + 1200^2} \; \angle\tan^{-1}\left(\frac{1200}{800}\right) = 1.4 \; \angle 56° \; k\Omega$$

FIGURE 12–4 Impedance triangle for Example 12.2.

The magnitude of Z is then 1.4 kΩ and the phase angle is 56°. The calculation from rectangular to polar form, and vice versa, can be done on the calculator if you have the $\boxed{R{\longrightarrow}P}$ and the $\boxed{P{\longrightarrow}R}$ keys. For example, one way to change 800 + j1200 to polar form, using a Casio fx or Radio Shack calculator, is

800 $\boxed{INV}$ $\boxed{R{\longrightarrow}P}$ 1200 $\boxed{=}$ → 1.44 kΩ $\boxed{x{\longrightarrow}y}$ → 56°

See after Example 11.10 for other calculator examples.

In ac circuits, impedance takes the place of resistance and Ohm's law applies for both *scalar* and *phasor* quantities:

$$I = \frac{V}{Z}$$

(12.4)

For example, if the total voltage in the *RL* circuit of Example 12.2 is $V_T = 15 \; \angle 56°$ V, then by dividing phasors in polar form, the phasor current I is

$$I = \frac{15 \; \angle 56° \; V}{1.44 \; \angle 56° \; k\Omega} = \frac{15}{1440} \; \angle(56° - 56°) = 10 \; \angle 0° \; mA$$

Remember, when you divide complex phasors you divide the magnitudes and subtract the angles. Observe that the phase angle of V_T and Z are the same, 56°, and the phase angle of the current and V_R are the same, 0°.

EXAMPLE 12.3

Given the *RL* series circuit in Figure 12–2 with $R = 15$ kΩ, $X_L = 11$ kΩ, and $I = 800$ μA,

1. Find V_R and V_L.

2. Find the total voltage V_T and the impedance Z in polar form.

EXAMPLE 12.3 (Cont.)

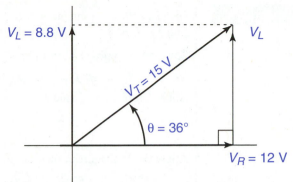

$$V_T = 12 + j8.8 = \sqrt{(12)^2 + (8.8)^2} \; \underline{/\tan^{-1}} \left(\tfrac{8.8}{12}\right) = 15 \; \underline{/36°} \; V$$

FIGURE 12–5 Voltage phasors in an *RL* series circuit for Example 12.3.

Solution   Figure 12–5 show the voltage phasor diagram for the *RL* circuit.

1. You can find the magnitudes of V_R and V_L by applying Ohm's law solved for *V*:

$$V_R = IR = (800 \ \mu A)(15 \ k\Omega) = 12 \ V$$

$$V_L = IX_L = (800 \ \mu A)(11 \ k\Omega) = 8.8 \ V$$

2. Then apply (12.2) to find V_T:

$$V_T = V_R + jV_L = 12 + j8.8 = \sqrt{12^2 + 8.8^2} \; \underline{/\tan^{-1}} \left(\frac{8.8}{12}\right)$$

$$= \sqrt{221} \; \underline{/\tan^{-1}} \; (0.733) = 15 \; \underline{/36°} \; V$$

The magnitude of V_T is then 15 V, and the phase angle $\theta = 36°$.

You can find the impedance *Z* from the voltage and the current using Ohm's law (12.4). In a series circuit, the phase angle of *Z* is the same as the phase angle of V_T. See Figure 12–3. Therefore, it is only necessary to calculate the magnitude of *Z*:

$$Z = \frac{V_T}{I} = \frac{15 \ V}{800 \ \mu A} = 19 \ k\Omega$$

The impedance in polar form is then

$$Z = 19 \; \underline{/36°} \; k\Omega$$

You can also find the impedance using the given values of *R* and X_L and formula (12.3).

$$Z = R + jX_L = 15 + j11 = \sqrt{15^2 + 11^2} \; \underline{/\tan^{-1}} \left(\frac{11}{15}\right)$$

$$= \sqrt{346} \; \underline{/\tan^{-1}} \; (0.733) = 19 \; \underline{/36°} \; k\Omega$$

Study the next example, which illustrates many of the ideas presented about *RL* series circuits.

EXAMPLE 12.4

Given an RL series circuit with $L = 500$ mH, $f = 400$ Hz, $V_L = 6.5$ V, and $Z = 3.0$ kΩ, find X_L, I, R, V_R, and V_T in polar form.

Solution   Apply (12.1) to find X_L:

$$X_L = 2\pi f L = 2\pi(400 \text{ Hz})(500 \text{ mH}) = 1.26 \text{ k}\Omega \approx 1.3 \text{ k}\Omega$$

Apply Ohm's law to find I:

$$I = \frac{V_L}{X_L} = \frac{6.5 \text{ V}}{1.26 \text{ k}\Omega} = 5.2 \text{ mA}$$

Apply the Pythagorean theorem to the impedance triangle to find R:

$$R = \sqrt{Z^2 - X_L{}^2} = \sqrt{3.0^2 - 1.26^2} = 2.7 \text{ k}\Omega$$

Then V_R can be found using Ohm's law:

$$V_R = IR = (5.2 \text{ mA})(2.7 \text{ k}\Omega) = 14 \text{ V}$$

Apply (12.2) to find V_T in polar form:

$$V_T = V_R + jV_L = 14 + j6.5 = \sqrt{14^2 + 6.5^2} \; \underline{/\tan^{-1}\left(\frac{6.5}{14}\right)} = 15 \; \underline{/25°} \text{ V}$$

EXERCISE 12.1

For all problems, round answers to two significant digits.

In problems 1 through 10, using the given values for an inductance L, find the indicated quantity.

1. $L = 1.0$ H, $f = 600$ Hz; X_L
2. $L = 1.5$ H, $f = 500$ Hz; X_L
3. $L = 20$ mH, $f = 5.0$ kHz; X_L
4. $L = 100$ mH, $f = 6.0$ kHz; X_L
5. $L = 800$ mH, $f = 2.0$ kHz; X_L

6. $L = 700$ mH, $f = 1.0$ kHz; X_L
7. $X_L = 10$ kΩ, $f = 15$ kHz; L
8. $X_L = 2.5$ kΩ, $f = 10$ kHz; L
9. $X_L = 200 \ \Omega$, $L = 50$ mH; f
10. $X_L = 3.5$ kΩ, $L = 75$ mH; f

In problems 11 through 14, given a circuit containing an inductance L and using the given values, find the indicated quantity.

11. $V_L = 12$ V, $I = 15$ mA; X_L
12. $V_L = 5.0$ V, $I = 3.5$ mA; X_L

13. $X_L = 25$ kΩ, $I = 100$ μA; V_L
14. $X_L = 600 \ \Omega$, $V_L = 1.5$ V; I

In problems 15 through 20, given an RL series circuit, find the impedance Z in rectangular and polar form for the given values of R and X_L.

15. $R = 750 \ \Omega$, $X_L = 1.5$ kΩ
16. $R = 1.0$ kΩ, $X_L = 800 \ \Omega$
17. $R = 3.3$ kΩ, $X_L = 4.5$ kΩ

18. $R = 6.8$ kΩ, $X_L = 3.9$ kΩ
19. $R = 2.0$ kΩ, $X_L = 1.2$ kΩ
20. $R = 1.5$ kΩ, $X_L = 1.5$ kΩ

In problems 21 through 24, given an RL series circuit with the given values,
(a) Find V_R and V_L.   (b) Find V_T and Z in polar form.

21. $R = 600 \ \Omega$, $X_L = 800 \ \Omega$, $I = 20$ mA
22. $R = 1.0$ kΩ, $X_L = 1.5$ kΩ, $I = 5$ mA

23. $R = 22$ kΩ, $X_L = 18$ kΩ, $I = 750$ μA
24. $R = 2.7$ kΩ, $X_L = 2.2$ kΩ, $I = 15$ mA

**In problems 25 through 28, given an *RL* series circuit with the given values,
(a) Find Z in polar form. (b) Find *I*, V_R and V_L.**

25. $R = 300 \ \Omega$, $X_L = 500 \ \Omega$, $V_T = 20$ V
26. $R = 3.0 \ k\Omega$, $X_L = 1.0 \ k\Omega$, $V_T = 12$ V
27. $R = 4.3 \ k\Omega$, $X_L = 2.4 \ k\Omega$, $V_T = 60$ V
28. $R = 5.6 \ k\Omega$, $X_L = 9.5 \ k\Omega$, $V_T = 9.0$ V

In problems 29 through 32 solve each applied problem.

29. Given an *RL* series circuit with $L = 150$ mH,
$f = 1.5$ kHz, $V_L = 7.5$ V, and $Z = 2.3 \ k\Omega$, find
$X_L, I, R, V_R,$ and V_T in polar form. (Note:
$R = \sqrt{Z^2 - X_L{}^2}$)

30. Given an *RL* series circuit with $f = 800$ Hz, $R = 12 \ k\Omega$, $V_R = 10$ V and $X_L = 5.5 \ k\Omega$, find $L, I, V_L, V_T,$ and Z in polar form.

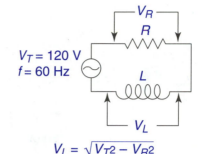

$$V_T = 120 \text{ V}$$
$$f = 60 \text{ Hz}$$

$$V_L = \sqrt{V_{T}{}^2 - V_{R}{}^2}$$

**FIGURE 12– 6 *RL* series
circuit for problems 31
and 32.**

31. Figure 12–6 shows a resistance in series with an inductance connected to a 120-V, 60-Hz generator. If $R = 100 \ \Omega$ and $L = 750$ mH, find $X_L, I, V_R, V_L,$ and Z in polar form.

32. In the circuit of Figure 12–6, if the resistance $R = 150 \ \Omega$ and V_R is measured to be 50 V, find $I, V_L, X_L, L,$ and Z in polar form. (Note: $V_L = \sqrt{V_T{}^2 - V_R{}^2}$.)

☰ 12.2

CAPACITIVE REACTANCE AND *RC* CIRCUITS

Capacitive Reactance

A capacitor consists of two conductors separated by an insulator, or dielectric, and is capable of storing charge. See Figure 12–7. The measure of the capacitor's ability to store charge is called capacitance *C* and is measured in farads (F). The capacitive reactance X_C is a measure of the capacitor's effect on the applied voltage and current. Reactance is measured in the same units as resistance, ohms, and depends on the capacitance *C* and the frequency f:

$$X_C = \frac{1}{2\pi f C} \tag{12.5}$$

For example, a capacitance $C = 10$ nF with a frequency $f = 5.0$ kHz has a capacitive reactance of

$$X_C = \frac{1}{2(3.14)(5.0 \text{ kHz})(10 \text{ nF})} = 3.18 \ k\Omega \approx 3.2 \ k\Omega$$

You can also use formula (12.5) to find *C* or f given the other quantities by solving the formula for *C* or f. For example, given $C = 200$ nF, the capacitive reactance will be $X_C = 500 \ \Omega$ at a frequency of

$$f = \frac{1}{2\pi X_C C} = \frac{1}{2(3.14)(500)(200 \text{ nF})} = 1.59 \text{ kHz} \approx 1.6 \text{ kHz}$$

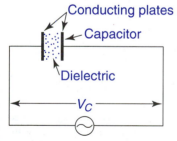

**FIGURE 12–7 Capacitor
and capacitive reactance
for Example 12.5.**

C = Capacitance

$$X_C = \frac{1}{2\pi f C} \qquad X_C = \frac{V_C}{I}$$

Capacitive reactance is also a type of impedance, and Ohm's law applies to X_C, the current I, and the voltage V_C across the capacitor in an ac circuit:

$$X_C = \frac{V_C}{I}$$

Study the next example, which applies these ideas.

EXAMPLE 12.5

Figure 12–7 shows a circuit containing a capacitor that consists of two conducting plates separated by a dielectric. If the voltage across the capacitor $V_C = 14$ V, $f = 2.0$ kHz, and $I = 15$ mA, find X_C and C.

Solution   Use Ohm's law to find X_C:

$$X_C = \frac{V_C}{I} = \frac{14\ \text{V}}{15\ \text{mA}} = 930\ \Omega$$

Then use (12.5) to find C:

$$C = \frac{1}{2\pi f X_C} = \frac{1}{2(3.14)(2.0\ \text{kHz})(933\ \Omega)} = 85\ \text{nF}$$

RC Circuits

When a capacitance and a resistance are connected in series to an ac source, the sine wave voltage across the capacitance V_C is out of phase with the sine wave voltage across the resistance V_R. The stored charge in the capacitor tends to oppose the applied voltage and V_C *lags* V_R by 90°. See Figure 12–8, which shows the RC circuit and the two voltages. The voltage V_R is a sine wave with a phase angle of 0°. The voltage V_C is a sine wave with a phase angle of –90° and therefore lags V_R by 90°. You can also say that V_R leads V_C by 90°.

In a series circuit, the current is the same throughout the circuit and is in phase with the voltage across the resistance. Therefore, the current also leads V_C by 90°. Figure 12–9(a) shows the phasor diagram of the voltages and the current in the RC series circuit. Here, V_R is the real component and $–V_C$ is the negative imaginary component of the resultant, or total, voltage V_T:

$$V_T = V_R - jV_C = \sqrt{V_R^2 + V_C^2}\ \angle \tan^{-1}\left(\frac{-V_C}{V_R}\right) \tag{12.6}$$

Observe that the phasor V_C has a phase angle of –90°. As a result, the phase angle θ of V_T will also be a negative angle in the fourth quadrant.

The resistance and the reactance in an RC series circuit have the same phase relationship as the voltages. The resistance R is the real component and $–X_C$ is the negative imaginary component of the total impedance Z:

$$Z = R - jX_C = \sqrt{R^2 + X_C^2}\ \angle \tan^{-1}\left(\frac{-X_C}{R}\right) \tag{12.7}$$

Figure 12–9(b) shows the impedance phasors. Note that Z has the same phase angle as V_T. Study the following example that shows how to apply these ideas.

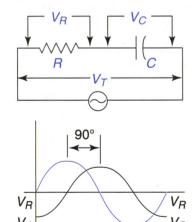

V_C lags V_R by 90°
or
V_R leads V_C by 90°

FIGURE 12– 8   Voltage waves in an RC series circuit for Examples 12.6 and 12.7.

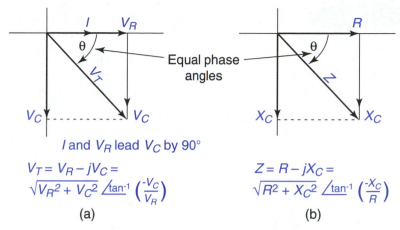

FIGURE 12– 9   Phasor relationships in an *RC* series circuit.

EXAMPLE 12.6

Given the *RC* circuit in Figure 12–8 with $R = 3.0 \text{ k}\Omega$ and $X_C = 1.0 \text{ k}\Omega$, find the total impedance *Z* of the circuit in rectangular and polar form.

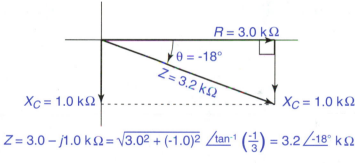

$$Z = 3.0 - j1.0 \text{ k}\Omega = \sqrt{3.0^2 + (-1.0)^2} \, \angle \tan^{-1}\left(\frac{-1}{3}\right) = 3.2 \, \angle\text{-18}° \text{ k}\Omega$$

FIGURE 12– 10   Impedance phasors for Example 12.6.

Solution   Draw the impedance phasors with *R* as the real component and $-X_C$ as the negative imaginary component of *Z*. See Figure 12–10. Then apply (12.7):

$$Z = R - jX_C = 3.0 - j1.0 \text{ k}\Omega = \sqrt{3.0^2 + 1.0^2} \, \angle \tan^{-1}\left(\frac{-1.0}{3.0}\right)$$

$$= \sqrt{10} \, \angle \tan^{-1}(-0.333) = 3.2 \, \angle -18° \text{ k}\Omega$$

The magnitude of *Z* is then 3.2 kΩ, and the phase angle is −18°.

Study the next two examples, which illustrate many of these ideas about *RC* series circuits.

EXAMPLE 12.7

Given the RC circuit in Figure 12–8 with $R = 7.5$ kΩ, $C = 33$ nF, $f = 500$ Hz, and $V_T = 20$ V, find X_C, I, V_C, V_R, and Z in polar form.

Solution   Apply (12.5) to find X_C:

$$X_C = \frac{1}{2\pi fC} = \frac{1}{2(3.14)(500\text{ Hz})(33\text{ nF})} = 9.6\text{ k}\Omega$$

Then find Z by applying (12.7):

$$Z = \sqrt{(7.5)^2 + (9.6)^2}\ \angle \tan^{-1}\left(\frac{-9.6}{7.5}\right) = 12.2\ \angle{-52°}\text{ k}\Omega$$

Find the current I by applying Ohm's law to Z and V_T:

$$I = \frac{V_T}{Z} = \frac{20\text{ V}}{12.2\text{ k}\Omega} = 1.6\text{ mA}$$

Now find V_C and V_R by applying Ohm's law again:

$$V_C = IX_C = (1.6\text{ mA})(9.6\text{ k}\Omega) = 15\text{ V}$$

$$V_R = IR = (1.6\text{ mA})(7.5\text{ k}\Omega) = 12\text{ V}$$

EXAMPLE 12.8

Given an RC series circuit with $R = 6.8$ kΩ, $I = 750$ μA, and $V_C = 8.5$ V, find X_C, V_R, V_T, and Z in polar form.

Solution   Use Ohm's law to find X_C and V_R:

$$X_C = \frac{V_C}{I} = \frac{8.5\text{ V}}{750\text{ mA}} = 11.3\text{ k}\Omega$$

$$V_R = IR = (750\text{ μA})(6.8\text{ k}\Omega) = 5.1\text{ V}$$

You can now find V_T and Z by applying (12.6) and (12.7):

$$V_T = \sqrt{5.1^2 + 8.5^2}\ \angle \tan^{-1}\left(\frac{-8.5}{5.1}\right) = 9.9\ \angle{-59°}\text{ V}$$

$$Z = \sqrt{6.8^2 + 11.3^2}\ \angle \tan^{-1}\left(\frac{-11.3}{6.8}\right) = 13.2\ \angle{-59°}\text{ k}\Omega$$

You can check some of the results by using the positive value of the phase angle θ (reference angle) and the sine and cosine functions. See Figure 12–11. The components of V_T and Z are given by

$$V_R = V_T \cos\theta \quad \text{and} \quad V_C = V_T \sin\theta$$

$$R = Z \cos\theta \quad \text{and} \quad X_C = Z \sin\theta$$

For example, to check the value of V_T, compute

$$V_T = \frac{V_C}{\sin\theta} = \frac{8.5}{\sin 59°} = 9.9\text{ V}$$

EXAMPLE 12.8 (Cont.)

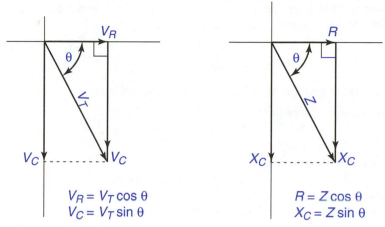

$$V_R = V_T \cos \theta$$
$$V_C = V_T \sin \theta$$

$$R = Z \cos \theta$$
$$X_C = Z \sin \theta$$

FIGURE 12–11   Phasor components for Example 12.8.

To check the value of X_C, you can compute

$$X_C = Z \sin \theta = 13.2 \sin 59° = 11.3 \text{ k}\Omega$$

EXERCISE 12.2

For all problems round answers to two significant digits.

In problems 1 through 10, using the given values for a capacitance C, find the indicated quantity.

1. $C = 20$ nF, $f = 6.0$ kHz; X_C

2. $C = 50$ nF, $f = 5.0$ kHz; X_C

3. $C = 500$ pF, $f = 15$ kHz; X_C

4. $C = 400$ pF, $f = 20$ kHz; X_C

5. $C = 1.0$ µF, $f = 300$ Hz; X_C

6. $C = 2.0$ µF, $f = 200$ Hz; X_C

7. $X_C = 10$ kΩ, $f = 10$ kHz; C

8. $X_C = 800$ Ω, $f = 2.0$ kHz; C

9. $X_C = 600$ Ω, $C = 200$ nF; f

10. $X_C = 15$ kΩ, $C = 800$ pF; f

In problems 11 through 14, given a circuit containing a capacitance C and using the given values, find the indicated quantity.

11. $V_C = 20$ V, $I = 12$ mA; X_C

12. $V_C = 12$ V, $I = 1.5$ mA; X_C

13. $X_C = 750$ Ω, $I = 7.5$ mA; V_C

14. $X_C = 25$ kΩ, $V_C = 10$ V; I

In problems 15 through 20, given an RC series circuit, find the impedance Z in rectangular and polar form for the given values of R and X_C.

15. $R = 1.6$ kΩ, $X_C = 1.2$ kΩ

16. $R = 8.2$ kΩ, $X_C = 16$ kΩ

17. $R = 680$ Ω, $X_C = 910$ Ω

18. $R = 30$ kΩ, $X_C = 20$ kΩ

19. $R = 1.5$ kΩ, $X_C = 750$ Ω

20. $R = 820$ Ω, $X_C = 1.0$ kΩ

In problems 21 through 24, given an RC series circuit with the given values,
(a) Find V_R and V_C.   (b) Find V_T and Z in polar form.

21. $R = 7.5$ kΩ, $X_C = 4.0$ kΩ, $I = 3.2$ mA

22. $R = 10$ kΩ, $X_C = 25$ kΩ, $I = 500$ µA

23. $R = 500$ Ω, $X_C = 800$ Ω, $I = 4.0$ mA

24. $R = 4.3$ kΩ, $X_C = 2.7$ kΩ, $I = 6.5$ mA

In problems 25 through 28, given an *RC* series circuit with the given values,
(a) Find *Z* in polar form. **(b)** Find *I*, V_R, and V_C.

25. $R = 680\ \Omega$, $X_C = 620\ \Omega$, $V_T = 20$ V

26. $R = 1.2$ kΩ, $X_C = 850\ \Omega$, $V_T = 24$ V

27. $R = 8.2$ kΩ, $X_C = 13$ kΩ, $V_T = 12$ V

28. $R = 20$ kΩ, $X_C = 30$ kΩ, $V_T = 10$ V

In problems 29 through 32, solve each applied problem.

29. Given an *RC* series circuit with $R = 3.3$ kΩ, $C = 100$ nF, $f = 1.0$ kHz, and $V_T = 15$ V, find X_C, I, V_C, V_R, and Z in polar form.

30. Given an *RC* series circuit with $R = 820\ \Omega$, $I = 9.5$ mA, and $V_C = 22$ V, find X_C, V_R, V_T, and Z in polar form.

31. Figure 12–12 shows a resistance in series with a capacitance connected to a 120-V, 60-Hz generator. If $R = 22$ kΩ and $C = 150$ nF, find X_C, V_R, V_C, I, and Z in polar form.

32. In the circuit of Figure 12–12, the capacitive reactance $X_C = 450\ \Omega$, and V_C is measured to be 100 V. Find R, C, I, V_R, and Z in polar form. (Note: $R^2 = \sqrt{Z^2 - X_C^2}$)

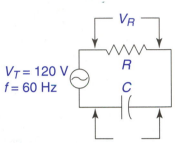

FIGURE 12–12 RC series circuit for problems 31 and 32.

≡ **12.3**

RLC CIRCUITS AND RESONANCE

When an inductance and a capacitance are connected in series in an ac circuit, the reactances tend to cancel each other. This is because inductive reactance has a phase angle of 90°, while capacitive reactance has a phase angle of −90°. The phasors have opposite direction and subtract from each other. Consider the *RLC* series circuit in Figure 12–13 containing a resistance, an inductance, and a capacitance in series.

The total impedance of the *RLC* series circuit in rectangular form is

$$Z = R + j(X_L - X_C) = R \pm jX \qquad (12.8)$$

where $X = |X_L - X_C|$ is the magnitude of the net reactance. When X_L is greater than X_C, the net reactance X is inductive, and the imaginary term jX will be positive. See Figure 12–14(a). When X_L is less than X_C, X is capacitive, and the imaginary term jX will be negative. See Figure 12–14(b).

For example, consider the *RLC* series circuit in Figure 12–15(a) where $R = 500\ \Omega$, $X_L = 600\ \Omega$, and $X_C = 1.0$ kΩ. Applying (12.8), the total impedance in rectangular form is

$$Z = 500 + j(1000 - 600)\ \Omega = 500 - j400\ \Omega$$

The magnitude of the net reactance $X = |600 - 1000| = 400\ \Omega$. Here, X is capacitive since X_L is less than X_C. The circuit is equivalent to an *RC* series circuit where $R = 500\ \Omega$ and $X_C = X = 400\ \Omega$. See Figure 12–15(b). This is called the *equivalent series circuit*.

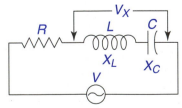

Net reactance $X_C = |X_L - X_C|$

FIGURE 12–13 RLC series circuit for Example 12.9.

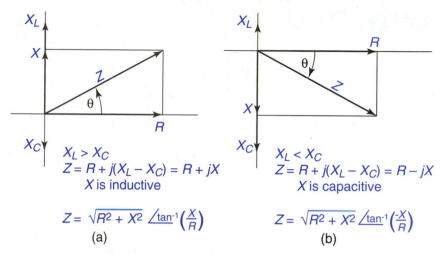

FIGURE 12–14   Impedance phasors in an *RLC* circuit.

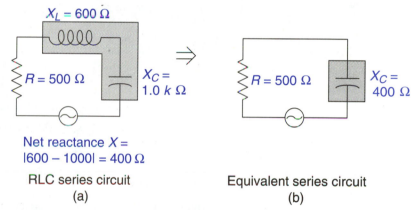

RLC series circuit
(a)

Equivalent series circuit
(b)

FIGURE 12–15   *RLC* series circuit and equivalent series circuit.

The total impedance Z in polar form can be obtained by applying formula (12.3) when X is inductive and by applying formula (12.7) when X is capacitive. See Figures 12–14(a) and 12–14(b). Applying (12.7) for the equivalent *RC* circuit, Z in polar form is

$$Z = 500 - j\,400\,\Omega = \sqrt{500^2 + 400^2}\ \angle \tan^{-1}\left(\frac{-400}{500}\right) = 640\ \angle -39°\ \Omega$$

Similarly, the total voltage of an *RLC* series circuit in polar form is given by

$$V_T = V_R + j(V_L - V_C) = V_R \pm jV_X \tag{12.9}$$

where $V_X = |\,V_L - V_C\,|$ is the magnitude of the net reactive voltage. To change V_T to polar form, use formula (12.2) when X is inductive and formula (12.6) when X is capacitive. Carefully study the next problem, which illustrates these concepts and some formulas from Sections 12.1 and 12.2.

EXAMPLE 12.9

Consider the *RLC* series circuit in Figure 12–13 where $R = 3.3$ kΩ, $L = 400$ mH, and $C = 20$ nF. If the applied voltage $V_T = 14$ V with a frequency $f = 3.0$ kHz, find the current I, X_L, X_C, V_R, V_L, and V_C. Also find the total impedance Z and V_T in rectangular and polar form.

Solution Find the reactances X_L and X_C using formulas (12.1) and (12.5):

$$X_L = 2\pi fL = 2(3.14)(3.0\,\text{kHz})(400\,\text{mH}) = 7.54\ \text{k}\Omega \approx 7.5\,\text{k}\Omega$$

$$X_C = \frac{1}{2\pi fC} = \frac{1}{2(3.14)(3.0\,\text{kHz})(10\,\text{nF})} = 5.31\,\text{k}\Omega \approx 5.3\,\text{k}\Omega$$

Then applying (12.8), the total impedance in rectangular form is

$$Z = R + j(X_L - X_C) = 3.3 + j(7.5 - 5.3)\ \text{k}\Omega = 3.3 + j\,2.2\ \text{k}\Omega$$

Here X_L is greater than X_C and jX is positive. The net reactance is inductive, and the equivalent series circuit is an *RL* series circuit with $R = 3.3$ kΩ and $X_L = 2.2$ kΩ.
    To change Z to polar form, apply (12.3):

$$Z = \sqrt{3.3^2 + 2.2^2}\ \angle \tan^{-1}\left(\frac{2.2}{3.3}\right) = 4.0\ \angle{-34°}\ \text{k}\Omega$$

Using Ohm's law the current is

$$I = \frac{V}{Z} = \frac{14\,\text{V}}{4.0\,\text{k}\Omega} = 3.5\,\text{mA}$$

Using Ohm's law again, the voltages across R, L, and C are

$$V_R = IR = (3.5\,\text{mA})(3.3\,\text{k}\Omega) = 11.6\,\text{V}$$

$$V_L = IX_L = (3.5\,\text{mA})(7.5\,\text{k}\Omega) = 26.3\,\text{V}$$

$$V_C = IX_C = (3.5\,\text{mA})(5.3\,\text{k}\Omega) = 18.6\,\text{V}$$

The voltages V_R, V_L, and V_C may appear to add up to more than the applied voltage, 14 V. However, V_L tends to cancel V_C, and their phasor sum equals 14 V as follows: Apply (12.9) to find V_T in rectangular form:

$$V_T = 11.6 + j(26.3 - 18.6)\ \text{V} = 11.6 + j\,7.7$$

The net reactive voltage V_X is then 7.7 V. Since X is inductive, apply (12.2) to change V_T to polar form:

$$V_T = \sqrt{V_R{}^2 + V_X{}^2}\ \angle \tan^{-1}\left(\frac{V_X}{V_R}\right) = \sqrt{11.6^2 + 7.7^2}\ \angle \tan^{-1}\left(\frac{7.7}{11.6}\right)$$

$$= 14\ \angle{34°}\ \text{V}$$

The magnitude of the total voltage is then 14 V, as it should be.

◼

When an ac series circuit contains more than one resistance, you can add the resistances, as in a dc circuit, to obtain the total resistance. When an ac series circuit contains more than one inductive reactance, or more than one capacitive reactance, you add the similar reactances to obtain the total inductive reactance and the total capacitive reactance:

$$X_{L_T} = X_{L_1} + X_{L_2} + X_{L_3} + \cdots$$

$$X_{C_T} = X_{C_1} + X_{C_2} + X_{C_3} + \cdots$$

You can, therefore, always reduce the circuit to an *RLC* series circuit with one resistance, one inductance, and one capacitance. Then, as in Example 12.9, you can find the equivalent *RL* or *RC* series circuit.

The power expended by the resistance in an ac circuit is called the *true power P* and is equal to the power dissipated in the resistive component I^2R. Since $R = Z \cos \theta = \frac{V}{I} \cos \theta$, the true power in terms of the applied voltage and current is given by

$$P = I^2R = I^2 \left(\frac{V}{I} \cos \theta \right) = VI \cos \theta \tag{12.10}$$

The value of $\cos \theta$ is called the power factor. For example, the true power in the circuit of Example 12.9 is:

$$P = (14 \text{ V})(3.5 \text{ mA}) \cos 34° = 40.6 \text{ mW}$$

Series Resonance

In an *RLC* circuit, the frequency at which the inductive reactance equals the capacitive reactance ($X_L = X_C$) is called the *resonant frequency* f_r. At this frequency, the net reactance $X = 0$, and the circuit is purely resistive. That is, the total impedance $Z = R$, and the phase angle $\theta = 0°$. The true power is at a maximum because when $\theta = 0°$, $\cos \theta = 1$ and $P = VI$. At the resonant frequency, $X_L = X_C$, and therefore,

$$2\pi f L = \frac{1}{2\pi f C}$$

which leads to

$$f^2 = \frac{1}{4\pi^2 LC}$$

Taking the square root of both sides, the formula for the resonant frequency f_r is

$$f_r = \frac{1}{2\pi\sqrt{LC}} \tag{12.11}$$

For example, if $L = 50$ mH and $C = 50$ nF, the resonant frequency is

$$f_r = \frac{1}{2(3.14)\sqrt{(50 \text{ mH})(50 \text{ nF})}} = 3.18 \text{ kHz}$$

One way this calculation can be done on a Casio fx or Radio Shack calculator is as follows:

1 ÷ 2 ÷ π ÷ (50 [EXP] 3 [+/-] × 50 [EXP] 9 [+/-])
[√] [=] → 3183

Resonance is important in tuning radio frequency (*RF*) circuits because at the resonant frequency, the impedance *Z* is at a minimum, and the current and power are at a maximum.

EXAMPLE 12.10

At a frequency of 4.0 kHz, how much inductance is required in series with a capacitance of 10 nF to obtain resonance?

Solution Solve formula (12.11) for *L*. Square both sides:

$$f^2 = \frac{1}{4\pi^2 LC}$$

Then multiply both sides by $\frac{L}{f_r^2}$:

$$\left(\frac{L}{f_r^2}\right)f^2 = \frac{1}{4\pi^2 LC}\left(\frac{L}{f_r^2}\right)$$

$$L = \frac{1}{4\pi^2 f_r^2 C}$$

Now substitute the given values to obtain the value for *L*:

$$L = \frac{1}{4\pi f_r^2 C} = \frac{1}{4(3.14)^2(4.0\,\text{kHz})^2(10\,\text{nF})} = 158\,\text{mH}$$

Study the next example, which applies many of the ideas in this chapter.

EXAMPLE 12.11

Given an *RLC* series circuit with $R = 300\ \Omega$, $L = 300$ mH, $C = 40$ nF and the applied voltage is 20 V, find the resonant frequency f_r. At the resonant frequency, show that $X_L = X_C$, and $V_L = V_C$, and find V_R, *I*, *Z*, and the true power *P*.

Solution Apply (12.11) to find f_r:

$$f_r = \frac{1}{2\pi\sqrt{LC}} = \frac{1}{2(3.14)(\sqrt{(300\,\text{mH})(40\,\text{nF})}} = 1.45\,\text{kHz}$$

Use formulas (12.1) and (12.5) to find X_L and X_C:

$$X_L = 2\pi f_r L = 2(3.14)(1.45\,\text{kHz})(300\,\text{mH}) = 2.7\,\text{k}\Omega$$

$$X_C = \frac{1}{2\pi f_r C} = \frac{1}{2(3.14)(1.45\,\text{kHz})(40\,\text{nF})} = 2.7\,\text{k}\Omega$$

Here, X_L may not always equal X_C exactly because the value of f_r may be rounded off and be approximate.

EXAMPLE 12.11 (Cont.)

Since $X_L = X_C$, the net reactance $X = 0$ and $Z = R = 300\ \Omega$. Applying Ohm's law, the current and voltages are then

$$I = \frac{V}{R} = \frac{20\ V}{300\ \Omega} = 67\ mA$$

$$V_R = IR = (67\ mA)(300\ \Omega) = 20\ V$$

$$V_L = IX_L = (67\ mA)(1.45\ kHz) = 97\ V$$

$$V_C = IX_C = (67\ mA)(1.45\ kHz) = 97\ V$$

Note that $V_L = V_C$ and that the voltages are both greater than the applied voltage of 14 V. However, they are 180° out of phase and cancel each other.

Apply (12.10) to find the true power with $\theta = 0°$:

$$P = VI \cos \theta = (20\ V)(67\ mA)\cos 0° = 1.3\ W$$

If the frequency is greater than f_r, $X_L > X_C$, and the circuit is inductive. If the frequency is less than f_r, $X_L < X_C$, and the circuit is capacitive. See problems 19 and 20.

ERROR BOX

A common error when working with *RLC* circuits is getting the wrong sign for the phase angle θ. The net reactance X is considered a positive quantity; however, you must know if it is inductive or capacitive and supply the correct sign for θ. If X_L is larger than X_C, θ is positive. If X_L is less than X_C, θ is negative. When using the calculator, if you enter a positive value and press $\boxed{\text{tan}^{-1}}$, you will get a positive angle. If you enter the *same* value with a negative sign and press $\boxed{\text{tan}^{-1}}$, you will get the *same* angle with a negative sign. Therefore, if X is capacitive, you can enter a positive value for $\frac{X}{R}$ and make the angle negative, or you can enter a negative value for $\frac{X}{R}$. See if you can get the correct angle in each of the practice problems.

Practice Problems: For each series *RLC* circuit, find the phase angle θ.

1. $R = 10\ \Omega$, $X_L = 20\ \Omega$, $X_C = 10\ \Omega$
2. $R = 10\ \Omega$, $X_L = 10\ \Omega$, $X_C = 20\ \Omega$
3. $R = 500\ \Omega$, $X_L = 600\ \Omega$, $X_C = 1.0\ k\Omega$
4. $R = 100\ \Omega$, $X_L = 1000\ \Omega$, $X_C = 1.0\ k\Omega$

Answers: 1. 45° 2. −45° 3. −39° 4. 0°

EXERCISE 12.3

For all problems round answers to two significant digits.

In problems 1 through 6, using the given values for the series RLC circuit in Figure 12–13, find the net reactance X and the impedance Z in rectangular and polar form.

1. $R = 600\ \Omega$, $X_L = 800\ \Omega$, $X_C = 500\ \Omega$
2. $R = 3.0\ k\Omega$, $X_L = 5.0\ k\Omega$, $X_C = 3.0\ k\Omega$
3. $R = 1.8\ k\Omega$, $X_L = 3.3\ k\Omega$, $X_C = 5.5\ k\Omega$
4. $R = 5.1\ k\Omega$, $X_L = 10\ k\Omega$, $X_C = 16\ k\Omega$
5. $R = 1.0\ k\Omega$, $X_L = 750\ \Omega$, $X_C = 1.2\ k\Omega$
6. $R = 500\ \Omega$, $X_L = 1.3\ k\Omega$, $X_C = 900\ \Omega$

In problems 7 through 12, using the given values for series resonance, find the indicated value.

7. $L = 100$ mH, $C = 20$ nF; f_r
8. $L = 50$ mH, $C = 100$ nF; f_r
9. $L = 40$ mH, $C = 800$ pF; f_r
10. $L = 350$ mH, $C = 100$ pF; f_r
11. $L = 400$ µH, $f_r = 10$ kHz; C
12. $C = 1.0$ µF, $f_r = 600$ Hz; L

In problems 13 through 20 solve each applied problem.

13. In the RLC series circuit in Figure 12–13 $R = 330\ \Omega$, $L = 50$ mH, $C = 100$ nF and the applied voltage $V = 12$ V with a frequency $f = 2.6$ kHz. Find I, X_L, X_C, V_R, V_L, V_C, and the impedance Z in rectangular and polar form. See Example 12.9.

14. In the RLC circuit in Figure 12–13 $R = 1.0\ k\Omega$, $L = 200$ mH, $C = 40$ nF and the applied voltage $V = 24$ V with a frequency $f = 1.5$ kHz. Find I, X_L, X_C, V_R, V_L, V_C, and the impedance Z in rectangular and polar form. See Example 12.9.

15. Figure 12–16 shows an ac series circuit containing two inductors, a capacitor, and a resistor connected to a voltage source V. If $V = 20$ V, $R = 600\ \Omega$, $X_{L_1} = 500\ \Omega$, $X_{L_2} = 750\ \Omega$, and $X_C = 800\ \Omega$, find Z, I, and the true power P. (Note: $X_L = X_{L_1} + X_{L_2}$.)

16. In the circuit in Figure 12–16, given $V = 10$ V, $R = 1.0\ k\Omega$, $X_{L_1} = 2.0\ k\Omega$, $X_{L_2} = 3.0\ k\Omega$, and $X_C = 7.5\ k\Omega$, find Z, I, and the true power P.

17. Given the RLC series circuit in Figure 12–17 with $R = 1.8\ k\Omega$, $X_L = 3.6\ k\Omega$, $X_C = 1.6\ k\Omega$ and $I = 10$ mA, find V_R, V_L, V_C, and V_T in rectangular and polar form.

18. In Figure 12.17, given $R = 510\ \Omega$, $X_L = 360\ \Omega$, $X_C = 750\ \Omega$ and $V_R = 5.0$ V, find I, V_L, V_C, and V_T in rectangular and polar form.

19. In Example 12.11, show that when $f = 1.8$ kHz and is greater than f_r, the circuit is inductive. Find the net reactance X and the impedance Z.

20. In Example 12.11, show that when $f = 1.2$ kHz and is less than f_r, the circuit is capacitive. Find the net reactance X and the impedance Z.

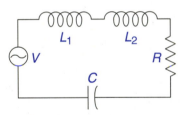

FIGURE 12–16 RLC circuit for problems 15 and 16.

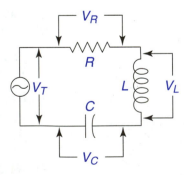

FIGURE 12–17 RLC circuit for problems 17 and 18.

≡ CHAPTER HIGHLIGHTS

12.1 INDUCTIVE REACTANCE AND *RL* CIRCUITS

Inductive reactance is measured in ohms in terms of the frequency f and the inductance L:

$$X_L = 2\pi f L \qquad (12.1)$$

In an RL series circuit, the voltage across the inductance V_L leads the current and the voltage across the resistance V_R by 90°. The voltage V_R is the real component, and V_L is the imaginary component of the total voltage:

$$V_T = V_R + jV_L = \sqrt{V_R{}^2 + V_L{}^2} \; \angle \tan^{-1}\left(\frac{V_L}{V_R}\right) \qquad (12.2)$$

For the impedance Z, R is the real component, and X_L is the imaginary component:

$$Z = R + jX_L = \sqrt{R^2 + X_L{}^2} \; \angle \tan^{-1}\left(\frac{X_L}{R}\right) \qquad (12.3)$$

Ohm's law for ac circuits applies to both scalar and phasor quantities:

$$I = \frac{V}{Z} \qquad (12.4)$$

12.2 CAPACITIVE REACTANCE AND *RC* CIRCUITS

Capacitive reactance is measured in ohms in terms of the frequency f and the capacitance C:

$$X_C = \frac{1}{2\pi f C} \qquad (12.5)$$

In an RC circuit, the current and the voltage across the resistance V_R lead the voltage across the capacitance V_C by 90°. The voltage, V_R is the real component, and $-V_C$ is the negative imaginary component of the total voltage:

$$V_T = V_R - jV_C = \sqrt{V_R{}^2 + V_C{}^2} \; \angle \tan^{-1}\left(\frac{-V_C}{V_R}\right) \qquad (12.6)$$

For the impedance Z, R is the real component, and $-X_C$ is the negative imaginary component:

$$Z = R - jX_C = \sqrt{R^2 + X_C{}^2} \; \angle \tan^{-1}\left(\frac{-X_C}{R}\right) \qquad (12.7)$$

12.3 *RLC* CIRCUITS AND RESONANCE

Inductive reactance and capacitive reactance in series tend to cancel each other. The total impendance of an RLC series circuit is

$$Z = R + j(X_L - X_C) = R \pm jX \qquad (12.8)$$

where $X = |X_L - X_C|$ is the magnitude of the net reactance.

When X_L is greater than X_C, X is inductive with a phase angle $\theta = 90°$. When X_L is less than X_C, X is capacitive with $\theta = -90°$. See Figure 12–14. To find Z, use (12.3) when X is inductive and (12.7) when X is capacitive.

The true power in an ac circuit is

$$P = VI\cos\theta \qquad (12.9)$$

Series resonance is when $X_L = X_C$, and the circuit is purely resistive with $Z = R$ and $\theta = 0°$. The resonant frequency is

$$f_r = \frac{1}{2\pi\sqrt{LC}} \qquad (12.10)$$

At the resonant frequency the current and the true power are at a maximum.

≡ REVIEW QUESTIONS

For all problems round answers to two significant digits.

In problems 1 through 6, using the given values for each series ac circuit, find the indicated quantity.

1. *RL* circuit: $L = 250$ mH, $f = 6.0$ kHz; X_L

2. *RL* circuit: $X_L = 740\ \Omega$, $L = 60$ mH; f

3. RC circuit: $C = 120$ nF, $f = 500$ Hz; X_C

4. RC circuit: $X_C = 4.0$ kΩ, $f = 10$ kHz; C

5. RLC circuit: $L = 200$ mH, $C = 500$ pF; f_r

6. RLC circuit: $L = 600$ mH, $C = 40$ nF; f_r

In problems 7 through 12, using the given values for each series ac circuit, find the impedance Z in rectangular and polar form.

7. RL circuit: $R = 820$ Ω, $X_L = 1.0$ kΩ

8. RL circuit: $R = 3.3$ kΩ, $X_L = 1.8$ kΩ

9. RC circuit: $R = 6.2$ kΩ, $X_C = 8.5$ kΩ

10. RC circuit: $R = 7.5$ kΩ, $X_C = 5.0$ kΩ

11. RLC circuit: $R = 470$ Ω, $X_L = 910$ Ω, $X_C = 360$ Ω

12. RLC circuit: $R = 1.1$ kΩ, $X_L = 700$ Ω, $X_C = 1.3$ kΩ

In problems 13 through 20 solve each applied problem.

13. In an RL series circuit, $R = 1.8$ kΩ, $f = 900$ Hz, $L = 200$ mH, and $I = 10$ mA. Find X_L, V_R, V_L and, V_T and Z in polar and rectangular form.

14. In an RL series circuit, $R = 20$ kΩ, $f = 2.0$ kHz, $L = 1.6$ H, and $V_T = 10$ V. Find X_L, V_R, V_L, I, and Z in polar and rectangular form.

15. In an RC series circuit, $R = 3.3$ kΩ, $f = 2.3$ kHz, $C = 15$ nF, and $I = 800$ µA. Find X_C, V_R, V_C, and V_T and Z in polar and rectangular form.

16. In an RC series circuit, $R = 510$ Ω, $f = 250$ Hz, $C = 1.0$ µF, and $V_T = 10$ V. Find X_C, V_R, V_C, I, and Z in polar and rectangular form.

17. In an RLC series circuit, $R = 3.0$ kΩ, $L = 150$ mH, and $C = 10$ nF. If the applied voltage $V = 32$ V with a frequency $f = 6.0$ kHz, find X_L, X_C, V_R, V_L, V_C, I, and Z in polar and rectangular form.

18. In an RLC series circuit, $R = 6.2$ kΩ, $L = 50$ mH, $C = 750$ pF, and $f = 20$ kHz. If $I = 5.5$ mA, find X_L, X_C, V_R, V_L, V_C, and V_T and Z in polar and rectangular form.

19. Figure 12–18 shows an ac series circuit containing two capacitors, an inductor, and a resistor connected to a voltage source. If $V = 9.0$ V, $R = 680$ Ω, $X_{C_1} = 500$ Ω, $X_{C_2} = 1.6$ kΩ, and $X_L = 1.2$ kΩ, find Z, I, and the true power P.

20. In the circuit in Figure 12–18, given $I = 5.0$ mA, $R = 5.6$ kΩ, $X_{C_1} = 2.7$ kΩ, $X_{C_2} = 2.2$ kΩ, and $X_L = 8.2$ kΩ, find Z, V, and the true power P.

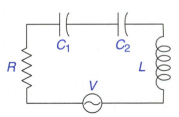

FIGURE 12–18 *RLC* series circuit for problems 19 and 20.

Parallel AC Circuits

Courtesy of Texas Instruments Inc.

*A technician works in the production process
of circuits and computer chips.*

Chapter 12 studied *RL*, *RC*, and *RLC* series ac circuits and how to analyze them. This chapter studies parallel ac circuits containing the same components. The basic difference between series and parallel ac circuits is the same as their difference in dc circuits. In an ac series circuit, the current is the *same* through each component but the voltage is *different* across each component. In an ac parallel circuit, the opposite is true. The voltage is the *same* across each component but the current is *different* through each component. The analysis of parallel ac circuits therefore focuses on the currents in the same way as the analysis of series circuits focuses on the voltages. Series-parallel ac circuits are analyzed in the last section of the chapter, which also shows how to reduce them to equivalent series circuits.

Chapter Objectives

In this chapter, you will learn:

- How to find the total current and impedance of a parallel *RL* circuit.

- How to find the total current and impedance of a parallel *RC* circuit.

- How to change an *RL* and *RC* parallel circuit to an equivalent series circuit.

- How to find the conductance, susceptance, and admittance of an ac parallel circuit.

- How to find the total current and impedance of a parallel *RLC* circuit.

- How to find the equivalent impedance of two impedances in parallel.

- How to find the total impedance of a series-parallel circuit and reduce it to an equivalent series circuit.

13.1
RL AND *RC* CIRCUITS

RL Circuits

Figure 13–1(a) shows an *RL* parallel circuit containing a resistance and an inductance in parallel. The voltage across the resistance and the voltage across the inductance are the same. The phase angle of the voltage *V* is chosen as the reference of 0° for the phasor diagram in Figure 13–1(b). Study the phasor diagram carefully.

The current through the resistance I_R is in phase with *V* and therefore has the same phase angle of 0°. The current through the inductance I_L is out of phase with the voltage, and lags *V* by 90°. Therefore, I_L also lags I_R by 90°. Since *V* and I_R have a phase angle of 0° in Figure 13–1(b), I_L has a phase angle of $-90°$.

The total current I_T is the phasor sum of I_R and I_L, where I_R is the real component and $-jI_L$ is the negative imaginary component:

$$I_T = I_R - jI_L = \sqrt{I_R{}^2 + I_L{}^2}\ \angle\tan^{-1}\left(\frac{-I_L}{I_R}\right) \tag{13.1}$$

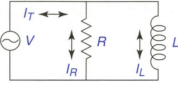

$R = 5.0\ \text{k}\Omega$
$X_L = 4.0\ \text{k}\Omega$

RL parallel circuit

(a)

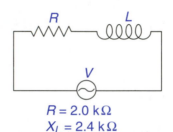

$$I_T = I_R - jI_L = \sqrt{I_R{}^2 + I_L{}^2}\ \angle\tan^{-1}\left(\frac{-I_L}{I_R}\right)$$

$$Z = \frac{V}{I_T}$$

Current phasors

(b)

$R = 2.0\ \text{k}\Omega$
$X_L = 2.4\ \text{k}\Omega$

Equivalent *RL* series circuit

(c)

FIGURE 13–1 *RL* parallel circuit and equivalent *RL* series circuit for Example 13.1.

From the total current and voltage, you can calculate the total impedance in an ac parallel circuit by applying Ohm's law:

$$Z = \frac{V}{I_T}$$

Using V and I_T in polar form, the phase angle of Z is found as follows. The phase angle of V is $0°$ and the phase angle of I_T is some negative angle θ. Applying the division rule for complex phasors (11.5), you subtract angles:

$$Z = \frac{V\ \angle 0°}{I_T\ \angle\theta} = \frac{V}{I_T}\ \angle(0° - \theta) = \frac{V}{I_T}\ \angle -\theta$$

The phase angle of Z is therefore $-\theta$. Since θ is a negative angle, $-\theta$ is a positive angle, and the impedance Z has a positive phase angle.

It is important to note that *there is no impedance triangle in a parallel circuit* as in a series circuit. Impedances are linked to voltages, not currents.

Carefully study the next example, which applies the above ideas.

EXAMPLE 13.1

Given the *RL* circuit in Figure 13–1 with $V = 10$ V, $R = 5.0$ kΩ, and $X_L = 4.0$ kΩ, find the total current I_T and the impedance Z of the circuit in polar and rectangular form.

Solution   First find the magnitude of I_R and I_L using Ohm's law:

$$I_R = \frac{V}{R} = \frac{10 \text{ V}}{5.0 \text{ k}\Omega} = 2.0 \text{ mA}$$

$$I_L = \frac{V}{X_L} = \frac{10 \text{ V}}{4.0 \text{ k}\Omega} = 2.5 \text{ mA}$$

Then apply (13.1) to find I_T, which is the phasor sum of I_R and I_L:

$$I_T = 2.0 - j2.5 \text{ mA} = \sqrt{2.0^2 + 2.5^2} \; \angle \tan^{-1}\left(\frac{-2.5}{2.0}\right) = 3.2 \; \angle{-51°} \text{ mA}$$

Applying Ohm's law using phasors, the total impedance in polar form is

$$Z = \frac{V}{I_T} = \frac{10 \; \angle 0° \text{ V}}{3.2 \; \angle{-51°} \text{ mA}} = \frac{10}{0.0032} \; \angle[0° - (-51°)]$$

$$= 3.1 \angle 51° \text{ k}\Omega$$

Note that the phase angle of the impedance is the negative of the phase angle of I_T.

To convert Z to rectangular form, use formula (11.3) for phasor components from Chapter 11:

$$Z \angle \theta = Z \cos \theta + j(Z \sin \theta)$$

Then

$$Z = 3.1 \; \angle 51° \text{ k}\Omega = 3.1 \cos 51° + j3.1 \sin 51° \text{ k}\Omega = 2.0 + j2.4 \text{ k}\Omega$$

Observe that the imaginary component of Z is positive. This represents an inductive reactance and means that the *RL* parallel circuit in Figure 13–1(a) *is equivalent to an RL series circuit* where $R = 2.0$ kΩ and $X_L = 2.4$ kΩ. See Figure 13–1(c). Any *RL* parallel circuit can be changed to an equivalent *RL* series circuit by this procedure.

▪

RC Circuits

Figure 13–2(a) shows an *RC* parallel circuit containing a resistance and a capacitance in parallel. The analysis is similar to that for an *RL* parallel circuit. The voltage V is the same across the resistance and the capacitance. The phase angle of V is chosen as the reference angle of 0° for the phasor diagram in Figure 13–2(b). The current through the resistance I_R is in phase with V and therefore also has a phase angle of 0°. The current through the capacitance I_C *leads V by 90°*. Therefore, I_C leads I_R by 90° in Figure 13–2(b).

The total current is the phasor sum of I_R and I_C, where I_R is the real component and jI_C is the positive imaginary component:

$$I_T = I_R + jI_C = \sqrt{I_R^2 + I_C^2} \; \angle \tan^{-1}\left(\frac{I_C}{I_R}\right) \tag{13.2}$$

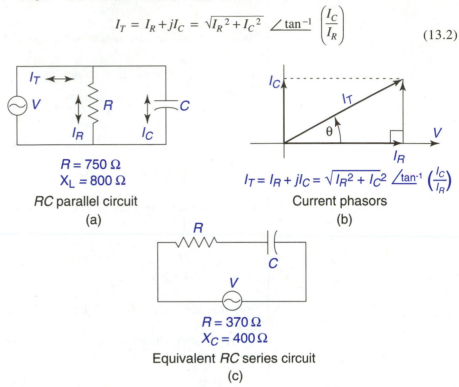

$R = 750\ \Omega$
$X_L = 800\ \Omega$

RC parallel circuit

(a)

$I_T = I_R + jI_C = \sqrt{I_R^2 + I_C^2} \; \angle\text{tan}^{-1}\left(\frac{I_C}{I_R}\right)$

Current phasors

(b)

$R = 370\ \Omega$
$X_C = 400\ \Omega$

Equivalent *RC* series circuit

(c)

FIGURE 13–2 *RC* parallel circuit and equivalent *RC* series circuit for Example 13.2.

As in the parallel *RL* circuit, the phase angle of *Z* is the negative of the phase angle of I_T. Since I_T has a positive phase angle, Z_T *has a negative phase angle* as in a series *RC* circuit. Study the next example.

EXAMPLE 13.2

Given the parallel *RC* circuit in Figure 13–2(a) with $V = 12$ V, $R = 750\ \Omega$, and $X_C = 800\ \Omega$, find I_T and *Z* in polar and rectangular form.

Solution First find the currents I_R and I_C:

$$I_R = \frac{12\text{ V}}{750\ \Omega} = 16\text{ mA}$$

$$I_C = \frac{12\text{ V}}{800\ \Omega} = 15\text{ mA}$$

Apply (13.2) to find the total current:

$$I_T = 16 + j15\text{ mA} = \sqrt{16^2 + 15^2} \; \angle \tan^{-1}\left(\frac{15}{16}\right) = 22 \; \angle 43°\text{ mA}$$

EXAMPLE 13.2 (Cont.)

Use Ohm's law and a phase angle of 0° for *V* to find *Z* in polar form:

$$Z = \frac{12 \angle 0° \text{ V}}{22 \angle 43° \text{ mA}} = \frac{12}{0.022} \angle (0° - 43°) = 545 \angle -43° \text{ } \Omega$$

Note that the phase angle of *Z* is the negative of the phase angle of I_T. In rectangular form, *Z* is then to two significant digits

$$Z = 545 \cos (-43°) + j545 \sin (-43°) \text{ } \Omega = 400 - j370 \text{ } \Omega$$

Since the imaginary component of *Z* is negative, it represents a capacitive reactance. The *RC* parallel circuit in Figure 13–2(a) is then equivalent to a series *RC* circuit with $R = 400$ Ω and $X_C = 370$ Ω. See Figure 13–2(c). Any *RC* parallel circuit can be changed to an equivalent *RC* series circuit.

▪

Susceptance and Admittance

In parallel dc circuits, the concept of conductance *G* is used to help analyze the circuit. Conductance is defined as the reciprocal of resistance and is measured in siemens (S):

$$G = \frac{1}{R}$$

Similarly, in parallel ac circuits the concepts of *susceptance B* and *admittance Y* are used to help analyze the ac circuit. They are phasors, defined as the reciprocals of reactance and impedance, respectively, and are measured in siemens:

$$B = \frac{1}{X}$$

$$Y = \frac{1}{Z} \tag{13.3}$$

Because of the reciprocal relation, *B* has a negative phase angle for an inductive reactance and a positive phase angle for a capacitive reactance. This is the same as the phase angles for inductive and capacitive currents.

For a resistance *R* in parallel with a reactance, the total admittance *Y* can be written as

$$Y = \frac{1}{R} \pm j\frac{1}{X} = G \pm jB \tag{13.4}$$

Note that *the imaginary component jB is negative for an inductive reactance and positive for a capacitive reactance.* For example, for the *RL* circuit in Figure 13–1(a), $R = 5.0$ kΩ and $X_L = 4.0$ kΩ. The admittance is

$$Y = \frac{1}{5.0 \text{ k}\Omega} - j\frac{1}{4.0 \text{ k}\Omega} = 200 - j250 \text{ } \mu\text{S}$$

Observe that the imaginary component is negative. The conductance $G = 200\ \mu S$ and the magnitude of the susceptance $B = 250\ \mu S$.

For the RC circuit in Figure 13–2(a), $R = 750\ \Omega$ and $X_C = 800\ \Omega$. The admittance is

$$Y = \frac{1}{750\ \Omega} - j\frac{1}{800\ \Omega} = 1.33 + j1.25\ \text{mS}$$

Observe that the imaginary component is positive. The conductance $G = 1.33\ \text{mS}$ and the susceptance $B = 1.25\ \text{mS}$.

Study the next example that shows two methods of changing a parallel circuit to an equivalent series circuit. The second method applies the concept of admittance.

EXAMPLE 13.3

Given an RC parallel circuit with $R = 1.3\ k\Omega$, $C = 10\ \text{nF}$, and $f = 10\ \text{kHz}$, find the equivalent RC series circuit.

Solution To find the equivalent series circuit, you need to find the impedance of the parallel circuit in rectangular form. Two different ways are shown. The first way is to find the total current as shown in Examples 13.1 and 13.2. *You can assume any convenient value for the voltage.* Suppose $V = 10\ \text{V}$. Then

$$X_C = \frac{1}{2\pi f C} = \frac{1}{2(3.14)(10\ \text{kHz})(10\ \text{nF})} = 1.6\ k\Omega$$

The currents are

$$I_R = \frac{V_R}{R} = \frac{10\ \text{V}}{1.3\ k\Omega} = 7.7\ \text{mA}$$

$$I_C = \frac{V_C}{X_C} = \frac{10\ \text{V}}{1.6\ k\Omega} = 6.3\ \text{mA}$$

The total current in rectangular and polar form is

$$I_T = 7.7 + j6.3\ \text{mA} = 9.9\ \underline{/39°}\ \text{mA}$$

The impedance in polar form is

$$Z = \frac{V}{I} = \frac{10\ \text{V}\ \underline{/0°}\ \text{V}}{9.9\ \underline{/39°}\ \text{mA}} = 1.0\ \underline{/-39°}\ k\Omega$$

The impedance in rectangular form is to two significant digits

$$Z = 1.0\ \cos(-39°) + j1.0\ \sin(-39°)\ k\Omega = 780 - j630\ \Omega$$

The equivalent RC series circuit then has $R = 780\ \Omega$ and $X_C = 630\ \Omega$. The equivalent value of the capacitance is

$$C = \frac{1}{2\pi f X_C} = \frac{1}{2(3.14)(10\ \text{kHz})(630\ \Omega)} = 25\ \text{nF}$$

EXAMPLE 13.3 (Cont.)

The second more direct way to find the impedance without finding the currents is to first find the admittance of the parallel circuit. Using $R = 1.3$ kΩ and $X_C = 1.6$ kΩ, apply formula (13.4) where jB is positive:

$$Y = \frac{1}{1.3 \text{ k}\Omega} + j\frac{1}{1.6 \text{ k}\Omega} = 770 + j630 \, \mu\text{S}$$

You can then find the impedance of the parallel circuit from formula (13.3). Since $Y = \frac{1}{Z}$, it follows that

$$Z = \frac{1}{Y} = \frac{1}{770 + j630 \, \mu\text{S}}$$

To perform the division, change 1 and Y to polar form:

$$Z = \frac{1 \angle 0°}{\sqrt{770^2 + 630^2} \ \angle \tan^{-1}\left(\frac{630}{770}\right)} = \frac{1 \angle 0°}{990 \angle 39° \, \mu\text{S}}$$

$$Z = \frac{1}{990} \angle (0° - 39°) = 1.0 \angle -39° \text{ k}\Omega$$

Note that the phase angle for 1 is 0°. You can then change the impedance to rectangular form and find the equivalent series circuit as just shown. ▪

ERROR BOX

A common error when working with parallel circuits is confusing the phase angles for impedance and current. An inductive reactance X_L has a *positive* phase angle, but an inductive current has a *negative* phase angle. A capacitive reactance X_C has a *negative* phase angle, but a capacitive current has a *positive* phase angle. See if you can get the correct phase angles in the practice problems.

Practice Problems: An ac parallel circuit with an applied voltage of 12 V contains a resistance $R = 7.5$ kΩ in parallel with a reactance X. For each given value of the reactance, find I_T and Z in polar form.

1. $X_L = 10$ kΩ 2. $X_C = 10$ kΩ 3. $X_L = 3.0$ kΩ 4. $X_C = 3.0$ kΩ

Answers:

4. $I_T = 4.3 \angle 68°$ mA, $Z = 4.3 \angle -68°$ kΩ

3. $I_T = 4.3 \angle -68°$ mA, $Z = 2.8 \angle 68°$ kΩ

2. $I_T = 2.0 \angle 37°$ mA, $Z = 6.0 \angle -37°$ Ω

1. $I_T = 2.0 \angle -37°$ mA, $Z = 6.0 \angle 37°$ Ω

EXERCISE 13.1

For all problems round answers to two significant digits.

In problems 1 through 8, using the given values for the parallel circuit, find the total current I_T and the impedance Z in rectangular and polar form.

1. *RL* circuit: $V = 12$ V, $R = 7.5$ kΩ, $X_L = 6.5$ kΩ **2.** *RL* circuit: $V = 20$ V, $R = 10$ kΩ, $X_L = 10$ kΩ

3. *RL* circuit: $V = 5.0$ V, $R = 1.2$ kΩ, $X_L = 820$ Ω

4. *RL* circuit: $V = 9.0$ V, $R = 910$ Ω, $X_L = 1.3$ kΩ

5. *RC* circuit: $V = 15$ V, $R = 750$ Ω, $X_C = 1.0$ kΩ

6. *RC* circuit: $V = 30$ V, $R = 620$ Ω, $X_C = 470$ Ω

7. *RC* circuit: $V = 24$ V, $R = 30$ kΩ, $X_C = 47$ kΩ

8. *RC* circuit: $V = 25$ V, $R = 3.0$ kΩ, $X_C = 2.0$ kΩ

For problems 9 through 16, find the admittance *Y* in rectangular and polar form for each of the parallel circuits in problems 1 through 8.

In problems 17 through 22 solve each applied problem.

17. In the *RL* parallel circuit in Figure 13–3, $R = 24$ kΩ, $L = 300$ mH, and the applied voltage $V = 12$ V with a frequency $f = 10$ kHz. Find
   (a) I_T and Z in polar form.
   (b) R and X_L of the equivalent *RL* series circuit.

18. In the *RL* parallel circuit in Figure 13–3, $R = 6.8$ kΩ, $X_L = 7.5$ kΩ, and $I_R = 800$ μA. Find V, I_L, I_T, and Z in polar form.

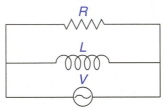

FIGURE 13–3 *RL* parallel circuit for problems 17 and 18.

19. In the *RC* parallel circuit in Figure 13–4, $R = 620$ Ω, $C = 40$ nF, and the applied voltage $V = 18$ V with a frequency $f = 5.0$ kHz. Find
   (a) I_T and Z in polar form.
   (b) R and X_C of the equivalent *RC* series circuit.

20. In the *RC* parallel circuit in Figure 13–4, $R = 8.2$ kΩ, $X_C = 5.6$ kΩ, and if $I_C = 5.0$ mA, find V, I_R, I_T, and Z in polar form.

21. Given an *RL* parallel circuit with $R = 1.8$ kΩ, $L = 100$ mH, and $f = 2.5$ kHz, what resistance and inductance connected in series is equivalent to this circuit? See Example 13.3.

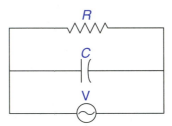

FIGURE 13–4 *RL* parallel circuit for problems 19 and 20.

22. Given an *RC* parallel circuit with $R = 10$ kΩ, $C = 500$ pF, and $f = 20$ kHz, what resistance and capacitance connected in series is equivalent to this circuit? See Example 13.3.

≡ 13.2

RLC CIRCUITS

When an inductance and a capacitance are connected in parallel, the inductive current tends to oppose the capacitive current. The currents oppose each other because the phase angle of I_L is –90° and the phase angle of I_C is 90°. The phasors have opposite direction and subtract from each other. Consider the *RLC* parallel circuit in Figure 13–5.

The total current of the *RLC* circuit in rectangular form is

$$I_T = I_R + j(I_C - I_L) = I_R \pm jI_X \tag{13.5}$$

where $I_X = |\,I_C - I_L\,|$ is the magnitude of the net reactive current. When I_L is greater than I_C ($I_L > I_C$), I_X is inductive, and the imaginary term jI_X is negative. See Figure 13–6(a). When I_C is greater than I_L ($I_C > I_L$), I_X is capacitive and jI_X is positive. See Figure 13–6(b).

The current I_T can be changed to polar form by using formula (13.1) when I_X is inductive, and by using formula (13.2) when I_X is capacitive. For example, suppose

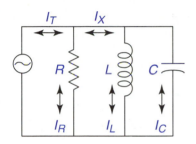

Net reactive current =
$I_X = |I_C - I_L|$

FIGURE 13–5 *RLC* parallel circuit.

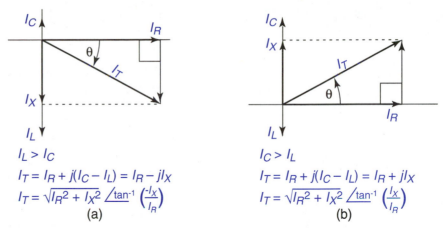

$I_L > I_C$
$I_T = I_R + j(I_C - I_L) = I_R - jI_X$
$I_T = \sqrt{I_R^2 + I_X^2} \angle \tan^{-1}\left(\frac{-I_X}{I_R}\right)$

(a)

$I_C > I_L$
$I_T = I_R + j(I_C - I_L) = I_R + jI_X$
$I_T = \sqrt{I_R^2 + I_X^2} \angle \tan^{-1}\left(\frac{I_X}{I_R}\right)$

(b)

FIGURE 13–6 Current phasors in an *RLC* parallel circuit.

in Figure 13–5 that $I_R = 10$ mA, $I_L = 5.0$ mA, and $I_C = 15$ mA. Applying (13.5), the total current is

$$I_T = 10 + j(15 - 5.0)\text{ mA} = 10 + j10\text{ mA}$$

Here, ($I_C > I_L$) so I_X is capacitive and the imaginary term jI_X is positive. The circuit is equivalent to an *RC* parallel circuit where $I_R = 10$ mA and $I_C = I_X = 10$ mA. Applying (13.2), using I_X for I_C, the total current in polar form is

$$I_T = 10 + j10\text{ mA} = \sqrt{10^2 + 10^2} \angle \tan^{-1}\left(\frac{10}{10}\right) = 14 \angle 45°\text{ mA}$$

Study the next example, which applies these ideas and shows two methods of changing a parallel *RLC* circuit to an equivalent series circuit. The second method applies the concept of admittance.

EXAMPLE 13.4

In the *RLC* parallel circuit in Figure 13–7(a), $R = 1.3$ kΩ, $L = 150$ mH, $C = 300$ nF, and if the applied voltage $V = 30$ V with a frequency $f = 600$ Hz, find I_T, Z, and the values of R and L or C in the equivalent series circuit.

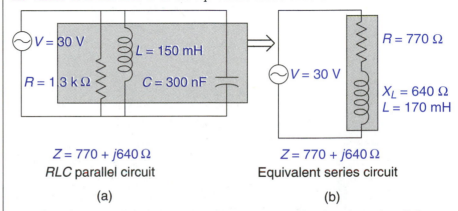

$Z = 770 + j640$ Ω
RLC parallel circuit

(a)

$Z = 770 + j640$ Ω
Equivalent series circuit

(b)

FIGURE 13–7 *RLC* parallel circuit and equivalent series circuit for Example 13.4.

EXAMPLE 13.4 (Cont.)

Solution Two methods are shown. The first method works with the branch currents as in the preceding illustration. The second method works with the admittance.

1. Find X_L and X_C using formulas (12.1) and (12.5):

$$X_L = 2\pi fL = 2(3.14)(600 \text{ Hz})(150 \text{ mH}) = 565 \, \Omega$$

$$X_C = \frac{1}{2\pi fC} = \frac{1}{2(3.14)(600 \text{ Hz})(300 \text{ nF})} = 884 \, \Omega$$

Apply Ohm's law to find the resistive, inductive, and capacitive currents:

$$I_R = \frac{V}{R} = \frac{30 \text{ V}}{1.3 \text{ k}\Omega} = 23 \text{ mA}$$

$$I_L = \frac{V}{X_L} = \frac{30 \text{ V}}{565 \, \Omega} = 53 \text{ mA}$$

$$I_C = \frac{V}{X_C} = \frac{30 \text{ V}}{884 \, \Omega} = 34 \text{ mA}$$

Apply (13.5) to find the total current in rectangular form:

$$I_T = I_R + j(I_C - I_L) = 23 + j(34 - 53) \text{ mA} = 23 - j19 \text{ mA}$$

Since I_L is greater than I_C, I_X is inductive and jI_X is negative. The RLC parallel circuit is then equivalent to an RL parallel circuit where $I_R = 23$ mA and $I_L = I_X = 19$ mA. Apply (13.1) to change I_T to polar form using I_X for I_L:

$$I_T = \sqrt{23^2 + 19^2} \; \angle \tan^{-1}\left(\frac{-19}{23}\right) = 30 \; \angle{-40°} \text{ mA}$$

To find the impedance use Ohm's law with phasors:

$$Z = \frac{V}{I_T} = \frac{30 \; \angle 0° \text{ V}}{30 \; \angle{-40°} \text{ mA}} = 1.0 \; \angle 40° \text{ k}\Omega$$

In rectangular form Z is

$$Z = 1.0 \cos 40° + j1.0 \sin 40° \text{ k}\Omega = 770 + j640 \, \Omega$$

The equivalent series circuit is then an RL series circuit where $R = 770 \, \Omega$ and $X_L = 640 \, \Omega$. See Figure 13–7(b). The value of the inductance L in the equivalent series circuit is found using formula (12.1) solved for L:

$$L = \frac{X_L}{2\pi f} = \frac{640}{2(3.14)(600)} = 170 \text{ mH}$$

2. In an RLC parallel circuit the admittance is given by

$$Y = \frac{1}{R} + j\left(\frac{1}{X_C} - \frac{1}{X_L}\right) = G + j(B_C - B_L) = G \pm jB \tag{13.6}$$

where $B = |B_C - B_L|$ is the magnitude of the net susceptance. When $B_C > B_L$, the imaginary term jB is positive, which means the net reactive current is capacitive. When $B_L > B_C$, the term jB is negative, and the net reactive current is inductive.

EXAMPLE 13.4 (Cont.)

Using X_L and X_C calculated earlier and applying (13.6), the admittance of the *RLC* circuit in Figure 13–7(a) is then

$$Y = \frac{1}{1.3\text{ k}\Omega} + j\left(\frac{1}{884\ \Omega} - \frac{1}{565\ \Omega}\right) = 770 - j640\ \mu\text{S}$$

The imaginary term is negative, which means the net reactive current is inductive. Find the impedance Z by changing Y to polar form and then taking the reciprocal of Y:

$$Y = \sqrt{770^2 + 640^2} \ \angle \tan^{-1}\left(\frac{-640}{770}\right) = 1.0 \ \angle -40° \ \text{mS}$$

$$Z = \frac{1}{Y} = \frac{1 \ \angle 0°}{1.0 \ \angle -40° \ \text{mS}} = 1.0 \ \angle 40° \ \text{k}\Omega$$

This is the same value of Z found by the first method. You can then change Z to rectangular form to find the same equivalent series circuit determined earlier. To find I_T note by Ohm's law that

$$I = \frac{V}{Z} = VY$$

Therefore,

$$I_T = (30\ \text{V})(770 - j640\ \mu\text{S}) = 23 - j19\ \text{mA}$$

which agrees with the value previously calculated.

▪

Applying the procedures shown in Example 13.4, any *RLC* parallel circuit can first be reduced to an *RL* or *RC* parallel circuit. Then, by finding the impedance, an equivalent *RL* or *RC* series circuit can be found for the *RL* or *RC* parallel circuit.

Figure 13–8 shows an *RLC* parallel circuit that contains two resistive branches. This type of *RLC* circuit can easily be converted to one with a simple resistive branch by finding the equivalent resistance of the two resistances in parallel. Study the next example, which shows how to analyze such a circuit.

EXAMPLE 13.5

In the parallel circuit in Figure 13–8(a) $R_1 = 3.3$ kΩ, $R_2 = 4.3$ kΩ, $X_L = 2.7$ kΩ, and $X_C = 1.3$ kΩ. Find the resistance and the reactance of the equivalent series circuit.

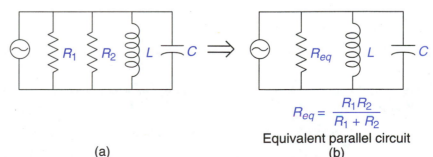

$$R_{eq} = \frac{R_1 R_2}{R_1 + R_2}$$

Equivalent parallel circuit

(a)   (b)

FIGURE 13–8 *RLC* parallel circuit for Example 13.5.

EXAMPLE 13.5 (Cont.)

Solution   First find the equivalent resistance of the two parallel resistances using the formula from dc circuits:

$$R_{EQ} = \frac{R_1 R_2}{R_1 + R_2} = \frac{(3.3)(4.3)}{3.3 + 4.3} = 1.87 \text{ k}\Omega$$

The circuit can then be reduced to an equivalent *RLC* circuit with one resistive branch. See Figure 13–8(b). The impedance of this reduced circuit can then be found by either of the methods shown in Example 13.4.

Using the admittance, the solution is as follows. Apply (13.6) to find *Y* in rectangular and polar form:

$$Y = \frac{1}{1.87} + j\left(\frac{1}{1.3} - \frac{1}{2.7}\right)\text{k}\Omega = 535 + 399 \text{ μS} = 670 \text{ } \angle 37° \text{ μS}$$

Since the imaginary term is positive, the circuit is capacitive. The total impedance in polar and rectangular form is

$$Z = \frac{1}{Y} = \frac{1}{670 \text{ } \angle 37° \text{ μS}} = 1.5 \text{ } \angle{-37°} \text{ k}\Omega = 1.2 - j0.90 \text{ k}\Omega$$

The equivalent series circuit is then an *RC* series circuit where $R = 1.2$ kΩ and $X_C = 900$ Ω.

▪

EXERCISE 13.2

For all problems round answers to two significant digits.

In problems 1 through 4, using the given values for an *RLC* parallel circuit, find I_T in rectangular and polar form.

1. $I_R = 30$ mA, $I_L = 80$ mA, $I_C = 40$ mA

2. $I_R = 50$ mA, $I_L = 100$ mA, $I_C = 60$ mA

3. $I_R = 55$ mA, $I_L = 30$ mA, $I_C = 90$ mA

4. $I_R = 500$ μA, $I_L = 600$ μA, $I_C = 300$ μA

In problems 5 through 8, using the given values for an *RLC* parallel circuit, find I_T and Z_T in rectangular and polar form.

5. $V = 12$ V, $R = 390$ Ω, $X_L = 430$ Ω, $X_C = 750$ Ω

6. $V = 24$ V, $R = 1.4$ kΩ, $X_L = 510$ Ω, $X_C = 1.5$ kΩ

7. $V = 36$ V, $R = 1.6$ kΩ, $X_L = 2.2$ kΩ, $X_C = 1.2$ kΩ

8. $V = 6.0$ V, $R = 10$ kΩ, $X_L = 20$ kΩ, $X_C = 5.6$ kΩ

For problems 9 through 12, find the admittance *Y* in rectangular and polar form for each of the parallel circuits in problems 5 through 8.

In problems 13 through 18 solve each applied problem.

13. For the *RLC* parallel circuit in Figure 13–5, $V = 10$ V, $f = 20$ kHz, $R = 4.3$ kΩ, $L = 10$ mH, and $C = 5.0$ nF. Find I_T, Z, and the equivalent series circuit including the values of R and L or C. See Example 13.4.

14. In the *RLC* parallel circuit in Figure 13–5, $V = 15$ V, $f = 50$ kHz, $R = 6.8$ kΩ, $L = 30$ mH, and $C = 500$ pF. Find I_T, Z, and the equivalent series circuit including the values of R and L or C. See Example 13.4.

15. For the parallel circuit in Figure 13–8(a), $R_1 = 8.2$ kΩ, $R_2 = 15$ kΩ, $X_L = 5.6$ kΩ, and $X_C = 10$ kΩ. Find Z in polar form and the resistance and reactance of the equivalent series circuit. See Example 13.5.

16. For the parallel circuit in Figure 13–8(a), $R_1 = 1.2$ kΩ, $R_2 = 1.5$ kΩ, $L = 100$ mH, $C = 0.50$ μF, and the frequency $f = 1.0$ kHz, find the admittance and the impedance of the circuit. See Example 13.5.

17. Given an *RLC* parallel circuit with $R = 1.5$ kΩ, $L = 50$ mH, $C = 10$ nF, $f = 5.0$ kHz, and the total current $I_T = 12$ mA, find the total impedance Z and the voltage V. (Note: First find the total admittance. See Example 13.4.)

18. Given an *RLC* parallel circuit with $R = 3.3$ kΩ, $L = 60$ mH, $C = 100$ nF, $f = 2.5$ kHz, and the total current $I_T = 15$ mA, find the total impedance Z and the voltage V. (Note: First find the total admittance. See Example 13.4.)

▤ 13.3
SERIES-PARALLEL CIRCUITS

Series-parallel ac circuits can be analyzed in many ways. For any series-parallel circuit, the following relationships can be used to simplify the circuit. Similar to series resistances in a dc circuit, the total impedance of two or more impedances in series is equal to their sum:

$$Z_T = Z_1 + Z_2 + Z_3 + \cdots \tag{13.7}$$

Also similar to two resistances in parallel in a dc circuit, the equivalent impedance of two impedances in parallel is equal to their product divided by their sum:

$$Z_{EQ} = \frac{Z_1 Z_2}{Z_1 + Z_2} \tag{13.8}$$

Formula (13.8) can also be expressed in terms of reciprocals:

$$\frac{1}{Z_{EQ}} = \frac{1}{Z_1} + \frac{1}{Z_2}$$

Note that formulas (13.7) and (13.8) apply to *complex phasors*. The operations of multiplication and division are done as shown in Chapter 11. Study the following examples, which show how to use these formulas.

EXAMPLE 13.6

Figure 13–9(a) shows a series-parallel circuit consisting of a resistance in series with an inductance and a capacitance in parallel. If $R = 1.5$ kΩ, $X_L = 470$ Ω, and $X_C = 620$ Ω, find the total impedance of the circuit in rectangular and polar form.

Solution First express each reactance as a complex phasor in rectangular form:

$$Z_1 = 0 + jX_L = 0 + j470 = j470 \, \Omega$$
$$Z_2 = 0 - jX_C = 0 - j620 = -j620 \, \Omega$$

EXAMPLE 13.6 (Cont.)

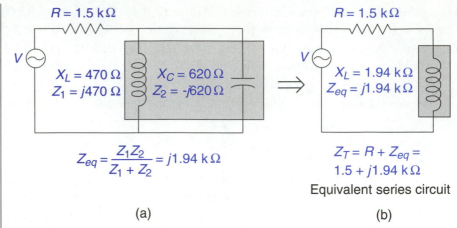

$$Z_{eq} = \frac{Z_1 Z_2}{Z_1 + Z_2} = j1.94 \text{ k}\Omega$$

$$Z_T = R + Z_{eq} =$$
$$1.5 + j1.94 \text{ k}\Omega$$

Equivalent series circuit

(a)   (b)

FIGURE 13–9   Series-parallel circuit for Example 13.6.

Then apply (13.8) to Z_1 and Z_2 in parallel, and find the equivalent impedance Z_{EQ}:

$$Z_{EQ} = \frac{Z_1 Z_2}{Z_1 + Z_2} = \frac{(j470)(-j620)}{(j470) + (-j620)} = \frac{-j^2\,291{,}400}{-j150}$$

$$= \frac{-(-1)\,291{,}400}{-j150} = \frac{291{,}400}{-j150} = \frac{1943}{-j}$$

Note that j^2 is replaced by -1, and the fraction is simplified by dividing the numerator and denominator by 150. The division is done by multiplying the numerator and denominator by j:

$$Z_{EQ} = \frac{1943(j)}{-j(j)} = \frac{j1943}{-j^2} = \frac{j1943}{-(-1)} = j1.94 \text{ k}\Omega$$

The equivalent impedance has a positive imaginary component and is inductive. Find the total impedance Z_T by adding R to Z_{EQ} in rectangular form and then changing to polar form:

$$Z_T = R + Z_{EQ} = 1.5 + j1.94 \text{ k}\Omega = 2.5 \underline{/52°} \text{ k}\Omega$$

The circuit is therefore equivalent to an RL series circuit where $R = 1.5$ kΩ and $X_L = 1.94$ kΩ. See Figure 13–9(b).

Observe that two phasors can be multiplied and divided in polar form but must be added in rectangular form. Therefore, when using formula (13.8), it may be necessary to work in both polar and rectangular forms as the next example shows.

Courtesy of Arthur Kramer and NYC Technical College.

Students in a microprocessor systems laboratory.

EXAMPLE 13.7

Figure 13–10(a) shows an important series-parallel circuit containing a capacitance in parallel with a resistance and an inductance in series. If $f = 10$ kHz, $R = 150$ Ω, $L = 10$ mH, and $C = 20$ nF, find the total impedance of the circuit.

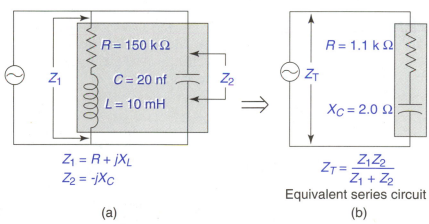

(a)

$Z_1 = R + jX_L$
$Z_2 = -jX_C$

$Z_T = \dfrac{Z_1 Z_2}{Z_1 + Z_2}$

Equivalent series circuit

(b)

FIGURE 13–10   Series-parallel circuit for Example 13.7.

Solution   Compute the reactances:

$$X_L = 2\pi f L = 2(3.14)(10 \text{ kHz})(10 \text{ mH}) = 628 \ \Omega$$

$$X_C = \frac{1}{2\pi f C} = \frac{1}{2(3.14)(10 \text{ kHz})(20 \text{ nF})} = 796 \ \Omega$$

EXAMPLE 13.7 (Cont.)

Then find the impedance of the series string Z_1 by adding R and X_L in phasor form:

$$Z_1 = R + jX_L = 150 + j628 \ \Omega$$

To find the total impedance, apply formula (13.8) with $Z_1 = 150 + j628$ and $Z_2 = -jX_C = -j796$:

$$Z_T = \frac{Z_1 Z_2}{Z_1 + Z_2} = \frac{(150 + j628)(-j796)}{(150 + j628) + (-j796)}$$

Add Z_1 and Z_2 in the denominator:

$$Z_T = \frac{(150 + j628)(-j796)}{150 - j168}$$

You can complete the calculation by changing each phasor to polar form and applying the rules for multiplying and dividing phasors:

$$Z_T = \frac{(646 \ \angle 77°)(796 \ \angle -90°)}{225 \ \angle 48°} = \frac{(646)(796)}{225} \ \angle(77° - 90° - 48°)$$

Then $Z_T = 2.3 \ \angle -61° \ k\Omega = 1.1 - j2.0 \ k\Omega$.
The series-parallel circuit is equivalent to an RC series circuit where $R = 1.1 \ k\Omega$ and $X_C = 2.0 \ k\Omega$. See Figure 13–10(b).

■

EXAMPLE 13.8

Figure 13–11(a) shows a common series-parallel network containing a resistance in series with a parallel circuit containing an inductive branch and a capacitive branch. If $R_1 = 100 \ \Omega$, $X_L = 4.0 \ k\Omega$, $R_2 = 400 \ \Omega$, $X_C = 5.0 \ k\Omega$, and $R_S = 2.7 \ k\Omega$, find the total impedance Z_T.

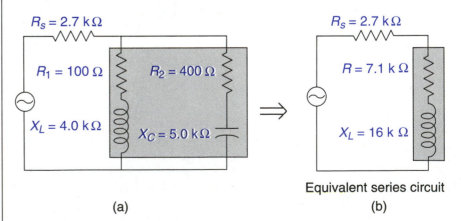

(a)   (b)

FIGURE 13–11   Series-parallel network for Example 13.8.

EXAMPLE 13.8 (Cont.)

Solution First find the equivalent impedance of the parallel circuit. Express each impedance as a complex phasor in the same units, ohms:

$$Z_1 = R_1 = 100\,\Omega$$

$$Z_2 = X_L = j4000\,\Omega$$

$$Z_3 = R_2 = 400\,\Omega$$

$$Z_4 = X_C = -j5000\,\Omega$$

Apply (13.7) to express the impedance of each parallel branch:

$$Z_{12} = Z_1 + Z_2 = 100 + j\,4000\,\,\Omega$$

$$Z_{34} = Z_3 + Z_4 = 400 - j\,5000\,\,\Omega$$

Then apply (13.8) to find the equivalent impedance of the two parallel branches:

$$Z_{EQ} = \frac{Z_{12}\,Z_{34}}{Z_{12} + Z_{34}} = \frac{(100 + j4000)(400 - j5000)}{(100 + j4000) + (400 - j5000)}$$

Add the impedances in the denominator:

$$Z_{EQ} = \frac{(100 + j4000)(400 - j5000)}{(500 - j1000)}$$

Now change to polar form and perform the operations:

$$Z_{EQ} = \frac{(4001\ \angle 88.6°\ )\,(5016\ \angle{-85.4°}\ )}{1118\ \angle{-63.4°}}$$

$$Z_{EQ} = \frac{(7001)(5016)}{1118}\angle[88.6° - 85.4° - (-63.4°)]$$

Then:

$$Z_{EQ} = 18\ \angle 67°\ \text{k}\Omega = 7.1 + j16\ \text{k}\Omega$$

The parallel circuit is then equivalent to an *RL* series circuit where $R = 7.1$ kΩ and $X_L = 16$ kΩ. See Figure 13–11(b).

Now add the series resistance R_S to Z_{EQ} to get the total impedance:

$$Z_T = R_S + Z_{EQ} = 2.7 + (7.1 + j\,16) = 9.8 + j\,16\ \text{k}\Omega$$

When ac networks become more complex, other analysis techniques need to be employed such as Kirchhoff's laws, the superposition theorem, Thevenin's theorem, and Norton's theorem.

EXERCISE 13.3

In problems 1 through 14 solve each applied problem to two significant digits.

1. In Example 13.6, find Z_T when $R = 1.3$ kΩ, $X_L = 510$ Ω, and $X_C = 680$ Ω.
2. In Example 13.6, find Z_T when $R = 2.0$ kΩ, $X_L = 2.4$ kΩ, and $X_C = 1.8$ kΩ.

3. In the series-parallel circuit in Figure 13–12, R = 750 Ω, X_L = 900 Ω, and X_C = 820 Ω. Find the impedance of the circuit.

4. In the circuit in Figure 13–12, R = 5.6 kΩ, X_L = 1.2 kΩ, and X_C = 3.3 kΩ. Find the impedance of the circuit.

5. In Example 13.7, find Z_T when f = 2.5 kHz, R = 330 Ω, L = 30 mH, and C = 100 nF.

6. In Example 13.7, find Z_T when f = 15 kHz, R = 470 Ω, L = 5.0 mH, and C = 30 nF.

7. Figure 13–13 shows a series-parallel circuit containing a capacitance in series with a resistance and an inductance in parallel. If X_C = 5.1 kΩ, R = 5.6 kΩ, and X_L = 4.7 kΩ, find the total impedance of the circuit.

8. In the circuit of Figure 13–13, X_C = 560 Ω, R = 430 Ω, and X_L = 750 Ω. Find the total impedance of the circuit.

9. Figure 13–14 shows a series-parallel network containing two parallel branches. If R = 10 kΩ, X_C = 20 kΩ, and X_L = 10 kΩ, find the total impedance of the network.

10. In the network of Figure 13–14, find Z_T when R = 5.0 kΩ, X_C = 10 kΩ, and X_L = 10 kΩ.

11. In Example 13.8 find Z_T when R_S = 5.1 kΩ, R_1 = 100 Ω, R_2 = 200 Ω, X_L = 5.0 kΩ, and X_C = 6.0 kΩ.

12. In Example 13.8 find Z_T when R_S = 18 kΩ, R_1 = 500 Ω, R_2 = 200 Ω, X_L = 15 kΩ, and X_C = 12 kΩ.

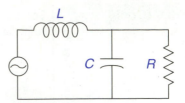

FIGURE 13–12
Series-parallel circuit for problems 3 and 4.

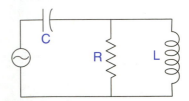

FIGURE 13–13
Series-parallel circuit for problems 7 and 8.

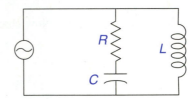

FIGURE 13–14
Series-parallel network for problems 9 and 10.

13. In Example 13.6, given that V_T = 15$\angle 0°$ V, find I_T, V_R, and V_L in rectangular and polar form. (Note: Kirchoff's law applies to a closed loop in an ac circuit for *phasor voltages*: $V_T = V_R + V_L$.)

14. In Example 13.7, given that V_T = 10$\angle 0°$ V, find I_R, V_R, and V_L in rectangular and polar form. (Note: Kirchoff's law applies to a closed loop in an ac circuit for *phasor voltages*: $V_T = V_R + V_L$.)

≡ CHAPTER HIGHLIGHTS

13.1 *RL* AND *RC* CIRCUITS

In an *RL* parallel circuit, I_L lags V_L and I_R by 90° and therefore has a negative phase angle of 90°. The total current in an *RL* parallel circuit is

$$I_T = I_R - jI_L = \sqrt{I_R{}^2 + I_L{}^2} \ \angle \tan^{-1}\left(\frac{-I_L}{I_R}\right) \quad (13.1)$$

In an *RC* parallel circuit, I_C leads V_C and I_C by 90° and therefore has a positive phase angle of 90°. The total current in an *RC* parallel circuit is

$$I_T = I_R + jI_C = \sqrt{I_R{}^2 + I_C{}^2} \ \angle \tan^{-1}\left(\frac{I_C}{I_R}\right) \quad (13.2)$$

See Examples 13.1 and 13.2.

In a parallel circuit, conductance G, susceptance B, and admittance Y are defined:

$$G = \frac{1}{R}$$

$$B = \frac{1}{X}$$

$$Y = \frac{1}{Z} \qquad (13.3)$$

For a resistance R in parallel with a reactance, the total admittance Y is:

$$Y = \frac{1}{R} \pm j\frac{1}{X} = G \pm jB \qquad (13.4)$$

The imaginary component jB is negative for an RL circuit and positive for an RC circuit. Study Example 13–3.

13.2 *RLC* CIRCUITS

The total current of an RLC circuit in rectangular form is

$$I_T = I_R + j(I_C - I_L) = I_R \pm jI_X \qquad (13.5)$$

where $I_X = |I_C - I_L|$ is the magnitude of the net reactive current.

When $I_L > I_C$, use formula (13.1) for I_T. When $I_L < I_C$, use formula (13.2) for I_T. See Figure 13–6.

In an RLC parallel circuit, the admittance is given by

$$Y = \frac{1}{R} + j\left(\frac{1}{X_C} - \frac{1}{X_L}\right) = G + j(B_C - B_L) = G \pm jB \qquad (13.6)$$

where $B = |B_C - B_L|$ is the magnitude of the net susceptance. Study Example 13.4.

13.3 SERIES-PARALLEL CIRCUITS

The total impedance of two or more impedances in series is equal to their sum:

$$Z_T = Z_1 + Z_2 + Z_3 + \cdots \qquad (13.7)$$

The equivalent impedance of two parallel impedances is equal to their product divided by their sum:

$$Z_{EQ} = \frac{Z_1 Z_2}{Z_1 + Z_2} \qquad (13.8)$$

Formulas (13.6) and (13.7) apply to *complex phasors* and their operations of multiplication and division. Study Examples 13.6, 13.7, and 13.8.

☰ REVIEW QUESTIONS

For all problems round answers to two significant digits.

In problems 1 through 4, using the given values for the parallel ac circuit, find the total current and the total impedance in rectangular and polar form.

1. *RL* circuit: $V = 9.0$ V, $R = 4.3$ kΩ, $X_L = 6.2$ kΩ

2. *RC* circuit: $V = 12$ V, $R = 820$ Ω, $X_C = 1.8$ kΩ

3. *RLC* circuit: $V = 30$ V, $R = 20$ kΩ, $X_L = 20$ kΩ, $X_C = 15$ kΩ

4. *RLC* circuit: $V = 24$ V, $R = 1.8$ kΩ, $X_L = 750$ Ω, $X_C = 1.0$ kΩ

In problems 5 through 8, find the admittance in rectangular and polar form for each of the parallel circuits in problems 1 through 4.

In problems 9 through 14 solve each applied problem.

9. In an RL parallel circuit, $R = 560$ Ω, $L = 20$ mH, and the applied voltage $V = 6.0$ V with a frequency $f = 6.0$ kHz. Find I_T, Z, and the equivalent series circuit values of R and L.

10. In an RC parallel circuit, $R = 2.0$ kΩ, $C = 25$ nF, and the applied voltage $V = 20$ V with a frequency $f = 2.0$ kHz. Find I_T, Z, and the equivalent series circuit values of R and C.

11. In an *RLC* parallel circuit, $R = 1.5$ kΩ, $L = 50$ mH, $C = 10$ nF, and the applied voltage $V = 24$ V with a frequency $f = 10$ kHz. Find I_T, Z, and the equivalent series circuit values of R and L or C.

12. In an *RLC* parallel circuit, $R = 3.3$ kΩ, $L = 100$ mH, $C = 50$ nF, and $f = 1.5$ kHz. If the total current $I_T = 20$ mA, find the admittance Y, the total impedance Z, and the voltage V.

13. In the series-parallel network in Figure 13–15, $R_1 = 150$ Ω, $X_L = 620$ Ω, $R_2 = 100$ Ω, and $X_C = 510$ Ω. Find the total impedance of the network.

14. Figure 13–16 shows a series-parallel network. If $R_1 = 1.0$ kΩ, $R_2 = 200$ Ω, $X_L = 2.0$ kΩ, and $X_C = 1.5$ kΩ, find the total impedance of the network.

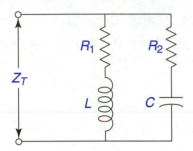

FIGURE 13–15 Series-parallel network for problem 13.

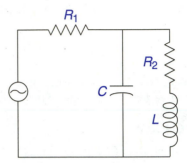

FIGURE 13–16 Series-parallel network for problem 14.

Logarithms

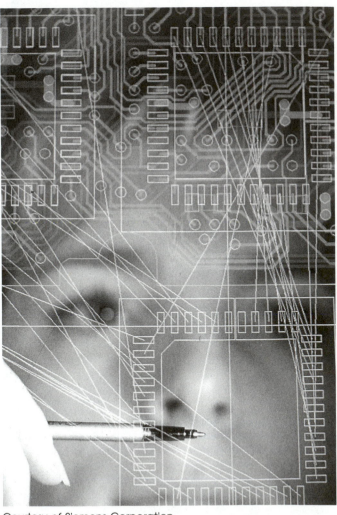

Courtesy of Siemens Corporation.

A technician examines a line card during testingof a central office switching system.

A logarithm is another name for an exponent. We use the term *logarithm* when we are working with the exponent in a formula or an equation. Logarithms simplify calculations and allow us to solve equations with exponents. There are many applications of exponents and logarithms in electronics, and they are presented throughout the chapter and especially in Section 14.3. This chapter begins with a study of exponential growth curves and exponential decay curves. These curves are important as they show the behavior of current and voltage in *RC* and *RL* circuits.

Chapter Objectives

In this chapter, you will learn:

- How to graph exponential growth and decay curves.
- How to find time constants in *RC* and *RL* circuits.
- How to find common and natural logarithms.
- How to solve exponential equations.
- How to use universal time constant curves.
- How to find instantaneous values of current and voltage in *RC* and *RL* circuits.
- How to compute power gain in electrical networks.

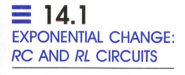

14.1
EXPONENTIAL CHANGE: *RC* AND *RL* CIRCUITS

An *exponential function* is defined as

$$y = b^x \quad b > 0 \tag{14.1}$$

where b is a positive constant called the *base*. Study the following examples that show how exponential functions give rise to growth and decay curves.

EXAMPLE 14.1

Graph the exponential function $y = 2^x$ from $x = -4$ to $x = 4$.

Solution   Substitute integral values of x, and compute the powers of 2. The table of values is as follows:

x	-4	-3	-2	-1	0	1	2	3	4
$y = 2^x$	0.0625	0.125	0.25	0.5	1	2	4	8	16

EXAMPLE 14.1 (Cont.)

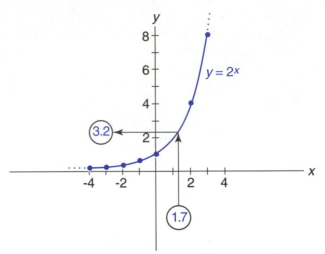

FIGURE 14–1   Exponential growth function for Example 14.1.

For example, when $x = -2$:

$$y = 2^{-2} = \frac{1}{2^2} = \frac{1}{4} = 0.25$$

Remember that a negative exponent means invert and change to a positive exponent (Section 3.2). The graph is shown in Figure 14–1. Study the graph. Observe that as x increases, y increases at a faster and faster rate and becomes infinitely large. The graph gets steeper and steeper and is said to increase exponentially. By using the graph of the exponential function 2^x in Figure 14–1, a value of y corresponding to *any value* of the exponent x can be defined. For example, $2^{1.7} \approx 3.2$ as shown on the graph by the arrows. The calculator provides a more precise answer:

$$2 \boxed{x^y} \; 1.7 \; = \; \rightarrow \; 3.2490 \; \ldots$$

The graph of an exponential function b^x, when $b > 1$ is called an *exponential growth function* such as $y = 2^x$ shown in figure 14–1. Many natural growth patterns behave in this way. Plants and animals in their early stages grow exponentially. Animal populations tend to grow exponentially when uncontrolled. The amount of money in a savings bank grows exponentially at a constant compound interest rate.

Observe also in Figure 14–1 that as x decreases, y decreases at a slower and slower rate and becomes infinitely small. The graph approaches the horizontal x axis as y gets closer and closer to zero. The x axis is called an *asymptote* of the graph. However, y never attains zero and never becomes negative. This decrease in y is the opposite of exponential growth and is called *exponential decay*. Study the next example, which shows exponential decay more directly using a negative exponent.

EXAMPLE 14.2

Graph the exponential function $y = 2^{-x}$ from $x = -4$ to $x = 4$.

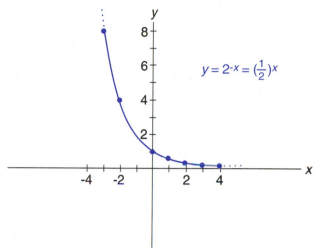

FIGURE 14–2 Exponential decay function for Example 14.2.

Solution   Substitute integral values of x and compute the powers of 2. The table of values is the reverse of the table in Example 14.1 as follows:

x	-4	-3	-2	-1	0	1	2	3	4
$y = 2^{-x}$	16	8	4	2	1	0.50	0.25	0.125	0.0625

This is because when x becomes $-x$, it reverses the values of x and, therefore, the values of y. Also note that the function can be written

$$y = 2^{-x} = \left(\frac{1}{2}\right)^{x}$$

The graph is shown in Figure 14–2 and is the mirror image of the graph in Figure 14–1 where the y axis is the mirror. Observe that as x increases, the value of y decreases at a slower and slower rate approaching a value of zero but never reaching it. This process is called exponential decay because natural decay patterns behave in this way, such as radioactive decay and biological decay. Exponential decay also occurs in electronic circuits as explained below.

▪

The Exponential Function e^x

In electronics, there are several important examples of exponential decay. When an inductive circuit is shorted, the current decays exponentially. When a capacitor is discharged, the voltage decays exponentially. Most important, *when a capacitor is charged or discharged*, the current decays exponentially. In these examples, it is the *rate of decay that is proportional to the amount of current or voltage.* That is, as the current or voltage decreases, the rate of decay also decreases proportionally. Exponential change is best described mathematically by the exponential function e^x. The number e is an irrational number like π and is equal to an infinite decimal:

$$e = 2.718281828459 \dots$$

The value of e is found using the methods of calculus. The following expression approaches e as n approaches infinity:

$$\left(1 + \frac{1}{n}\right)^n \rightarrow e \text{ as } n \rightarrow \infty \text{ (infinity)}$$

You can approximate e on the calculator by letting n equal a large number such as 10^6. Then

$$1 + \frac{1}{n} = 1 + \frac{1}{10^6} = 1.000001$$

and $\left(1 + \frac{1}{n}\right)^n = (1.000001)^{1,000,000}$ on the calculator is

$$1.000001 \;\boxed{y^x}\; 1,000,000 \;=\; \rightarrow 2.71828$$

This is e to six figures. Values of e^x are obtained on the calculator directly by pressing either $\boxed{\text{INV}}$, $\boxed{\text{2nd F}}$, or $\boxed{\text{SHIFT}}$, and then $\boxed{\text{lnx}}$ or $\boxed{\text{ln}}$. For example, e^3 is

$$3 \;\boxed{\text{INV}}\; \boxed{\text{ln}} \;\rightarrow\; 20.086$$

and $e^{-1.5}$ is

$$1.5 \;\boxed{+/-}\; \boxed{\text{SHIFT}}\; \boxed{\text{lnx}} \;\rightarrow\; 0.2231$$

The letter e was chosen in honor of the Swiss mathematician Leonhard Euler (1707–83), who was a great contributor to mathematics.

RC Time Constant

The following example closes the circuit and shows exponential decay in an *RC* circuit using the exponential function e^x. The example also explains the important concept of time constant.

EXAMPLE 14.3 Close the Circuit 	Figure 14–3 shows an *RC* circuit containing a constant voltage source *V* and a single pole double throw (SPDT) switch *S*. When *S* is switched to position 1, the capacitor is connected to the voltage source. The capacitor charges and the current *t* seconds after the switch is closed is given by $$i = \frac{V}{R}e^{-(t/RC)} \qquad (14.2)$$ After the capacitor is fully charged and the capacitor voltage v_C equals the applied voltage *V*, if *S* is switched to position 2, the capacitor will discharge through the resistance. The current *t* seconds after *S* is switched to position 2 is also given by equation (14.2).

EXAMPLE 14.3 (Cont.)

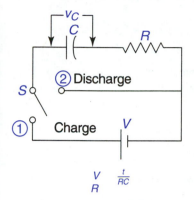

FIGURE 14–3 *RC* circuit, charge and discharge current, for Example 14.3.

Draw the graph of i versus t when $V = 100$ V, $R = 500$ Ω, and $C = 2.0$ μF.

Solution The *time constant* for an *RC* circuit is given by

$$\tau = RC \quad (\tau = \text{tau}) \qquad (14.3)$$

The time constant provides a standard time unit to compare different circuits. The time constant for the given circuit is

$$\tau = (500 \text{ }\Omega)(2.0 \text{ }\mu\text{F}) = (500)(2.0 \times 10^{-6}) = 0.0010 \text{ s} = 1.0 \text{ ms}$$

The time constant tells you how long it takes for the current to decrease to 63% of its initial value. See Figure 14–4. After five time constants, the current decreases more than 99% to practically zero.

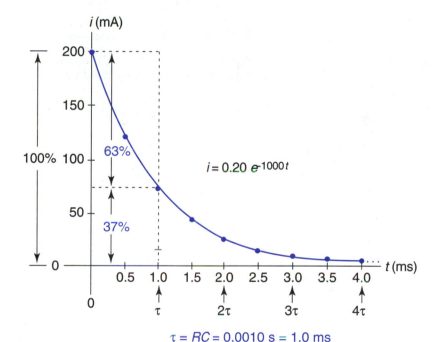

$$\tau = RC = 0.0010 \text{ s} = 1.0 \text{ ms}$$

FIGURE 14–4 Exponential decay of current in an *RC* circuit for Example 14.3.

To draw the graph, substitute the given values in (14.2). The equation for i in amps is then

$$i = \frac{100}{500} e^{(-t/0.001)} = 0.20 \, e^{-1000t}$$

For the graph, it is sufficient to choose values of t up to four time constants $= 4(0.0010) = 4.0$ ms. Choosing values of t every 0.5 ms, the table of values for i is as follows:

t (ms)	0	0.5	1.0	1.5	2.0	2.5	3.0	3.5	4.0
i (mA)	200	120	74	45	27	16	10	6.0	3.7

EXAMPLE 14.3 (Cont.)

For example, when $t = 1.5$ ms $= 0.0015$ s,

$$i = 0.20e^{-1000(0.0015)} = 0.20e^{-1.5} = 0.045 \text{ A} = 45 \text{ mA}$$

This can be done on the calculator as follows:

$$0.20 \; \boxed{\times} \; 1.5 \; \boxed{+/-} \; \boxed{\text{INV}} \; \boxed{\text{In}} \; \boxed{=} \; \rightarrow 0.0446$$

The graph in Figure 14–4 shows the exponential decay of current. Theoretically, the current never reaches zero; however, practically it reaches zero after five time constants. ▪

RL Time Constant

Figure 14–5(a) shows an RL circuit containing a constant voltage source V. When the switch is closed, the current increase is given by the exponential function:

$$i = \frac{V}{R}(1 - e^{-Rt/L}) \tag{14.4}$$

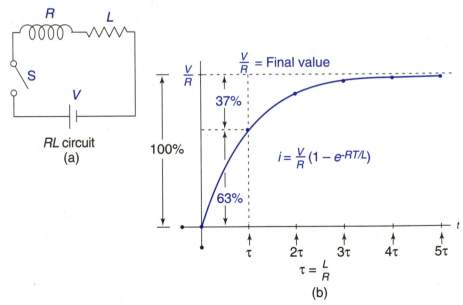

FIGURE 14–5 *RL circuit and exponential graph of current.*

The current is initially zero when the switch is closed and increases at a *slower and slower* rate to a final value of $\frac{V}{R}$. The rate of change of current decays to zero in the same way as the rate of change of current in an *RC* circuit. The time constant is given by

$$\tau = \frac{L}{R} \tag{14.5}$$

The time constant tells you how long it takes for the current to increase to 63% of its final value. See the graph in Figure 14–5(b). It is the same exponential curve as Figure 14–4, but it is now turned upside down. The final value is practically reached after five time constants.

For example, if $R = 330 \ \Omega$, $L = 500$ mH, and $V = 30$ V, the time constant is

$$\tau = \frac{L}{R} = \frac{500 \text{ mH}}{330 \ \Omega} = \frac{0.50}{330} = 0.00152 \text{ s} = 1.52 \text{ ms}$$

The current after 2 time constants ($t = 3.04$ ms) is

$$i = \frac{30 \text{ V}}{330 \ \Omega} (1 - e^{-(330)(0.00304) / 0.50})$$

$$= 0.0909(1 - e^{-2.0}) = 0.0909(0.865) = 0.0786 \text{A} \approx 79 \text{ mA}$$

EXERCISE 14.1

In all problems, round answers to two significant digits.

In problems 1 through 8, graph each exponential function plotting at least six points.

1. $y = 2.5^x$

2. $y = 1.5^x$

3. $y = 1.5^{-x}$

4. $y = 2.5^{-x}$

5. $y = e^x$

6. $y = e^{-x}$

7. $y = 1 - e^{-x}$

8. $y = 1 - e^x$

Applications to Electronics In problems 9 through 16, find the time constant of the given circuit.

9. *RC* circuit: $R = 200 \ \Omega$, $C = 5.0 \ \mu\text{F}$

10. *RC* circuit: $R = 1.0 \ \text{k}\Omega$, $C = 2.0 \ \mu\text{F}$

11. *RC* circuit: $R = 10 \ \text{k}\Omega$, $C = 400$ nF

12. *RC* circuit: $R = 2.0 \ \text{M}\Omega$, $C = 200$ pF

13. *RL* circuit: $R = 30 \ \Omega$, $L = 60$ mH

14. *RL* circuit: $R = 75 \ \Omega$, $L = 500$ mH

15. *RL* circuit: $R = 1.0 \ \text{k}\Omega$, $L = 0.5$ H

16. *RL* circuit: $R = 120 \ \Omega$, $L = 100$ mH

In problems 17 through 20 solve each problem by graphing the indicated curve.

17. Given the *RC* circuit in Figure 14–3 with $V = 50$ V, $R = 200 \ \Omega$, and $C = 10 \ \mu\text{F}$,
   (a) Find the time constant of the circuit.
   (b) Graph the exponential decay of current when *S* is switched to position 2 after the capacitor is charged. Show points for one, two, three, and four time constants. See Example 14.3.

18. Do the same as in problem 17 for $V = 100$ V, $R = 2.0 \ \text{k}\Omega$, and $C = 5.0 \ \mu\text{F}$.

19. Given the *RL* circuit in Figure 14–5 with $V = 60$ V, $R = 75 \ \Omega$, and $L = 300$ mH,
   (a) Find the time constant of the circuit.
   (b) Graph the exponential curve of current when the switch is closed showing points for one, two, three, and four time constants. See *RL* time constant.

20. Do the same as in problem 19 when $V = 20$ V, $R = 1.0 \ \text{k}\Omega$, and $L = 0.10$ H.

≣ 14.2
COMMON AND NATURAL LOGARITHMS

A logarithmic function is the inverse of an exponential function in the same way that division is the inverse of multiplication and $\tan^{-1}$ is the inverse of tan. A *logarithmic function* is defined as

$$y = \log_b x, \text{ where } b^y = x \; [b > 0 \text{ and } b \neq 1] \tag{14.6}$$

The definition says: "The logarithm of x to the base b is the exponent y," or simply, "log x to the base $b = y$." *A logarithm is therefore another way of looking at an exponent.* For example,

$$\log_2 16 = 4 \text{ means } 2^4 = 16$$

The first equation is called *logarithmic form,* and the second equation is called *exponential form.* Other examples of logarithmic and exponential form are shown in the following example.

EXAMPLE 14.4

Give some examples of logarithmic and exponential form.

Solution Examples of the two forms are shown in the following table.

Logarithmic Form	Exponential Form
$\log_5 125 = 3$	$5^3 = 125$
$\log_4 8 = 1.5$	$4^{1.5} = 8$
$\log_{10} 0.001 = -3$	$10^{-3} = 0.001$
$\log_e 20 = 3$	$e^3 = 20$
$\log_e 0.368 = -1$	$e^{-1} = 0.368$

Common Logarithms

The number 10 is used as a base for logarithms because our number system is based on 10 and it is therefore easy to work with. Base 10 logarithms are called *common logs.* We do not write the base when it is 10. It is understood, as shown in the following table of common logs:

$$\log 0.001 = \log 10^{-3} = -3$$
$$\log 0.01 = \log 10^{-2} = -2$$
$$\log 0.1 = \log 10^{-1} = -1$$
$$\log 1.0 = \log 10^0 = 0$$
$$\log 10 = \log 10^1 = 1$$
$$\log 100 = \log 10^2 = 2$$
$$\log 1000 = \log 10^3 = 3$$

Observe that the power of 10 *is* the logarithm. That is, $\log(10^x) = x$. Also note that the logarithm of a number between two numbers in the table lies between the two powers of ten. For example, log 50 = 1.70 is between 1 and 2, and log 0.5 = −0.301 is between −1 and 0. You should memorize this table. It will give you a better understanding of logarithms and provide a quick check on results.

Natural Logarithms

The exponential number e is the other number that is used as a base for logarithms. The reason for using e is because many natural exponential functions are easy to work with when expressed in terms of e, such as formula (14.2) for an *RC* circuit and formula (14.4) for an *RL* circuit. Base e logarithms are called *natural logs*. When the base is e, we use "ln" instead of log and also do not write the base:

$$\ln 7.39 = 2$$

$$\ln 0.368 = -1$$

In particular, note that $\ln e = 1$, $\ln e^x = x$, and $\ln 1 = 0$.

Your calculator has keys for common and natural logarithms. Press $\boxed{\log}$ to find the common log of a number:

$$3.52 \ \boxed{\log} \ \to \ 0.5465$$

$$0.0511 \ \boxed{\log} \ \to \ -1.292$$

Observe that for a positive number smaller than one, the logarithm is negative. If you enter zero or a negative number and press $\boxed{\log}$, the display will show error since these logarithms are not defined.

Press $\boxed{\ln}$ or $\boxed{\ln x}$ to find the natural log:

$$0.587 \ \boxed{\ln} \ \to \ -0.5327$$

$$4.39 \ \boxed{\ln x} \ \to \ 1.479$$

The exponential key $\boxed{10^x}$ is the inverse of the common log key $\boxed{\log}$. For example, if you are given that log $x = 0.5021$ and want to find x, use the key $\boxed{10^x}$:

$$0.5021 \ \boxed{10^x} \ \to \ 3.178$$

The inverse of the logarithm is also called the antilog. Similarly, $\boxed{e^x}$ is the inverse of the natural log key $\boxed{\ln}$. If ln $x = -0.8679$, then x is found by using $\boxed{e^x}$:

$$0.8679 \ \boxed{+/-} \ \boxed{e^x} \ \to \ 0.4198$$

Exponential Equations

Logarithms were discovered in 1614 by John Napier of Scotland to simplify calculations by using exponents. The rules for logarithms *for any base* follow from the rules for exponents:

$$\log (AB) = \log A + \log B \quad \text{(Addition rule)} \tag{14.7}$$

$$\log \left(\frac{A}{B}\right) = \log A - \log B \quad \text{(Subtraction rule)} \tag{14.8}$$

$$\log (A^x) = x(\log A) \quad \text{(Multiplication rule)} \tag{14.9}$$

The addition and subtraction rules can be used to simplify multiplication and division calculations. However, they are rarely used for that today because of the calculator. The multiplication rule, however, enables you to solve equations with exponents as shown in the following example.

EXAMPLE 14.5

Solve the exponential equation: $2^x = 50$

Solution   Take the common logarithm (or natural logarithm) of both sides of the equation:

$$\log (2^x) = \log 50$$

Apply the multiplication rule (14.9) and make the x a coefficient:

$$x(\log 2) = \log 50$$

Then divide by log 2 to find x and evaluate on the calculator:

$$x = \frac{\log 50}{\log 2} = \frac{1.6990}{0.3010} = 5.64$$

This can be done in one series of steps on the calculator:

$$50 \; \boxed{\log} \; \boxed{\div} \; 2 \; \boxed{\log} \; \boxed{=} \; \rightarrow 5.64$$

This means that $2^{5.64} = 50$. You can quickly check if this answer is approximately correct. Observe that the value of x should be between 5 and 6 since $2^5 = 32$ and $2^6 = 64$. You can also use natural logs instead of common logs to solve this equation.

ERROR BOX

A common error when working with logarithms is not seeing a simple relationship between a logarithm and an exponent. As shown with powers of 10, $\log 10^x = x$. You can also see this from the multiplication rule: $\log (10^x) = x(\log 10) = x(1) = x$. Similarly, you should understand that $\ln (e^x) = x$. See if you can apply this idea, and do the practice problems *without the calculator* to reinforce the concepts.

Practice Problems: Find the value of x or y *without the calculator:* Give answers to 8 and 10 in terms of e.

1. $\log 10^3 = y$ 2. $\ln e^2 = y$ 3. $\log \left(\frac{1}{10}\right) = y$ 4. $\ln \left(\frac{1}{e^2}\right) = y$

5. $\log 10^{1.5} = y$ 6. $\ln \sqrt{e} = y$ 7. $\log x = 2$ 8. $\ln x = 1$

9. $\log x = -1$ 10. $\ln x = 0.5$

Answers: 7. 100   8. e 9. $\frac{1}{10}$ 10. $\sqrt{e}$

1. 3   2. 2   3. −1   4. −2   5. 1.5   6. 0.5

The next example shows an exponential equation where natural logs should be used.

EXAMPLE 14.6

Solve the following exponential equation:

$$1 - e^{-10t} = 0.70$$

Solution   First solve for the exponential term by moving e^{-10t} and 0.70 to opposite sides of the equation:

$$e^{-10t} = 1 - 0.70 = 0.30$$

Take the natural logarithm of both sides since the base is e:

$$\ln(e^{-10t}) = \ln 0.30$$

The left side simplifies to $-10t$ since, by the multiplication rule,

$$\ln(e^{-10t}) = -10t(\ln e) = -10t(1) = -10t$$

You can also use the fact that $\ln(e^x) = x$ for any value of x. (See the Error Box.) The solution for t is then:

$$-10t = \ln 0.30$$

$$t = \frac{\ln 0.30}{-10} = \frac{-1.20}{-10} = 0.12$$

Study the next example, which closes the circuit and applies these ideas to a problem in electronics.

EXAMPLE 14.7

Close the Circuit

The charge or discharge current for the RC circuit in Figure 14–3 is given by the following exponential equation:

$$i = \frac{V}{R} e^{-(t/RC)}$$

If $V = 12$ V, $R = 100\ \Omega$, and $C = 5.0\ \mu F$, find the time t when:

1. The current $i = 10$ mA.

2. The current has decayed to 30% of its initial value.

Solution

1. To find t when $i = 10$ mA $= 0.010$ A, first find the time constant:

$$\tau = RC = (100\ \Omega)(5.0\ \mu F) = (100)(5.0 \times 10^{-6})$$

$$= 0.0005\ s = 0.50\ ms = 500\ \mu s$$

Then substitute the values of V, R, i, and τ into the given exponential equation:

$$0.010 = \frac{12}{100} e^{-(t/0.0005)} = 0.12\, e^{(-2000t)}$$

Note that $\frac{t}{0.0005} = 2000t$. To solve for t, first solve for the exponential term. Divide both sides by 0.12:

$$e^{(-2000t)} = \frac{0.010}{0.12} = 0.0833$$

EXAMPLE 14.7 (Cont.)

Take the natural logarithm of both sides since the base is e:

$$\ln[e^{(-2000t)}] = \ln 0.0833$$

The left side is simply $-2000t$ and the solution for t is

$$-2000t = \ln 0.0833$$

$$t = \frac{\ln 0.0833}{-2000} = \frac{-2.485}{-2000} = 1.24\,\text{ms}$$

2. To find when the current has decayed to 30% of its initial value, you can let $i_0 = \frac{V}{R}$ be the initial value of the current. The exponential equation can then be written:

$$i = i_0\,e^{-(t/RC)} = i_0\,e^{(-2000t)}$$

$$\frac{i}{i_0} = e^{(-2000t)}$$

When the current has decayed to 30% of its initial value,

$$\frac{i}{i_0} = 0.30$$

Therefore,

$$0.30 = e^{(-2000t)}$$

Taking the natural logarithm of both sides, the solution for t is

$$\ln(e^{-2000t}) = \ln 0.30$$

$$-2000t = \ln 0.30$$

$$t = \frac{\ln 0.30}{-2000} = \frac{-1.204}{-2000} = 0.60\,\text{ms} = 600\,\mu\text{s}$$

Therefore, the current reduces to 30% of its initial value after 0.60 ms or 600 μs.

EXERCISE 14.2

In problems 1 through 10, write each exponential equation in logarithmic form.

1. $2^3 = 8$

2. $5^{-1} = 0.2$

3. $10^4 = 10{,}000$

4. $10^{-2} = 0.01$

5. $10^{-0.2} = 0.631$

6. $10^{1.6} = 39.8$

7. $e^2 = 7.39$

8. $e^{-1} = 0.368$

9. $e^{-1.5} = 0.223$

10. $e^{0.5} = 1.649$

In problems 11 through 20, write each logarithmic equation in exponential form.

11. $\log_5 25 = 2$

12. $\log_2 0.25 = -2$

13. $\log 1000 = 3$

14. $\log 0.1 = -1$

15. $\log 0.316 = -0.5$

16. $\log 200 = 2.30$

17. $\ln 0.135 = -2$

18. $\ln 54.6 = 4$

19. $\ln 3 = 1.10$

20. $\ln 0.60 = -0.511$

In problems 21 through 30, find each logarithm to three significant digits.

21. $\log 86.1$

22. $\log 5.33$

23. $\log 0.105$

24. $\log 0.0484$

25. $\log (5.23 \times 10^3)$

26. $\log (2.55 \times 10^{-6})$

27. $\ln 25$

28. $\ln 1.43$

29. $\ln (6.52 \times 10^{-3})$

30. $\ln (0.45 \times 10^3)$

In problems 31 through 38, find the value of x to three significant digits.

31. $\log x = 0.700$

32. $\log x = 2.32$

33. $\ln x = 6.00$

34. $\ln x = 1.35$

35. $\log x = -3.12$

36. $\log x = -0.831$

37. $\ln x = -0.345$

38. $\ln x = -2.12$

In problems 39 through 50, solve each exponential equation to three significant digits.

39. $4^t = 23$

40. $5^n = 2$

41. $10^{x+1} = 20$

42. $10^{2n} = 0.15$

43. $e^{5t} = 5.2$

44. $e^{1.7t} = 0.36$

45. $e^{-1.3t} = 0.135$

46. $e^{-0.356t} = 4.32$

47. $1 - e^{-300t} = 0.90$

48. $1 - e^{-0.75t} = 0.50$

49. $1 - e^{0.25t} = -1.62$

50. $1 - e^{5.34t} = 0.98$

In problems 51 and 52 solve each problem to three significant digits.

51. When interest is compounded daily in a bank account at a rate of r percent a year, the principal (initial amount) will double in approximately n years where $n = \dfrac{\log 2}{\log (1 + r)}$. Find n when r is

(a) 2%

(b) 3%

(c) 5%.

52. The German lightning calculator Zacharias Dase (1824–61) was able to *mentally* calculate the product of two n-digit numbers in an incredibly small amount of time given approximately by the formula:

$$\log t = 2.72 \log n + \log 0.102$$

where t is in seconds. For example, he could multiply two 5-digit numbers in his head in less than one second. Find how long it took him to mentally multiply two

(a) 10-digit numbers.

(b) 20-digit numbers.

Applications to Electronics

In problems 53 through 58 solve each applied problem to two significant digits.

53. In Example 14.7, find the time t when

(a) The current $i = 50$ mA.

(b) The current has decayed to 25% of its original value.

54. In the RC circuit in Figure 14–3, after the capacitor is charged, if S is switched to position 2 causing the capacitor to discharge through the resistance, the voltage across the capacitor is given by

$$v_C = Ve^{-(t/RC)}$$

If $V = 60$ V, $R = 300$ Ω, and $C = 5.0$ μF, find the time t when
(a) The capacitor voltage $v_C = 15$ V.
(b) v_C has decayed to 50% of its initial value.

55. The capacitance of a cylindrical capacitor is given by

$$C = \frac{cl}{(18 \times 10^9)\ln(R_2/R_1)}$$

where c = dielectric constant, l = length, R_1 = inside radius, and R_2 = outside radius. Find C when $c = 5.8$, $l = 0.15$ m, $R_2 = 1.1$ cm, and $R_1 = 1.0$ cm.

56. A discharged battery is being charged by a solar cell. The voltage after t hours is given by

$$v = 13.2(1 - e^{-0.10t})$$

Find t when $v = 12.6$ V.

57. In the RL circuit in Figure 14–5(a), the current after the switch S is closed is given by

$$i = \frac{V}{R}\left(1 - e^{-Rt/L}\right)$$

If $V = 100$ V, $R = 200$ Ω, and $L = 0.5$ H, find the time t when
(a) The current $i = 400$ mA.
(b) The current has increased to 50% of its final value.

58. In the RC circuit shown in Figure 14–6, the voltage across the capacitor after the switch S is closed is given by the following exponential function:

$$v_C = V\left(1 - e^{(-t/RC)}\right)$$

If $V = 20$ V, $R = 750$ Ω, and $C = 10$ μF, find the time t when
(a) The voltage $v_C = 5.0$ V.
(b) v_C has increased to 50% of its final value.

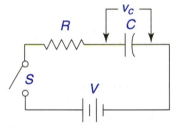

FIGURE 14–6 Charging voltage in an RC circuit for problem 58.

☰ 14.3

APPLICATIONS OF LOGARITHMS

Close the Circuit

All the examples in this section close the circuit and show applications of logarithms in electronics.

RC and RL Circuits

An important application of logarithms, finding values of current and voltage in RC and RL circuits, was shown in Section 14.2. The exponential curves for these circuits were shown in Section 14.1. We now study these ideas closer and examine the general picture for any RC or RL circuit.

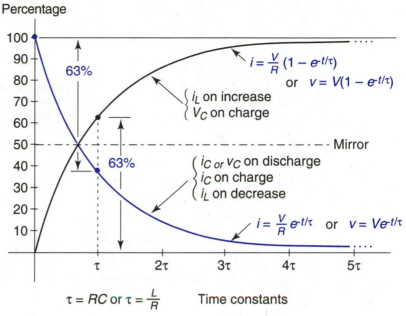

Percentage

$\tau = RC \text{ or } \tau = \dfrac{L}{R}$ Time constants

FIGURE 14–7 Universal time constant curves for *RC* and *RL* circuits.

Figure 14–7 shows the two universal exponential curves that apply to all *RC* and *RL* circuits. Study Figure 14–7 carefully. The curves are based on multiples of the time constant and the percentage of current or voltage. They can, therefore, be applied to any *RC* or *RL* circuit. The curve of exponential decay, which starts at 100% and decreases toward zero, has an equation of the type:

$$i = \frac{V}{R}e^{-t/\tau}$$

$$v = Ve^{-t/\tau} \tag{14.10}$$

where $\dfrac{V}{R}$ = initial current.
V = initial voltage.
t = time.
τ = time constant.

Note $\tau = RC$ for an *RC* circuit and $\tau = \dfrac{L}{R}$ for an *RL* circuit. This decay curve illustrates the following four situations:

▪ *RC circuit:* Capacitor current i_C on charge. Capacitor current i_C on discharge. Capacitor voltage v_C on discharge.

▪ *RL circuit:* Inductor current i_L on decrease.

The other rising exponential curve that starts at zero and increases toward 100% has an equation of the type:

$$i = \frac{V}{R}\left(1 - e^{-t/\tau}\right)$$

$$v = V\left(1 - e^{-t/\tau}\right) \tag{14.11}$$

where $\frac{V}{R}$ = steady state or maximum current.

V = steady-state or maximum voltage.

t = time.

τ = time constant.

This curve is mathematically the mirror image of the decay curve. The horizontal line passing through the point of intersection of the curves at 50% represents the mirror. This curve illustrates the following two situations:

▪ *RC circuit:* Capacitor voltage v_C on charge.

▪ *RL circuit:* Inductor current i_L on increase.

The interpretation of the curves is as follows. Consider the *RC* circuit in Figure 14–8(a). When *S* is switched to position 1 to charge the capacitor, the initial current = $\frac{V}{R}$. Then i_C decays according to (14.10), while v_C rises according to (14.11) to reach a maximum value = V. After v_C reaches V, if *S* is switched to position 2 to discharge the capacitor, the initial current is again $\frac{V}{R}$ and the initial voltage = V. Then i_C and v_C both decay according to (14.10).

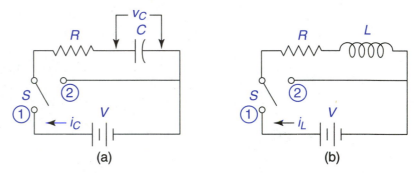

FIGURE 14–8   Exponential change in *RC* and *RL* circuits for Example 14.8
(a) Charge and discharge in   (b) Current increase and
    an *RC* circuit.   decrease in an *RL* circuit.

For the *RL* circuit in Figure 14–8(b), when *S* is switched to position 1, i_C increases according to (14.11) to reach a maximum value of $\frac{V}{R}$. After i_L reaches $\frac{V}{R}$, if *S* is switched to position 2, the initial current is $\frac{V}{R}$ and i_L decays according to (14.10). Since the curves have the same shape it follows that as a function of time, the percent decrease from the initial value for the decay curve equals the percent increase toward the final value for the other curve. This percent change is given by the following formula:

$$\text{Percent change} = \left(1 - e^{-t/\tau}\right) \times (100\%)$$

where t/τ = number of time constants. The values for t/τ and the percent change are shown in Table 14.1. After 0.70 time constant, the percent change is 50%, and after 1 time constant, the percent change is 63%. After 5 time constants, the percent change is 99%, and the circuit reaches steady state for all practical purposes.

TABLE 14.1

Number of Time Constants	Percent Change
0.25	22%
0.50	39%
0.70	50%
1.0	63%
2.0	87%
3.0	95%
4.0	98%
5.0	99%

Study the following example, which shows how to apply the above ideas.

EXAMPLE 14.8

For the RC circuit in Figure 14–8(a) $V = 20$ V, $R = 250$ Ω, and $C = 10$ μF. If S is switched to position 1 to charge the capacitor,

1. Find i_C and v_C after 1.5 ms.

2. Find t when $i_C = 20$mA.

3. How much time does it take for v_C to increase to 95% of its final value?

Solution

1. To find values of i_C and v_C, first find the time constant:

$$\tau = RC = (250\ \Omega)\ (10\ \mu F) = (250)(10 \times 10^{-6}) = 0.0025\ \text{s} = 2.5\ \text{ms}$$

Then to find i_C after 1.5 ms, substitute $t = 1.5$ ms, $V = 20$ V, $R = 250$ Ω, and $\tau = 2.5$ ms in (14.10):

$$i_C = \frac{V}{R}e^{-t/\tau} = \frac{20}{250}e^{-1.5/2.5} = 0.08\,(0.5488) = 0.0439\,\text{A} \approx 44\,\text{mA}$$

To find v_C after 1.5 ms, substitute $t = 1.0$ ms, $V = 20$ V, and $\tau = 2.5$ ms in (14.11):

$$v_C = V\left(1 - e^{-t/\tau}\right) = 20\left(1 - e^{-1.5/2.5}\right) = 20\,(0.4512) = 9.0\,\text{V}$$

2. To find t when $i_C = 20$ mA, substitute the given values in (14.10) and solve for t using natural logarithms:

$$0.020 = \frac{20}{250}e^{-t/2.5} = 0.08\,e^{-t/2.5}$$

$$e^{-t/2.5} = \frac{0.02}{0.08} = 0.25$$

$$\ln\,(e^{-t/2.5}) = \ln 0.25$$

$$\frac{-t}{2.5} = \ln 0.25$$

$$t = -2.5\,(\ln 0.25) = 3.5\ \text{ms}$$

EXAMPLE 14.8 (Cont.)

Note that the calculations are done using milliseconds for t and τ.

3. To find how long it takes for v_C to increase to 95% of its final value, note from Table 14.1 that this percent change corresponds to 3.0 time constants. Therefore, the time is

$$t = 3.0\,\tau = 3.0(2.5\ \text{ms}) = 7.5\ \text{ms}$$

Decibels and Power Ratios

Human senses are not as precise as electronic measurements. We detect changes in sound levels not based on the actual power increase but based on a *ratio* of the power increase. To illustrate, suppose the power output of an amplifier changes from 5 W to 10 W. This is an *actual* increase of 5 W but a *power ratio* increase of $\frac{10}{5} = \frac{2}{1}$. The increase in sound intensity that the human ear detects will *not* be the same if the power increases another 5 W to 15 W. However, the increase *will* sound the same if the power increases by the *same* ratio from 10 W to 20 W. That is, the power change from 5 W to 10 W will sound the same as the power change from 10 W to 20 W, or from 20 W to 40 W, and so on.

As a result of this phenomenon, sound intensity levels are defined in terms of logarithms of power ratios. This places sound intensity levels corresponding to power values such as 5 W, 10 W, 20 W, and 40 W equally apart on a scale. This is because, by the subtraction rule for logarithms (14.8), log (10/5) = log 10 – log 5 is the same value as log (20/10) = log 20 – log 10 or the log of any two numbers with the same ratio. The original unit of the gain in sound intensity is called the bel in honor of Alexander Graham Bell:

$$\text{bel} = \log\left[\frac{P_2}{P_1}\right]$$

where P_2 = output power and P_1 = input power. The bel turned out to be too large a unit for practical use, so we use the decibel (dB) to define the gain in sound intensity. The decibel is one-tenth the size of a bel as follows:

$$\text{dB} = 10 \log\left[\frac{P_2}{P_1}\right] \tag{14.12}$$

Note that the decibel is defined in terms of common logarithms. For example, if the power input to an amplifier $P_1 = 0.25$ W and the power output $P_2 = 4.0$ W, the gain in sound intensity is

$$\text{dB} = 10 \log\frac{4.0}{0.25} = 10 \log 16 = 10\,(1.204) = 12.0\ \text{dB gain}$$

The human ear can just detect an increase of sound intensity of 1 dB.

Study the following example, which shows an application of logarithms using (14.12).

EXAMPLE 14.9

The input power of an amplifier is 5.0 mW. What output power is needed for a decibel gain of 20 dB?

Solution Substitute the given values in (14.12):

$$20 = 10 \log\left(\frac{P_2}{5.0}\right)$$

Solve for $\log\left(\frac{P_2}{5.0}\right)$:

$$\log\left(\frac{P_2}{5.0}\right) = \frac{20}{10} = 2$$

Rewrite in exponential form and solve for P_2:

$$10^2 = \frac{P_2}{5.0}$$

$$P_2 = 500\,\text{mW}$$

Therefore, an output power of 500 mW represents an increase in sound intensity of 20 dB.

▪

In an amplifier, if the input resistance R is the same as the output resistance R, then by the power law $P = \frac{V^2}{R}$, and you can write formula (14.12) in terms of voltage:

$$dB = 10 \log\left[\frac{V_2^2/R}{V_1^2/R}\right]$$

where V_1 = input voltage and V_2 = output voltage. Inverting and multiplying the fractions, the R's divide out and the fraction becomes:

$$dB = 10 \log\left[\frac{V_2^2}{V_1^2}\right] = 10 \log\left[\frac{V_2}{V_1}\right]^2$$

Applying the multiplication rule (14.9) and making the exponent 2 a coefficient, you have

$$dB = 2(10) \log\left[\frac{V_2}{V_1}\right] = 20 \log\left[\frac{V_2}{V_1}\right] \qquad (14.13)$$

The decibel gain can, therefore, be based on a voltage ratio unit as well. For example, suppose the input voltage of an amplifier is $V_1 = 40$ mV and the output voltage is $V_2 = 5.5$ V. If the input and output resistances are the same, the power gain to two significant digits is

$$dB = 20 \log\left(\frac{5.5}{0.040}\right) = 20\,(2.14) = 43\,dB\text{ gain}$$

EXAMPLE 14.10

The voltage gain of a network is 22 dB. If the input and output resistances are the same and the output voltage is measured at 16 V, what is the input voltage?

Solution Substitute in formula (14.13) and solve for V_1 using logarithms:

$$22 = 20 \log \left(\frac{16}{V_1} \right)$$

$$\log \left(\frac{16}{V_1} \right) = \frac{22}{20} = 1.1$$

Change to exponential form:

$$10^{1.1} = \frac{16}{V_1}$$

$$V_1 = \frac{16}{10^{1.1}} = 1.3 \text{ V}$$

Decibel levels must be in reference to some standard level which corresponds to 0-dB. One of the common references used in telecommunications is $P_0 = 1$ mW. For this reference, the symbol dBm is generally used. A level of 10 dBm then means an input power $P_0 = P_1 = 1$ mW and an output power $P_2 = 10$ mW. This corresponds to a decibel gain of $10 \log \left(\frac{10}{1} \right) = 10$ dB.

Antenna Gain

The *gain* of an antenna compares the power input of the antenna with the power input required by a standard reference antenna to produce the same signal strength. If P_1 is the power input of the given antenna and P_2 is the power input of the standard antenna, the *power gain* of the given antenna is given by formula (14.12) for decibel gain. Study the following example, which applies this idea.

EXAMPLE 14.11

A multielement transmitting antenna produces a 150 mV/m (millivolts per meter) signal at a receiving station when supplied with 500 W. A standard dipole antenna requires 2.0 kW to produce the same signal strength at the receiving station. What is the power gain of the multielement antenna?

Solution Substitute $P_2 = 2.0$ kW = 2000 W and $P_1 = 500$ W into (14.12) and compute the power gain:

$$dB = 10 \log \frac{2000}{500} = 6.0 \text{ dB}$$

Therefore, the power gain of the multielement antenna is 6.0 dB.

EXERCISE 14.3

For all problems round answers to two significant digits. In problems 1 through 8 solve each applied problem.

1. In the RC circuit in Figure 14–8(a), $V = 60$ V, $R = 750$ Ω, and $C = 20$ μF. If S is switched to position 1 to charge the capacitor,
   (a) Find i_C and v_C after 20 ms.
   (b) How much time does it take for $i_C = 50$ mA?, $v_C = 25$ V?

2. In problem 1 after the capacitor is fully charged, S is switched to position 2, discharging the capacitor.
   (a) Find i_C and v_C after 10 ms.
   (b) How much time does it take for $i_C = 10$ mA?, $v_C = 10$ V?

3. In problem 1, how much time does it take for
   (a) i_C to decrease to 50% of its initial value?
   (b) v_C to increase to 98% of its final value?

4. In problem 2, how much time does it take for
   (a) v_C to decrease to 39% of its initial value?
   (b) i_C to decrease to 87% of its initial value?

5. In the RL circuit in Figure 14–8(b), $V = 100$ V, $R = 100$ Ω, and $L = 500$ mH. If S is switched to position 1 causing the current to increase,
   (a) Find i_L after 10 ms.
   (b) How much time does it take for $i_L = 500$ mA?
   (c) How much time does it take for i_L to increase to 22% of its final value?

6. In problem 5, after the circuit reaches steady state, S is switched to position 2 and the current decreases.
   (a) Find i_L after 4 ms.
   (b) How much time does it take for $i_L = 250$ mA?
   (c) How much time does it take for i_L to decrease to 99% of its initial value?

7. In Figure 14–8(a) when S is switched to position 1, the charge on the capacitor in coulombs (C) is given by the exponential equation:
$$q_C = CV(1 - e^{-t/RC})$$
   If $V = 40$ V, $C = 100$ μF, and $R = 1.0$ kΩ, find
   (a) q_C after 250 ms
   (b) The time t for the capacitor to be 95% charged.
   (Note: This equation is the same curve as (14.11) where CV represents 100% charge.)

8. In Figure 14–8(a) when S is switched to position 1, the voltage across the resistance is given by the exponential equation:
$$v_R = Ve^{-t/RC}$$
   If $V = 20$ V, $C = 500$ nF, and $R = 2.0$ kΩ, find
   (a) v_R after 500 μs.
   (b) The time t for v_R to decrease to 63% of its initial value.
   (Note: This equation is the same as (14.10).)

In problems 9 through 12, P_1 = input power and P_2 = output power of an amplifier. Find the decibel gain.

9. $P_1 = 50$ mW, $P_2 = 5$ W

10. $P_1 = 100$ mW, $P_2 = 15$ W

11. $P_1 = 1.0$ W, $P_2 = 80$ W

12. $P_1 = 60$ mW, $P_2 = 20$ W

In problems 13 through 16, V_1 = input voltage and V_2 = output voltage of an amplifier. Find the decibel gain.

13. $V_1 = 40$ mV, $V_2 = 9$ V

14. $V_1 = 10$ mV, $V_2 = 6$ V

15. $V_1 = 200$ mV, $V_2 = 10$ V

16. $V_1 = 100$ mV, $V_2 = 25$ V

In problems 17 through 22 solve each applied problem.

17. The input power of an audio frequency amplifier is 10 mW. What output power is needed for a decibel gain of 10 dB?

18. The output power of an amplifier is 40 W with a decibel gain of 24 dB. What is the input power of the amplifier?

19. The voltage gain of an amplifier is 30 dB, and the output voltage is measured at 12 V. What is the input voltage if the input and output resistances are the same?

20. The voltage input of a network is 0.5 V, and the decibel gain is 25 dB. What is the output voltage if the input and output resistances are the same?

21. A medium-frequency directional antenna produces a signal of 110 mV/m at a receiving station when supplied with 1 kW of power. A standard vertical antenna requires 2 kW of power to produce the same signal strength at the receiving station. What is the power gain of the directional antenna?

22. A multielement antenna produces 25 μV/m at a receiving station when supplied with 75 W of power. A standard dipole half-wave antenna requires 500 W of power to produce the same signal strength at the receiving station. What is the power gain of the multielement antenna?

≡ CHAPTER HIGHLIGHTS

14.1 EXPONENTIAL CHANGE: *RC* AND *RL* CIRCUITS

An *exponential function* is defined as

$$y = b^x \quad b > 0 \qquad (14.1)$$

When $b > 1$, b^x is an exponential growth function. As x increases, y increases at a faster and faster rate and becomes infinitely large. When $0 < b < 1$, b^x is an exponential decay function. As x increases, b^x decreases at a slower and slower rate and decays to zero. See Figures 14–1 and 14–2.

The exponential function e^x, where $e \approx 2.718$, is used to describe exponential current and voltage changes in *RC* and *RL* circuits. To obtain values of e^x on the calculator, press $\boxed{\text{INV}}$, $\boxed{\text{SHIFT}}$, or $\boxed{\text{2nd F}}$, and then $\boxed{\ln x}$ or $\boxed{\ln}$. This is the same as $\boxed{e^x}$.

In the *RC* circuit in Figure 14–3, the current t seconds after S is switched to position 1 for charging the capacitor, or position 2 for discharging when the capacitor is fully charged, is given by

$$i = \frac{V}{R} e^{(-t/RC)} \qquad (14.2)$$

The time constant for an *RC* circuit is

$$\tau = RC \qquad (14.3)$$

The current decreases 63% in one time constant. In five time constants, the current decays to practically zero. See Example 14.3.

In the *RL* circuit in Figure 14–5, the current t seconds after the switch is closed is given by

$$i = \frac{V}{R}(1 - e^{-Rt/L}) \qquad (14.4)$$

The time constant for an *RL* circuit is

$$\tau = \frac{L}{R} \qquad (14.5)$$

The current increases at a slower and slower rate to a final steady-state value of $\frac{V}{R}$ after five time constants.

14.2 COMMON AND NATURAL LOGARITHMS

A *logarithmic function* is defined as

$$y = \log_b x, \text{ where } b^y = x \; [b > 0 \text{ and } b \neq 1]$$
$$(14.6)$$

See Example 14–4.

Common logarithms use base 10. When the base is not written, it is understood to be 10.

Natural logarithms use base e, where ln is used instead of log and the base is understood.

On the calculator, common logarithms are obtained by pressing $\boxed{\log}$. To find the inverse of a logarithm (antilog), press $\boxed{\text{INV}}$, $\boxed{\text{2nd F}}$, or $\boxed{\text{SHIFT}}$, and then $\boxed{\log}$. This is the same as $\boxed{10^x}$.

Natural logarithms are obtained by pressing $\boxed{\ln x}$ or $\boxed{\ln}$. To find the inverse of a natural logarithm, press the same as given for $\boxed{e^x}$.

The rules for logarithms for any base follow from the rules for exponents:

$$\log (AB) = \log A + \log B \quad \text{(Addition rule)}$$
$$(14.7)$$

$$\log \left(\frac{A}{B}\right) = \log A - \log B \quad \text{(Subtraction rule)}$$
$$(14.8)$$

$$\log (A^x) = x (\log A) \quad \text{(Multiplication rule)}$$
$$(14.9)$$

To solve an exponential equation, isolate the exponential term and take the logarithm of both sides applying the multiplication rule (14.9). See Examples 14.6 and 14.7.

14.3 APPLICATIONS OF LOGARITHMS

The universal exponential curves for voltage and current change in an *RC* circuit, and current change in an *RL* circuit, are shown in Figure 14–7.

The curve of exponential decay has an equation of the type

$$i = \frac{V}{R} e^{-t/\tau}$$

$$v = V e^{-t/\tau} \quad (14.10)$$

where $\frac{V}{R}$ = initial current.
V = initial voltage.
t = time.
and τ = time constant.

The decay curve illustrates the four situations:

- *RC circuit:* Capacitor current i_C on charge. Capacitor current i_C on discharge. Capacitor voltage v_C on discharge.
- *RL circuit:* Inductor current i_L on decrease.

The other rising exponential curve has an equation of the type:

$$i = \frac{V}{R}\left(1 - e^{-t/\tau}\right)$$

$$v = V\left(1 - e^{-t/\tau}\right) \quad (14.11)$$

where $\frac{V}{R}$ = steady-state or maximum current.
V = steady-state or maximum voltage.
t = time
and τ = time constant.

This curve illustrates the two situations:

- *RC circuit:* Capacitor voltage v_C on charge.
- *RL circuit:* Inductor current i_L on increase.

Study Example 14.8.

Power gain of an amplifier or any electrical network is defined in terms of decibels, which is a measure of the increase in sound intensity:

$$dB = 10 \log \frac{P_2}{P_1}$$
$$(14.12)$$

where P_2 = output power and P_1 = input power. Also, in an amplifier if the input resistance equals the output resistance,

$$dB = 20 \log \frac{V_2}{V_1}$$
$$(14.13)$$

where V_1 = input voltage and V_2 = output voltage. See Examples 14.9 and 14.10.

Decibel levels must be based on a standard power level that corresponds to 0 dB such as $P_0 = 1$ mW.

Antenna gain is also defined in terms of (14.12) where P_1 is the power input of the given antenna and P_2 is the power input of a standard antenna that produces the same signal strength.

≡ REVIEW QUESTIONS

In problems 1 through 4, write each equation in logarithmic form.

1. $10^3 = 1000$

2. $10^{-2.3} = 0.00501$

3. $e^{-3.5} = 0.0302$

4. $e^{0.28} = 1.32$

In problems 5 through 8, write each equation in exponential form.

5. $\log 0.0776 = -1.11$

6. $\log 5.33 = 0.727$

7. $\ln 3.69 = 1.31$

8. $\ln (4.55 \times 10^{-3}) = -5.39$

In problems 9 through 12, find each logarithm to three significant digits.

9. $\log 25$

10. $\log (5.23 \times 10^{-6})$

11. $\ln 0.675$

12. $\ln 36.5$

In problems 13 through 16, find x to three significant digits.

13. $\log x = 0.859$

14. $\log x = -8.25$

15. $\ln x = 4.21$

16. $\ln x = -0.434$

In problems 17 and 18, solve the exponential equation to three significant digits.

17. $10^{3x} = 30$

18. $1 - e^{-10t} = 0.50$

In problems 19 and 20 solve each applied problem to two significant digits.

19. A virus population increases according to the exponential equation:

$$N = N_0 \, (2)^{0.20 \, t}$$

where the initial number of viruses $N_0 = 10^6$ and $N =$ number after t hours. Graph N versus t for $t = -10$ to $t = 20$.

20. In problem 19,

   (a) How many hours does it take for the number of viruses to increase from one million to two million?

   (b) How many hours does it take for the population to increase ten times?

Applications to Electronics   In problems 21 through 26 solve each applied problem to two significant digits.

21. In the *RC* circuit in Figure 14–8(a) $V = 40$ V, $R = 1.5$ kW, and $C = 2.0$ μF. If S is switched to position 1 to charge the capacitor,

   (a) Find the time constant of the circuit.

   (b) Give the equation for the exponential decay of current i_C in milliamps.

   (c) Find i_C after 10 ms.

   (d) Find the time when $i_C = 5.0$ mA.

22. In the *RL* circuit in Figure 14–8(b) $V = 100$ V, $R = 100$ Ω, and $L = 500$ mH. If S is switched to position 1 causing the current i_L to increase,

   (a) Find the time constant of the circuit.

   (b) Give the exponential equation for the increase of current.

   (c) Find i_L after 8 ms.

   (d) Find the time when $i_L = 500$ mA.

23. In Figure 14–8(a) when S is switched to position 1, the voltage across the resistance v_R is given by the voltage decay curve in Figure 14–7. If $V = 9.0$ V, $C = 10$ μF, and $R = 1.2$ kΩ, find
   (a) v_R after 20 ms.
   (b) the time for v_R to decay to 40% of its initial value.

24. Given the voltage charge curve for a capacitor in Figure 14–7, how many time constants does it take for the voltage to increase to
   (a) 50% of its final value?
   (b) 75% of its final value?

25. An audio frequency amplifier has an input power of 2 mW and an input voltage of 1.5 V. If the decibel gain is 30 dB and the input and output resistances are the same, what are the output power and output voltage?

26. A directional antenna produces a signal of 100 μV/m at a receiving station when supplied with 50 W of power. A standard vertical antenna requires 250 W of power to produce the same signal strength at the receiving station. What is the power gain of the directional antenna?

CHAPTER

15

Computer Number Systems

Photo Courtesy of Hewlett-Packard Company.

A technician custom designs a local area network (LAN).

Our number system is called the decimal number system because it is based on the number 10. It evolved from the use of our fingers for counting. However, other systems can be used just as effectively. In the electronics of a microprocessor or computer, switches are either open or closed, current is either flowing in one direction or the other, magnetic cores are magnetized either clockwise or counterclockwise, and so on. Because of this *bistable* nature of electronic states, the base 2 or binary system is the primary system used in computers. Other systems compatible with the binary system, the base 8 or octal system, and the base 16 or hexadecimal system, are also used in conjunction with the binary system. This chapter studies these three computer number systems and computer codes that use these systems to represent data.

Chapter Objectives

In this chapter, you will learn:

- How to convert binary numbers to decimal numbers and vice versa.
- How to add and subtract binary numbers.
- How to convert octal numbers to decimal numbers and vice versa.
- How to convert octal numbers to binary numbers and vice versa.
- How to convert hexadecimal numbers to decimal numbers and vice versa.
- How to convert hexadecimal numbers to binary numbers and vice versa.
- The binary and hexadecimal codes used in computers.

15.1
BINARY NUMBER SYSTEM

The binary number or base 2 system uses only two digits, 0 and 1, to express all numbers. Data in a computer is represented by a series of binary digits or *bits*, and calculations are done by use of binary arithmetic. To better understand the base 2 system, and the base 8 and base 16 systems, it is helpful to first take a closer look at our base 10 decimal system.

Decimal Number System

The decimal number system contains 10 digits:

$$0, \ 1, \ 2, \ 3, \ 4, \ 5, \ 6, \ 7, \ 8, \ 9$$

which represent the first 10 positive integers in the system. The *base* or *radix* of the decimal system is therefore 10. A number larger than ten is expressed using these digits, where each digit represents a multiple of a *power of 10*. The powers of

10 increase from left to right. For example, the number 3789 in expanded form is equal to

$$3789 = 3 \times 10^3 + 7 \times 10^2 + 8 \times 10^1 + 9 \times 10^0$$
$$= 3 \times 1000 + 7 \times 100 + 8 \times 10 + 9 \times 1 = 3789$$

The powers of 10 are called the place values of the digits and are shown in Table 15.1. Observe in Table 15.1 that digits to the right of the decimal point correspond to negative powers of 10.

TABLE 15.1   Place Values of the Decimal Digits

				Decimal Point ↓			
Power	10^3	10^2	10^1	10^0	.	10^{-1}	10^{-2}	10^{-3}
Value	1000	100	10	1		0.1	0.01	0.001

EXAMPLE 15.1

Write the decimal number 52.34 in expanded form.

Solution   Using the place values of the digits from Table 15.1, the number in expanded form is

$$52.34 = 5 \times 10^1 + 2 \times 10^0 + 3 \times 10^{-1} + 4 \times 10^{-2}$$
$$= 5 \times 10 + 2 \times 1 + 3 \times 0.1 + 4 \times 0.01$$

We can now look at the binary system and compare it to the decimal system.

Binary Number System

The binary number system contains only two digits, 0 and 1, which represent the first two positive integers. The base or radix of the system is 2. A number larger than one is expressed using the digits 0 and 1, as shown in Table 15.2. Each digit represents a power of 2.

For example, the binary number 1101_2 in expanded form is

$$1101_2 = 1 \times 2^3 + 1 \times 2^2 + 0 \times 2^1 + 1 \times 2^0$$
$$= 1 \times 8 + 1 \times 4 + 0 \times 2 + 1 \times 1 = 13$$

The subscript 2 is used to identify the base when working with numbers from different bases. *If there is no subscript, the base is understood to be 10.*

The place values of the binary digits, including negative powers of 2 to the right of the *binary point*, are shown in Table 15.3. Note that the negative powers of 2 are fractions whose numerators are 1 and whose denominators are 2, 4, 8, and so on.

TABLE 15.2 Binary and Decimal Numbers

Binary	Decimal
0	0
1	1
10	2
11	3
100	4
101	5
110	6
111	7
1000	8
1001	9
1010	10
1011	11
1100	12
1101	13
1110	14
1111	15
10000	16

TABLE 15.3 Place Values of the Binary Digits

Binary Point
↓

Power	2^6	2^5	2^4	2^3	2^2	2^1	2^0	.	2^{-1}	2^{-2}	2^{-3}
Value	64	32	16	8	4	2	1		0.5	0.25	0.125

Study the next example, which shows how to convert binary numbers to decimal numbers.

EXAMPLE 15.2

Convert the following to decimal numbers:

1. 110100_2

2. 10.101_2

Solution

1. To convert a binary number to a decimal number, multiply each digit by its place value in Table 15.3 and add:

$$110100_2 = 1 \times 2^5 + 1 \times 2^4 + 0 \times 2^3 + 1 \times 2^2 + 0 \times 2^1 + 0 \times 2^0$$

$$= 32 + 16 + 0 + 4 + 0 + 0 = 52$$

EXAMPLE 15.2 (Cont.)

Observe that you only need to add the place values that have digits of 1. You do not need to consider the zero digits.

2. The binary number 10.101 contains a binary point and is not an integer. It contains a binary fraction, and you need to add negative powers of 2:

$$10.101 = 1 \times 2^1 + 1 \times 2^{-1} + 1 \times 2^{-3}$$
$$= 1 \times 2 + 1 \times 0.5 + 1 \times 0.125 = 2.625$$

To do the reverse of this process, that is, to convert a decimal integer to a binary number, you can apply the following algorithm called the *method of remainders:*

1. Divide the decimal integer by 2.
2. Record the remainder (0 or 1).
3. Divide the quotient by 2.
4. Record the remainder again.
5. Repeat steps 3 and 4 until the quotient is zero. The remainders in reverse order represent the binary integer.

The flowchart for the method of remainders algorithm is shown in Figure 15–1. Study the following example.

EXAMPLE 15.3

Convert the decimal number 46 to a binary integer.

Solution Apply the method of remainders. Divide 46 by 2 and record the remainder:

$$\frac{23}{2\overline{)46}} \quad \text{remainder} = 0$$

Continue dividing the quotients and writing down the remainders until the quotient is zero. Work from the top down as follows:

$$
\begin{array}{ccc}
& & \text{Remainders} \\
2\,\overline{)46} & & 0 \\
2\,\overline{)23} & & 1 \\
2\,\overline{)11} & & 1 \\
2\,\overline{)5} & & 1 \\
2\,\overline{)2} & & 0 \\
2\,\overline{)1} & & 1 \\
0 & &
\end{array}
$$

The binary digits or bits are the remainders reading from the bottom up: $46 = 101110_2$.

Another algorithm for converting a decimal integer to a binary integer is by working with powers of 2:

1. Choose the largest power of 2 not exceeding the decimal integer.

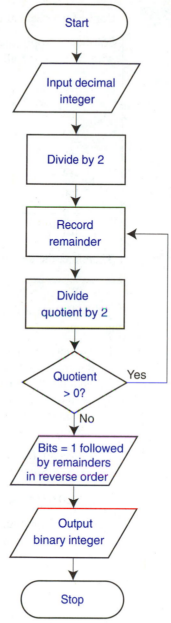

FIGURE 15–1 Flowchart for converting a decimal integer to a binary integer for Example 15.3.

2. Add smaller and smaller powers of 2 until you equal the number exactly without exceeding the number.

3. The powers of 2 that you add are the 1 bits of the binary integer and the powers of 2 that you do not add are the 0 bits.

Study the next example, which applies this algorithm.

EXAMPLE 15.4

Convert the decimal number 101 to a binary integer.

Solution Apply the preceding algorithm and add powers of 2 beginning with $2^6 = 64$ as follows:

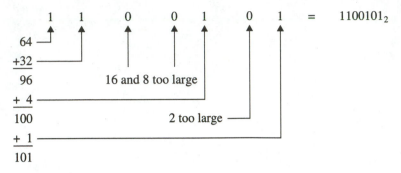

Therefore, the decimal number $101 = 1100101_2$.

To convert a decimal fraction to a binary fraction, you apply the following algorithm:

1. Multiply the fraction by 2.
2. Record the integral digit to the left of the decimal point (0 or 1).
3. Multiply the *fractional part of the result by 2.*
4. Record the integral digit again.
5. Continue steps 3 and 4 until the fractional part of the result is zero. The integral digits in order represent the binary fraction.

The flowchart for converting a decimal fraction to a binary fraction is shown in Figure 15–2. Study the following example.

EXAMPLE 15.5

Convert 0.625 to a binary fraction.

Solution Apply the preceding algorithm. Multiply 0.625 and subsequent fractional parts by 2. Write down the integral digits until the fractional part is zero:

$$
\begin{array}{ll}
& \text{Integral digits} \\
2(0.625) = 1.250 & 1 \\
2(0.250) = 0.50 & 0 \\
2(0.50) \ \ = 1.00 & 1 \\
\end{array}
$$

The binary fraction is the integral digits from the top down:

$$0.625 = 0.101_2$$

Study the next example, which shows how to convert a decimal number containing a fraction to a binary number.

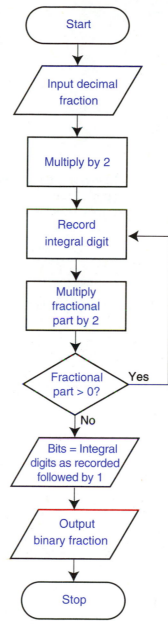

FIGURE 15–2 Flowchart for
converting a decimal fraction to
a binary fraction for Example 15.5.

EXAMPLE 15.6

Convert 44.25 to a binary number.

Solution   Convert the integral part, 44, and the fractional part, 0.25, separately
applying the preceding algorithms:

EXAMPLE 15.6 (Cont.)

	Remainders		Integral digits	
2)44	0			
2)22	0		$2(0.25) = 0.50$	0
2)11	1		$2(0.50) = 1.00$	1
2)5	1			
2)2	0			
2)1	1			
0				

Therefore, the decimal number $44.25 = 101\ 100.01_2$. Observe that the conversion of the fractional part will not always end, but can result in a repeating binary fraction. For example, the decimal fraction $0.8 = 0.1100\ 1100\ 1100\ \ldots$ (see problem 44).

■

EXERCISE 15.1

In problems 1 through 6, write each decimal number in expanded form.

1. 1328

2. 45,217

3. 38.96

4. 21.152

5. 10.01

6. 101.11

In problems 7 through 22, convert each binary number to a decimal number.

7. 11

8. 10

9. 100

10. 111

11. 1110

12. 1001

13. 10001

14. 10101

15. 110011

16. 100100

17. 10.11

18. 11.01

19. 1.101

20. 10.001

21. 11.1011

22. 1.1111

In problems 23 through 40, convert each decimal number to a binary number.

23. 4

24. 6

25. 11

26. 27

27. 57

28. 63

29. 99

30. 111

31. 128

32. 200

33. 0.5

34. 0.75

35. 0.875

36. 0.375

37. 2.25

38. 7.125

39. 33.625

40. 5.5625

In problems 41 through 44 solve each applied problem.

41. Complete Table 15.2 for the decimal numbers 17 through 32.

42. How many binary digits are needed to write the following decimal numbers:
   (a) 100.
   (b) 1000.
   (c) 10,000.
   (d) 1,000,000.

43. One way a computer stores data is by a binary-coded decimal (BCD), where 4 bits are used to represent each decimal digit. For example, 0010 1001 is a BCD for the decimal number 29. What decimal numbers do each of the following BCD's represent?
   (a) 0100 1001.
   (b) 0100 0000.
   (c) 0000 0111.
   (d) 0001 0010.

44. Show that the decimal fraction 0.8 is equal to the repeating binary fraction 0.1100 1100 1100 … Apply the method shown in Example 15.5 and show that the integral digits repeat.

15.2
BINARY ARITHMETIC

The algorithms for binary arithmetic are essentially the same as those for decimal numbers that you are familiar with. However, since there are only two digits in the binary system, addition is much simpler. The subscript 2 for a binary number is omitted in this section and is understood.

Binary Addition

Table 15.4 shows the four basic binary sums plus two other useful sums for adding binary numbers. The table tells the digit to record and the digit to carry to the next column.

TABLE 15.4   Binary Sums

$0 + 0 = \ \ 0$	
$0 + 1 = 1 + 0 = \ \ 1$	
$1 + 1 = \ \ 10$	(0 and carry 1)
$1 + 1 + 1 = 10 + 1 = \ \ 11$	(1 and carry 1)
$1 + 1 + 1 + 1 = 11 + 1 = 100$	(0 and carry 10)

Study the following examples, which show how to add binary numbers.

EXAMPLE 15.7

Add the binary numbers:

$$\begin{array}{r} 110 \\ \underline{111} \end{array}$$

Solution   Add each column starting with the units column and working to the left applying the sums in Table 15.4 and carrying digits:

$$
\begin{array}{ccccc}
 & & 1 & & 1 \quad \text{(carried digit)} \\
110 & & 110 & & 110 \\
\underline{111} & \Rightarrow & \underline{111} & \Rightarrow & \underline{111} \\
1 & & 01 & & 1101
\end{array}
$$

EXAMPLE 15.7 (Cont.)

Column two adds up to 10. Therefore 1 is carried to the third column as shown on the top. The decimal equivalent of the addition is:

$$
\begin{array}{r}
6 \\
\underline{7} \\
13
\end{array}
$$

▪

EXAMPLE 15.8

Add the binary numbers:

$$
\begin{array}{r}
10 \\
11 \\
111 \\
\underline{101}
\end{array}
$$

Solution The addition is as follows with the carried digits on the top:

$$
\begin{array}{r}
101 \qquad \text{(carried digits)} \\
\hline
10 \\
11 \\
111 \\
\underline{101} \\
10001
\end{array}
$$

The first column adds up to 11 and carry 1. The second column adds up to 100 with a carry of 10 to the third column. The third column adds up to 100. You can also do this addition by adding partial sums of two numbers at a time:

$$
\begin{array}{r}
10 \\
\underline{+11} \\
101 \\
\underline{+111} \\
1100 \\
\underline{+101} \\
10001
\end{array}
$$

The method of partial sums is used in computers. To add numbers with binary points, you align the points, as in decimal arithmetic, and add the same way as shown in Examples 15.7 and 15.8.

▪

It is common to add zeros in front of a binary number so that all columns being used are represented. For example, if four columns are used, 11 is written 0011 and 101 is written 0101. Most computers store numbers using a fixed number of digits and supply leading zeros if necessary.

Binary Subtraction Using Complements

Subtraction in a computer is usually done by adding complements since it simplifies the microprocessor circuitry. It is helpful to first see how subtraction is done with decimal numbers by adding complements. The *9's complement* of a decimal number is the number obtained by subtracting each digit from 9. The *10's complement* of a number is the number obtained by adding one to its 9's complement, or by subtracting the original number from a power of 10. For example, three numbers and their complements are:

Decimal Number	9's Complement	10's Complement
3	9 − 3 = 6	6 + 1 = 7 or 10 − 3 = 7
18	99 − 18 = 81	81 + 1 = 82 or 100 − 18 = 82
515	999 − 515 = 484	484 + 1 = 485 or 1000 − 515 = 485

The following example illustrates how subtraction is done by adding complements for decimal numbers.

EXAMPLE 15.9

Perform the following decimal subtraction by
1. Adding the 9's complement.
2. Adding the 10's complement.

$$817$$
$$-321$$

Solution

1. To subtract by adding the 9's complement, first obtain the 9's complement of the number to be subtracted (subtrahend): 999 − 321 = 678. Then use the following addition algorithm:

$$817$$
$$\underline{678}$$
$$495$$
$$\underline{\quad} \quad \text{(end-around carry)}$$
$$496$$

Take the leading 1 of the sum and add it to the remaining digits to obtain the answer to the subtraction. This is called *end-around carry* in computer terminology.

2. To subtract by adding the 10's complement, first obtain the 10's complement of the subtrahend: 678 + 1 = 679 or 1000 − 321 = 679. Then use the following addition algorithm:

$$817$$
$$\underline{679}$$
$$\cancel{1}496 = 496 \quad \text{(delete leading 1)}$$

Delete the leading 1 of the sum to obtain the answer.

Binary subtraction is done in a similar way to that shown for decimal subtraction in Example 15.9 by adding the 1's complement or the 2's complement. The 1's complement of a binary number is the number obtained by subtracting each digit from 1. *This is the same as changing each 0 to 1 and each 1 to 0.* The 2's complement of a number is the number obtained by adding 1 to its 1's complement. For example, three binary numbers and their complements are:

Binary Number	1's Complement	2's Complement
101	010	010 + 1 = 011
1011	0100	0100 + 1 = 0101
0110	1001	1001 + 1 = 1010

Observe for the first and second number that a leading zero is used to express the complement. For the third number, observe that a leading zero is added and becomes a 1 in the complement. Adding a leading zero or zeros does not affect the results when adding complements. Study the next two examples that explain how binary subtraction is done by adding complements, similar to the decimal subtraction shown in Example 15.9.

EXAMPLE 15.10

Subtract the following binary numbers by

1. Adding the 1's complement.

2. Adding the 2's complement.

$$1011$$
$$-1001$$

Solution

1. To subtract by adding the 1's complement, first obtain the 1's complement of the subtrahend by changing 0's to 1's and 1's to 0's: $1001 \rightarrow 0110$. Then use the following addition algorithm:

$$11 \quad \text{(carried digits)}$$
$$1011$$
$$0110$$
$$\overline{10001}$$
$$+1 \quad \text{(end-around carry)}$$
$$\overline{0010} = 10$$

Take the leading one of the sum and add it to the remaining digits, similar to when adding the 9's complement.

2. To subtract by adding the 2's complement, first obtain the 2's complement of the subtrahend: $0110 + 1 = 0111$. Then use the following addition algorithm:

$$111 \quad \text{(carried digits)}$$
$$1011$$
$$0111$$
$$\overline{10010} = 10 \quad \text{(delete leading one)}$$

Delete the leading 1 as is done when adding the 10's complement.

EXAMPLE 15.11

Subtract the following binary numbers:

$$10110$$
$$\underline{-1101}$$

Solution Add a leading zero to the subtrahend, then add either the 1's complement or the 2's complement using the algorithm shown in Example 15.10. The subtraction using the 1's complement is as follows:

$$
\begin{array}{lllll}
 & & & 11 & \text{(carried digits)} \\
10110 & & 10110 & & 10110 \\
\underline{-01101} & \Rightarrow & \underline{10010} & \Rightarrow & \underline{10010} \\
 & & & 101000 & \\
 & & & \underline{+1} & \text{(end-around carry)} \\
 & & & \overline{01001} = 1001 &
\end{array}
$$

See problem 31, which asks you to check this example using the 2's complement.

EXERCISE 15.2

In problems 1 through 14, add the binary numbers.

1. 11
   $\underline{10}$

2. 11
   $\underline{100}$

3. 100
   $\underline{111}$

4. 111
   $\underline{111}$

5. 110
   011
   $\underline{110}$

6. 011
   011
   $\underline{100}$

7. 11.1
   10.1
   $\underline{10.1}$

8. 1.10
   1.10
   $\underline{11.11}$

9. 1
   10
   101
   $\underline{111}$

10. 1
   111
   101
   $\underline{101}$

11. 010
   101
   101
   $\underline{111}$

12. 001
   100
   011
   $\underline{110}$

13. 101.01
   11.01
   $\underline{111.11}$

14. 111.10
   100.11
   $\underline{110.01}$

In problems 15 through 18, subtract the decimal numbers by adding the

(a) 9's complement.
(b) 10's complement.

15. 78
    −13

16. 132
    − 56

17. 1001
    − 111

18. 999
    −111

In problems 19 through 22, subtract the binary numbers by adding the

(a) 1's complement.
(b) 2's complement.

19. 111
    −101

20. 1101
    −1010

21. 1111
    − 110

22. 10110
    − 101

In problems 23 through 30, subtract the binary numbers by adding the 1's complement or the 2's complement.

23. 110
    −101

24. 111
    −110

25. 1101
    −1010

26. 1100
    − 111

27. 11000
    − 110

28. 10111
    − 1111

29. 100.01
    − 1.11

30. 1110.1
    − 101.1

In problems 31 and 32 perform the indicated operations.

31. Subtract the binary numbers in Example 15.11 by adding the 2's complement, and check that you get the same answer.

32. Many computers use storage registers that are set for a fixed number of digits. Consider a computer that stores the first 10 binary digits. A binary number such as 10011 will then be stored as 0 000 010 011. Show that if this number is subtracted from 0 000 110 111 by adding the 1's complement the computer will obtain the correct answer, which is 100 100.

≡ 15.3
OCTAL NUMBER SYSTEM

After studying binary numbers, you realize that they can be awkward to work with unless you are a computer. A binary number is generally more than three times as long as a decimal number. A long string of 0's and 1's can be difficult to recognize or remember. Number systems with a larger base that can easily be converted to the binary system are therefore used in computers also. Octal (base 8) and hexadecimal (base 16) numbers are used to identify storage locations in a computer memory and to code and represent data. The octal number system contains the following eight digits:

$$0, 1, 2, 3, 4, 5, 6, 7$$

The place values of the octal digits are powers of 8 and are shown in Table 15.5. Table 15.6 shows the first 16 octal numbers with their decimal and binary equivalents. Note that after the number 7 in the octal system, the next place must be used to represent higher numbers so that the decimal number $8 = 10_8$.

TABLE 15.5 Place Values of the Octal Digits

					Octal Point ↓		
Power	8^4	8^3	8^2	8^1	8^0	8^{-1}	8^{-2}
Value	4096	512	64	8	1	0.125	0.015625

TABLE 15.6 Decimal, Octal, and Binary Numbers

Decimal	Octal	Binary
0	0	0
1	1	1
2	2	10
3	3	11
4	4	100
5	5	101
6	6	110
7	7	111
8	10	1000
9	11	1001
10	12	1010
11	13	1011
12	14	1100
13	15	1101
14	16	1110
15	17	1111
16	20	10000

Conversions among Octal, Decimal, and Binary

Conversions between octal and decimal numbers use procedures which are similar to those for binary and decimal numbers. The first example shows how to convert an octal number to a decimal number.

EXAMPLE 15.12

Convert the octal number 372_8 to a decimal number.

Solution Multiply each digit by its place value in Table 15.5 and add:

$$372_8 = 3(64) + 7(8) + 2(1) = 192 + 56 + 2 = 250$$

The next example shows how to convert a decimal number to an octal number.

EXAMPLE 15.13

Convert the decimal number 281 to an octal number.

Solution Use the method of remainders as with binary numbers. Divide the number by 8, and the quotients repeatedly by 8, until the quotient is zero. Record the remainders:

Remainders

$8\,)\overline{281}$ 1 ↑

$8\,)\overline{35}$ 3

$8\,)\overline{4}$ 4

0

The digits of the octal number are the remainders reading from the bottom up:

$$281 = 431_8.$$

Conversions between octal and binary numbers can be done very directly. Since $2^3 = 8$, every three binary bits is equivalent to one octal digit. *To convert an octal number to a binary number*, change each digit into its three-bit binary equivalent, adding a zero or zeros if necessary. For example, to convert the octal number 561_8 to a binary number, write the three bits corresponding to each digit:

5   6   1

101   110   001

Then $561_8 = 101\ 110\ 001_2$.

To convert a binary number to an octal number, reverse the procedure and change each three bits to an octal digit working from left to right. For example, to convert the binary number 1 111 010$_2$ to an octal number, arrange in groups of three bits, adding leading zeros, and write the corresponding octal digits:

$$001 \qquad 111 \qquad 010$$
$$1 \qquad\quad 7 \qquad\quad 2$$

Therefore, 1 111 010$_2$ = 172$_8$.

Study the next example, which shows conversion among octal, binary, and decimal numbers.

EXAMPLE 15.14

Express each number as a binary, octal, and decimal number:

1. 763$_8$

2. 11 101 100$_2$

Solution

1. To express 763$_8$ as a binary number, change each digit into its three-bit binary equivalent:

$$7 \qquad\quad 6 \qquad\quad 3$$
$$111 \qquad 110 \qquad 011$$

Then 763$_8$ = 111 110 011$_2$. To express as a decimal number, multiply each digit by its place value:

$$763_8 = 7(8^2) + 6(8) + 3(1) = 448 + 48 + 3 = 499$$

2. To express 11 101 100$_2$ as an octal number, change each three bits to an octal digit:

$$011 \qquad 101 \qquad 100$$
$$3 \qquad\quad 5 \qquad\quad 4$$

Then 11 101 100$_2$ = 354$_8$. To change to a decimal number, add up the place values of the 1's:

$$11\ 101\ 100_2 = 2^7 + 2^6 + 2^5 + 2^3 + 2^2 = 128 + 64 + 32 + 8 + 4 = 236$$

If your calculator has a BASE N key, you can do the preceding conversions on the calculator. The procedure is shown in the next section following Example 15.17.

EXERCISE 15.3

In problems 1 through 8, convert each octal number to a decimal number.

1. 12

2. 17

3. 65

4. 70

5. 324

6. 635

7. 7777

8. 1001

In problems 9 through 16, convert each decimal number to an octal number.

9. 9

10. 13

11. 38

12. 60

13. 212

14. 481

15. 1100

16. 3030

In problems 17 through 24, convert each octal number to a binary number.

17. 10

18. 15

19. 23

20. 34

21. 761

22. 520

23. 1054

24. 4076

In problems 25 through 32, convert each binary number to an octal number.

25. 1100

26. 1011

27. 11 011

28. 10 111

29. 101 111

30. 101 000

31. 11 001 010

32. 1 110 100

In problems 33 through 36 perform the necessary conversions.

33. Express each number as a binary, octal, and decimal number:
   (a) 623_8.
   (b) $10\ 100\ 101_2$.

34. Express each decimal number as an octal and a binary number:
   (a) 400.
   (b) 100.

35. Complete Table 15.6 for the decimal numbers 17 through 32.

36. In a computer, data is represented by use of a binary code. For example, one code, which uses six bits for each character, represents the letter A as the binary number 110 001. Find the octal and decimal equivalents of this number.

15.4
HEXADECIMAL NUMBER SYSTEM

The hexadecimal number system, like the octal number system, is compatible with the binary number system. However, it is used much more often than the octal system in computer coding. The hexadecimal, or hex, number system contains the following 16 digits:

0, 1, 2, 3, 4, 5, 6, 7, 8, 9, A, B, C, D, E, F

The letters A through F are equal to the decimal numbers 10 through 15, respectively. Table 15.7 shows the first 21 hexadecimal numbers with their decimal and binary equivalents. Note that after the number F (15) in the hexadecimal system, the next place must be used to represent higher numbers. The place values of the hexadecimal digits are powers of 16 and are shown in Table 15.8.

TABLE 15.7 Decimal, Hexadecimal, and Binary Numbers

Decimal	Hexadecimal	Binary
0	0	0
1	1	1
2	2	10
3	3	11
4	4	100
5	5	101
6	6	100
7	7	110
8	8	111
9	9	1000
10	A	1001
11	B	1010
12	C	1100
13	D	1101
14	E	1110
15	F	1111
16	10	10000
17	11	10001
18	12	10010
19	13	10011
20	14	10100

TABLE 15.8 Place Values of the Hexadecimal Digits

					Hex Point ↓	
Powers	16^3	16^2	16^1	16^0	.	16^{-1}
Value	4096	256	16	1		0.0625

Conversions among Hexadecimal, Decimal, Binary, and Octal

Conversions between hexadecimal and decimal numbers use the same procedures as with binary and octal numbers except that the hexadecimal numbers A through F must be substituted for the decimal numbers 10 through 15, respectively. Study the following examples, which illustrate the procedures.

EXAMPLE 15.15

Convert each hexadecimal number to its decimal equivalent.

1. E5

2. 2B.4

Solution

1. To convert E5 to a decimal number, multiply each digit by its place value shown in Table 15.8 and add:

$$E5 = 14(16) + 5(1) = 224 + 5 = 229$$

Therefore, $E5_{16} = 229$. Note that 14 must be used for E when working with decimal numbers.

2. To convert 2B.4 to a decimal number, do the same as in part 1 using 11 for B:

$$2B.4 = 2(16) + 11(1) + 4(0.0625) = 32 + 11 + 0.2500 = 43.25$$

Therefore, $2B.4_{16} = 43.25$

■

EXAMPLE 15.16

Convert 380 to a hexadecimal number.

Solution   Use the method of remainders as with binary and octal numbers. Divide the number by 16, and the quotients repeatedly by 16, until the quotient is zero. Record the remainders:

Remainders

$$
\begin{array}{rl}
16\ )\overline{380} & \quad 12 = C \\
16\ )\overline{23} & \quad 7 \\
16\ )\overline{1} & \quad 1 \\
0 &
\end{array}
$$

Note that a remainder of 12 is changed to a C. You must change remainders of 10 through 15 to A through F, respectively. The hexadecimal number is given by the remainders reading from the bottom up:

$$380 = 17C_{16}$$

■

Figure 15–3 shows a flowchart for converting a decimal number to a hexadecimal number as shown in Example 15.16. Division by 16 can be done on a calculator and the remainder found by multiplying the decimal part of the quotient by 16. For example, dividing 380 by 16, you obtain

$$\frac{380}{16} = 23.75 \text{ and } 0.75(16) = 12 \text{ remainder}$$

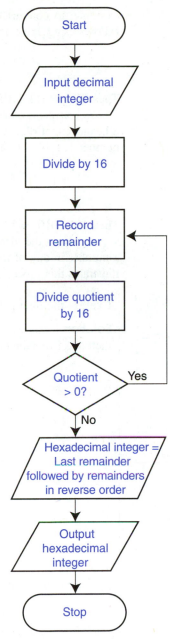

FIGURE 15–3 Flowchart for converting
a decimal number to a hexadecimal
number for Example 15.16.

Conversions between hexadecimal and binary numbers are done very directly, similar to octal-binary conversions. Each hexadecimal digit is equivalent to four binary bits since $2^4 = 16$. *To convert a hexadecimal number to a binary number*, change each digit to its four-bit binary equivalent, adding zeros if necessary. For example, to convert the hexadecimal number 7D to a binary number, write it as follows using Table 15.7:

$$\underbrace{7}_{0111} \quad \underbrace{D}_{1101}$$

Then $7D_{16} = 111\ 1101_2$.

To convert a binary number to a hexadecimal number, change each four bits to a hexadecimal digit working from left to right. For example, to convert the binary number 101 1010 to a hexadecimal number, write it as follows using Table 15.7:

$$\underbrace{0101}_{5} \quad \underbrace{1010}_{A}$$

Then $101\ 1010_2 = 5A_{16}$.

Applying the ideas presented in this chapter you should now be able to express a number in any of the four bases 2, 8, 10, and 16. Study the next example, which illustrates this process.

EXAMPLE 15.17

Express the decimal number 252 in base 2, 8, and 16.

Solution You can start with base 2, 8, or 16. Starting with base 2, use the method of remainders:

$$
\begin{array}{r|r}
 & \text{Remainders} \\
2\ \overline{)252} & 0 \\
2\ \overline{)126} & 0 \\
2\ \overline{)63} & 1 \\
2\ \overline{)31} & 1 \\
2\ \overline{)15} & 1 \\
2\ \overline{)7} & 1 \\
2\ \overline{)3} & 1 \\
2\ \overline{)1} & 1 \\
0 &
\end{array}
$$

Then $252 = 11111100_2$. Change this binary number to an octal number by using three-bit equivalents:

$$\underbrace{011}_{3} \quad \underbrace{111}_{7} \quad \underbrace{100}_{4}$$

Then $252 = 374_8$. Change the binary number to a hexadecimal number by using four-bit equivalents:

$$\underbrace{1111}_{F} \quad \underbrace{1100}_{C}$$

Then $252 = FC_{16} = 374_8 = 11111100_2$.

You can do this example on your calculator if it has a BASE N mode. You must first put it in base *n* mode by pressing MODE and the key that corresponds to BASE N. One way to solve Example 15.17 on a Casio fx or Radio Shack calculator is

$$\boxed{\text{MODE}} \quad \boxed{1} \quad 252 \quad \boxed{\text{BIN}} \rightarrow 11111100 \quad \boxed{\text{OCT}} \rightarrow 374 \quad \boxed{\text{HEX}} \rightarrow \text{FC}$$

ERROR BOX

A common error when working with hexadecimal numbers is incorrectly using or identifying the digits A through F for 10 through 15. See if you can do the practice problems *without a table or calculator* to better learn the hexadecimal system.

Practice Problems: Change each hexadecimal number to a decimal number.

1. 1A   2. B2   3. 3C   4. D4   5. 5E   6. F6

Change each decimal number to a hexadecimal number.

7. 161   8. 43   9. 195   10. 77   11. 229   12. 111

Answers: 7. A1   8. 2B   9. C3   10. 4D   11. E5   12. 6F
1. 26   2. 178   3. 60   4. 212   5. 94   6. 246

EXERCISE 15.4

In problems 1 through 14, convert each hexadecimal number to a decimal number.

1. 1B
2. 2E
3. A0
4. C3
5. CD
6. FF
7. 123
8. 285
9. A0B
10. BE7
11. 10.2
12. 2C.A
13. 1001
14. 10000

In problems 15 through 24, convert each decimal number to a hexadecimal number.

15. 21
16. 24
17. 40
18. 35
19. 85
20. 72
21. 142
22. 116
23. 343
24. 421

In problems 25 through 32, convert each hexadecimal number to a binary number.

25. 32
26. 89
27. AB
28. FE
29. 6C7
30. B81
31. 1E0B
32. C0D0

In problems 33 through 40, convert each binary number to a hexadecimal number.

33. 1110

34. 1011

35. 11011

36. 101101

37. 101111

38. 1111111

39. 1011100

40. 11101001

In problems 41 through 44, perform the necessary conversions.

41. Express the decimal number 160 in base 2, 8, and 16.

42. Express the decimal number 341 in base 2, 8, and 16.

43. In a computer, data is represented by use of a binary code. For example, the XS-3 code represents the decimal number 267 as the binary number 0101 1001 1010. The XS-3 is a four-bit code since four bits represent each character. Find the hexadecimal and decimal equivalents of this binary number.

44. One standard six-bit binary computer code encodes the word GO as the binary number 110111 100110. Find the hexadecimal and decimal equivalents of this number.

≡ 15.5
COMPUTER CODES

Alphanumeric data is data consisting of letters or numbers or both, such as words, addresses, equations, and tables. Such data is represented in a computer using a binary code. Several binary codes have been used throughout the development of computers, but today the one that is primarily used is ASCII (pronounced "AS-KEY"), which stands for American Standard Code for Information Interchange. Another code, less frequently used, is EBCDIC (pronounced "EBB-SE-DIC"), which stands for Extended Binary-Coded Decimal Interchange Code. Both these codes are extensions of a four-bit binary coded decimal (BCD) where four *numeric bits* are used to represent the digits 0 through 9. For example, the BCD representation of the decimal number 278 uses 12 bits:

$$
\begin{array}{ccc}
2 & 7 & 8 \\
\overbrace{0010} & \overbrace{0111} & \overbrace{1000}
\end{array}
$$

Each group of four bits represents one numeric *character* and is the binary equivalent of the digit. This is different from the straight binary representation of 278, which uses 9 bits:

$$278 = 1\ 0001\ 0110_2$$

Straight binary representation of numbers is used in computers when performing arithmetic operations. Binary-coded decimal representation is used for arithmetic operations and for information processing when *only* numeric characters are involved. When alphanumeric characters are used, more than four bits are necessary, and codes have been developed that use six, seven, and eight bits to represent each character. The computer codes ASCII and EBCDIC use four *numeric bits* (least significant bits) and four leading *zone bits* (more significant bits), to code a character:

Zone Bits   **Numeric Bits**

For example, in ASCII, A is coded as 0100 0001. The zone bits are 0100 and the numeric bits are 0001. In EBCDIC, A is coded as 1100 0001. The zone bits are 1100 and the numeric bits are 0001. Observe that the numeric bits are the same for A in both codes.

A code of n bits contains 2^n combinations of bits. An eight-bit code such as ASCII can therefore accommodate $2^8 = 256$ characters. This includes the basic 36-character set consisting of the 10 digits and 26 letters and as many as 220 other characters such as $, =, ?, and so on.

ASCII

The ASCII code was developed by the American National Standards Institute (ANSI) and is used in microcomputers and minicomputers and in non-IBM mainframe systems. Table 15.9 shows the ASCII code for the 36 basic characters and 12 other characters. The hexadecimal, or hex, equivalents of each binary representation are also given and are used more often than the binary representation. They are a convenient way of referring to a character since eight bits can readily be converted to two hex digits and vice versa. For example, the character K is coded as 0100 1011, which in hex is 4B. Observe in Table 15.9 that the numeric bits for the integers 0 through 9 are the binary equivalents of the integers.

TABLE 15.9 American Standard Code for Information Interchange (ASCII)

| | Hex | Binary | Zone Bits | | |
| | | | 3 | 4 | 5 |
			0011	0100	0101
	0	0000	0	@	P
	1	0001	1	A	Q
	2	0010	2	B	R
	3	0011	3	C	S
	4	0100	4	D	T
	5	0101	5	E	U
	6	0110	6	F	V
	7	0111	7	G	W
Numeric Bits	8	1000	8	H	X
	9	1001	9	I	Y
	A	1010	:	J	Z
	B	1011	;	K	[
	C	1100	<	L	\
	D	1101	=	M	]
	E	1110	>	N	^
	F	1111	?	O	_

A computer almost always stores an extra bit for each character called a *check bit* or *parity bit* that precedes the zone bits. The check bit is chosen so that the sum of the 1-bits is either even or odd depending on whether the computer operates on even parity or odd parity. For example, if the computer operates on even parity, the check bit for K would be 0 since the sum of the 1-bits is an even number 4. If the computer operates on odd parity, the check bit for K would be 1:

$$\text{Check bit}$$
$$\downarrow$$

| K even parity | 0 0100 1011 |
| K odd parity | 1 0100 1011 |

The check bit may be used to detect an error when a character is transmitted. If a bit has been erased or changed, the parity changes, and the computer, detecting the wrong parity, transmits the data again. An odd parity bit has the advantage that all zeros (0 0000 0000) will never be transmitted and thus will not be confused with "no information."

EXAMPLE 15.18

Given the name SAM,

1. What is the ASCII code for this name in binary, hexadecimal, and decimal code?

2. Suppose the computer uses an odd parity check. What would the binary code for SAM become?

Solution

1. Using Table 15.9 the binary and hexadecimal ASCII code is

S		A		M	
0101	0011	0100	0001	0100	1101
5	3	4	1	4	D

You can find the decimal code by converting the hexadecimal (or binary) numbers:

$$\text{S: } 53_{16} = 5(16) + 3(1) = 83$$
$$\text{A: } 41_{16} = 4(16) + 1(1) = 65$$
$$\text{M: } 4D_{16} = 4(16) + 13(1) = 77$$

2. An odd-parity check means the bits in each byte must add up to an odd number. Since the 1-bits for S, A, and M each add up to even numbers, a 1 would have to be added for each character:

S	A	M
1 0101 0011	1 0100 0001	1 0100 1101

▪

The number of bits used to represent a character is called a *byte*. A byte in ASCII contains eight bits, or nine bits including the check bit. A computer word is a set of bytes that are processed as a group. Words vary in length. For example, in certain computers, a word consists of four 8-bit bytes or 32 bits.

EXAMPLE 15.19

What is the following computer word in ASCII using four 8-bit bytes?

00110001001110010011100100110101

Solution   Partition the bits into 8-bit bytes and use Table 15.9 to obtain

0011 0001	0011 1001	0011 1001	0011 0101
1	9	9	5

▪

EBCDIC

The EBCDIC code was developed by IBM and is the extension of an earlier six-bit code. It is used mainly in IBM mainframe systems and in IBM compatible systems. Table 15.10 shows the EBCDIC code for the 36 basic characters. Observe that in ASCII and EBCDIC the numeric bits for the integers 0 through 9 and the letters A through I are the same. The alphabetical characters use three sets of zone bits in EBCDIC, whereas ASCII uses only two sets.

TABLE 15.10   Extended Binary-Coded Decimal Interchange Code (EBCDIC)

			Zone Bits			
			C	**D**	**E**	**F**
	Hex	**Binary**	**1100**	**1101**	**1110**	**1111**
	0	0000				Ø
	1	0001	A	J		1
	2	0010	B	K	S	2
	3	001·1	C	L	T	3
	4	0100	D	M	U	4
	5	0101	E	N	V	5
	6	0110	F	O	W	6
Numeric Bits	7	0111	G	P	X	7
	8	1000	H	Q	Y	8
	9	1001	I	R	Z	9

EXERCISE 15.5

In problems 1 through 4, express each number in

  (a) Four-bit BCD code.
  (b) Straight binary representation.

1. 18   **3.** 122

2. 37   **4.** 253

In problems 5 through 10, express each computer word using ASCII code in

  (a) binary code.
  (b) hexadecimal code.
  (c) decimal code.

5. X7   **8.** JOE

6. 3M   **9.** WABC

7. CIA   **10.** 122B

In problems 11 through 16, do the same as for problems 5 through 10, respectively, but using the EBCDIC code.

In problems 17 through 24 solve each applied problem.

17. (a) What is a *byte*?
    (b) How long is a byte in ASCII including a parity bit?

18. What is a computer *word?*

19. What would be the binary ASCII code for the word FBI if the computer uses an even-parity check?

20. What would be the binary EBCDIC code for the word KGB if the computer uses an odd-parity check?

21. What is the following computer word in ASCII?

$$00110111001100000110111$$

22. What is the following computer word in EBCDIC?

$$110000101100100111100011$$

23. The ASCII code for the character "$" in octal code is 44. Express this in
    (a) Eight-bit binary code.
    (b) Hexadecimal code.

24. Complete the following word in ASCII, where each byte has a check bit added for odd parity, by supplying the last byte:

$$1\ 0100\ 0001$$

$$1\ 0101\ 0011$$

$$0\ 0100\ 0011$$

$$0\ 0100\ 1001$$

▬ CHAPTER HIGHLIGHTS

15.1 BINARY NUMBER SYSTEM

The binary number system contains two digits: 0 and 1. The place values of the binary digits are as follows:

Binary Point ↓

Power	2^6	2^5	2^4	2^3	2^2	2^1	2^0	.	2^{-1}	2^{-2}	2^{-3}
Value	64	32	16	8	4	2	1		0.5	0.25	0.125

To convert a binary number to a decimal number, multiply each digit by its place value and add.

To convert a decimal number to a binary number, divide the number by 2 and the quotients repeatedly by 2 until you get zero. Record the remainders, and write them in the reverse order to obtain the binary digits. See Example 15.3.

15.2 BINARY ARITHMETIC

The basic binary sums are as follows:

$$0 + 0 = \quad 0$$
$$0 + 1 = 1 + 0 = \quad 1$$
$$1 + 1 = \quad 10 \quad (0 \text{ and carry } 1)$$
$$1 + 1 + 1 = 10 + 1 = \quad 11 \quad (1 \text{ and carry } 1)$$
$$1 + 1 + 1 + 1 = 11 + 1 = 100 \quad (0 \text{ and carry } 10)$$

Add binary numbers the same as decimal numbers applying the preceding sums, and carry the indicated digits. See Examples 15.7 and 15.8.

The 1's complement of a binary number is the number obtained by subtracting each digit from 1. This is the same as changing each 0 to 1 and each 1 to 0. The 2's complement of a number is the number obtained by adding 1 to its 1's complement.

To subtract binary numbers either:

1. Add the 1's complement of the subtrahend (number to be subtracted) and end-around carry the leading 1.
2. Add the 2's complement and delete the leading 1. See Examples 15.10 and 15.11.

15.3 OCTAL NUMBER SYSTEM

The octal number system contains eight digits:

$$0,1,2,3,4,5,6,7$$

The place values of the octal digits are as follows:

Octal Point ↓

Power	8^4	8^3	8^2	8^1	8^0	.	8^{-1}	8^{-2}
Value	4096	512	64	8	1		0.125	0.015625

To convert an octal number to a decimal number, multiply each digit by its place value and add.

To convert a decimal number to an octal number, divide the number by 8 and the quotients repeatedly by 8 until you get zero. Record the remainders, and write them in the reverse order to obtain the octal digits. See Example 15.13.

To convert an octal number to a binary number, change each digit into its three-bit binary equivalent, adding zeros if necessary.

To convert a binary number to an octal number, reverse this procedure, and change each three bits to an octal digit working from left to right.

15.4 HEXADECIMAL NUMBER SYSTEM

The hexadecimal or hex number system contains the following 16 digits:

0, 1, 2, 3, 4, 5, 6, 7, 8, 9, A, B, C, D, E, F

The letters A through F are equal to the decimal numbers 10 through 15, respectively.

The place values of the hexadecimal digits are as follows:

Hex Point ↓

Powers	16^3	16^2	16^1	16^0	.	16^{-1}
Value	4096	256	16	1		0.0625

To convert a hexadecimal number to a decimal number, multiply each digit by its place value and add.

To convert a decimal number to a hexadecimal number, divide the number by 16 and the quotients repeatedly by 16 until you get zero. Record the remainders, and write them in the reverse order to obtain the hex digits. See Example 15.16.

To convert a hexadecimal number to a binary number, change each digit into its four-bit binary equivalent, adding zeros if necessary.

To convert a binary number to a hexadecimal number, reverse this procedure, and change each four bits to a hex digit working from left to right. See Example 15.17.

15.5 COMPUTER CODES

Two binary codes in use today in computers are ASCII and EBCDIC. ASCII is the primary one and is used in microcomputers, minicomputers, and non-IBM mainframe systems. EBCDIC is used in IBM mainframe systems. Both codes use eight bits for alphanumeric characters and may or may not use an extra parity bit. A byte is the number of bits used to represent a character. See Tables 15.9 and 15.10 and Example 15.18.

≡ REVIEW QUESTIONS

In problems 1 through 4, convert each binary number to a decimal number.

1. 1111

2. 1101

3. 11011

4. 100.11

In problems 5 through 8, convert each decimal number to a binary number.

5. 19

6. 34

7. 125

8. 5.5

In problems 9 through 12, calculate each binary sum.

9. 101
 111

10. 11
 110
 101

11. 11
 111
 100
 101

12. 1.01
 0.11
 11.01

In problems 13 through 16, perform each binary subtraction.

13. 1010
 − 101

14. 1100
 −1001

15. 10000
 − 1111

16. 11001
 − 1110

In problems 17 through 22, convert each number to a decimal number.

17. 75_8

18. 167_8

19. $3A_{16}$

20. EB_{16}

21. $17A_{16}$

22. DAD_{16}

In problems 23 through 28, convert each number to a binary number.

23. 46_8

24. 103_8

25. $2A_{16}$

26. $8B_{16}$

27. $9CD_{16}$

28. $FE5_{16}$

In problems 29 through 34, convert each decimal number to

    (a) An octal number.
    (b) A hexadecimal number.

29. 27

30. 63

31. 111

32. 347

33. 1056

34. 2112

In problems 35 through 40, convert each binary number to

    (a) An octal number.
    (b) A hexadecimal number.

35. 111

36. 1101

37. 10011

38. 11110

39. 101 0101

40. 1100 1100

In problems 41 through 44, express each computer word in binary code using

    (a) ASCII.
    (b) EBCDIC.

41. GO

42. 101

43. XS3

44. SEA

In problems 45 through 48 solve each applied problem.

45. Complete the computer message in hexadecimal ASCII code by supplying the last word:

$$4841535445$$

$$4D414B4553$$

46. Complete the following message in binary EBCDIC code:

$$1101 \ 0011$$

$$1101 \ 0110$$

$$1110 \ 0101$$

47. The ASCII code for the character "+" in hex is 2B. Using Table 15.9, write the binary ASCII code for the mathematical equation $1 + 1 = 2$ if the computer uses even parity.

48. Given a binary number, the 1's complement of the 1's complement is the original number since each digit is reversed twice back to the original. Show that the 2's complement of the 2's complement is also the original number for the following binary numbers:

    (a) 1100.
    (b) 1011000.

CHAPTER
16
Boolean Algebra

Courtesy of Arthur Kramer and NYC Technical College.

Students learning the operation and installation of local area networks (LANs).

In the nineteenth century the English mathematician George Boole (1815–64) developed a mathematical system called Boolean algebra, or Boolean logic, which has proved to be very useful in the design of circuits for information processing. The laws of Boolean algebra are similar in some ways to those of arithmetic and algebra, but in other ways are quite different.

The first section looks at the mathematical structure of Boolean algebra. The second section shows the relationship between Boolean algebra and the basic logic gates and how to use Boolean algebra to analyze logic circuits. The last section shows how these ideas are used to simplify and design logic circuits using Karnaugh maps.

Chapter Objectives

In this chapter, you will learn:

- The basic laws and theorems of Boolean algebra.
- How to use truth tables to prove Boolean statements.
- The three basic logic gates AND, OR, and NOT.
- How to construct truth tables for logic circuits.
- How to express the output of a logic circuit using Boolean algebra.
- How to simplify a logic circuit using Karnaugh maps.

16.1 BOOLEAN ALGEBRA

Boolean algebra consists of a set of elements that satisfy three basic operations:

1. The OR operation "+".
2. The AND operation "·".
3. The *complement* of an element A, called NOT A and written $\overline{A}$.

The operations are defined in terms of *truth tables*, where each element is assigned a truth value of 1 ("true") or 0 ("not true"). These are shown in Table 16.1 for any two elements A and B. Observe that there are four possible combinations of 0 and 1 to consider when there are two elements in a statement.

TABLE 16.1   Truth Tables for AND, OR, and NOT

		AND	OR	NOT
A	B	$A \cdot B$	$A + B$	$\overline{A}$
0	0	0	0	1
0	1	0	1	1
1	0	0	1	0
1	1	1	1	0

Note in Table 16.1 that

1. *A* AND *B* is true, only if *both A* and *B* are true.
2. *A* OR *B* is not true, only if *both A* and *B* are not true.
3. NOT *A* has the opposite truth value of *A*.

The truth values for AND, OR, and NOT come from their application in mathematical logic. For example, suppose *A* and *B* represent the following statements:

> *A*: Computers process information with binary numbers
>
> *B*: 8 − 5 = 2

Here, *A* is a true statement about computers, and *B* is *not* true in arithmetic. The statement *A* AND *B* is

> Computers process information with binary numbers and 8 − 5 = 2.

Based on the truth table, this statement is *not* true because *B* is not true. The statement *A* OR *B* is

> Computers process information with binary numbers or 8 − 5 = 2.

Based on the truth table this statement *is* true because *A* is true. The statement NOT *A* is

> Computers do not process information with binary numbers.

This statement is *not* true because *A* is true. The statement NOT *B* is

$$8 - 5 \neq 2$$

This statement *is* true because *B* is not true. These results make sense logically and provide a basis for the truth tables.

The 0 and 1 shown in the truth tables represent essential elements in Boolean algebra. The set of elements and the operations AND, OR, and NOT satisfy laws like those of arithmetic shown in Chapter 1:

- Commutative laws:

$$A + B = B + A$$

$$A \cdot B = B \cdot A$$

- Distributive laws:

$$A + (B \cdot C) = (A + B) \cdot (A + C)$$

$$A \cdot (B + C) = A \cdot B + A \cdot C$$

- Identity laws:

$$A + 0 = A$$

$$A \cdot 1 = A$$

- Complement laws:

$$A + \overline{A} = 1$$

$$A \cdot \overline{A} = 0$$

Observe that the first distributive law does not apply in arithmetic, where + is addition and · is multiplication, while the second distributive law does. The complement laws are related to the following rules in arithmetic and algebra:

$$x + (-x) = 0$$

$$x \cdot x^{-1} = x \cdot \frac{1}{x} = 1$$

However, the complement laws are not the same as these rules.

The dot symbol for AND in Boolean algebra is usually omitted, like the dot for multiplication in algebra. That is, AB means A · B. The AND operation is done before OR and does not have to be enclosed in parentheses. The preceding distributive laws can, therefore, be written:

$$A + BC = (A + B)(A + C)$$

$$A(B + C) = AB + AC$$

The distributive laws and each of the other laws above are dual statements. The *dual* of a Boolean statement is the statement obtained by interchanging the operations of + and · and by interchanging 0 and 1. In Boolean algebra,

Rule If a statement is true, the dual statement is also true.

Several basic theorems (true statements) follow from these laws. They are shown in Table 16.2, where each statement is written next to its dual.

TABLE 16.2 Basic Theorems in Boolean Algebra

$\overline{0} = 1$	$\overline{1} = 0$
$A + 1 = 1$	$A \cdot 0 = 0$
$A + A = A$	$A \cdot A = A$
$A + AB = A$	$A(A + B) = A$
$(A + B) + C = A + (B + C)$	$(A \cdot B) \cdot C = A \cdot (B \cdot C)$

Observe in Table 16.2 that

1. The complement of 1 is 0 and vice versa.
2. An element operated on itself results in the same element. This is very different from arithmetic.
3. The last theorem and its dual are associative laws and are the same as those in arithmetic.

Statements in Boolean algebra can be shown to be equivalent by the use of truth tables. If two expressions have the same truth values, then they are equal. The truth values are based on the defined truth tables for the operations of AND, OR, and NOT shown in Table 16.1. Study the following examples that show how to prove a theorem in Boolean algebra.

EXAMPLE 16.1

Prove the following Boolean theorem:

$$A + AB = A$$

Solution   Construct a truth table for A, B, AB and $A + AB$. Show that the truth values for A and $A + AB$ are the same. This is done as follows:

A	B	AB	A + AB
0	0	0	0
0	1	0	0
1	0	0	1
1	1	1	1

Same

The first three columns in the truth table are the same as the given truth table for A AND B in Table 16.1. The last column compares the column for A and the column for AB and applies the defined truth values for the operation of OR given in Table 16.1. Here, A is treated as "A" and AB is treated as "B" in Table 16.1. Then, since the column for A and the column for $A + AB$ have the same truth values, the two statements are equivalent, and the theorem is true.

▪

EXAMPLE 16.2

Prove the following Boolean theorem:

$$\overline{AB} = \overline{A} + \overline{B}$$

Solution   This statement says NOT (A AND B) is equivalent to (NOT A) OR (NOT B). To show this equivalence, construct two truth tables, one for $\overline{AB}$ and one for $\overline{A} + \overline{B}$. Then show that the truth values for each statement are the same. In each truth table, one column is used for each new element, or operation, building up to the complete statement:

A	B	AB	$\overline{AB}$
0	0	0	1
0	1	0	1
1	0	0	1
1	1	1	0

A	B	$\overline{A}$	$\overline{B}$	$\overline{A} + \overline{B}$
0	0	1	1	1
0	1	1	0	1
1	0	0	1	1
1	1	0	0	0

The columns for A and B must be the same in each table. To obtain the truth values for the last column, apply the defined truth values for AND, OR, and NOT. Then compare the last column in each table. Since the truth values are the same, they are equivalent, and the theorem is true.

▪

The theorem in Example 16.2 is one of two important laws in Boolean algebra known as *De Morgan's laws*. The two laws are duals of each other:

$$\overline{AB} = \overline{A} + \overline{B}$$

$$\overline{A + B} = \overline{A} \cdot \overline{B}$$

EXAMPLE 16.3

Prove the following Boolean theorem:

$$(AB)C = A(BC)$$

Solution There are *eight* possible combinations of truth values for three elements. The truth tables, therefore, have eight rows:

A	B	AB	C	(AB)C
0	0	0	0	0
0	0	0	1	0
0	1	0	0	0
0	1	0	1	0
1	0	0	0	0
1	0	0	1	0
1	1	1	0	0
1	1	1	1	1

A	B	C	BC	A(BC)
0	0	0	0	0
0	0	1	0	0
0	1	0	0	0
0	1	1	1	0
1	0	0	0	0
1	0	1	0	0
1	1	0	0	0
1	1	1	1	1

The last column in each table is the same, which proves the theorem.

Note that the 0s and 1s in the columns for A, B, and C represent the first eight binary numbers. There is a direct relation between binary numbers that have two digits and Boolean algebra that has two elements 0 and 1.

■

EXERCISE 16.1

1. Given the following statements:

   A: Copper is a poor conductor.
   B: $\sqrt{0.01} = 0.1$

   write the statements A AND B, A OR B, NOT A, and NOT B, and tell whether each is true, or not true, based on the actual truth of A and B.

2. Do the same as in problem 1 for the following statements:

   A: $V = IR$
   B: New York is in California.

3. Which of the laws for Boolean algebra also apply in arithmetic and algebra?

4. Explain why the complement laws for Boolean algebra cannot apply in algebra.

In problems 5 through 10, give the dual of each Boolean statement.

5. $A \cdot 1 = A$

6. $A \cdot \overline{A} = 0$

7. $A + 1 = A + \overline{A}$

8. $A + 0 = A \cdot A$

9. $A + \overline{A}B = A + B$

10. $AB + BC = (A + C)B$

In problems 11 through 18, prove each Boolean theorem by use of truth tables.

11. $A + A = A$

12. $\overline{\overline{A}} = \overline{A}$

13. $A(A + B) = A$

14. $(A + B)(A + B) = A + B$

15. $\overline{A} + \overline{A} \cdot \overline{B} = \overline{A}$

16. $\overline{A + B} = \overline{A} \cdot \overline{B}$

17. $(A + B) + C = A + (B + C)$

18. $(A + C)(A + B) = A + B(A + C)$

≡ 16.2
LOGIC CIRCUITS

Boolean algebra was first applied to switching circuit problems by Claude Shannon in 1938. His mathematical methods of working with switches and relays proved very useful in circuit design and are used today in the design of logic circuits and computers.

Series and Parallel Switching Circuits

See Figure 16–1, which shows the two basic kinds of switching circuits. A series circuit contains two or more switches connected end to end in the same path. In a series circuit, current will flow only if all switches are closed. A parallel circuit contains two or more switches connected in parallel paths. In a parallel circuit, current will flow if either switch is closed. The properties of each circuit are shown in a truth table where 0 = Open and 1 = Closed.

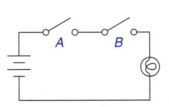

Series switching circuit

A	B	Output A AND B
0	0	0
0	1	0
1	0	0
1	1	1

(a)

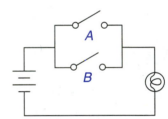

Parallel switching circuit

A	B	Output A OR B
0	0	0
0	1	1
1	0	1
1	1	1

(b)

FIGURE 16–1 Basic switching circuits.

Note the following about the truth tables:

1. The truth table for the series circuit is the same as the truth table for the Boolean operation AND.
2. The truth table for the parallel circuit is the same as the truth table for the Boolean operation OR.

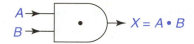

AND gate

A	B	X = A • B
0	0	0
0	1	0
1	0	0
1	1	1

FIGURE 16–2   AND gate.

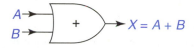

OR gate

A	B	X = A + B
0	0	0
0	1	1
1	0	1
1	1	1

FIGURE 16–3   OR gate.

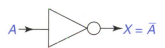

NOT gate

A	X = $\overline{A}$
0	1
1	0

(a)

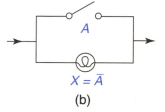

(b)

FIGURE 16– 4
NOT gate.

Because of this, the concepts of Boolean algebra can be applied to the analysis of switching circuits, or logic circuits, as follows.

Logic Gates

Logic gates control the transmission of input pulses and are the building blocks of electronic circuits. There are three basic logic gates corresponding to the Boolean operations AND, OR, and NOT.

The AND Gate   An AND gate corresponds to the *series* switching circuit shown in Figure 16–1. The output X is expressed in terms of the inputs A and B using Boolean symbols: $X = A \cdot B$. The AND gate and its truth table are shown symbolically in Figure 16–2. It will allow a pulse or signal to be transmitted, denoted by $X = 1$, *only* when $A = 1$ and $B = 1$; that is, only when there is an input to both A and B. When there is an output from a gate, it is said to be *enabled*.

The OR Gate   An OR gate corresponds to the *parallel* switching circuit shown in Figure 16–1. The output X is expressed using Boolean symbols: $X = A + B$. The OR gate and its truth table are shown symbolically in Figure 16–3. It allows a pulse or signal to be transmitted, $X = 1$, when either $A = 1$ or $B = 1$. That is, when there is an input at either A or B.

   The AND gate and the OR gate may have more than two inputs. An AND gate with three inputs is $X = A \cdot B \cdot C$, and an OR gate with three inputs is $X = A + B + C$.

The NOT Gate   A NOT gate corresponds to the complement operation and is called an inverter. The NOT gate has only one input A, and its output is expressed in Boolean symbols: $X = \overline{A}$. The NOT gate and its truth table are shown symbolically in Figure 16–4(a).

   A NOT gate can be thought of as an electrical switch that allows current to flow through a circuit when it is open and does not allow current to flow when it is closed. In Figure 16–4(b), the current flows through the lamp when the switch is open. When the switch is closed, the lamp is short circuited and no current flows through it. A NOT gate can also be thought of as a voltage inverter that changes high voltage to low voltage and vice versa.

Logic Circuits

A *logic circuit* is a circuit containing one or more of the three basic logic gates. A logic circuit has a direct relationship to a switching circuit or a computer circuit. The logic gates are its basic elements. Since the three basic logic gates have the same truth tables as the corresponding Boolean operations, *logic circuits have the properties of Boolean algebra.*

   There are certain common logic circuits that are used often. Study the following example, which shows a common type consisting of two or more AND gates feeding into an OR gate.

EXAMPLE 16.4

Figure 16–5 shows an AND–OR logic circuit consisting of two AND gates feeding into an OR gate. Find the Boolean expression for the output X in terms of the inputs A and B, and construct the truth table for the circuit.

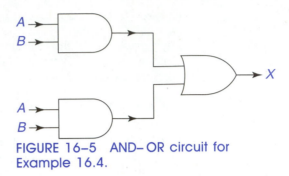

FIGURE 16–5   AND–OR circuit for Example 16.4.

Solution   To find the output X, find the output of each gate separately, moving in the direction of the signal from left to right. The output of each AND gate is $A \cdot B$. See Figure 16–6. These become the inputs of the OR gate and the output of the OR gate is then

$$X = A \cdot B + A \cdot B$$

This is known as a *Boolean sum of products.*

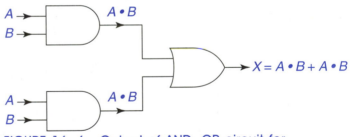

FIGURE 16–6   Output of AND–OR circuit for Example 16.4.

To construct the truth table, work with the Boolean expression for the output X and proceed as in Section 16.1 with the four combinations of the inputs A and B:

A	B	$A \cdot B$	$X = A \cdot B + A \cdot B$
0	0	0	0
0	1	0	0
1	0	0	0
1	1	1	1

In the truth table, note that that the output X is also equivalent to the product $A \cdot B$ since the columns are the same. This can also be shown by applying the laws of

EXAMPLE 16.4 (Cont.)

Boolean algebra as follows. Apply the theorem $A + A = A$ from Table 16.2 to the term $A \cdot B$:

$$X = A \cdot B + A \cdot B = A \cdot B$$

This means that the logic circuit in Figure 16–5 can be simplified to a basic AND gate. Simplifying switching circuits is one of the important applications of Boolean algebra.

▪

NAND and NOR Gates

Two useful logic circuits are the NAND and NOR gates. A NAND gate is equivalent to an AND gate followed by a NOT gate. A NOR gate is equivalent to an OR gate followed by a NOT gate. Figure 16–7 shows the two gates and their truth tables. Note that the bubble shown in the NOT gate is also used for the NAND and NOR gates and represents the NOT operation.

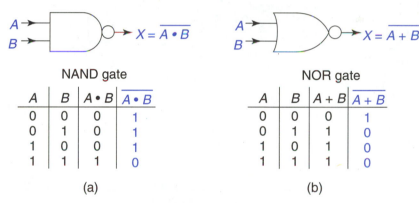

A	B	$A \bullet B$	$\overline{A \bullet B}$
0	0	0	1
0	1	0	1
1	0	0	1
1	1	1	0

(a)

A	B	$A + B$	$\overline{A + B}$
0	0	0	1
0	1	1	0
1	0	1	0
1	1	1	0

(b)

FIGURE 16–7   NAND and NOR gates.

The NAND gate has a high output when either or both inputs are low, whereas the NOR gate has a low output when either or both inputs are high.

EXAMPLE 16.5

Given the logic circuit in Figure 16–8 with inputs A, B, and C,

1. Find the output in terms of A, B, and C.

2. Show that the output is equivalent to $B \cdot C$ by truth tables and by applying the laws of Boolean algebra.

Solution

1. Moving from left to right, the output of each gate is shown in Figure 16–9. The output X is made up of the two AND gates feeding into the OR gate and is equal to

$$X = \overline{A} \cdot B \cdot C + B \cdot C$$

EXAMPLE 16.5 (Cont.)

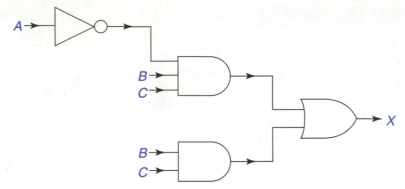

FIGURE 16–8 Logic circuit for Example 16.5.

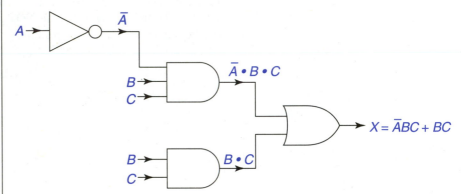

FIGURE 16–9 Output of the logic circuit for Example 16.5.

2. The truth table for the output contains eight possible combinations because of the three inputs:

A	B	C	$\overline{A}$	$\overline{A}B$	$\overline{A}BC$	BC	$X = \overline{A}BC + BC$
0	0	0	1	0	0	0	0
0	0	1	1	0	0	0	0
0	1	0	1	1	0	0	0
0	1	1	1	1	1	1	1
1	0	0	0	0	0	0	0
1	0	1	0	0	0	0	0
1	1	0	0	0	0	0	0
1	1	1	0	0	0	1	1

SAME

Observe that the last two columns are identical, which shows that:

$$X = \overline{A}BC + BC = BC$$

EXAMPLE 16.5 (Cont.)

A truth table with n inputs contains 2^n possible combinations of 0s and 1s. For example, four inputs has $2^4 = 16$ combinations in the truth table.

To show this equivalence using the laws of Boolean algebra, first apply the distributive law to the term BC in X:

$$X = \overline{A}BC + BC = BC(\overline{A} + 1)$$

Then from Table 16.2, $A + 1 = 1$ for any element A. Applying this to $\overline{A}$, you have

$$X = BC(\overline{A} + 1) = BC(1) = BC$$

▪

ERROR BOX

A common error when working with logic circuits is confusing the AND and OR gates. An AND gate has an output only when *all* inputs are sending signals. The OR gate has an output when *any one* input is sending a signal. See if you can do the practice problems without looking at the truth tables.

Practice Problems: Given the following inputs for an AND gate and an OR gate, tell the output for each gate.

1. $A = 0$, $B = 1$ 2. $A = 1$, $B = 0$ 3. $A = 0$, $B = 0$, $C = 0$
4. $A = 1$, $B = 1$, $C = 0$ 5. $A = 1$, $B = 1$, $C = 1$

Answers (AND,OR):   1̇ '1̇ '5̇   1̇ '0̇ '4̇   0̇ '0̇ '3̇   1̇ '0̇ '2̇   1̇ '0̇ '1̇

Two other logic circuits that are common are the XOR (exclusive OR) and XNOR (exclusive NOR) gates. The XOR and XNOR gates are special forms of the OR and NOR gates, respectively. The exclusive means that only certain combinations of inputs produce outputs. For the XOR gate, only an *odd* number of 1 inputs produce a 1 output. For the XNOR gate, only an *even* number of 1 inputs produce a 1 output. These gates and their truth tables are shown symbolically in Figure 16–10. Note that the Boolean symbol for the operation of XOR is a circled plus sign. The ability of these gates to distinguish between an odd and even number of inputs makes them useful in computer circuits to recognize even and odd parity in binary transmission.

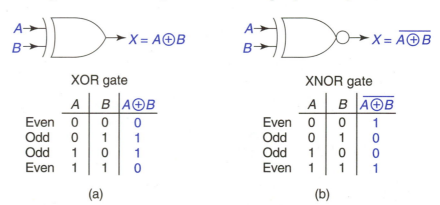

	A	B	$A \oplus B$
Even	0	0	0
Odd	0	1	1
Odd	1	0	1
Even	1	1	0

(a)

	A	B	$\overline{A \oplus B}$
Even	0	0	1
Odd	0	1	0
Odd	1	0	0
Even	1	1	1

(b)

FIGURE 16–10   XOR and XNOR gates.

EXERCISE 16.2

1. Figure 16–11 shows a combination of a series and parallel switching circuit with three switches. Which switches need to be closed to produce an output at the lamp?

2. Express the output of the circuit in Figure 16–11 in Boolean algebra when all three switches are closed.

In problems 3 through 16, give the output of each logic gate for the given inputs.

3. AND gate; $A = 1$, $B = 0$
4. AND gate; $A = 1$, $B = 1$
5. OR gate; $A = 0$, $B = 1$
6. OR gate; $A = 0$, $B = 0$
7. NOT gate; $A = 1$
8. NOT gate; $A = 0$
9. NAND gate; $A = 0$, $B = 1$
10. NAND gate; $A = 1$, $B = 1$
11. NOR gate; $A = 1$, $B = 1$

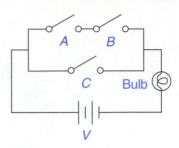

FIGURE 16–11   Switching circuit for problems 1 and 2.

12. NOR gate; $A = 0$, $B = 0$
13. XOR gate; $A = 1$, $B = 0$
14. XOR gate; $A = 1$, $B = 1$
15. XNOR gate; $A = 0$, $B = 0$
16. XNOR gate; $A = 1$, $B = 1$

For each logic circuit in problems 17 through 20,

(a) Determine the output X.
(b) Construct the truth table.
    (See Example 16.4).

17.

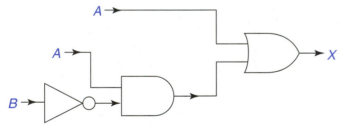

FIGURE 16–12   Logic circuit for problems 17 and 21.

18.

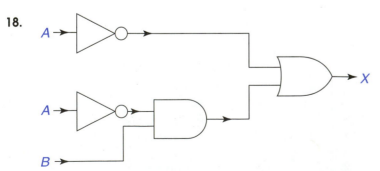

FIGURE 16–13   Logic circuit for problems 18 and 22.

19.

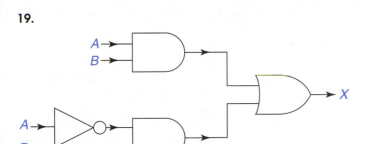

FIGURE 16–14 Logic circuit for problems 19 and 23.

20.

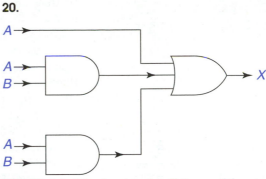

FIGURE 16–15 Logic circuit for problems 20 and 24.

In problems 21 through 24, show, by applying the laws of Boolean algebra, that the output of the given circuit is equivalent to the output that is given. (See Examples 16.4 and 16.5.)

21. Figure 16–12; $X = A$

22. Figure 16–13; $X = \overline{A}$

23. Figure 16–14; $X = B$

24. Figure 16–15; $X = A$

For each logic circuit in problems 25 and 26,

(a) Determine the output X.

Show that the output is equivalent to the given output

(b) By Truth tables.

(c) By applying the laws of Boolean algebra.

25.

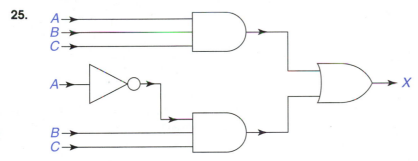

FIGURE 16–16 Logic circuit for problem 25.

26.

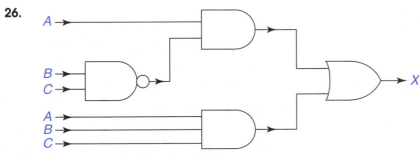

FIGURE 16–17 Logic circuit for problem 26.

≡ 16.3
KARNAUGH MAPS

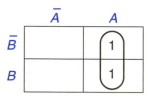

	$\overline{A}$	A
$\overline{B}$	$\overline{A}\,\overline{B}$	$A\overline{B}$
B	$\overline{A}B$	AB

FIGURE 16–18
Karnaugh map for
two variables.

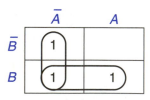

$$AB + A\overline{B} = A$$

FIGURE 16–19
Adjacent squares in
a two-variable
Karnaugh map.

In Boolean algebra and logic circuits it is shown that certain Boolean sums of products are equivalent to simpler expressions. Example 16.4 shows that $AB + AB = AB$. Example 16.5 shows that $\overline{A}BC + BC = BC$.

A useful way to simplify Boolean sums of products is by using a Karnaugh map. These are checkerboard-like diagrams in which squares represent products of Boolean elements. For example, Figure 16–18 shows a Karnaugh map for two variables A, B.

In a truth table, each variable has two states: 0 or 1. In a Karnaugh map, each variable appears as itself and its complement. The complement is like the other state of the variable. If A is 0 then $\overline{A}$ is 1, and vice versa. Each row and each column in the two-variable map represents one variable. The first column represents $\overline{A}$, the first row $\overline{B}$, the second column A, and the second row B. Each small square in the map represents the product of the variables in that column and that row. For example, the upper-left square is $\overline{A}\,\overline{B}$, the upper-right $A\overline{B}$, the lower-left $\overline{A}B$ and the lower-right AB. These four products are the four fundamental Boolean products for two variables.

Consider now a Boolean sum of products containing two variables such as

$$AB + A\overline{B}$$

This sum can be shown to be equal to a simpler expression containing fewer products as follows. Place a 1 in each square corresponding to one of these products as shown in Figure 16–19. If two adjacent squares have 1s, circle them together as shown. The circled squares are then used to determine a simpler expression. Since the circled squares constitute the column A, it follows that

$$AB + A\overline{B} = A$$

This means that the sum of all the products in a row, or a column, is equivalent to the variable that represents that row or that column.

This result can also be shown using truth tables or the theorems of Boolean algebra. For example, applying the distributive law and the complement law:

$$AB + A\overline{B} = A(B + \overline{B}) = A(1) = A$$

A Karnaugh map, however, shows this statement in a more direct and simpler way. Study the next example, which further illustrates Karnaugh maps.

EXAMPLE 16.6

	$\overline{A}$	A
$\overline{B}$	1	
B	1	1

$$AB + \overline{A}B + \overline{A}\,\overline{B} = \overline{A} + B$$

FIGURE 16–20 Adjacent
squares in a Karnaugh
map for Example 16.6.

Simplify the following Boolean sum of products by using a Karnaugh map:

$$AB + \overline{A}B + \overline{A}\,\overline{B}$$

Solution   Place a 1 in each square corresponding to one of the products as shown in Figure 16–20. Circle adjacent squares containing 1s. One of the circles represents the variable $\overline{A}$ and the other circle the variable B. The sum of these two variables then represents the equivalent expression:

$$AB + \overline{A}B + \overline{A}\,\overline{B} = \overline{A} + B$$

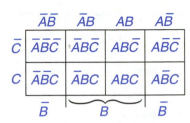

FIGURE 16-21 Karnaugh map for three variables.

Three Variables

When there are two variables, there are four possible products in the Karnaugh map, corresponding to the combinations of 0s and 1s in the truth table. Similarly, when there are three variables, there are eight possible products to consider in a Karnaugh map. Figure 16–21 shows the Karnaugh map for three variables. The columns represent the four products from the Karnaugh map for two variables A and B. The rows represent the variables $\overline{C}$ and C. Each square represents the product of three variables, and there are eight possible combinations of products. Note also that the first two columns constitute $\overline{A}$, the last two columns A, the middle two columns B, and the first and last column $\overline{B}$.

Now consider the following Boolean sum of products:

$$ABC + A\overline{B}C + AB\overline{C}$$

Place a 1 in each square corresponding to a product. Circle together any 1s that are adjacent to each other as shown in Figure 16–22. Each circled group of products can then be replaced by the product common to the group. Here, AB is common to ABC and $AB\overline{C}$, and AC is common to ABC and $A\overline{B}C$. Therefore, the original expression is equivalent to the sum of two products:

$$ABC + A\overline{B}C + AB\overline{C} = AB + AC$$

The first column and the last column are also considered adjacent columns in a Karnaugh map as they contain common products. You can think of the map as being cut out and wrapped onto a cylinder so that the first and last columns are touching. Study the next example, which illustates this situation.

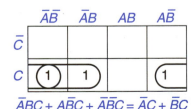

$$ABC + A\overline{B}C + AB\overline{C} = AB + AC$$

FIGURE 16-22 Adjacent squares in a three-variable Karnaugh map.

EXAMPLE 16.7

Simplify the following Boolean expression:
$$\overline{A}BC + A\overline{B}C + \overline{A}\,\overline{B}\,C$$

Solution   Place a 1 in each square corresponding to a product. One of the products is in the first column and one is in the last column, but they are adjacent as both are in the bottom row. See Figure 16–23. Therefore, there are two groups of 1s that should be circled. Here, $\overline{A}C$ is common to $\overline{A}BC$ and $\overline{A}\,\overline{B}\,C$, and $\overline{B}C$ is common to $A\overline{B}C$ and $\overline{A}\overline{B}C$. The Boolean expression is, therefore, equivalent to

$$\overline{A}BC + A\overline{B}C + \overline{A}\,\overline{B}\,C = \overline{A}C + \overline{B}C$$

As shown in Figure 16–21, two columns or four adjacent squares constitute each of the single variables A, B, $\overline{A}$, and $\overline{B}$. Also, each row of four adjacent squares constitutes C or $\overline{C}$. This means that if any group of four adjacent squares are circled, they correspond to one of these single variables, and these four products of three variables can then be replaced by a single variable. For example, as shown in Figure 16–21, the following Boolean expression is equivalent to $\overline{B}$:

$$\overline{A}\,\overline{B}\,\overline{C} + \overline{A}\,\overline{B}\,C + A\,\overline{B}\,\overline{C} + A\overline{B}C = \overline{B}$$

$$\overline{A}BC + A\overline{B}C + \overline{A}\overline{B}C = \overline{A}C + \overline{B}C$$

FIGURE 16-23 Adjacent squares on the edges of a Karnaugh map for Example 16.7.

■

Four Variables

A Karnaugh map of four variables contains 16 products as shown in Figure 16–24. Each column, each row, or any other four adjacent squares correspond to a product of two variables. Two adjacent squares correspond to a product of three variables. Consider the Boolean expression containing four variables:

$$ABCD + \overline{A}BCD + AB\overline{C}D + \overline{A}B\overline{C}D$$

	$\overline{A}\overline{B}$	$\overline{A}B$	AB	$A\overline{B}$
$\overline{C}\overline{D}$	$\overline{A}\overline{B}\overline{C}\overline{D}$	$\overline{A}B\overline{C}\overline{D}$	$AB\overline{C}\overline{D}$	$A\overline{B}\overline{C}\overline{D}$
$\overline{C}D$	$\overline{A}\overline{B}\overline{C}D$	$\overline{A}B\overline{C}D$	$AB\overline{C}D$	$A\overline{B}\overline{C}D$
CD	$\overline{A}\overline{B}CD$	$\overline{A}BCD$	$ABCD$	$A\overline{B}CD$
$C\overline{D}$	$\overline{A}\overline{B}C\overline{D}$	$\overline{A}BC\overline{D}$	$ABC\overline{D}$	$A\overline{B}C\overline{D}$

FIGURE 16–24   Karnaugh map for four variables.

To simplify this expression using the Karnaugh map, place a 1 in each square corresponding to a product as shown in Figure 16–25. Circle adjacent squares that form a group of four adjacent squares. The expression is then equivalent to the product of the two variables common to each square, which is *BD*:

$$ABCD + \overline{A}BCD + AB\overline{C}D + \overline{A}B\overline{C}D = BD$$

$$\overline{A}BCD + \overline{A}BCD + AB\overline{C}D + \overline{A}B\overline{C}D = BD$$

FIGURE 16–25   Four adjacent squares in a four-variable Karnaugh map.

In a four-variable map, not only are the first and last columns considered adjacent but also the first and last rows are considered adjacent. Study the next example which illustrates this situation.

EXAMPLE 16.8

Simplify the following Boolean expression:

$$ABC\overline{D} + AB\overline{C}\,\overline{D} + A\overline{B}C\overline{D} + \overline{A}\,\overline{B}C\overline{D}$$

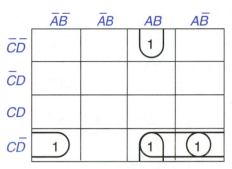

$$ABC\overline{D} + AB\overline{C}\overline{D} + A\overline{B}C\overline{D} + \overline{A}\,\overline{B}C\overline{D} =$$

$$AB\overline{D} + AC\overline{D} + \overline{B}C\overline{D}$$

FIGURE 16–26 Adjacent squares on the edges of a Karnaugh map for Example 16.8.

Solution   Place a 1 in each square corresponding to a product as shown in Figure 16–26. There are then *two* pairs of adjacent squares in the last row and one pair of adjacent squares in the third column. Each pair of adjacent squares corresponds to the product of three variables that is common to the two squares. The Boolean expression is then equivalent to

$$ABC\overline{D} + AB\overline{C}\,\overline{D} + A\overline{B}C\overline{D} + \overline{A}\,\overline{B}C\overline{D} = AB\overline{D} + AC\overline{D} + \overline{B}C\overline{D}$$ ▪

Karnaugh maps can also be drawn for five or more variables and are very useful for simplifying complex Boolean expressions that represent logic circuits.

EXERCISE 16.3

In problems 1 through 10, simplify each Boolean sum of products by using a Karnaugh map.

1. $AB + \overline{A}B$
2. $A\overline{B} + \overline{A}\,\overline{B}$
3. $\overline{A}B + A\overline{B} + \overline{A}\,\overline{B}$
4. $AB + A\overline{B} + \overline{A}B$
5. $ABC + \overline{A}BC + AB\overline{C}$
6. $\overline{A}BC + \overline{A}B\overline{C} + \overline{A}\,\overline{B}\,\overline{C}$
7. $\overline{A}\,\overline{B}C + A\overline{B}C + \overline{A}\,\overline{B}\,\overline{C}$
8. $\overline{A}\,\overline{B}\,\overline{C} + AB\overline{C} + A\overline{B}\,\overline{C}$
9. $A\overline{B}\,\overline{C} + \overline{A}\,\overline{B}\,\overline{C} + \overline{A}BC$
10. $ABC + \overline{A}BC + \overline{A}\,\overline{B}C$

In problems 11 through 14 prove each statement true.

11. Show that the Boolean expression

$$ABC + A\overline{B}C + AB\overline{C} + A\overline{B}\,\overline{C}$$

can be simplified to a single variable.

12. Show that the Boolean expression

$$\overline{A}B\overline{C} + \overline{A}BC + \overline{A}\,\overline{B}C + \overline{A}\,\overline{B}\,\overline{C}$$

can be simplified to a single variable.

13. Show that the Boolean expression

$$A\overline{B} + \overline{A}B$$

cannot be simplified.

14. Show that the Boolean expression

$$ABC + \overline{A}B\overline{C} + A\,\overline{B}\,\overline{C}$$

cannot be simplified.

In problems 15 through 18, simplify each Boolean expression containing four variables:

15. $\overline{A}\,\overline{B}\,\overline{C}\,\overline{D} + \overline{A}BC\overline{D} + AB\overline{C}\,\overline{D} + A\overline{B}\,\overline{C}\,\overline{D}$

16. $ABCD + A\overline{B}CD + AB\overline{C}D + A\overline{B}\,\overline{C}D$

17. $\overline{A}\,\overline{B}CD + \overline{A}\,\overline{B}\,\overline{C}D + A\overline{B}CD$

18. $AB\overline{C}\,\overline{D} + A\overline{B}C\overline{D} + ABC\overline{D}$

In problems 19 and 20 simplify the Boolean expression.

19. Show that the Boolean expression

$$AB + A\overline{B}C + A\overline{B}\,\overline{C}$$

can be simplified to a single variable.

20. Show that the Boolean expression

$$\overline{A}BC + ABCD + ABC\overline{D}$$

can be simplified to an expression containing two variables.

▤ CHAPTER HIGHLIGHTS

16.1 BOOLEAN ALGEBRA

The basic truth tables for the operations of AND, OR, and NOT are as follows:

		AND	**OR**	**NOT**
A	*B*	*A · B*	*A + B*	*$\overline{A}$*
0	0	0	0	1
0	1	0	1	1
1	0	0	1	0
1	1	1	1	0

The laws of Boolean algebra are as follows:

▪ Commutative laws:

$$A + B = B + A$$
$$A \cdot B = B \cdot A$$

▪ Distributive laws:

$$A + (B \cdot C) = (A + B) \cdot (A + C)$$
$$A \cdot (B + C) = A \cdot B + A \cdot B$$

▪ Identity laws:

$$A + 0 = A$$
$$A \cdot 1 = A$$

▪ Complement laws:

$$A + \overline{A} = 1$$
$$A \cdot \overline{A} = 0$$

Study the basic theorems in Table 16.2. To prove a Boolean statement is true, construct a truth table, and show that the columns for each side of the statement are the same for all possible truth values of the variables. Study Examples 16.1 through 16.3.

16.2 LOGIC CIRCUITS

The three basic logic gates are AND gate, OR gate, and NOT gate. The truth tables for the these gates are the same as the corresponding operations in Boolean algebra shown in Section 16.1.

The AND gate and the OR gate may have more than two inputs.

A logic circuit produces an output that is a combination of the basic gates. The truth table for the circuit shows the truth values of the output for all possible truth values of the inputs. Study Examples 16.4 and 16.5.

The NAND gate is an AND gate followed by a NOT gate. The NOR gate is an OR gate followed by a NOT gate. See Figure 16–7.

16.3 KARNAUGH MAPS

Karnaugh maps are checkerboard-like diagrams where each square represents a term in a Boolean expression.

A Karnaugh map for two variables is shown in Figure 16–18, for three variables in Figure 16–21, and for four variables in Figure 16–24.

Place a 1 in each square containing a term in a Boolean expression and circle adjacent squares. Two adjacent squares can be replaced by a term containing one less variable. Four adjacent squares can be replaced by a term containing two less variables. Study Examples 16.6 through 16.8. The vertical edges of a three-variable map and both the vertical and the horizontal edges of a four-variable map are considered adjacent.

≡ REVIEW QUESTIONS

In problems 1 and 2,

(a) Prove each Boolean statement by use of truth tables.
(b) Give the dual of each Boolean statement.

1. $A(\overline{A} + B) = AB$

2. $A(A + BC) = A$

In problems 3 through 6, give the output of each logic gate for the given inputs:

3. AND gate; $A = 0$, $B = 1$

5. NOR gate; $A = 1$, $B = 0$

4. OR gate; $A = 1$, $B = 0$

6. NAND gate; $A = 0$, $B = 0$

For each logic circuit in problem 7 and 8,

(a) Determine the output X.
(b) Construct the truth table.
(c) Show by applying the laws of Boolean algebra that the output is equivalent to the output that is given.

7. $X = A + B$

8. $X = AB$

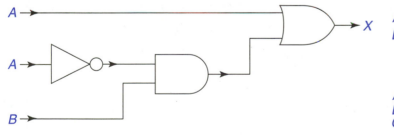

FIGURE 16–27 Logic circuit for problem 7.

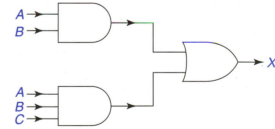

FIGURE 16–28 Logic circuit for problem 8.

In problems 9 through 12, simplify each Boolean expression using a Karnaugh map.

9. $\overline{A}B + \overline{A}\overline{B}$

11. $\overline{A}B\overline{C} + AB\overline{C} + \overline{A}BC + ABC$

10. $ABC + AB\overline{C} + \overline{A}BC$

12. $ABCD + ABC\overline{D} + AB\overline{C}\,\overline{D}$

Scientific and Graphing Calculator Functions

Basic scientific calculator operations are shown in Sections 1.5 and 2.5. Other functions and operations are shown in examples throughout the text. This appendix lists the key operations found on most scientific calculators with references in brackets to examples in the text that illustrate the operation.

≡ SCIENTIFIC CALCULATOR FUNCTIONS

Key	Function	Examples
MODE	Sets the operational mode	[9.9, 11.5]
DRG	Sets the angle measurement for degrees, radians or gradient	[9.9, 10.1]
INV, 2nd F, or SHIFT	Accesses the function above the key	[2.17, 2.24, 2.25, 9.11, 9.13, 9.18, 9.19, 12.2, 14.3]
FIX	Fixes the number of significant digits	
x^2	Squares a number	[2.20, 2.21, 2.22]
$\sqrt{}$	Finds the square root of a number	[2.5, 2.14, 2.22, 2.23, 3.11]
M$_{in}$, STO, or X→M	Stores a number in memory	[1.25]
MR, RCL, or RM	Recalls a number from memory	[1.25]
(	Opens parentheses	[1.25, 1.27, 1.28, 2.1, 2.20, 11.5]
)	Closes parentheses	[1.25, 1.27, 1.28, 2.1, 2.20, 11.5]
π	Enters the number π	[2.14, 3.11, 9.9, 10.1]

$\boxed{1/x}$ or $\boxed{x^{-1}}$	Finds the reciprocal of a number	[2.14, 2.21, 2.24, 3.11, 4.7, 5.10, 5.11, 5.12]
$\boxed{y^x}$ or $\boxed{x^y}$	Raises to a power	[2.1, 2.20, 2.21, 2.24, 7.1, 14.1]
$\boxed{\sqrt[3]{}}$	Finds the cube root of a number	[2.24]
$\boxed{\sqrt[x]{y}}$ or $\boxed{x^{\frac{1}{y}}}$	Finds the nth root of a number	[2.24]
$\boxed{+/-}$	Changes the sign of a number	[2.13, 2.14, 2.24, 2.25, 3.11, 7.1, 11.5, 14.3]
$\boxed{\text{EXP}}$ or $\boxed{\text{EE}}$	Sets the display to enter a power of 10 for scientific notation	[2.13, 2.14, 2.24, 2.25, 3.11, 7.1]
$\boxed{\mu}$, $\boxed{m}$, $\boxed{k}$, and $\boxed{M}$	Electrical units keys found on electrical engineering calculators	[2.17, 2.25]
$\boxed{\sin}$, $\boxed{\cos}$, $\boxed{\tan}$	Finds the sine, cosine, or tangent of an angle	[9.9, 10.1]
$\boxed{\sin^{-1}}$, $\boxed{\cos^{-1}}$, $\boxed{\tan^{-1}}$	Finds the inverse sine, inverse cosine, or inverse tangent of a number	[9.11, 9.13, 9.18, 9.19]
$\boxed{10^x}$	Raises 10 to the x power	
$\boxed{e^x}$	Raises e to the x power	[14.3]
$\boxed{\log}$	Finds the common logarithm of a number to the base 10	[14.5]
$\boxed{\ln}$ or $\boxed{\ln x}$	Finds the natural logarithm of a number to the base e	
$\boxed{\text{CPLX}}$	Used to set the calculator to complex mode	[11.5]
$\boxed{\text{Re}\leftrightarrow\text{Im}}$	Allows entry of the real and imaginary parts of a complex number for calculations with complex numbers	[11.5]
$\boxed{i}$	Used to enter the imaginary part of a complex number	[11.5]
$\boxed{x\longleftrightarrow y}$	Changes the display between two memory registers	[11.10, 12.2]
$\boxed{R\longrightarrow P}$	Changes from rectangular to polar form	[11.10, 12.2]
$\boxed{P\longrightarrow R}$	Changes from polar to rectangular form	[11.10]

BASE N	Used to set the calculator to a number base mode	[15.17]
BIN, OCT, or HEX	Used to change a decimal number to binary, octal, or hexadecimal form	[15.17]

≡ GRAPHING CALCULATOR FUNCTIONS

There are several graphing calculators available, including the Sharp EL9200 and EL9300, Casio fx-7700 and fx-8700, and the Texas Instruments series TI-81, TI-82, and TI-85. Most of the keys shown previously for scientific calculators also apply to graphing calculators with a few exceptions.

A function key such as $\sqrt{}$ or sin is pressed *before* a value is entered.

An ENTER or EXE key is used instead of an = key to display the result after any calculation or function key is used.

An up arrow key ∧ (like a computer) is used for an exponent.

For example, with TI graphing calculators the following calculations are done as follows:

Calculation	Keystrokes
4^2	4 x^2 ENTER
$\sqrt[3]{27}$	27 ∧ 3 x^{-1} ENTER
tan 40°	TAN 40 ENTER
ln 10	LN 10 ENTER

The graphing calculators have many computerlike capabilities. This section introduces some of the basic graphing functions of the popular TI-82.

Before drawing a graph, you need to press MODE to set the conditions for the graph. There are seven choices:

- Notation: normal, scientific, or engineering.
- Accuracy: number of decimal places displayed.
- Angle measure: radians or degrees.
- Type of graph: function, parametric, polar, or sequence.
- Type of display: connected line or dot.
- Type of plot: sequential or simultaneous plotting.
- Type of screen: full or split screen.

To make choices for any of these conditions, you move the cursor around similar to a computer keyboard. There are arrow keys < and > to move the cursor left and right, and arrow keys ∧ and ∨ to move the cursor up and down. Press ENTER when the cursor is on the condition you want. The examples given in the next few paragraphs use normal notation, radian measure, function mode, connected line, sequential plotting, and full or split screen modes.

To set the range of values on the x and y axes, you can use the [ZOOM] key. Some of the choices are as follows:

- [ZOOM] 4: Sets a small window for a split screen; x ranges from -4.7 to 4.7 and y from -3.1 to 3.1.

- [ZOOM] 5: Sets a large window for a full screen; x ranges from -15 to 15 and y from -10 to 10.

- [ZOOM] 6: Sets a square window for a full screen; x ranges from -10 to 10 and y from -10 to 10.

- [ZOOM] 7: Sets a window for trigonometric functions; x ranges from -2π to 2π and y from -3 to 3.

You can also set any values you want for the x and y axes by using the [WINDOW] key. Make entries by moving the cursor and using the [DEL] and [INS] keys to delete or insert characters.

Now, to enter the equation of a graph, press [Y=] and enter the expression using the [X, T, Θ] key for the variable. The variable is X when in function mode, T in parametric mode, and θ in polar mode. Use the [CLEAR] key to clear a line or the entire display. For example, to enter $y = \sin x$, key in the following:

[ZOOM] 7 [Y=] [CLEAR] [SIN] [X, T, Θ] [ENTER]

The display shows the first function, and the cursor moves to the second function:

$$y_1 = \sin x$$

$$y_2 =$$

Now press [GRAPH] to display the graph. The calculator graphs $y = \sin x$ from -2π to 2π.

You can enter a second function by pressing [Y=] again. It is possible to enter and graph as many as 10 functions. For example, to graph the voltage wave $y = 2 \cos x$ and the current wave $y = 3 \sin x$ on the same set of axes, you key in

[ZOOM] 7 [Y=] [CLEAR] 2 [X] [COS] [X, T, Θ] [ENTER] 3 [X] [SIN] [X, T, Θ] [GRAPH]

You can select which functions you want the calculator to graph by moving the cursor to the equals sign and pressing [ENTER] to highlight the sign. If you do not want the function to be graphed, press [ENTER] again to remove the highlight.

You can draw a vertical line on a graph by first pressing [2nd] [DRAW] 4, then positioning the cursor where you want the line and pressing [ENTER].

Another very useful feature is to see the values of x and y for any point on the graph. Press [TRACE] to locate the cursor on the graph. The coordinates of the point are shown at the bottom of the display. Then by using the left and right arrow keys you can move the cursor along the graph and see the coordinates of the points. If there are two graphs, use the up and down arrow keys to move the point between the graphs.

You can also see a table of values of x and y by pressing [2nd] [TABLE].

APPENDIX

B

Notation and Formulas

≡ ELECTRICAL NOTATION

Quantity	Letter(s) Used	Unit	Symbol
Admittance	Y	Siemen	S
Angular velocity	ω (omega)	Radians per second	rad/s
Capacitance	C	Farad	F
Charge	Q	Coulomb	C
Conductance	G	Siemen	S
Current	I, i	Ampere	A
Energy	E	Kilowatt-hour	kWh
Frequency	f	Hertz	Hz
Impedance	Z	Ohm	Ω (omega)
Inductance	L	Henry	H
Period	T	Second	s
Power	P	Watt	W
Power gain	dB	Decibel	dB
Reactance	X	Ohm	Ω
Resistance	R	Ohm	Ω
Resonant frequency	f_r	Hertz	Hz
Susceptance	B	Siemen	S
Time constant	τ (tau)	Second	s
Voltage	V, v	Volt	V

≡ METRIC OR SI PREFIXES

Power of 10	Prefix	Symbol
10^{-15}	femto	f
10^{-12}	pico	p
10^{-9}	nano	n
10^{-6}	micro	μ (mu)
10^{-3}	milli	m
10^{-2}	centi	c
10^{3}	kilo	k
10^{6}	mega	M
10^{9}	giga	G
10^{12}	tera	T

≡ MATHEMATICAL FORMULAS

Exponent Rules

$$(x^m)(x^n) = x^{m+n} \tag{3.6}$$

$$\frac{x^m}{x^n} = x^{m-n} \quad (x \neq 0) \tag{3.7}$$

$$(x^m)^n = x^{mn} \tag{3.8}$$

$$(xy)^n = x^n y^n \quad \text{and} \quad \left(\frac{x}{y}\right)^n = \frac{x^n}{y^n} \tag{3.9}$$

$$x^0 = 1 \quad (x \neq 0) \tag{3.11}$$

$$x^{-n} = \left(\frac{1}{x}\right)^n = \frac{1}{x^n} \quad \text{and} \quad \left(\frac{x}{y}\right)^{-n} = \left(\frac{y}{x}\right)^n \quad (x, y \neq 0) \tag{3.12}$$

$$x^{1/n} = \sqrt[n]{x}$$

$$x^{m/n} = \left(\sqrt[n]{x}\right)^m \tag{7.1}$$

Quadratic Formula

$$x = \frac{-b \pm \sqrt{b^2 - 4ac}}{2a} \tag{7.7}$$

Trigonometry

$$\pi \text{ radians} = 180°$$

$$1 \text{ rad} = \left(\frac{180}{\pi}\right)° = 57.3°$$

$$1° = \frac{\pi}{180} \text{ rad} = \frac{1}{57.3} \text{ rad} = 0.0175 \text{ rad}$$

For a point (x,y) on a circle of radius r,

$$x^2 + y^2 = r^2$$

$$\sin \theta = \frac{y}{r}$$

$$\cos \theta = \frac{x}{r}$$

$$\tan \theta = \frac{y}{x}$$

Vector or Phasor Components

$$V_x = V \cos \theta \ (x \text{ component})$$

$$V_y = V \sin \theta \ (y \text{ component}) \tag{9.7}$$

Law of Sines

$$\frac{a}{\sin A} = \frac{b}{\sin B} = \frac{c}{\sin C} \tag{9.8}$$

Law of Cosines

$$c^2 = a^2 + b^2 - 2ab(\cos C) \tag{9.10}$$

Complex Phasors

$$x + jy = \sqrt{x^2 + y^2} \ \angle \tan^{-1}\left(\frac{y}{x}\right) = Z \ \angle \theta \tag{11.2}$$

$$Z \ \angle \theta = Z \cos \theta + jZ \sin \theta = x + jy \tag{11.3}$$

$$(Z_1 \angle \theta_1)(Z_2 \angle \theta_2) = Z_1 Z_2 \angle (\theta_1 + \theta_2) \qquad (11.4)$$

$$\frac{Z_1 \angle \theta_1}{Z_2 \angle \theta_2} = \frac{Z_1}{Z_2} \angle (\theta_1 - \theta_2) \qquad (11.5)$$

Logarithms

$$y = \log x \text{ means } 10^y = x$$

$$y = \ln x \text{ means } e^y = x \text{ where } e \approx 2.718$$

$$\log (AB) = \log A + \log B \qquad (14.7)$$

$$\log \left(\frac{A}{B}\right) = \log A - \log B \qquad (14.8)$$

$$\log (A^x) = x(\log A) \qquad (14.9)$$

☰ ELECTRICAL FORMULAS

DC Circuit

Ohm's law:

$$I = \frac{V}{R}$$

$$V = IR$$

$$R = \frac{V}{I} \qquad (5.1)$$

Power formulas:

$$P = VI$$

$$I = \frac{P}{V}$$

$$V = \frac{P}{I} \qquad (5.2)$$

$$P = I^2 R \quad \text{and} \quad P = \frac{V^2}{R} \qquad (5.3)$$

$$R = \frac{P}{I^2} \quad \text{and} \quad R = \frac{V^2}{P} \qquad (5.3a)$$

$$I = \sqrt{\frac{P}{R}} \quad \text{and} \quad V = \sqrt{PR} \qquad (5.3b)$$

Series DC Circuit

Total resistance:
$$R_T = R_1 + R_2 + R_3 + \cdots + R_n \qquad (5.4)$$

Voltage divider:
$$V_1 = \frac{R_1}{R_T}(V_T) \qquad (8.3)$$

Parallel DC Circuit

Total current:
$$I_T = I_1 + I_2 + I_3 + \cdots + I_n \qquad (5.5)$$

Total resistance:
$$\frac{1}{R_T} = \frac{1}{R_1} + \frac{1}{R_2} + \frac{1}{R_3} + \cdots + \frac{1}{R_n} \qquad (5.6)$$

Conductance:
$$G = \frac{1}{R} \qquad (5.10)$$

Total conductance:
$$G_T = G_1 + G_2 + G_3 + \cdots + G_n \qquad (5.11)$$

Current divider:
$$I_1 = \frac{G_1}{G_T}(I_T) \qquad (8.5)$$

Two Parallel Resistances

$$R_T = \frac{R_1 R_2}{R_1 + R_2} \qquad (5.8)$$

$$R_1 = \frac{R_2 R_T}{R_2 - R_T} \qquad (5.9)$$

$$I_1 = \frac{R_2}{R_1 + R_2}(I_T) \qquad (8.4)$$

Kirchhoff's Laws

The algebraic sum of all currents entering and leaving any point in
a circuit equals zero. $\qquad (8.1)$

The algebraic sum of the voltages around any closed path equals zero. $\qquad (8.2)$

AC Circuit

Ohm's law:

$$I = \frac{V}{Z} \qquad (12.4)$$

True power:

$$P = VI \cos \theta \qquad (12.9)$$

RL Series Circuit

Inductive reactance:

$$X_L = 2\pi fL \qquad (12.1)$$

Total voltage:

$$V_T = V_R + jV_L = \sqrt{V_R^2 + V_L^2} \angle \tan^{-1}\left(\frac{V_L}{V_R}\right) \qquad (12.2)$$

Impedance:

$$Z = R + jX_L = \sqrt{R^2 + X_L^2} \angle \tan^{-1}\left(\frac{X_L}{R}\right) \qquad (12.3)$$

RC Series Circuit

Capacitive reactance:

$$X_C = \frac{1}{2\pi fC} \qquad (12.5)$$

Total voltage:

$$V_T = V_R - jV_C = \sqrt{V_R^2 + V_C^2} \angle \tan^{-1}\left(\frac{-V_C}{V_R}\right) \qquad (12.6)$$

Impedance:

$$Z = R - jX_C = \sqrt{R^2 + X_C^2} \angle \tan^{-1}\left(\frac{-X_C}{R}\right) \qquad (12.7)$$

RLC Series Circuit

Impedance:

$$Z = R + j(X_L - X_C) \qquad (12.8)$$

Use (12.3) when $X_L > X_C$. Use (12.7) when $X_L < X_C$.

Resonant frequency:

$$f_r = \frac{1}{2\pi\sqrt{LC}} \qquad (12.10)$$

RL Parallel Circuit

Total current:

$$I_T = I_R - jI_L = \sqrt{I_R{}^2 + I_L{}^2} \;\; \angle \tan^{-1} \left(\frac{-I_L}{I_R} \right) \tag{13.1}$$

Conductance:

$$G = \frac{1}{R} \tag{13.3}$$

Susceptance:

$$B = \frac{1}{X} \tag{13.3}$$

Admittance:

$$Y = \frac{1}{Z} \tag{13.3}$$

$$Y = \frac{1}{R} - j\frac{1}{X_L} = G - jB_L \tag{13.4}$$

RC Parallel Circuit

Total current:

$$I_T = I_R + jI_C = \sqrt{I_R{}^2 + I_C{}^2} \;\; \angle \tan^{-1} \left(\frac{I_C}{I_R} \right) \tag{13.2}$$

Admittance:

$$Y = \frac{1}{R} + j\frac{1}{X_C} = G + jB_C \tag{13.4}$$

RLC Parallel Circuit

Total current:

$$I_T = I_R + j(I_C - I_L) \tag{13.5}$$

Use (13.1) when $I_L > I_C$. Use (13.2) when $I_L < I_C$.
Admittance:

$$Y = \frac{1}{R} + j\left(\frac{1}{X_C} - \frac{1}{X_L} \right) = G + j(B_C - B_L) \tag{13.6}$$

Series Impedances

$$Z_T = Z_1 + Z_2 + Z_3 + \cdots \tag{13.7}$$

Two Parallel Impedances

$$Z_{EQ} = \frac{Z_1 Z_2}{Z_1 + Z_2} \tag{13.8}$$

RC and RL Circuits

Time constants:

$$RC: \quad \tau = RC \tag{14.3}$$

$$RL: \quad \tau = \frac{L}{R} \tag{14.5}$$

The decay curves

$$i = \frac{V}{R}e^{-t/\tau} \quad \text{or} \quad v = Ve^{-t/\tau} \tag{14.10}$$

illustrate the following four situations:

- ▪ RC circuit: Capacitor current i_c on charge. Capacitor current i_c on discharge. Capacitor voltage v_c on discharge.
- ▪ RL circuit: Inductor current i_L on decrease.

The exponential curves

$$i = \frac{V}{R}\left(1 - e^{-t/\tau}\right) \quad \text{or} \quad v = V\left(1 - e^{-t/\tau}\right) \tag{14.11}$$

illustrate the following two situations:

- ▪ RC circuit: Capacitor voltage v_C on charge.
- ▪ RL circuit: Inductor current i_L on increase.

Power Gain

$$\text{dB} = 10\log\left(\frac{P_2}{P_1}\right) \tag{14.12}$$

Logic Gates

		AND	OR	NOT
A	*B*	*A* · *B*	*A* + *B*	$\overline{A}$
0	0	0	0	1
0	1	0	1	1
1	0	0	1	0
1	1	1	1	0

APPENDIX C

Answers to Odd-Numbered Problems

≡ CHAPTER 1

EXERCISE 1.1

1. 16	**3.** 120	**5.** 41
7. 36	**9.** 26	**11.** 6
13. 6	**15.** 3	**17.** 8

19. 2 mi/gal   **21.** 57,600,000 mi

23. 1 A   **25.** 15 mV   **27.** 7

EXERCISE 1.2

1. $\frac{3}{5}$ **3.** $\frac{4}{5}$ **5.** $\frac{3}{4}$

7. $\frac{2}{15}$ **9.** $\frac{4}{3}$ **11.** 2

13. 9   **15.** 12   **17.** 1

19. $\frac{5}{8}$ **21.** $\frac{17}{60}$ **23.** $\frac{19}{20}$

25. $\frac{85}{24}$ **27.** $\frac{41}{36}$ **29.** $\frac{13}{6}$

31. $\frac{14}{15}$ **33.** $\frac{39}{10}$ **35.** 5

37. $200   **39.** 1 ft 4 in.   **41.** $\frac{1}{4}$ V

43. 1 Ω   **45.** $\frac{29}{4}$ Ω or $7\frac{1}{4}$ Ω   **47.** 4

EXERCISE 1.3

1. 10.09	**3.** 4.05	**5.** 0.69
7. 30	**9.** 0.08	**11.** 12.3
13. 4	**15.** 7	**17.** $\frac{1}{5} = 0.20 = 20\%$

19. $\frac{17}{100} = 0.17 = 17\%$ **21.** $\frac{7}{125} = 0.056 = 5.6\%$

23. $\frac{1}{250} = 0.004 = 0.4\%$ **25.** $\frac{3}{4} = 0.75 = 75\%$

27. $\frac{3}{20} = 0.15 = 15\%$ **29.** $\frac{3}{2} = 1.50 = 150\%$

31. 30 g   **33.** **(a)** $42; **(b)** $44.10

35. **(a)** $40.00; **(b)** $43.00   **37.** 10% for 2 years

39. 110%, $\frac{1120}{1000}$, $\frac{9}{8}$, $1\frac{1}{6}$, 1.19   **41.** 40.12 V

43. **(a)** 5%; **(b)** 12.6 V   **45.** 20%

EXERCISE 1.4

1. 3	**3.** 2	**5.** 3
7. 3	**9.** 4	**11.** **(a)** 5.15; **(b)** 5.15
13. **(a)** 31.26; **(b)** 31.3		**15.** **(a)** 321.87; **(b)** 322
17. **(a)** 29.00; **(b)** 29.0	**19.** 5,570,000 m³	

21. 100 in²   **23.** **(a)** 1.3 Ω; **(b)** 1.28 Ω

25. $R_1 = 38$ Ω, $R_2 = 24$ Ω

27. **(a)** 3; **(b)** −22; **(c)** −1; **(d)** 0;

     (e) When a number is negative

EXERCISE 1.5

1. 2	**3.** 21.1	**5.** 0.5		**19.** 0.00266	**21.** 197,000,000 mi^2	
7. 2.4	**9.** 0.68	**11.** 0.0143		**23.** (a) 4.67%; (b) 19.1%	**25.** 0.405 (SOHO)	
13. 35.5	**15.** 1.90	**17.** 8.27		**27.** 667 Ω = 0.667 kΩ	**29.** 5.71 Ω	

REVIEW QUESTIONS

1. 22	**3.** 27	**5.** 30	**19.** 31.0	**21.** 66.9	**23.** 54.5 mi/h
7. $\frac{23}{30}$	**9.** $\frac{7}{12}$	**11.** 4.04	**25.** 4.24 in ≈ $4\frac{1}{4}$ in	**27.** $34.56	**29.** 20 Ω
13. 20	**15.** 33	**17.** 0.95	**31.** 27 mA	**33.** 60 Ω	**35.** 21 mA

≡ CHAPTER 2

EXERCISE 2.1

1. 81	**3.** $\frac{1}{4} = 0.25$	**5.** 0.343	**31.** $10\sqrt{2}$	**33.** 14	**35.** 2
7. $\frac{1}{9}$	**9.** $\frac{11}{250} = 0.044$	**11.** $\frac{1}{20} = 0.05$	**37.** $\frac{\sqrt{7}}{4}$	**39.** 0.1	**41.** 1.728 cm^3
13. $\frac{5}{64}$	**15.** 4	**17.** $\frac{5}{2} = 2.5$	**43.** 100 mm, 10 cm		**45.** 17 W
19. 0.8	**21.** 0.09	**23.** 3	**47.** 20 Ω	**49.** 0.20 A	**51.** 2
25. $\frac{1}{2} = 0.5$	**27.** 0.4	**29.** $2\sqrt{2}$			

EXERCISE 2.2

1. 10^3	**3.** 10^{-6}	**5.** 10^{-9}	**29.** 2.5×10^{-5}	**31.** 0.75×10^6	**33.** 16×10^3
7. 100,000	**9.** 0.000000001, $\frac{1}{1,000,000,000}$		**35.** 7×10^{-9}	**37.** 125×10^6	**39.** 16×10^{12}
			41. 4×10^3	**43.** 2×10^4	**45.** 1×10^4
11. 0.001, $\frac{1}{1000}$	**13.** 10^6	**15.** 10^{-5}	**47.** 9×10^1	**49.** 6.5×10^4	**51.** 3.0 V
17. 10^{-6}	**19.** 10^3	**21.** 10^2	**53.** 5.0×10^{-3} A		**55.** 2.0×10^3 Ω
23. 12×10^2	**25.** $6 \times 10^0 = 6$	**27.** 3×10^3	**57.** 6.4 W	**59.** 50×10^{-3} A	**61.** 3.1×10^{-3} Ω

EXERCISE 2.3

1. (a) 4.26×10^{10};	(b) 42.6×10^9	**17.** 0.0000320	**19.** 4,440,000
3. (a) 9.30×10^{-8};	(b) 93.0×10^{-9}	**21.** 1.01	**23.** 9.42×10^{-3}
5. (a) & (b) 2.35×10^0		**25.** 0.899×10^6	**27.** 53.3×10^{-3}
7. (a) 6.22×10^4;	(b) 62.2×10^3	**29.** 40.0×10^6	**31.** 76.0%
9. (a) 5.64×10^7;	(b) 56.4×10^6	**33.** 2.40×10^{-6} m	**35.** 5.32×10^{-23} grams
11. (a) 1.14×10^{-7};	(b) 0.114×10^{-6} or 114×10^{-9}	**37.** 439×10^9 kWh	**39.** 5.41×10^3 Ω = 5.41 kΩ
13. 26,400,000,000	**15.** 0.0000000000931	**41.** 1.32×10^{-9} F	**43.** 2.91×10^3 Hz

EXERCISE 2.4

1. 2300 V	**3.** 0.310 A	**5.** 5200 mW	**19.** 373 K	**21.** 54.4 kg	**23.** 41.1 m
7. 0.860 mA	**9.** 130 pF	**11.** 3.88 MHz	**25.** 0.181 hp	**27.** 4.72 ft-lb	**29.** 115 ft/s
13. 4.10 m	**15.** 600 cm^2	**17.** 3500 μA	**31.** 35°C	**33.** 650 lb	**35.** 0.0630 in

37. 1.85 km   **39.** 0.128 V = 128 mV   **45.** 0.295 in^2 **47.** 1470 Ω = 1.47 kΩ

41. 6.20 kΩ   **43.** (a) $\dfrac{0.628\ \Omega}{1000\ \text{ft}}$;   (b) 0.0130 in.^{2}

EXERCISE 2.5

1. 58.1	**3.** 0.0101	**5.** 0.655
7. 119	**9.** 0.242	**11.** 57.0
13. 15.0	**15.** 10.1	**17.** 57.4
19. 93.7	**21.** 539	**23.** 0.250×10^9

25. 0.215×10^0 **27.** 2.08×10^6 **29.** 78.9
31. 6.46 cm   **33.** 4.20 V   **35.** 747 Ω
37. 30.3 kΩ

REVIEW QUESTIONS

1. 4	**3.** 9	**5.** 2.5
7. 0.25×10^6	**9.** $5\sqrt{3}$	**11.** 0.2
13. 277	**15.** 0.136	**17.** 33.0×10^3
19. 2.05×10^{-3}	**21.** 200×10^{-6}	**23.** 3.18

25. 510 μs   **27.** 1.50 MΩ   **29.** 2.30 MW
31. 0.762 m   **33.** 6,370 km   **35.** 201,000 hp
37. 53.8 ft   **39.** 56.4 V   **41.** 11.0 mA
43. 669 Ω   **45.** 33.3 mH   **47.** 50.4×10^{-6} H = 50.4 μH

☰ CHAPTER 3

EXERCISE 3.1

1. −3	**3.** 4	**5.** 21
7. −44	**9.** −7	**11.** 0.06
13. −2	**15.** −36	**17.** 48
19. $\frac{-1}{5}$	**21.** 0.5	**23.** $\frac{-23}{6}$
25. −0.5	**27.** $-8x + 3y$	**29.** $-1.5V_1 - 2.3V_2$
31. $2I^2R$	**33.** $-by - 1$	**35.** $5X_C{}^2 - X_C - 7$

37. $\dfrac{R_1}{3}$ **39.** $5V_0 - 2V_i$ **41.** $n^2 + n + 2$

43. 8,848 m − (−11,034 m) = 19,882 m
29,028 ft − (−36,201 ft) = 65,229 ft

45. $0° - 45° + 90° = +45°$ **47.** $3\pi r^2 h - 2\pi r$

49. −158°C   **51.** 1.3 + 0.8 − 1.7 − 0.4 = 0

53. $3.8IR + 0.4$ **55.** $\dfrac{1}{12}W + 2$

EXERCISE 3.2

1. x^6	**3.** R^2	**5.** $27p^3 v^6$
7. 10^{12}	**9.** $\dfrac{x^2}{9b^4}$	**11.** I^2
13. $\frac{1}{25} = 0.04$	**15.** $\frac{9}{2} = 4.5$	**17.** $10^2 = 100$
19. $9T^{-2} = \dfrac{9}{T^2}$	**21.** $Y^{-1} = \dfrac{1}{Y}$	**23.** $-14Q^5$
25. $3y^3$	**27.** $2m^3 - 2m^2 + 6m$	
29. $2a - b$	**31.** $3.4\pi f_r - 2.2$	

33. $0.04 \times 10^6 = 4 \times 10^4$ **35.** $2 \times 10^{-2} = 0.02$
37. $7.20 \times 10^{-7} = 0.720 \times 10^{-6}$ **39.** 2.03×10^{-3}
41. $1.98 \times 10^{-7} = 198 \times 10^{-9}$ **43.** $1 - \left(\dfrac{P_2}{P_1}\right)^3$
45. $\dfrac{13V^2}{4}$ **47.** 38 Ω   **49.** 1.74×10^{-3} Ω
51. 982 kHz   **53.** 111 pF   **55.** 2*X^3− 6*X^2+10*X

EXERCISE 3.3

1. $x^2 + 6x + 8$	**3.** $4P^2 - 2P - 20$
5. $6I^2 - 13IR + 6R^2$	**7.** $V_1{}^2 - 9$
9. $50L^2 - 8C^2$	**11.** $t^2 - 8t + 16$
13. $x^2y^2 + 2xy + 1$	**15.** $AX + AY - BX - BY$
17. $E^3 - E^2 - E + 1$	**19.** $G^3 + G^2 + G + 6$
21. $3Y(2Y + 1)$	**23.** $2I^2(2R_1 + R_2)$

25. $5(2s^2 + 4s - 3)$	**27.** $xyz(x + y + z)$
29. $(a - 2b)(a + 2b)$	**31.** $50(m - 2)(m + 2)$
33. $(X_C + 0.4)(X_C - 0.4)$	**35.** $(Z + 1)(Z + 5)$
37. $(v - 1)(v - 3)$	**39.** $(C_1 - 3)^2$
41. $(3G + 1)(G + 1)$	**43.** $(5x + 6y)(x - y)$
45. $4(R_T + 4)(2R_T - 1)$	**47.** $5P(2V + 5)(V - 1)$

49. $(s_R - 0.3)^2$ **51.** $(L^2 + 1)^2$

53. $(T^2 + 10^2)\,(T + 10)\,(T - 10)$

55. $(n + 3)\,(n - 2) = n^2 + n - 6$ **57.** $4(t + 4)\,(t + 3)$

59. $(3t + 2)\,(2t - 3) = 6t^2 - 5t - 6$ **61.** $\dfrac{e^{-1}}{R}\,(V - V_0)$

63. $0.25V^2\left(\dfrac{2}{R_1} + \dfrac{1}{R_2}\right)$

65. $\dfrac{V_1^2}{R} - \dfrac{V_2^2}{R} = \dfrac{1}{R}\,(V_1 + V_2)\,(V_1 - V_2)$

EXERCISE 3.4

1. $\dfrac{7y^2}{9}$ **3.** $\dfrac{C}{B + D}$ **5.** $\dfrac{(R_1 - R_2)}{5}$

7. $\dfrac{(\alpha + 3t)}{(3\alpha - t)}$ **9.** Not reducible **11.** $\dfrac{7X_L - 1}{9X_L - 4}$

13. $\dfrac{1}{V - R}$ **15.** $\dfrac{1}{f_c - 1}$ **17.** $\dfrac{Q}{Q + 2}$

19. $\dfrac{6}{x}$ **21.** $\dfrac{1}{2}$ **23.** $\dfrac{7}{uv}$

25. $\dfrac{s - t}{s}$ **27.** $\dfrac{2(I + 1)}{3}$ **29.** $5(2D - 1)$

31. $\dfrac{V_0 + 1}{V_0}$ **33.** $10^3(K - 1)$ **35.** $\dfrac{2(A - 2B)}{(A - 2)(A + 2t)}$

37. 1 **39.** 1 **41.** $2(F_x + F_y)$ **43.** $\dfrac{4r}{3\pi}$

45. $P = V\left(\dfrac{V}{R}\right) = \dfrac{V^2}{R}$ **47.** $\dfrac{3I_T R_1}{4}$ **49.** $5t(t + 1)$

EXERCISE 3.5

1. $\dfrac{23}{24}$ **3.** $\dfrac{37x}{45}$ **5.** $\dfrac{5X_L + 42X_C}{180}$

7. $\dfrac{3I_1^2 + 2I_2^2}{I_1 I_2}$ **9.** 3.5×10^{-3} or 0.0035

11. $\dfrac{20}{1 \times 10^4} = 2 \times 10^{-3}$ **13.** $\dfrac{18 - 8\omega}{15r\omega^2}$

15. $\dfrac{2d^2 + 6cd - 5c^2}{2d^2}$ **17.** $\dfrac{1}{V_R + 2}$

19. $\dfrac{8.1 + 2.3k}{I(1 + k)}$ **21.** $\dfrac{2}{3}$ **23.** $\dfrac{1}{2V(V - 1)}$

25. $\dfrac{-1}{t - 1} = \dfrac{1}{1 - t}$ **27.** $\dfrac{X - Y}{2(X + Y)}$

29. $\dfrac{3m_0^2 + 4m_0 - 3}{3(3m_0 + 1)(m_0 + 1)}$ **31.** $\dfrac{-2I - 13R}{(I + R)(I - R)(I + 2R)}$

33. $\dfrac{7I_1^2 + 9I_1 I_2 - I_2^2}{I_1 I_2 (I_1 + I_2)}$ **35.** $\dfrac{6C^2 + 3C - 1}{(C - 1)^2(C + 1)}$

37. $\dfrac{1}{i + 1}$ **39.** $1 - \alpha$ **41.** **(a)** $R_{avg} = \dfrac{2r_1 r_2}{r_1 + r_2}$;

(b) 24 mi/h

43. $\dfrac{f_1 f_2}{f_1 + f_2 - d}$ **45.** $\dfrac{R_1 R_2 R_3}{R_1 R_2 + R_2 R_3 + R_1 R_3}$

47. $\dfrac{1}{G_{eq}} = \dfrac{\dfrac{1}{G_1 G_2}}{\dfrac{1}{G_1} + \dfrac{1}{G_2}} = \dfrac{1}{G_1 + G_2} \Rightarrow G_{eq} = G_1 + G_2$

49. $\dfrac{V(2R_1 + 100)}{R_1(R_1 + 100)}$

REVIEW QUESTIONS

1. 24 **3.** -1 **5.** $3x - 2x^2$

7. $5CV - 4$ **9.** $k^4 P^4$ **11.** 10

13. $-21X_C^{\,4}$ **15.** $1.1V^2$

17. $0.192 \times 10^3 = 1.92 \times 10^2$ **19.** $3\phi^3 - 6\phi^2 + 3\phi$

21. $W^2 - 2W - 8$ **23.** $12k^2 - 36k + 27$

25. $I^3 - 2I^2 - 4I + 8$ **27.** $4AB(2A - 3B)$

29. $(V_R + 7)\,(V_R - 3)$ **31.** $(x + 0.1)\,(x + 0.2)$

33. $(2u - 3v)\,(3u - 4v)$ **35.** $\dfrac{11r}{5} = 0.73r$

37. $\dfrac{9V}{10}$ **39.** $\dfrac{R_1 + 2R_2}{R_1 + R_2}$ **41.** $\dfrac{1}{n - d}$

43. 5 **45.** $\dfrac{29E_1}{24}$ **47.** $\dfrac{5}{2 \times 10^6} = 2.5 \times 10^{-6}$

49. $\dfrac{4Z - 1}{(Z + 1)(Z - 1)}$ **51.** $\dfrac{r^2 - 10r}{3(r + 2)^2}$ **53.** $+49°F$

55. $h - 2$ **57.** $4R(I_1^2 + I_2^2)$ **59.** 0.24 A

61. $0.4(R_0 - x)\,(5R_0 + 2x)$ **63.** $\dfrac{4(t + 1)}{t - 1}$

CHAPTER 4

EXERCISE 4.1

1. 5	**3.** 1	**5.** −4		**31.** 12	**33.** 4	**35.** $\frac{2}{5}$
7. −1	**9.** 4	**11.** 6		**37.** 30	**39.** 3	**41.** $\frac{1}{2}$
13. −1	**15.** −11	**17.** 1.5		**43.** $500, $300, $200		**45.** −25°C
19. 3	**21.** $\frac{1}{2}$	**23.** 2.3		**47.** 22.5 kg	**49.** 80 mA	**51.** 25 mΩ
				53. 12 Ω	**55.** 60 μF	**57.** 2 mH
25. 2.5	**27.** $\frac{1}{5}$	**29.** 5		**59.** 90 Ω		

EXERCISE 4.2

1. $R = \dfrac{P}{I^2}$ **3.** $f = \dfrac{1}{2\pi C X_C}$ **5.** $r_g = \dfrac{V - IR}{I}$

7. $V_D = V - IR$ **9.** $I = \dfrac{V_T}{R_1 + R_2 + R_3}$

11. $\beta = \dfrac{I_C}{I_E - I_C}$ **13.** $L_T = \dfrac{L_1 L_2}{L_1 + L_2}$

15. $L = \dfrac{Z}{\omega(1 + \omega C Z)}$ **17. (a)** $h = 116{,}600 - 550T$; **(b)** 1100 ft

19. (a) $C = \dfrac{5}{9}(F - 32)$; **(b)** 25°C

21. (a) $R_{avg} = \dfrac{2R_1 R_2}{R_1 + R_2}$; **(b)** 48 mi/h

23. $P = (IR)I = I^2 R$

25. (a) $R_1 = \dfrac{V - V_0}{I}$; **(b)** 6.0 Ω

27. $(R_T R_1 R_X)\dfrac{1}{R_T} = (R_T R_1 R_X)\dfrac{1}{R_1} + (R_T R_1 R_X)\dfrac{1}{R_X} \Rightarrow R_1 R_X = R_T R_X + R_T R_1 \Rightarrow R_X(R_1 - R_T) = R_T R_1 \Rightarrow R_X = \dfrac{R_T R_1}{R_1 - R_T}$

29. (a) $\alpha = \dfrac{R_1 - R_0}{R_0 t}$; **(b)** 0.0031 per °C

31. (a) $\rho = \dfrac{RA}{l}$; **(b)** 2.1×10^{-8} Ω–m

33. (a) $\Delta V = -16$ V; **(b)** $\Delta V = R \Delta I$

35. (a) $\Delta V = IR$; **(b)** $\Delta V = -0.1 IR$

EXERCISE 4.3

1. $38,500, $27,500   **3.** $35,000

5. $42   **7.** 89   **9.** $260.87

11. 1925   **13.** 2 nmi/h   **15. (a)** West; **(b)** 10 mi/h

17. 5 mH, 6 mH   **19.** $I_1 = 50$ mA, $I_2 = 75$ mA, 150 mA

21. $V_1 = 6.0$ V, $V_2 = 5.7$ V   **23.** 200 cm, 150 cm

25. $V_1 = 21$V, $V_2 = 42$V, $V_3 = 7.0$V

27. Bulb 72 V, resistor 4.8 V

REVIEW QUESTIONS

1. 6   **3.** $\frac{1}{2}$ **5.** 0.60

7. 4   **9.** −0.1   **11.** 19 m/s

13. $500   **15.** 42V   **17.** $Q = \dfrac{Er^2}{k}$

19. $I = \dfrac{P}{V_1 + V_2}$ **21.** 3 A

23. (a) $C_2 = 2(C_T - C_1)$; **(b)** 60 μF

25. (a) $\Delta I_T = \dfrac{\Delta V}{R_0}$; **(b)** 30 mA   **27.** 600 W, 2400 W

CHAPTER 5

EXERCISE 5.1

1. $I_1 = 330$ mA **3.** $R_2 = 750$ Ω **5.** $V_T = 15$ V

7. $I_T = 360$ mA **9.** 17 V

11. (a) 500 mA; (b) 250 mA

13. (a) 400 mA; (b) 200 mA **15.** 50 mA

17. 100 mA **19.** 45 mΩ **21.** $\frac{2}{1}$

EXERCISE 5.2

1. 7.0 W **3.** 1.5 A **5.** 13 V

7. 14 W **9.** 6.4 W **11.** 10 Ω

13. 22 Ω **15.** 2.0 A **17.** 37 V

19. 1.7 mW **21.** 3.0 W **23.** 20 Ω

25. 19 mA **27.** 1.7 A, 5.0 A

29. 840 mA **31.** (a) 15,000 W = 15 kW (b) 810 Ω

33. (a) 170 mW; (b) 1.1 V

35. $P_1 = 3.1$ W, $P_2 = 2.4$ W

37. (a) 81 V; (b) 1.2 kWh = 4.3 MJ

39. 2.7 kW **41.** $P = VI = V\left(\dfrac{V}{R}\right) = \dfrac{V^2}{R}$

EXERCISE 5.3

1. (a) $I = 80$ mA; (b) $V_1 = 80$ V, $V_2 = 16$ V

3. $R_1 = 75$ Ω, $R_2 = 225$ Ω

5. $I = 100$ mA, $P = 200$ mW

7. (a) $V_2 = 3.0$ V, $V_3 = 6.0$ V; (b) $V_T = 12$ V

9. $R_D = 240$ Ω **11.** $V_1 = 15$ V, $V_2 = 30$ V, $P = 27$ W

13. 2.7 kΩ

EXERCISE 5.4

1. 7.5 Ω **3.** 28 Ω **5.** 460 Ω

7. 130 mS, 7.5 Ω **9.** 11 mS, 94 Ω

11. 2.5 mS, 390 Ω **13.** 150 Ω

15. 1.2 kΩ **17.** 20 Ω

19. $I_1 = 500$ mA, $I_2 = 400$ mA, $I_T = 900$ mA, $R_T = 13$ Ω

21. $I_1 = 330$ mA, $I_2 = 250$ mA, $I_3 = 280$ mA, $I_T = 860$ mA,
$R_T = 58$ Ω

23. $R_1 R_2 = R_T R_2 + R_T R_1 \Rightarrow R_1 R_2 = R_T(R_1 + R_2) \Rightarrow R_T = \dfrac{R_1 R_2}{R_1 + R_2}$

25. $R_{eq} = \dfrac{(R_1)(2R_1)}{R_1 + 2R_1} = \dfrac{2R_1^2}{3R_1} = \dfrac{2R_1}{3}$

EXERCISE 5.5

1. 30 Ω **3.** 30 Ω

5. $R_T = \dfrac{R_1 R_2}{R_1 + R_2} + R_3$, $R_T = 180$ Ω

7. $I_T = 26$ mA, $V_2 = 6.2$ V, $V_4 = 2.6$ V, $I_4 = 8.7$ mA

9. $R_T = 240$ Ω **11.** $R_1 = 20$ Ω **13.** $R_T = 50$ Ω

15. 46 mA, $V_1 = 4.6$ V, $V_2 = 3.7$ V **17.** 2.9 W

REVIEW QUESTIONS

1. (a) 500 mA;
(b) 750 mA, 50%; I is directly proportional to V

3. (a) 9.0 V; (b) 18 V **5.** 16 Ω

7. 88 W **9.** 1.5 W **11.** (a) 200 mA; (b) 3.6 kWh

13. (a) 470 mA; (b) $V_1 = 16$ V, $V_2 = 8.5$ V

15. (a) $R_D = 3.4$ Ω; (b) 15 Ω

17. (a) 99 Ω; (b) 99 Ω

19. (a) $R_T = \dfrac{(R_1 + R_2)R_3}{(R_1 + R_2) + R_3}$; (b) 94 Ω

21. $R_T = 50$ Ω, $V_1 = 25$ V, $V_3 = 14$ V

23. $R_T = 1.2$ kΩ, $V_T = 42$ V

≡ CHAPTER 6

EXERCISE 6.1

1. 3.
5. 7.
9.

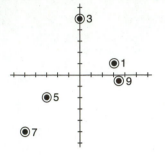

11.

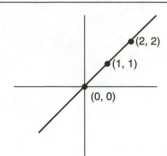

13.

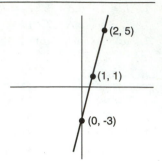

15.

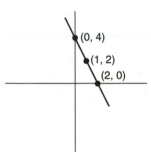

17.

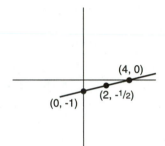

19.

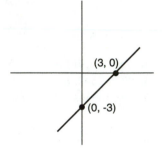

21.

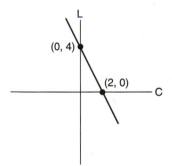

23.

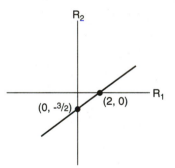

25.

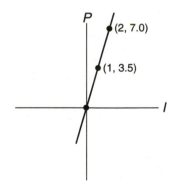

27.

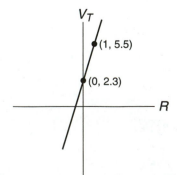

29.

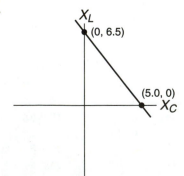

31.

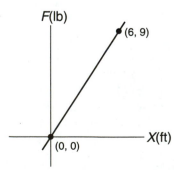

33.

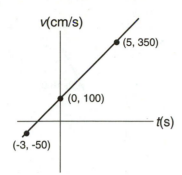

35. (a)

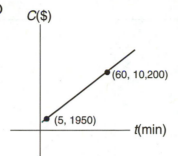

37.

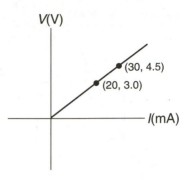

35. (b) $1200

39.

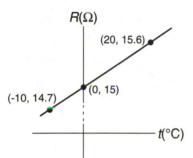

41.

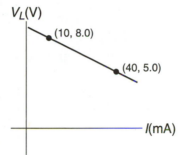

43.

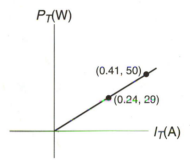

EXERCISE 6.2

1. $x = 3, y = 2$

3. $I_1 = 2, I_2 = 1$

5. $R = \frac{-2}{3}, X = \frac{1}{2}$

7. $I_a = -3, I_b = -3$

9. $P_1 = 0.8, P_2 = 0.2$

11. $r = 0.8, t = 0.3$

13. $A = 2, B = 5$

15. $x = 3, y = -1$

17. $G_1 = -2.8, G_2 = -3.2$

19. $V_x = -2.4, V_y = 5.2$

21. 46 real, 57 false

23. 18 m (59 ft), 36 m (118 ft)

25. 15–min walk, 45–min ride

27. Mr. Lee $300, wife $200

29. $E_x = 5.5$ N/C, $E_y = 3.5$ N/C

31. $I_1 = 110$ mA, $I_2 = 74$ mA

33. $G_1 = 50$ mS, $G_2 = 40$ mS; $R_1 = 20\ \Omega, R_2 = 25\ \Omega$

35. $V_1 = 22$ V, $V_2 = 12$ V

37. 3 ns (1st), 5 ns (2nd)

39. $R_x = 80\ \Omega, R_Y = 40\ \Omega$

EXERCISE 6.3

1. 5

3. 21

5. -7.8

7. 2

9. 1

11. 13

13. $X = 3, Y = 1$

15. $R = \frac{-1}{2}, V = \frac{1}{3}$

17. $I_1 = -2, I_2 = -4$

19. $R_A = 0.6, R_T = 0.4$

21. $a = 1.3, \beta = 1.1$

23. $I_A = -2.2, I_B = -4.4$

25. $X_C = 3.5, X_L = -1.5$

27. $x = 1, y = 2, z = 3$

29. $I_1 = 4, I_2 = -1, I_3 = 3$

31. $I_A = 0.1, I_B = 0.5, I_C = -0.3$

33. $p = 3, q = -5, r = -1$

35. $V_1 = \frac{1}{2} = 0.5, V_2 = \frac{2}{3} = 0.67, V_3 = \frac{3}{5} = 0.60$

37. $I_1 = 0.5, I_2 = 1.0, I_3 = -0.5$

39. $T_1 = 300, T_2 = 350$

41. Plane = 120 mi/h, wind = 30 mi/h

43. 0.25 gal, 0.75 gal, 0.50 gal

45. $I_1 = 190$ mA, $I_2 = 100$ mA

47. $V_A = 50$ V, $V_B = 200$ V, $V_C = 150$ V

49. $I_1 = 720$ mA, $I_2 = 880$ mA, $I_3 = 160$ mA

51. $19.5\ \Omega$, 0.00390/°C

53. $V_1 = 32$ V, $V_2 = 24$ V

55. $R_1 = 30\ \Omega, R_2 = 15\ \Omega$

57. $I_1 = 1.8$ A, $I_2 = 1.1$ A, $I_3 = 1.6$ A

REVIEW QUESTIONS

1.

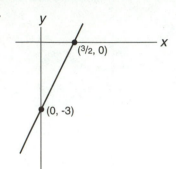

3.

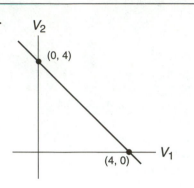

5.

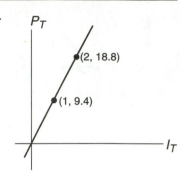

7. −534
9. 9
11. $x = -1, y = 2$
13. $I_1 = 0.4, I_2 = 0.9$

15. $X_L = 2.5, X_C = 3.5$
17. $x_1 = 3, x_2 = -1, x_3 = 1$
19. $I_1 = 1.5, I_2 = 2.0, I_3 = 3.5$

21.

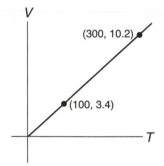

27.

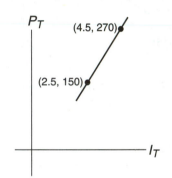

23. $11,500
25. $a = 5$ mi/h/s, $v = 200$ mi/h

29. 6.2 V, 0.30 Ω = 300 mΩ
31. $\lambda 1 = 0.6, \lambda_2 = 0.4, \lambda_3 = 0.2$
33. $I_1 = 200$ mA, $I_2 = 100$ mA

≡ CHAPTER 7

EXERCISE 7.1

1. 5
3. 9
5. 0.1
25. $10^3 V_x$
27. $108s^3$
29. $\dfrac{2\alpha \beta^2}{3}$

7. 27
9. 0.3
11. $1.2 \times 10^1 = 12$
31. 0.0084 m
33. 3.1 kHz
35. 0.019

13. 100
15. $\dfrac{125}{64}$
17. 2
37. 12 μA

19. $I_T{}^{0.4}$
21. $-3\omega^2$
23. $\dfrac{5}{L^2 X}$

EXERCISE 7.2

1. $6x$
3. $0.2I$
5. $2C_1{}^2C_2$
25. $2\sqrt{3} \times 10^3$
27. 1.2×10^{-3}
29. $\dfrac{\sqrt{3}}{3}$

7. $2\sqrt{3}$
9. $6\sqrt{3}$
11. $2\sqrt{2}$
31. $\dfrac{\sqrt{30}}{6}$
33. $\dfrac{2\sqrt{3}}{3}$
35. $\dfrac{3\sqrt{2}}{4}$

13. $r^2 \theta\sqrt{r \theta}$
15. $4X_L X_T{}^2\sqrt{2X_T}$
17. $2\sqrt[3]{3}$
23. $0.1I^2R_3$

19. 15
21. 3
37. $\dfrac{0.2\sqrt{V_x V_y}}{V_y}$
39. $2\sqrt{5} + 2\sqrt{7}$
41. $7\sqrt{5}$

43. $8\sqrt{3}$ **45.** $-3\sqrt{5}$ **47.** $3\sqrt{2}$

49. $6\sqrt{3} \times 10^6$ **51.** $7\sqrt{2} \times 10^{-3}$ **53.** $4\sqrt[3]{2}$

55. $5Y\sqrt{2Y}$ **57.** $\dfrac{3\sqrt{2}}{4}$ **59.** $\dfrac{17\sqrt{3}}{6G}$

61. 4.5, 3.139, 2.844, 2.828

63. 26, 13.981, 8.814, 7.300, 7.143, 7.141

65. 8.3 mi/h   **67.** 1.5, 1.296, 1.261, 1.260

69. 2.77×10^2 N/C   **71.** $\dfrac{\sqrt{4\pi^2 f^2 C^2 R^2 + 1}}{2\pi f C}$

73. $\dfrac{X_L R \sqrt{R^2 + X_L^2}}{R^2 + X_L^2}$

EXERCISE 7.3

1. $2, -2$ **3.** $5, -2$ **5.** $2, \dfrac{2}{3}$

7. $0, \dfrac{1}{7}$ **9.** $0, -0.433$ **11.** $1.2, -1.2$

13. $0.74, -0.74$ **15.** $\dfrac{1}{2}, -1$ **17.** $\dfrac{1}{2}, -3$

19. $5, -8$ **21.** $\dfrac{1}{4}, \dfrac{1}{2}$ **23.** $\pm\sqrt{\dfrac{P_1 + P_2}{R_T}}$

25. $R_2, \dfrac{R_2}{4}$ **27.** v_0, v_0 **29.** 40 s

31. 7.0 m, 10 m, 12 m   **33.** 2.2 mm, 4.5 mm

35. $\left(\dfrac{f}{2\pi L}\right) 2\pi f L = \dfrac{1}{2\pi f C}\left(\dfrac{f}{2\pi L}\right) \Rightarrow f^2 = \dfrac{1}{4\pi^2 LC} \Rightarrow$

$$f = \dfrac{1}{2\pi\sqrt{LC}} = \dfrac{\sqrt{LC}}{2\pi LC}$$

37. $X_L = 4.6\ \Omega, R = 2.2\ \Omega$ **39.** 2.0 A, 4.0 A

41. 1.0 A

EXERCISE 7.4

1. $-1, -3$ **3.** $2, \dfrac{-5}{3}$ **5.** $4, \dfrac{1}{2}$

7. $1 \pm \sqrt{2}; 2.41, -0.41$ **9.** $-2 \pm \sqrt{5}; 0.24, -4.24$

11. $1 \pm \sqrt{7}; 3.65, -1.65$ **13.** $\dfrac{3 \pm \sqrt{5}}{4}; 1.31, 0.19$

15. $\dfrac{1}{3}, -1$ **17.** $6.13, -0.46$ **19.** $\dfrac{-2 \pm \sqrt{-4}}{2}$

21. $\pm\sqrt{-1}$

23. 6 ft, 16 ft

25. 3.1 mm, 5.1 mm, 6.0 mm   **27.** 2.6 h

29. 5.0 A, 8.0 A   **31.** $13\ \Omega, 15\ \Omega$

33. 590 mA, 290 Ω; 3.4 A, 8.7 Ω

35. 1.3 A, 12 V; 390 mA, 38 V

REVIEW QUESTIONS

1. 9   **3.** 10   **5.** $2ab^2$

7. $R + 1$ **9.** $3\sqrt{2}$ **11.** $10X^2\sqrt{3X}$

13. 0.11   **15.** $\dfrac{\sqrt{2}}{4}$ **17.** $5\sqrt{6} - \sqrt{3}$

19. $\sqrt{2} \times 10^3$ **21.** $\dfrac{5f\sqrt{3}}{6}$ **23.** $t R_a\sqrt{t R_a}$

25. 4; 2.875; 2.655; 2.646   **27.** $5, -1$

29. $0, \dfrac{7}{8}$ **31.** $\dfrac{1}{3}, -1$ **33.** $\dfrac{-3 \pm \sqrt{5}}{2}; -0.38, -2.62$

35. $\dfrac{-1 \pm \sqrt{19}}{2}; 1.68, -2.68$ **37.** $\dfrac{-5 \pm \sqrt{97}}{6}; 2.47, -0.81$

39. $0, \dfrac{V_1 + V_2}{R_T}$ **41.** (a) $\dfrac{\sqrt{e(e+1)}}{2e}$; (b) 1.1; (c) 1.1

43. 2.8×10^8 m/S   **45.** (a) $V_{\text{rms}} = \dfrac{V_{\text{max}}\sqrt{2}}{2}$; (b) 120 V

47. $10\ \Omega, 15\ \Omega$ **49.** 380 mA, 34 Ω; 2.6 A, 730 mΩ

☰ CHAPTER 8

EXERCISE 8.1

1. See statement in text   **3.** $I_1 - I_2 - I_3 = 0$

5. $V_1 - I_1 R_1 - I_3 R_3 = 0$ **7.** $I_3 - I_1 - I_2 = 0$

9. $V_1 - I_1 R_1 - I_3 R_3 = 0$

11. $I_1 = 1.6$ A, $I_2 = 860$ mA, $I_3 = 710$ mA

13. $I_1 = 1.4$ A, $I_2 = 630$ mA, $I_3 = 2.0$ A

15. $I_1 = 1.7$ A, $I_2 = 1.9$ A, $I_3 = 230$ mA

17. 4.0 V, 10 Ω

19. $I_1 = 330$ mA, $I_2 = 1.2$ A, $I_3 = 390$ mA, $I_4 = 1.2$ A

21. $I_1 = 200$ mA, $I_2 = 500$ mA, $I_3 = 500$ mA, $I_4 = 800$ mA

EXERCISE 8.2

1. $V_1 = 3.2$ V, $V_1 = 1.9$ V, $V_3 = 3.9$ V
3. $I_1 = 4.3$ mA, $I_2 = 5.7$ mA
5. $I_1 = 100$ mA, $I_2 = 33$ mA, $I_3 = 67$ mA
7. **(a)** $V_2 = 8.0$ V, $V_3 = 4.0$ V;
   (b) $I_1 = 1.0$ A, $I_2 = 0.50$ A $= 500$ mA

9. $I_1 = 750$ mA, $V_2 = 3.8$ V
11. **(a)** $I_1 = 570$ mA, $I_2 = 290$ mA, $I_3 = 1.1$ A;
    (b) $V_2 = 200$ V, $V_3 = 80$ V, $V_4 = 120$ V
13. 25 Ω
15. $V_2 = 20$ V, $V_3 = 7.5$ V, $V_4 = 23$ V

EXERCISE 8.3

1. $I_T = 190$ mA, $V_1 = 4.6$ V
3. $I_1 = 100$ mA, $I_2 = 13$ mA

5. $I_1 = 600$ mA, $I_2 = 600$ mA, $I_3 = 1.2$ A
7. $I_1 = 4.6$ mA, $I_2 = 4.3$ mA, $I_3 = 380$ μA

EXERCISE 8.4

1. $V_{TH} = 4.9$ V, $R_{TH} = 5.7$ Ω, $I_L = 330$ mA, $V_L = 3.0$ V
3. $V_{TH} = 80$ V, $R_{TH} = 670$ Ω, $I_L = 22$ mA, $V_L = 65$ V
5. $V_{TH} = 13$ V, $R_{TH} = 2.2$ Ω, $I_L = 1.6$ A, $V_L = 9.7$ V

7. $V_{TH} = 33$ V, $R_{TH} = 1.2$ kΩ, $I_L = 12$ mA, $V_L = 18$ V
9. $V_{TH} = 1.0$ V, $R_{TH} = 32$ Ω
11. $V_{TH} = 6.8$ V, $R_{TH} = 12$ Ω, $I_L = 180$ mA, $V_L = 4.6$ V

EXERCISE 8.5

1. $I_N = 2.3$ A, $R_N = 3.4$ Ω, $I_L = 590$ mA, $V_L = 5.9$ V
3. $I_N = 100$ mA, $R_N = 170$ Ω, $I_L = 36$ mA, $V_L = 11$ V

5. $I_L = 400$ mA, $V_L = 2.0$ V

REVIEW QUESTIONS

1. $I_1 = 600$ mA, $I_2 = 600$ mA, $I_3 = 1.2$ A
3. $I_1 = 1.4$ A, $I_2 = 400$ mA, $I_3 = 900$ mA, $I_4 = 900$ mA
5. **(a)** $V_2 = 6.0$ V, $V_3 = 8.0$ V, $V_4 = 12$ V;
   (b) $I_1 = 52$ mA, $I_2 = 40$ mA

7. $I_1 = 280$ mA, $I_2 = 240$ mA, $I_3 = 520$ mA
9. $V_{TH} = 56$ V, $R_{TH} = 100$ Ω, $I_L = 340$ mA, $V_L = 20$ V
11. $I_3 = 35$ mA, $V_3 = 170$ V

≡ CHAPTER 9

EXERCISE 9.1

1. $\dfrac{\pi}{3} = 1.05$ rad
3. $\dfrac{\pi}{2} = 1.57$ rad
5. $\dfrac{7\pi}{4} = 5.50$ rad
7. 0.873 rad
9. $30°$
11. $270°$
13. $330°$
15. $69.3°$
17. $A = B = 70°$, $C = 110°$
19. $A = 49.7°$, $B = 40.3°$, $C = 139.7°$

21. $A = 35°$, $B = 55°$
23. $A = 0.97$ rad, $B = 2.17$ rad
25. 10
27. 2.2
29. 0.86
31. $x = 3.6$, $y = 4.8$
33. $x = 3.5$, $y = 3$
35. 6.4 ft
37. 15 mi
39. 90 ft
41. 230 mA
43. 29 kΩ
45. 169 mm

EXERCISE 9.2

1. $\sin A = \dfrac{8}{17} = 0.471$
   $\cos A = \dfrac{15}{17} = 0.882$
   $\tan A = \dfrac{8}{15} = 0.533$

3. $\sin A = 0.387$
   $\cos A = 0.922$
   $\tan A = 0.420$

5. $\sin A = 0.836$
   $\cos A = 0.549$
   $\tan A = 1.52$

7. sin A = 0.943
cos A = 0.333
tan A = 2.83

9. sin A = 0.707
cos A = 0.707
tan A = 1.00

11. 0.500 **13.** 0.987 **15.** 0.638

17. 0.951 **19.** 18.5° **21.** 65.0°

23. 22.6° **25.** 70.5°

27. B = 40°, a = 2.3, b = 1.9

29. A = 64.5°, b = 4.9, c = 11.4

31. A = 38.7°, B = 5.3°, b = 6.2

33. A = 0.84 rad, b = 4.0, c = 6.0

35. B = 45°, b = 2, c = 2.8

37. 89 m **39.** 180 ft **41.** 1.9°

43. E_x = 6.0 N/C, E_y = 6.4 N/C

45. Z = 4.5 kΩ, θ = 35° **47.** V = 212 V

49. 56° = 0.98 rad

EXERCISE 9.3

(For 1 through 23, answers appear in the following order: sin, cos, tan)

1. $\frac{4}{5}$ = 0.80, $\frac{3}{5}$ = 0.60, $\frac{4}{3}$ = 1.33

3. $\frac{-12}{13}$ = −0.923, $\frac{5}{13}$ = 0.385, $\frac{-12}{5}$ = −2,4

5. $\frac{-1}{\sqrt{2}}$ = −0.707, $\frac{-1}{\sqrt{2}}$ = −0.707, 1.0

7. 0.80,−0.60,−0.75 **9.** −0.868,−0.496, 1.75

11. 0.60, 0.80, 0.75 **13.** $\frac{-3}{5}$ = −0.60, $\frac{-4}{5}$ = −0.80, $\frac{3}{4}$ = 0.75

15. 0.923, 0.385, 2.40 **17.** −0.80,−0.60, 1.33

19. 0.40, 0.917, 0.436 **21.** 0.707,−0.707,−1.00

23. sin 50° = 0.767 **25.** − tan 30° = −0.577

27. cos 25° = 0.906 **29.** sin 180° = 0

31. − tan 53° = 1.33 **33.** tan 1.10 rad = 1.96

35. sin 20° = 0.342 **37.** cos 0° = 1.00

39. −tan $\frac{\pi}{8}$ rad = − 0.414

41. sin θ = 0.422, cos θ = −0.906, tan θ = −0.467

43. 9.00 knots **45.** 18.4 V, tan θ = −0.857

47. tan−23° = −0.424, 1.95 kΩ

49. (a) 850 mA; (b) 736 mA; (c) −601 mA

51. (a) −120 V; (b) −97.1 V

EXERCISE 9.4

1. 45.0° **3.** 50.0° **5.** 141.9°

7. −72.8° **9.** 0.5774 **11.** 25.0 °, 335.0°

13. 40.2°, 220.2° **15.** 213.8°, 326.2°

17. 153.3°, 333.3° **19.** 57.8° = 1.01 rad, 302.2° = 5.27 rad

21. 45.0° = 0.785 rad, 225° = 3.93 rad

23. 210° = 3.67 rad, 330° = 5.76 rad

25. 45.0° = 0.785 rad, 135° = 2.36 rad, 225° = 3.93 rad, 315° = 5.50 rad

27. 125° **29.** −56.3° **31.** 333°

33. (a) 0.629 rad, 2.51 rad; (b) 1.67 ms, 6.67 ms

EXERCISE 9.5

1. 26 N, 67° **3.** 59 ft/s, 154°

5. 6.4 × 10⁻⁶ N, 231° **7.** 40 ∠34° kΩ

9. 5.3 ∠−63° kΩ **11.** 12 ∠90° V

13. 1.3 ∠−39° kΩ **15.** 297 ∠57° μA

17. 46 N, 300° **19.** 7.1 mi/h, 10°

21. 71 ∠−62° V **23.** 86 ∠−54° mA

25. 18 ∠34° kΩ **27.** 30 ∠−29° kΩ

29. 3.4 × 10³ N/C, 169°

EXERCISE 9.6

1. C = 40°, a = 7.5, c = 5.3

3. A = 42°, b = 21, c = 13

5. B = 47°, C = 66°, c = 5.0

7. A = 51°, B = 24°, b = 0.24

9. c = 7.1, A = 42°, B = 68°

11. b = 0.42, A = 10°, C = 25°

13. C = 83°, A = 39°, B = 58°

15. C = 130°, A = 24°, B = 26°

17. $AB = 1500$ ft, $AC = 1200$ ft   **19.**   450 m   **27.**   17 k Ω, 30°

21.   30°   **23.** 15×10^{-6} N   **25.**   43 A

REVIEW QUESTIONS

1. $\frac{\pi}{6} = 0.524$ rad   **3.**   60°   **5.** $A = C = 70°$, $B = 110°$

(For 7 through 13, answers appear in the following order: sin, cos, tan)

7. $\frac{4}{5} = 0.80$, $\frac{3}{5} = 0.60$, $\frac{4}{3} = 1.3$

9. $\frac{5}{13} = 0.38$, $\frac{12}{13} = 0.92$, $\frac{5}{12} = 0.42$ **11.** $\frac{12}{13}$, $\frac{-5}{13}$, $\frac{-12}{5}$

13. -0.30, 0.95, -0.31

15.   30°, 150°; $\frac{\pi}{6} = 0.52$ rad, $\frac{5\pi}{6} = 2.62$ rad

17.   108°, 288°; 1.89 rad, 5.03 rad   **19.**   20 N, 37°

21.   12.5 $\angle -37°$ kΩ **23.** $c = 5.8$, $A = 31°$, $B = 59°$

25. $A = 41°$, $B = 79°$, $c = 13$ **27.**   36 mi

29.   75 m   **31.**   10 lb, 210°

33.   8.5 $\angle 16°$ kΩ **35.**   0.057°

37. $\omega t = 0.72$ rad at $t = 1.9$ ms and 5.6 rad at $t = 15$ ms

39.   32 $\angle 1.8°$ kΩ

CHAPTER 10

EXERCISE 10.1

1.
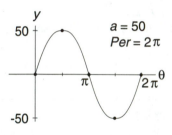
$a = 50$
$Per = 2\pi$

3.
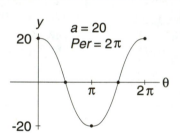
$a = 20$
$Per = 2\pi$

5.
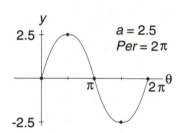
$a = 2.5$
$Per = 2\pi$

7.
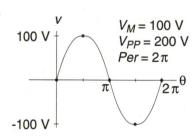
$V_M = 100$ V
$V_{PP} = 200$ V
$Per = 2\pi$

9.
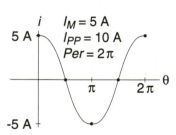
$I_M = 5$ A
$I_{PP} = 10$ A
$Per = 2\pi$

11.
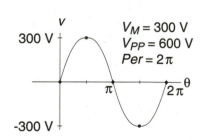
$V_M = 300$ V
$V_{PP} = 600$ V
$Per = 2\pi$

13.
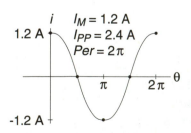
$I_M = 1.2$ A
$I_{PP} = 2.4$ A
$Per = 2\pi$

15.
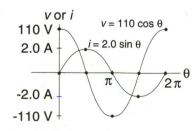
$v = 110 \cos \theta$
$i = 2.0 \sin \theta$

17.   **(a)**   39 V;

(b) $\frac{\pi}{6} = 0.524$ rad, $\frac{5\pi}{6} = 2.62$ rad

EXERCISE 10.2

1. 240 V
3. 1.6 A
5. 40 V
7. 160 V
9. 1.7 A

11. 35 mA
13. $I = 2.2$ A, $I_M = 3.1$ A, $P = 240$ W
15. $V = 3.3$ V, $P_1 = 61$ mW
17. $I_1 = 67$ mA, $P_2 = 4.3$ W

EXERCISE 10.3

1. 55 Hz, 18 ms
3. 500 Hz, 2 ms
5. 64 Hz, 16 ms
7. 17 ms
9. 10 μs
11. 14 ns
13. 4.0 A, 120π rad/s, 60 Hz, 17 ms

15. 310 V, 300 rad/s, 48 Hz, 21 ms
17. 390 rad/s, 63 Hz, 16 ms
19. $v = 180 \sin 120\pi t = 180 \sin 380t$
21. $i = 4.0 \sin 100\pi t = 4.0 \sin 310t$

23.

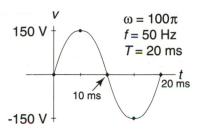

25.

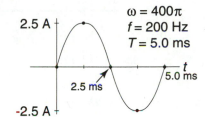

27. **(a)** 77 V; **(b)** 1.8 ms, 6.5 ms

29. **(a)** 1.2 A; **(b)** 12 ms, 18 ms

EXERCISE 10.4

1. 1.5 A, 50 Hz, 20 ms, π rad
3. 170 V, 55 Hz, 18 ms, $\dfrac{\pi}{4}$ rad
5. $i = 3.0 \sin\left(120\pi t + \dfrac{\pi}{2}\right)$

7. $v = 320 \sin\left(100\pi t + \dfrac{\pi}{4}\right)$
9. $v = 170 \sin (110\pi t + \pi)$

11.

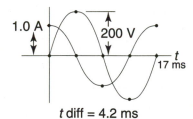

13. 50 Hz, $T = 20$ ms, t diff $= 5.0$ ms, v leads i by 90°
15. **(a)** -2.0 A; **(b)** 1.7 A

REVIEW QUESTIONS

1. $V_M = 160$ V, $V = 110$ V, 120π rad/s, 60 Hz, 17 ms, 0 rad

3. $I_M = 1.4$ A, $I = 850$ mA, 240π rad/s, 120 Hz, 8.3 ms, $\frac{\pi}{2}$ rad

5. $i = 2.4 \sin 200\pi t$

7. $v = 300 \sin\left(120\pi t + \frac{\pi}{2}\right)$

9. 75 V, 110 V, 3.8 W **11.** 83 Hz, 12 ms

13. 100 Hz, 10 ms, 2.5 ms

15. **(a)** 400 mA; **(b)** 1.1 ms, 5.6 ms

≡ CHAPTER 11

EXERCISE 11.1

1. $j3$

3. $j0.2$

5. $-j4$

7. $j0.2$

9. $\frac{j}{2} = j0.5$

11. $j2 \times 10^3$

13. -12

15. $j0.1$

17. 10

19. $j0.6$

21. $-j800$

23. $j4 \times 10^3$

25. $j48$

27. $\frac{-j}{2} = -j0.5$

29. $-j2$

31. $-j0.6$

33. $-j2 \times 10^{-3}$

35. 35 V

37. $-j220$ Ω

EXERCISE 11.2

1. $4 + j2$

3. $j5$

5. $8.4 + j2.7$

7. $-j4$

9. $17 - j7$

11. $j8$

13. $39.9 - j17.5$

15. 61

17. $39 + j54$

19. -4

21. $\frac{1}{10} - \frac{j}{5} = 0.1 - j0.2$

23. $-1 - j3$

25. $1 - j$

27. $1 - j2$

29. $1.4 - j1.8$

31. $4.5 + j0.50$

33. $\frac{9}{13} - \frac{j7}{13} = 0.69 - j0.54$

35. $0 + j2.5$

37. $x^2 - x + jx - x - jx + 1 - j^2 = x^2 - 2x + 2$

39. $13.7 - j3.2$ kΩ

41. $1.4 - j0.2$ kΩ

43. $100 - j50$ V

EXERCISE 11.3

1.

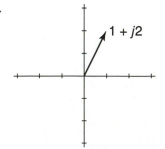

$1 + j2$

3.

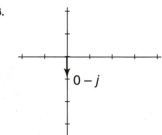

$0 - j$

5.

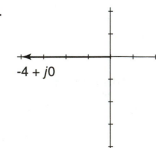

$-4 + j0$

7.

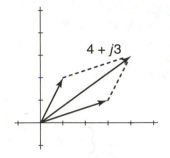

9.

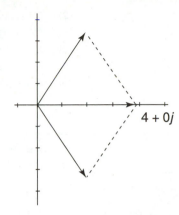

11.

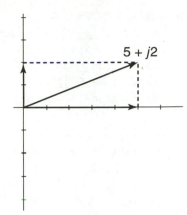

13.

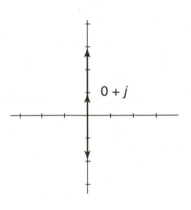

15. $2.8 \angle 45°$	**17.** $4.5 \angle -63°$	**19.** $55 \angle -90°$
21. $75 \angle 0°$	**23.** $29 \angle 59°$	**25.** $1.3 \angle -67°$
27. $2.1 + j2.1$	**29.** $43 - j25$	**31.** $100 + j0$
33. $0 - j10$	**35.** $2.9 - j1.4$	**37.** $0 + j15$
39. $89 \angle 40°$	**41.** $145 \angle -10°$	**43.** $10 \angle -90°$
45. $8.0 \angle 60°$	**47.** $200 \angle -50°$	**49.** $180 \angle 90°$
51. $3.0 \angle 50°$	**53.** $0.20 \angle -60°$	**55.** $1.5 \angle 90°$
57. $4 \angle 0°$	**59.** $1.5 \angle 90°$	**61.** $2.2 \angle -27°$
63. $6.0 \angle 24°$ kΩ	**65.** $4.8 \angle 48°$ mA	
67. $3.6 \angle 65°$ mA	**69.** $32 \angle 63°$ kΩ	

REVIEW QUESTIONS

1. -32	**3.** $-j0.5$	**5.** $6 + j2$
7. $22 - j7$	**9.** $0.2 + j0.4$	**11.** $2 - j4$
13. $3.2 \angle 72°$	**15.** $87 + j50$	**17.** $40 \angle -31°$

19. $14 \angle 40°$ **21.** $2 + j0 = 2 \angle 0°$ **23.** $22 \angle 56°$ V

25. $0.05 \angle 45°$ μS

CHAPTER 12

EXERCISE 12.1

1. 3.8 kΩ	**3.** 630 Ω	**5.** 10 kΩ
7. 110 mH	**9.** 640 Hz	**11.** 800 Ω
13. 2.5 V	**15.** $750 + j1500$ $\Omega = 1.7 \angle 63°$ kΩ	

17. $3.3 + j4.5$ k$\Omega = 5.6 \angle 54°$ kΩ

19. $2 + j1.2$ k$\Omega = 2.3 \angle 31°$ kΩ

21. **(a)** $V_R = 12$ V, $V_L = 16$ V;
  (b) $V_T = 20 \angle 53°$ V, $Z = 1.0 \angle 53°$ kΩ

23. **(a)** $V_R = 17$ V, $V_L = 14$ V;

  (b) $V_T = 21 \angle 39°$ V, $Z = 28 \angle 39°$ kΩ

25. **(a)** $580 \angle 59°$ Ω;
  (b) $I = 34$ mA, $V_R = 10$ V, $V_L = 17$ V

27. **(a)** $4.9 \angle 29°$ kΩ;
  (b) $I = 12$ mA, $V_R = 52$ V, $V_L = 29$ V

29. $X = 1.4$ kΩ, $I = 5.3$ mA, $R = 1.8$ kΩ, $V_R = 9.6$ V,
  $V_T = 12 \angle 38°$ V

31. $X = 280$ Ω, $I = 400$ mA, $V_R = 40$ V, $V_L = 110$ V,
  $Z = 300 \angle 71°$

EXERCISE 12.2

1. 1.3 kΩ	**3.** 21 kΩ	**5.** 530 Ω
7. 1.6 nF	**9.** 1.3 kHz	**11.** 1.7 kΩ
13. 5.6 V	**15.** $1.6 - j1.2$ k$\Omega = 2.0 \angle -37°$ kΩ	

17. $680 - j910$ $\Omega = 1.1 \angle -53°$ kΩ

19. $1.5 - j0.75$ k$\Omega = 1.7 \angle -27°$ kΩ

21. **(a)** $V_R = 24$ V, $V_C = 13$ V;
 (b) $V_T = 27$ $\angle -28°$ V, $Z = 8.5$ $\angle -28°$ kΩ
23. **(a)** $V_R = 2.0$ V, $V_C = 3.2$ V;
 (b) $V_T = 3.8$ $\angle -58°$ V, $Z = 940$ $\angle -58°$ Ω
25. **(a)** $Z = 920$ $\angle -42°$ Ω;
 (b) $I = 22$ mA, $V_R = 15$ V, $V_C = 13$ V

27. **(a)** $Z = 15$ $\angle -58°$ kΩ;
 (b) $I = 780$ μA, $V_R = 6.4$ V, $V_C = 10$ V
29. $X_C = 1.6$ kΩ, $I = 4.1$ mA, $V_C = 6.5$ V, $V_R = 14$ V,
 $Z = 3.7$ $\angle -26°$ kΩ
31. $X_C = 18$ kΩ, $V_R = 94$ V, $V_C = 76$ V, $I = 4.3$ mA,
 $Z = 28$ $\angle -39°$ kΩ

EXERCISE 12.3

1. $X = 300$ Ω, $Z = 600 + j\,300$ Ω $= 670$ $\angle 27°$ Ω
3. $X = 2.2$ kΩ, $Z = 1.8 - j\,2.2$ kΩ $= 2.8$ $\angle -51°$ kΩ
5. $X = 450$ Ω, $Z = 1000 - j\,450$ Ω $= 1100$ $\angle -24°$ Ω
7. 3.6 kHz   **9.** 28 kHz   **11.** 630 nF
13. $X_L = 820$ Ω, $X_C = 610$ Ω, $Z = 330 + j\,210$ Ω $= 390$ $\angle 32°$ Ω,
 $I = 31$ mA, $V_R = 10$ V, $V_L = 25$ V, $V_C = 19$ V

15. $Z = 600 + j\,450$ Ω $= 750$ $\angle 37°$ Ω, $I = 27$ mA,
 $P = 430$ mW
17. $V_R = 18$ V, $V_L = 36$ V, $V_C = 16$ V,
 $V_T = 18 + j\,20$ V $= 27$ $\angle 48°$ V
19. $X = 1.2$ kΩ, $Z = 0.30 + j\,1.2$ kΩ $= 1.2$ $\angle 76°$ kΩ

REVIEW QUESTIONS

1. 9.4 kΩ   **3.** 2.7 kΩ   **5.** 16 kHz
7. $820 + j\,1000$ Ω $= 1.3$ $\angle 51°$ kΩ
9. $6.2 - j\,8.5$ kΩ $= 11$ $\angle -54°$ kΩ
11. $470 + j\,550$ Ω $= 720$ $\angle 49°$ Ω
13. $X_L = 1.1$ kΩ, $V_R = 18$ V, $V_L = 11$ V,
 $V_T = 18 + j\,11$ V $= 21$ $\angle 31°$ V;
 $Z = 1.8 + j\,1.1$ kΩ $= 2.1$ $\angle 31°$ kΩ

15. $X_C = 4.6$ kΩ, $V_R = 2.6$ V, $V_C = 3.7$ V,
 $V_T = 2.6 - j\,3.7$ V $= 4.5$ $\angle -54°$ V,
 $Z = 3.3 - j\,4.6$ kΩ $= 5.7$ $\angle -54°$ kΩ
17. $X_L = 5.7$ kΩ, $X_C = 2.7$ kΩ, $V_R = 23$ V, $V_L = 43$ V,
 $V_C = 20$ V, $I = 7.5$ mA, $Z = 3.0 + j\,3.0$ kΩ $= 4.2$ $\angle 45°$ kΩ
19. $Z = 680 - j\,900$ Ω $= 1.1$ $\angle -53°$ kΩ, $I = 8.0$ mA,
 $P = 43$ mW

☰ CHAPTER 13

EXERCISE 13.1

1. $1.6 - j\,1.8$ mA $= 2.4$ $\angle -49°$ mA,
 $3.2 + j\,3.7$ kΩ $= 4.9$ $\angle 49°$ kΩ
3. $4.2 - j\,6.1$ mA $= 7.4$ $\angle -56°$ mA,
 $380 + j\,560$ Ω $= 680$ $\angle 56°$ Ω
5. $20 + j\,15$ mA $= 25$ $\angle 37°$ mA,
 $480 - j\,360$ Ω $= 600$ $\angle -37°$ Ω
7. $800 + j\,510$ μA $= 950$ $\angle 33°$ μA,
 $21 - j\,14$ kΩ $= 25$ $\angle -33°$ kΩ
9. $130 - j\,150$ μS $= 200$ $\angle -49°$ μS

11. $0.83 - j\,1.2$ mS $= 1.5$ $\angle -56°$ mS
13. $1.3 + j\,1.0$ ms $= 1.7$ $\angle 37°$ mS
15. $33 + j\,22$ μS $= 40$ $\angle 33°$ μS
17. **(a)** $I_T = 810$ $\angle -52°$ μA, $Z = 15$ $\angle 52°$ kΩ;
 (b) $R = 9.1$ kΩ, $X_L = 12$ kΩ
19. **(a)** $I_T = 37$ $\angle 38°$ mA, $Z = 490$ $\angle -38°$ Ω;
 (b) $R = 390$ Ω, $X_C = 300$ Ω
21. $R = 780$ Ω, L $= 57$ mH

EXERCISE 13.2

1. $30 - j\,40$ mA $= 50$ $\angle -53°$ mA
3. $55 + j\,60$ mA $= 81$ $\angle 47°$ mA
5. $I_T = 31 - j\,12$ mA $= 33$ $\angle -21°$ mA,
 $Z_T = 340 + j\,130$ Ω $= 360$ $\angle 21°$ Ω
7. $I_T = 23 + j\,14$ mA $= 26$ $\angle 31°$ mA,
 $Z_T = 1.2 - j\,0.71$ kΩ $= 1.4$ $\angle -31°$ kΩ
9. $Y = 2.7$ $\angle -21°$ mS $= 2.6 - j\,0.99$ mS

11. 730 $\angle 31°$ μS $= 630 + j\,380$ μS
13. $I_T = 2.3 - j\,1.7$ mA $= 2.9$ $\angle -36°$,
 $Z = 2.8 + j\,2.0$ kΩ $= 3.5$ $\angle 36°$ kΩ,
 ESC: $R = 2.8$ kΩ, L $= 16$ mH
15. $Z = 4.9$ $\angle 23°$ kΩ, ESC: $R = 4.5$ kΩ, $X_L = 1.9$ kΩ
17. $Z = 1.2 + j\,0.59$ kΩ $= 1.4$ $\angle 26°$ kΩ, $V = 16$ V

EXERCISE 13.3

1. $Z_T = 1.3 + j2.0$ k$\Omega = 2.4 \angle 57°$ kΩ
3. $Z_T = 410 + j530$ $\Omega = 670 \angle 52°$ Ω
5. $990 \angle -8.4°$ $\Omega = 980 - j140$ Ω
7. $2.3 - j2.3$ k$\Omega = 3.3 \angle -45°$ kΩ
9. $5.0 + j15$ k$\Omega = 16 \angle 72°$ kΩ

11. $13 - j28$ k$\Omega = 31 \angle -64°$ kΩ
13. $I_T = 6.1 \angle -52°$ mA $= 3.7 - j4.8$ mA,
    $V_R = 9.1 \angle -52°$ V $= 5.6 - j7.2$ V,
    $V_L = 12 \angle 37°$ V $= 9.4 + j7.2$ V

REVIEW QUESTIONS

1. $I_T = 2.1 - j1.5$ mA $= 2.5 \angle -35°$ mA,
   $Z = 2.9 + j2.0$ k$\Omega = 3.5 \angle 35°$ kΩ
3. $I_T = 1.5 + j0.50$ mA $= 1.6 \angle 18°$ mA,
   $Z = 18 - j6.0$ k$\Omega = 19 \angle -18°$ kΩ
5. $Y = 280 \angle -35°$ μS $= 230 - j160$ μS
7. $Y = 53 \angle 18°$ μS $= 50 + j17$ μS

9. $I_T = 11 - j8.0$ mA $= 13 \angle -37°$ mA,
   $Z = 360 + j270$ $\Omega = 450 \angle 37°$ Ω ; $R = 360$ Ω,
   $L = 7.1$ mH
11. $I_T = 16 + j7.5$ mA $= 18 \angle 25°$ mA,
    $Z = 1.2 - j0.57$ k$\Omega = 1.4 \angle -25°$ kΩ; $R = 1.2$ kΩ,
    $C = 28$ nF
13. $Z = 1.2 \angle -26°$ k$\Omega = 1.1 - j0.54$ kΩ

CHAPTER 14

EXERCISE 14.1

1.
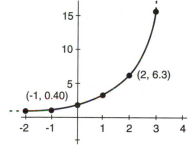

3.
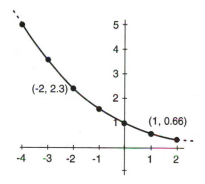

5.
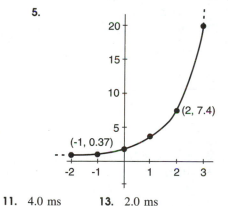

9. 1.0 ms   11. 4.0 ms   13. 2.0 ms
15. 0.50 ms = 500 μS

7.
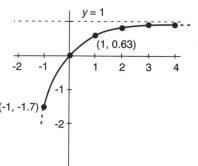

17.
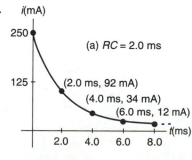

19.
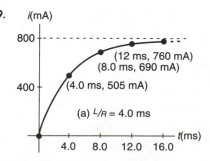

EXERCISE 14.2

1. $\log_2 8 = 3$

3. $\log 10{,}000 = 4$

5. $\log 0.631 = -0.2$

7. $\ln 7.39 = 2$

9. $\ln 0.223 = -1.5$

11. $5^2 = 25$

13. $10^3 = 1000$

15. $10^{-0.5} = 0.316$

17. $e^{-2} = 0.135$

19. $e^{1.10} = 3$

21. 1.94

23. -0.979

25. 3.72

27. 3.22

29. -5.03

31. 5.01

33. 403

35. $0.000759 = 0.759 \times 10^{-3}$

37. 0.708

39. 2.26

41. 0.301

43. 0.330

45. 1.54

47. $0.00768 = 7.68 \times 10^{-3}$

49. 3.85

51. **(a)** 35.0 yr; **(b)** 23.4 yr; **(c)** 14.2 yr

53. **(a)** 0.44 ms = 440 µS; **(b)** 0.69 ms = 690 µS

55. 0.51 nF = 510 pF

57. **(a)** 4.0 ms; **(b)** 1.7 ms

EXERCISE 14.3

1. **(a)** 21 mA, 44 V; **(b)** 7.1 ms, 8.1 ms

3. **(a)** 10 ms; **(b)** 60 ms

5. **(a)** 860 mA; **(b)** 3.5 ms; **(c)** 1.3 ms

7. **(a)** 3.7 mC; **(b)** 300 ms

9. 20 dB

11. 19 dB

13. 47 dB

15. 34 dB

17. 100 mW

19. 380 mV

21. 3.0 dB

REVIEW QUESTIONS

1. $\log 1000 = 3$

3. $\ln 0.0302 = -3.5$

5. $10^{-1.11} = 0.0776$

7. $e^{1.31} = 3.69$

9. 1.40

11. -0.3930

13. 7.23

15. 67.4

17. 0.492

21. **(a)** 3.0 ms; **(b)** $i_c = 0.27\, e^{-330t}$; **(c)** 950 µA;
(d) 5.0 ms

23. **(a)** 1.7 V; **(b)** 11 ms   **25.** $P_2 = 2.0$ W, $V_2 = 47$ V

19.

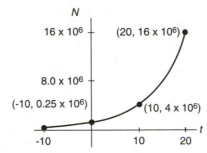

≡ CHAPTER 15

EXERCISE 15.1

1. $1(1000) + 3(100) + 2(10) + 8(1)$

3. $3(10) + 8(1) + 9(0.1) + 6(0.01)$

5. $1(10) + 0(1) + 0(0.1) + 1(0.01)$

7. 3

9. 4

11. 14

13. 17

15. 51

17. 2.75

19. 1.625

21. 3.6875

23. 100

25. 1011

27. 111 001

29. 1 100 011

31. 10 000 000

33. 0.1

35. 0.111

37. 10.01

39. 100 001.101

41.

17	10	001	25	11	001
18	10	010	26	11	010
19	10	011	27	11	011
20	10	100	28	11	100
21	10	101	29	11	101
22	10	110	30	11	110
23	10	111	31	11	111
24	11	000	32	100	000

43. **(a)** 49; **(b)** 40; **(c)** 7; **(d)** 12

EXERCISE 15.2

1. 101

3. 1011

5. 1111

7. 1000.1

9. 1111

11. 10011

13. 10 000.01

15. 65

17. 890

19. 10

21. 1001

23. 1

25. 11

27. 100 10

29. 10.10

31. 10110
     <u>10011</u>
     101001

EXERCISE 15.3

1. 10 **3.** 53 **5.** 212
7. 4095 **9.** 11 **11.** 46
13. 324 **15.** 2114 **17.** 1000
19. 10011 **21.** 111 110 001 **23.** 1 000 101 100
25. 14 **27.** 33 **29.** 57
31. 312 **33.** (a) $110\ 010\ 011_2 = 403$; (b) $245_8 = 165$

35.

17	21	10001		25	31	11001
18	22	10010		26	32	11010
19	23	10011		27	33	11011
20	24	10100		28	34	11100
21	25	10101		29	35	11101
22	26	10110		30	36	11110
23	27	10111		31	37	11111
24	30	11000		32	40	100000

EXERCISE 15.4

1. 27 **3.** 160 **5.** 205
7. 291 **9.** 2571 **11.** 16.125
13. 4097 **15.** 15 **17.** 28
19. 55 **21.** 8E **23.** 157
25. 11 0010 **27.** 1010 1011 **29.** 110 1100 1101
31. 1 1110 0000 1011 **33.** E
35. 1B **37.** 2F **39.** 5C
41. $1010\ 0000_2 = 240_8 = A0_{16}$ **43.** $59A_{16} = 1434$

EXERCISE 15.5

1. (a) 0001 1000; (b) 10010
3. (a) 0001 0010 0010; (b) 111 1010
5. (a) 0101 1000 0011 0111; (b) 58 37; (c) 88 55
7. (a) 0100 0011 0100 1001 0100 0001; (b) 43 49 41;
   (c) 67 73 65
9. (a) 0101 0111 0100 0001 0100 0010 0100 0011;
   (b) 57 41 42 43; (c) 87 65 66 67
11. (a) 1110 0111 1111 0111; (b) E7 F7; (c) 231 247

13. (a) 1100 0011 1100 1001 1100 0001;
   (b) C3 C9 C1; (c) 195 201 193
15. (a) 1110 0110 1100 0001 1100 0010 1100 0011;
   (b) E6 C1 C2 C3; (c) 230 193 194 195
17. (a) Number of bits used to denote a character;
   (b) 9 bits
19. 1 0100 0110 0 0100 0010 1 0100 1001 **21.** 707
23. (a) 0010 0100; (b) 24

REVIEW QUESTIONS

1. 15 **3.** 27 **5.** 10011
7. 111 1101 **9.** 1100 **11.** 10011
13. 101 **15.** 1 **17.** 61
19. 58 **21.** 378 **23.** 100 110
25. 10 1010 **27.** 1001 1100 1101
29. (a) 33; (b) 1B **31.** (a) 157; (b) 6F

33. (a) 2040; (b) 420 **35.** (a) 7; (b) 7
37. (a) 23; (b) 13 **39.** (a) 125; (b) 55
41. (a) 0100 0111 0100 1111; (b) 1100 0111 1101 0110
43. (a) 0101 1000 0101 0011 0011 0011;
   (b) 1110 0111 1110 0010 1111 0011
45. 57 41 53 54 45
47. 1 0011 0001 0 0010 1011 1 0011 0001 1 0011 1101 1 0011 0010

≡ CHAPTER 16

EXERCISE 16.1

1. Copper is a poor conductor and $\sqrt{0.01} = 0.1$. NOT TRUE
   Copper is a poor conductor or $\sqrt{0.01} = 0.1$. TRUE
   Copper is not a poor conductor. TRUE
   $0.01 \neq 0.1$. NOT TRUE

3. Commutative laws, identity laws and the distributive law: $A(B + C) = AB + AC$
5. $A + 0 = A$ **7.** $A \cdot 1 = A \cdot \overline{A}$ **9.** $A(\overline{A} + B) = AB$

11.

A	$A+A$
1	1
0	0

13.

A	B	$A+B$	$A(A+B)$
0	0	0	0
0	1	1	0
1	0	1	1
1	1	1	1

15.

A	B	$\overline{A}$	$\overline{B}$	$\overline{A}\cdot\overline{B}$	$\overline{A}+\overline{A}\cdot\overline{B}$
0	0	1	1	1	1
0	1	1	0	0	1
1	0	0	1	0	0
1	1	0	0	0	0

17.

A	B	C	$A+B$	$(A+B)+C$	$B+C$	$A+(B+C)$
0	0	1	0	1	1	1
0	0	0	0	0	0	0
0	1	1	1	1	1	1
0	1	0	1	1	1	1
1	0	1	1	1	1	1
1	0	0	1	1	1	1
1	1	1	1	1	1	1
1	1	0	1	1	1	1

EXERCISE 16.2

1. A and B or C **3.** 0   **5.** 1

7. 0   **9.** 1   **11.** 0

13. 1   **15.** 1

17. (a) $A+A\overline{B}$;   (b)

A	B	$\overline{B}$	$A\overline{B}$	$A+A\overline{B}$
0	0	1	0	1
0	1	0	0	0
1	0	1	1	1
1	1	0	0	1

19. (a) $AB+\overline{A}B$;   (b)

A	B	$\overline{A}$	AB	$\overline{A}B$	$AB+\overline{A}B$
0	0	1	0	0	0
0	1	1	0	1	1
1	0	0	0	0	0
1	1	0	1	0	1

21. $A+A\overline{B} = A(1+\overline{B}) = A(1) = A$

23. $AB+\overline{A}B = B(A+\overline{A}) = B(1) = B$

25. (a) $ABC+\overline{A}BC$;

(b)

A	B	C	$\overline{A}$	BC	ABC	$\overline{A}BC$	$ABC+\overline{A}BC$
0	0	0	1	0	0	0	0
0	0	1	1	0	0	0	0
0	1	0	1	0	0	0	0
0	1	1	1	1	0	1	1
1	0	0	0	0	0	0	0
1	0	1	0	0	0	0	0
1	1	0	0	0	0	0	0
1	1	1	0	1	1	0	1

(c) $ABC+\overline{A}BC = BC(A+\overline{A}) = BC(1) = BC$

EXERCISE 16.3

1. B **3.** $\overline{A} + \overline{B}$ **5.** $AB + BC$

7. $\overline{A}\,\overline{B} + \overline{B}C$ **9.** $\overline{B}\,\overline{C} + \overline{A}BC$ **11.** A

13. No adjacent squares   **15.** $\overline{C}\,\overline{D}$

17. $\overline{A}\,\overline{B}C + \overline{B}CD$ **19.** A

REVIEW QUESTIONS

1. (a)

A	B	$\overline{A}$	$\overline{A}+B$	$A(\overline{A}+B)$	AB
0	0	1	1	0	0
0	1	1	1	0	0
1	0	0	0	0	0
1	1	0	1	1	1

(b) $A+\overline{A}B = A+B$

3. 0   **5.** 0

7. (a) $A + \overline{A}B$;

(b)

A	B	$\overline{A}$	$\overline{A}B$	$A+\overline{A}B$
0	0	1	0	0
0	1	1	1	1
1	0	0	0	1
1	1	0	0	1

(c) $A+\overline{A}B = A+AB+\overline{A}B = A+B(A+\overline{A}) = A+B(1) = A+B$

9. $\overline{A}$ **11.** B

Index